DIGITAL SIGNAL PROCESSING APPLICATIONS USING THE ADSP-2100 FAMILY

VOLUME 2

ANALOG DEVICES TECHNICAL REFERENCE BOOKS

Published by Prentice Hall

Analog-Digital Conversion Handbook
Digital Signal Processing in VLSI
Digital Signal Processing Applications Using the ADSP-2100 Family,
   Volumes 1 and 2
Digital Signal Processing Laboratory Using the ADSP-2101 Microcomputer
ADSP-2100 Family User's Manual

Published by Analog Devices

Nonlinear Circuits Handbook
Transducer Interfacing Handbook
Synchro & Resolver Conversion
High-Speed Design Seminar
Mixed-Signal Design Seminar

DIGITAL SIGNAL PROCESSING APPLICATIONS USING THE ADSP-2100 FAMILY

VOLUME 2

by
The Applications Engineering Staff of
Analog Devices, DSP Division
Edited by Jere Babst

P T R PRENTICE HALL
Englewood Cliffs, New Jersey 07632

Published by P T R Prentice Hall
Prentice-Hall, Inc.
A Paramount Communications Company
Englewood Cliffs, New Jersey 07632

The publisher offers discounts on this book when ordered in bulk quantities.
For more information, contact: Corporate Sales Department, P T R Prentice Hall,
113 Sylvan Ave., Englewood Cliffs, NJ 07632.
Phone (201) 592-2863   Fax (201) 592-2249

Printed in the United States of America

10 9 8 7 6 5 4 3 2 1

ISBN 0-13-178567-2

Prentice-Hall International (UK) Limited, *London*
Prentice-Hall of Australia Pty. Limited, *Sydney*
Prentice-Hall Canada Inc., *Toronto*
Prentice-Hall Hispanoamericana, S.A., *Mexico*
Prentice-Hall of India Private Limited, *New Delhi*
Prentice-Hall of Japan, Inc., *Tokyo*
Simon & Schuster Asia Pte. Ltd., *Singapore*
Editora Prentice-Hall do Brasil, Ltda., *Rio de Janeiro*

Contents ■

CHAPTER 1   INTRODUCTION

1.1	OVERVIEW	1
1.2	ADSP-2100 FAMILY PROCESSORS	1
1.2.1	ADSP-2100 Family Base Architecture	4
1.2.2	ADSP-2101 Architecture	7
1.2.3	ADSP-2111 Architecture	9
1.2.4	ADSP-21msp50 Architecture	10
1.3	ASSEMBLY LANGUAGE OVERVIEW	11
1.4	DEVELOPMENT SYSTEM	13
1.5	CONVENTIONS OF NOTATION	14
1.6	PROGRAMS ON DISK	15
1.7	FOR FURTHER SUPPORT	15

CHAPTER 2   MODEMS

2.1	OVERVIEW	17
2.2	V.32 MODEM DEFINITION	17
2.2.1	Transmitter Algorithms	18
2.2.2	Receiver Algorithms	20
2.2.3	Scrambler	21
2.2.4	Descrambling	22
2.2.5	ADSP-2100 Family Implementation	24
2.2.6	Scrambler/Descrambler Programs	25
2.2.7	Raised Cosine Filter	32
2.2.8	ADSP-2100 Family Implementation	33
2.2.9	Trellis Encoding	37
2.2.10	ADSP-2100 Family Implementation	39
2.2.11	Viterbi Decoding	47
2.2.12	Data Constellation	50
2.2.13	Viterbi Algorithm	50
2.2.14	ADSP-2100 Family Implementation	52
2.2.15	Shortest Path Through Trellis Diagram	53

Contents

2.2.16	Viterbi Program	55
2.2.16.1	Initialization	55
2.2.16.2	Data Input & Euclidean Distance	55
2.2.16.3	Shortest Path	55
2.2.16.4	Last Surviving Path	55
2.2.16.5	Determination Of Error Corrected Data	56
2.3	QUADRATURE AMPLITUDE MODULATION	75
2.3.1	QAM Methodology	75
2.3.2	ADSP-2100 Family Implementation	78
2.4	ECHO CANCELLATION	81
2.4.1	Echo Cancellation Algorithm	82
2.4.2	ADSP-2100 Family Implementation Of LMS Algorithm	84
2.4.3	Frequency Offset Compensation	88
2.4.4	Family Implementation Of Hilbert Transform	91
2.4.5	V.32 Modem Implementation	96
2.5	ADAPTIVE EQUALIZATION	98
2.5.1	History Of Adaptive Filters	98
2.5.2	Applications Of Adaptive Filters	99
2.5.3	Channel Equalization In A Modem	101
2.5.3.1	Equalization	102
2.5.3.2	Performance Index	105
2.5.4	Equalizer Architectures	105
2.5.4.1	Real Or Complex	106
2.5.4.2	Sampling Rates	107
2.5.5	Least Mean Squared (LMS) Algorithm	109
2.5.6	Program Structure	112
2.5.6.1	Input New Sample	113
2.5.6.2	Filtering (Equalizing)	113
2.5.6.3	Training Sequence	114
2.5.6.4	Decision-Directed Adaptation	115
2.5.6.5	Tap Update (LMS Algorithm)	117
2.5.6.6	Output	118
2.5.7	Practical Considerations	119
2.5.7.1	Viterbi Decoder	119
2.5.7.2	Pseudo-Random Training Sequence	119
2.5.7.3	Delay Line Length	119
2.6	CONTINUOUS PHASE MODULATION	120
2.6.1	CPFSK Methodology	120
2.6.2	ADSP-2100 Family Implementation	121
2.7	V.27 *ter* & V.29 MODEM TRANSMITTERS	126
2.7.1	V.27 *ter* Transmitter	126
2.7.2	V.29 Transmitter	141
2.8	REFERENCES	155

Contents

CHAPTER 3   **LINEAR PREDICTIVE CODING**

3.1   OVERVIEW 157
3.2   LINEAR PREDICTION 157
3.3   7.8 kbits/s LPC 159
3.4   2.4 kbits/s LPC 163
3.5   LPC SUBROUTINES 169

CHAPTER 4   **GSM CODEC**

4.1   OVERVIEW 205
4.1.1   Speech Codec 205
4.1.2   Software Comments 206
4.1.2.1   Multiply With Rounding 206
4.1.2.2   Arithmetic Saturation Results 206
4.1.2.3   Temporary Arrays 207
4.1.2.4   Shared Subroutines 207
4.2   ENCODER 207
4.2.1   Down Scaling & Offset Compensation Of The Input 208
4.2.2   Pre-Emphasis Filtering 208
4.2.3   Auto-Correlation 209
4.2.4   The Schur Recursion 209
4.2.5   Transformation Of The Reflection Coefficients 211
4.2.6   Quantization & Coding Of Logarithmic-Area-Ratios 212
4.2.7   Decoding Of Logarithmic-Area-Ratios 212
4.2.8   Short Term Analysis Filtering 213
4.2.8.1   Transformation Of The LARs Into Coefficients 214
4.2.8.2   Short Term Analysis Filtering 215
4.2.9   Calculation Of The Long Term Parameters 215
4.2.9.1   Long Term Analysis Filtering 216
4.2.9.2   Long Term Synthesis Filtering 217
4.2.10   Residual Pulse Excitation Encoding Section 217
4.2.10.1   Weighting Filter 217
4.2.10.2   Adaptive Sample Rate Decimation 218
4.2.10.3   APCM Quantization Of Selected Sequence 218
4.2.10.4   APCM Inverse Quantization 219
4.2.10.5   Update Of The Short Term Residual Signal 219
4.3   DECODER 220
4.3.1   Short Term Synthesis Filtering 220
4.3.1.1   Short Term Synthesis Filter 221
4.3.2   Long Term Synthesis Filtering 221
4.3.3   Post Processing 222
4.4   BENCHMARKS & MEMORY REQUIREMENTS 222
4.5   LISTINGS 223

vii

Contents

CHAPTER 5   **SUB-BAND ADPCM**

5.1	OVERVIEW	293
5.2	SUB-BAND ADPCM ALGORITHM	294
5.3	TRANSMIT PATH	294
5.3.1	Transmit Quadrature Mirror Filter	294
5.3.2	Higher Sub-Band Encoder	296
5.3.3	Lower Sub-Band Encoder	296
5.4	RECEIVE PATH	298
5.4.1	Higher Sub-Band Decoder	298
5.4.2	Lower Sub-Band Decoder	299
5.4.3	Receive Quadrature Mirror Filter	300
5.5	ADSP-2100 FAMILY IMPLEMENTATION	300
5.6	SUBROUTINE DESCRIPTIONS	301
5.6.1	reset_mem	301
5.6.2	filtez	301
5.6.3	filtep	301
5.6.4	quantl	301
5.6.5	invqxl	302
5.6.6	logscl	302
5.6.7	scalel	302
5.6.8	upzero	302
5.6.9	uppol2	302
5.6.10	uppol1	302
5.6.11	limit	303
5.6.12	quanth	303
5.6.13	invqah	303
5.6.14	logsch	303
5.7	BENCHMARKS	328

CHAPTER 6   **SPEECH RECOGNITION**

6.1	OVERVIEW	329
6.2	SPEECH RECOGNITION SYSTEMS	330
6.2.1	Voice Production & Modeling	330
6.2.2	Training Phase	332
6.2.3	Recognition Phase	333
6.3	SOFTWARE IMPLEMENTATION	334
6.3.1	Word Acquisition & Analysis	335
6.3.1.1	Receive Shell	335
6.3.1.2	Frame Analysis	336
6.3.1.3	Endpoint Detection	337

Contents

6.3.1.4   Coefficient Conversion 338
6.3.2   Isolated Word Recognition 340
6.3.2.1   Library Routines 340
6.3.2.2   Comparison 341
6.3.2.3   Dynamic Time Warping 342
6.3.2.4   Ranking 346
6.3.3   Main Shell Routines 346
6.3.3.1   Executive Shell 347
6.3.3.2   Demonstration Shell 348
6.4   HARDWARE IMPLEMENTATION 349
6.5   LISTINGS 349
6.6   REFERENCES 440

CHAPTER 7   DISCRETE COSINE TRANSFORM

7.1   OVERVIEW 443
7.2   BACKGROUND 444
7.3   COMPUTATIONAL METHODS 449
7.4   HOU'S FAST DISCRETE COSINE ALGORITHM 449
7.5   ZIG-ZAG SCANNING OF DCT COEFFICIENTS 453
7.6   ZIG-ZAG SCANNING & ADSP-21XX PROCESSORS 454
7.7   LISTINGS 455
7.8   REFERENCES 480

CHAPTER 8   DIGITAL TONE DETECTION

8.1   OVERVIEW 481
8.2   IMPLEMENTATION 482
8.2.1   Choosing A Sampling Frequency 482
8.2.2   Picking The Best Value Of N 483
8.2.2.1   Leakage Loss 483
8.2.2.2   Frequency Resolution... 484
8.2.2.3   Detection Time 484
8.2.2.4   Tone Detection Categories... 484
8.2.2.5   Tone Detection Example 485
8.3   BENCHMARKS FOR THE EXAMPLE PROGRAM 489
8.4   LISTINGS 490

Contents

CHAPTER 9   DIGITAL CONTROL SYSTEM DESIGN

9.1   OVERVIEW 503
9.2   DIGITAL CONTROL SYSTEMS OVERVIEW 503
9.3   DIGITAL CONTROL SYSTEM MODEL 504
9.4   DIGITAL CONTROL SYSTEM HARDWARE 505
9.5   DIGITAL CONTROL SYSTEM SOFTWARE 507
9.6   DIGITAL PID CONTROLLER DESIGN 508
9.7   PID CONTROLLER IMPLEMENTATION 511
9.8   N'TH ORDER DIGITAL CONTROLLER DESIGN 513
9.8.1   Analog-Controller-Based Digital Design 513
9.8.2   Direct Digital Design 514
9.8.3   State-Space Design 515
9.9   N'TH ORDER DIGITAL CONTROLLER STRUCTURES 515
9.10   N'TH ORDER CONTROLLER IMPLEMENTATION 517
9.11   NOTCH FILTER EXAMPLE FOR THE ADSP-2100A 520
9.12   REFERENCES 523

CHAPTER 10   VARIATIONS ON IIR BIQUAD FILTERS

10.1   OVERVIEW 525
10.1.1   IIR Biquad Filter 525
10.1.2   Biquad Filter Subroutine 525
10.2   MULTIPRECISION FILTERS 527
10.2.1   Multiprecision Mult On ADSP-2100 Family DSPs 528
10.2.2   Double-Precision Biquad 531
10.2.3   Half, Double-Precision Biquad 537
10.2.4   Half, Triple-Precision Biquad 540
10.3   OPTIMIZED 16-BIT BIQUADS 544
10.4   CONCLUSION 549

CHAPTER 11   SOFTWARE UART

11.1   OVERVIEW 551
11.2   HARDWARE 551
11.3   SOFTWARE 552
11.3.1   Program Flow 553
11.3.2   Initialization & Timer Interrupt Routines 554
11.3.3   Transmit & Receive Subroutines 563
11.4   BAUD RATES 564
11.5   AUTOBAUD FEATURE 564
11.6   CHARACTER ECHO EXAMPLE 567
11.7   PROGRAM FILES 569

Contents

CHAPTER 12   HARDWARE INTERFACING

12.1   OVERVIEW 571
12.2   SOUNDPORT INTERFACES 571
12.2.1   ADSP-2111/AD1849 SoundPort Interface 572
12.2.2   ADSP-2105/AD1849 SoundPort Interface 578
12.2.3   ADSP-2101/AD1847 SoundPort Interface 585
12.3   INTERFACING DRAMS WITH THE ADSP-2100 FAMILY 603
12.3.1   DRAM Configuration 606
12.3.2   Multiplexed Memory Addressing 607
12.3.3   DSP & DRAM Control Signals 607
12.3.3.1   DSP Read/Write Timing 607
12.3.3.2   DRAM Read/Write Timing 608
12.3.3.3 $\overline{RAS}$ Generation 609
12.3.3.4 $\overline{CAS}$ Generation 610
12.3.3.5 $\overline{WRITE}$ & $\overline{OE}$ Generation 611
12.3.4   DSP To DRAM Interface Timing 611
12.3.4.1   DRAM Read Timing 611
12.3.4.2   DRAM Write Timing 612
12.3.5   Memory Access Modes 613
12.3.5.1   Page Mode 613
12.3.5.2   Enhanced Or Fast Page Mode 613
12.3.6   DRAM Refresh 614
12.3.7   DRAM Refresh Timing 615
12.3.8   EZ-LAB Implementation 617
12.3.9   DRAM Program Listings 618
12.3.10   DRAM Interfacing References 630
12.4   LOADING AN ADSP-2101 PROGRAM/SERIAL PORT 631
12.4.1   A Monitor 631
12.4.2   Implementation 632
12.5   MEMORY INTERFACING FOR THE ADSP-2105 637
12.5.1   Ex System 1: Boot Pages For Program Memory 637
12.5.2   Ex System 2: Booting With the -loader Option 638
12.5.3   Ex System 3: Internal & External PM RAM 639
12.5.4   Ex System 4: Using External PM ROM 640
12.5.5   Hardware Implications 640
12.5.6   Use Of The C-Compiler With ADSP-2105 Systems 641
12.5.7   Linking Modules Generated By The C-Compiler 642
12.5.8   Additional Suggestions 642
12.5.9   About The Example Programs 643
12.5.10   Appendix: Example System 1 645

Contents

FIGURES

Figure 1.1   ADSP-2100 Family Base Architecture 5
Figure 1.2   ADSP-2101 Architecture 8
Figure 1.3   ADSP-2111 Architecture 10
Figure 1.4   ADSP-21msp50 Architecture 11

Figure 2.1   Transmitter Block Diagram 19
Figure 2.2   Receiver Block Diagram 20
Figure 2.3   Call Mode Scrambler 22
Figure 2.4   Answer Mode Scrambler 23
Figure 2.5   Call Mode Descrambler 23
Figure 2.6   Answer Mode Descrambler 23
Figure 2.7   Circular Buffer Implementation For Scrambler 24
Figure 2.8   Raised Cosine Pulse Shaping Filter 34
Figure 2.9   Modem Transmitter 35
Figure 2.10   Encoder Block Diagram 38
Figure 2.11   V.32 Signal Constellation 39
Figure 2.12   Convolutional Encoder Block Diagram 41
Figure 2.13   Trellis Diagram For Convolutional Encoding 49
Figure 2.14   Signal Constellation Showing Encoder Output 51
Figure 2.15   Accumulated Distance Table Update Example 54
Figure 2.16   QAM Modulator Block Diagram 76
Figure 2.17   QAM Demodulator Block Diagram 77
Figure 2.18   Telephone Channel Block Diagram 81
Figure 2.19   Echo Canceller 83
Figure 2.20   LMS Adaptive Filter 84
Figure 2.21   Flowchart For LMS Stochastic Gradient Algorithm 85
Figure 2.22   Block Diagram Of Echo Canceller 89
Figure 2.23   Block Diagram Of Hilbert Transform 90
Figure 2.24   Spectrum Of Hilbert Frequency Shift 91
Figure 2.25   V.32 Modem Block Diagram 97
Figure 2.26   Example Short Impulse Response 102
Figure 2.27   Pure Delay Impulse Response 103
Figure 2.28   Equalizer Impulse Response 103
Figure 2.29   Transversal (FIR) Delay Line 106
Figure 2.30   IIR Delay Line 106
Figure 2.31   Fractionally Spaced Delay Line (FSE) 107
Figure 2.32   Adaptive Equalizer Flowchart 112
Figure 2.33   CPFSK Flow Diagram 122
Figure 2.34   Modem Transmitter Block Diagram 127
Figure 2.35   8-Point V.27 *ter* Constellation 128
Figure 2.36   4-Point V.27 *ter* Constellation 129

Contents

Figure 2.37   V.29 Constellation 142
Figure 2.38   V.29 Constellation For 7200 bits/s Fallback Mode 142
Figure 2.39   V.29 Constellation For 4800 bits/s Fallback Mode 143

Figure 5.1   Sub-Band ADPCM Algorithm Block Diagram 295
Figure 5.2   Higher Sub-Band Encoder Block Diagram 296
Figure 5.3   Lower Sub-Band Encoder Block Diagram 297
Figure 5.4   Higher Sub-Band Decoder Block Diagram 298
Figure 5.5   Lower Sub-Band Decoder Block Diagram 299

Figure 6.1   Speech Training System Block Diagram 332
Figure 6.2   Speech Recognition System Block Diagram 333
Figure 6.3   Distance Matrix With Slope Constraints 343
Figure 6.4   Time Warping Paths Between Sums & Distances 345
Figure 6.5   EXECSHEL.DSP Link File Menu Tree 347
Figure 6.6   DEMOSHEL.DSP Link File Menu Tree 348
Figure 6.7   Speech Recognition System Circuit Board Schematic 350

Figure 7.1   A Two-Dimensional Discrete Cosine Transform 444
Figure 7.2   The DCT Reduces The Blocking Artifact 446
Figure 7.3   Implementation Of An N=16 DCT 451
Figure 7.4   Signal Flow Graph For A Fast DCT 452
Figure 7.5   Zig-Zag Scanning Of Quantized Addresses 454

Figure 9.1   General Digital Control System 504
Figure 9.2   Digital Control System Model 505
Figure 9.3   ADSP-2101-Based Actuator Controller 506
Figure 9.4   PID Block Diagram 509
Figure 9.5   PD, PI, & PID Controllers 510
Figure 9.6   Second-Order Biquad Structure 515
Figure 9.7   Cascaded Biquad Sections 516
Figure 9.8   Fourth-Order Direct Form Controller 517

Figure 10.1   Second-Order Biquad IIR Filter Section 526
Figure 10.2   Multiprecision Multiplication Of 32-Bit Numbers 530
Figure 10.3   Modulo Addressing & Delay Line Data 545

Figure 11.1   General System Configuration 551
Figure 11.2   Example System Configuration 552
Figure 11.3   Receive Data Timing 562

xiii

Contents

Figure 12.1 Functional Block Diagram Of DRAM Interface 605
Figure 12.2 DSP Read/Write Timing 608
Figure 12.3 DRAM Read Cycle Timing 608
Figure 12.4 DRAM Delayed-Write Cycle Timing 609
Figure 12.5 $\overline{RAS}$ & $\overline{CAS}$ Timing For DRAM Read 611
Figure 12.6 $\overline{RAS}$ & $\overline{CAS}$ Timing For DRAM Write 612
Figure 12.7 EZ-LAB/DRAM Interface Board Connection 617
Figure 12.8 Boot Program Flow Diagram 633

LISTINGS

Listing 2.1   Call Mode Scrambler Main Routine 26
Listing 2.2   Call Mode Scrambler Scrambling Routine 28
Listing 2.3   Call Mode Descrambler Routine 30
Listing 2.4   Raised Cosine Filter 35
Listing 2.5   Trellis Encoder Program 42
Listing 2.6   Convolutional Encoder Routine 44
Listing 2.7   Signal Mapping Routine 46
Listing 2.8   Viterbi Decoder 56
Listing 2.9   Modulator Code 78
Listing 2.10 Demodulator Code 80
Listing 2.11 LMS Stochastic Gradient Implementation 86
Listing 2.12 Hilbert Transform Implementation 93
Listing 2.13 Delay Line Routine, Complex Tap Weights 108
Listing 2.14 LMS Routine 111
Listing 2.15 Input Routine 113
Listing 2.16 Filter Routine 114
Listing 2.17 Training Sequence Routine 115
Listing 2.18 Decision-Directed Adaptation Routine 116
Listing 2.19 Tap Update Routine 118
Listing 2.20 Output Routine 118
Listing 2.21 CPFSK Program (ADSP-2101) 123
Listing 2.22 Main V.27 *ter* Routine (MAIN27.DSP) 129
Listing 2.23 Data Acquisition Routine (GET27.DSP) 134
Listing 2.24 Data Scrambler Routine (SCRAM27.DSP) 135
Listing 2.25 IQ Generator Routine (IQ27.DSP) 136
Listing 2.26 Pulse Shape Filter Routine (PSF.DSP) 138
Listing 2.27 Random Number Generator Routine (RAND.DSP) 139
Listing 2.28 Signal Modulation Routine (MODULATE.DSP) 140
Listing 2.29 Main V.29 Routine (MAIN29.DSP) 144
Listing 2.30 Data Acquisition Routine (GET29.DSP) 151
Listing 2.31 Data Scrambler Routine (SCRAM29.DSP) 152
Listing 2.32 IQ Generator Routine (IQ29.DSP) 153

Contents

Listing 3.1   7.8 kbits/s LPC Routine 159
Listing 3.2   2.4 kbits/s LPC Routine 164
Listing 3.3   AUTOCOR.DSP Subroutine 169
Listing 3.4   DECODE.DSP Subroutine 172
Listing 3.5   DEEMP.DSP Subroutine 174
Listing 3.6   DURBIN.DSP Subroutine 175
Listing 3.7   DURBIN2.DSP Subroutine 180
Listing 3.8   ENCODE.DSP Subroutine 186
Listing 3.9   GAIN.DSP Subroutine 189
Listing 3.10   OVERFLOW.DSP Subroutine 190
Listing 3.11   PITCH.DSP Subroutine 191
Listing 3.12   POLY.DSP Subroutine 195
Listing 3.13   PREEMP.DSP Subroutine 196
Listing 3.14   RANDOM.DSP Subroutine 197
Listing 3.15   SQRT.DSP Subroutine 198
Listing 3.16   SSYNTH.DSP Subroutine 200

Listing 4.1   Initialization Routine (GSM_RSET.DSP) 224
Listing 4.2   Codec Routine (GSM0610.DSP) 228
Listing 4.3   Voice Activity Detection Routine (GSM0632.DSP) 256
Listing 4.4   Comfort Noise Insertion Routine (GSM_SID.DSP) 273
Listing 4.5   Discontinuous Trans Routine (GSM_DTX.DSP) 276
Listing 4.6   Data Acquisition Shell Routine (DMR21xx.DSP) 280

Listing 5.1   Implementation Of The G.722 Algorithm 304

Listing 6.1   Executive Shell Subroutine (EXECSHEL.DSP) 351
Listing 6.2   Demonstration Shell Subroutine (DEMOSHEL.DSP) 354
Listing 6.3   Data Variable Initialization Routine (INITIZE.DSP) 359
Listing 6.4   Receive Word Routine (RECVSHEL.DSP) 362
Listing 6.5   Frame Analysis Routine (ANALYZE.DSP) 369
Listing 6.6   Endpoint Detection Routine (ENDPOINT.DSP) 377
Listing 6.7   Coefficient Conversion Routine (CONVERT.DSP) 380
Listing 6.8   Library Functions Routine (LIB_FUNC.DSP) 398
Listing 6.9   Word Comparison Routine (COMPLIB.DSP) 402
Listing 6.10   Word Ranking Routine (RANKDIST.DSP) 407
Listing 6.11   Library Template Routine (WARPSHEL.DSP) 409
Listing 6.12   Y Coordinate Range Routine (YMINMAX.DSP) 416
Listing 6.13   Dynamic Time Warping Routine (TIMEWARP.DSP) 418
Listing 6.14   Vector Distance Routine (VECTDIST.DSP) 422
Listing 6.15   Display Driver Routine (DEMOBOX.DSP) 425
Listing 6.16   DTMF Signal Generator Routine (DTMF.DSP) 431
Listing 6.17   Automatic Dialing Routine (DTMFMAIN.DSP) 435

Contents

Listing 7.1   One-Dimensional Fast DCT (16 Points) Routine 456
Listing 7.2   DIF16 Subroutine 458
Listing 7.3   DIF8 Subroutine 459
Listing 7.4   DIF4 Subroutine 460
Listing 7.5   DIF2 Subroutine 462
Listing 7.6   RLR4 Subroutine 463
Listing 7.7   RLR8 Subroutine 464
Listing 7.8   RLR16 Subroutine 465
Listing 7.9   DC_AND_BREV Subroutine 466
Listing 7.10   Two-Dimensional Fast DCT (16 X 16 Points) Routine ... 467
Listing 7.11   One-Dimensional Fast DCT (8 Points) Routine 470
Listing 7.12   DIF8_8 Subroutine 472
Listing 7.13   DIF4_8 Subroutine 473
Listing 7.14   DIF2_8 Subroutine 474
Listing 7.15   RLR4_8 Subroutine 475
Listing 7.16   RLR8_8 Subroutine 476
Listing 7.17   DC_AND_BREV_8 Subroutine 477
Listing 7.18   Two-Dimensional Fast DCT (8 X 8 Points) Routine 478

Listing 8.1   Prime Factors Routine (FACTOR.C) 491
Listing 8.2   Prime Numbers Routine (PRIMES.C) 492
Listing 8.3   Best Sampling Frequency Routine (BESTFS.C) 493
Listing 8.4   Best Number Of Samples Routine (BESTN.C) 495
Listing 8.5   Coefficient Generating Routine (COEFGEN.C) 497
Listing 8.6   Tone Detection Routine (EXAMPLE.DSP) 499

Listing 9.1   PID_CONTROLLER Routine 512
Listing 9.2   BIQUAD_CONTROLLER Routine 519
Listing 9.3   NOTCH_FILTER Routine 521

Listing 10.1   Basic Biquad Filter Subroutine 526
Listing 10.2   Double-Precision Multiply Routine 529
Listing 10.3   Double-Precision IIR Biquad Subroutine 532
Listing 10.4   Optimized Double-Precision IIR Biquad Subroutine 535
Listing 10.5   Half, Double-Precision IIR Biquad Subroutine 537
Listing 10.6   Half, Triple-Precision IIR Biquad Subroutine 540
Listing 10.7   Optimized Basic Biquad Filter Subroutine 545
Listing 10.8   Second-Level Optimization Of Basic Biquad Filter 547

Listing 11.1   UART.DSP Code 554
Listing 11.2   Autobaud Example Program 565
Listing 11.3   Character Echo Program 568

xvi

Contents

Listing 12.1   ADSP-2111/AD1849 Talk-Through Routine 573
Listing 12.2   ADSP-2105/AD1849 Talk-Through Routine 579
Listing 12.3   ADSP-2101/AD1847 Talk-Through Routine 586
Listing 12.4   ADSP-2101/AD1847 Demonstration Routine 590
Listing 12.5   DRAM Read Program 618
Listing 12.6   DRAM Write Program 619
Listing 12.7   DRAM Refresh Program 620
Listing 12.8   DRAM Test Program 621
Listing 12.9   DRAM Speech Sample Record/Playback Program 625
Listing 12.10   Monitor Program Listing 635

TABLES

Table 1.1   ADSP-2100 Family Functional Differences 3

Table 2.1   Differential Encoder Lookup Table.................................... 40
Table 2.2   State Table For Convolutional Encoder 48
Table 2.3   Lookup Table Of X & Y Coordinates 53
Table 2.4   ADSP-2100 Benchmarks For Echo Cancellation 97
Table 2.5   8-Point V.27 *ter* Phase Changes 127
Table 2.6   4-Point V.27 *ter* Phase Changes 128
Table 2.7   8-Point V.29 Phase Changes 141

Table 3.1   Parameter Set For The Sound Synthesis Model 158

Table 4.1   GSM Implementation Benchmarks 223

Table 5.1   Decoder Modes Of Operation 298
Table 5.2   Inverse Adaptive Quantizer Modes Of Operation 300
Table 5.3   Typical Benchmark Performance 328

Table 6.1   Time Warping Boundaries 344

Table 7.1   Cosine vs. Fourier Transform Characteristics 445
Table 7.2   Benchmark Times For Executing The DCT 447

Table 8.1   Sample Frequencies & Prime Factors 485
Table 8.2   Sorted Sampling Frequencies (BESTFS.ERR) 486
Table 8.3   Sorted Values For N (BESTN.ERR) 487
Table 8.4   Goertzel Coefficients 489
Table 8.5   Typical Benchmark Performance 490

Contents

Table 10.1   Filter Routine Characteristics Summary 549

Table 12.1   Test System Components 606

INDEX

.... 647

Preface

This book is the second volume of applications for the ADSP-2100 Family of Digital Signal Processors, and it is intended to complement, rather than replace the information contained in *Digital Signal Processing Applications Using the ADSP-2100 Family, Volume 1*. Each chapter embraces a single application topic, briefly describes the algorithm, and discusses its implementation on ADSP-2100 Family Processors. Although several topics contained in this book are addressed in Volume 1, the information presented here will provide you with a new perspective when approaching these topics.

If you want to understand how processors optimized for digital signal processing, such as the ADSP-2100 Family, are used to solve particular problems, you will find this book informative. The topics explored in this volume include, but are not limited to, telecommunications, hardware interfaces, and data encoding, decoding, and transmission.

This book does not provide full explanations of the signal processing theory behind the applications. The contributors and editor assumed that you already understand the theory and practice applying to your area of interest. *Digital Signal Processing in VLSI**, a companion book in the Analog Devices technical reference set, provides much of the necessary basics. The references listed at the end of many of the chapters provide a wealth of additional information.

This volume includes solutions that vary in length and complexity. Here is a brief summary of each chapter's contents:

- *Introduction*

Overviews of the ADSP-2100 Family base architecture, additional peripherals on the ADSP-2101, ADSP-2111, and ADSP-21msp50, assembly language, and development system.

- *Modems*

Implementations for V.32, V.27 *ter*, and V.29 modems.

- *Linear Predictive Coding*

Techniques used to analyze, encode, and decode 7.8 kbits/s and 2.4 kbits/s speech signals.

---

*Higgins, Richard J., *Digital Signal Processing in VLSI*. Englewood Cliffs, NJ: Prentice Hall 1990

- *GSM Codec*

Implementation of the Pan-European Digital Mobile Radio (DMR) Speech Codec Specification 06.10. This chapter also includes subroutines for Voice Activity Detection (VAD, Specification 06.32) and Comfort Noise Insertion (CNI, Specification 06.12).

- *Sub-Band Adaptive Differential Pulse Code Modulation*

Implementation of the CCITT Sub-band ADPCM Recommendation G.722.

- *Speech Recognition*

A design example and demonstration that implements a speech recognition system using the ADSP-2101 EZ-LAB Demonstration Board and an expansion board.

- *Discrete Cosine Transform*

Implementation of an algorithm that performs a Discrete Cosine Transform.

- *Digital Tone Detection*

Techniques for detecting digital representations of sinusoidal tones.

- *Digital Control System Design*

Several algorithms and software and hardware design methods and guidelines for high-speed digital control systems.

- *Variations On IIR Biquad Filters*

Several variations on the basic IIR biquad filter that include multiprecision filters and optimized filter subroutines.

- *Software UART*

Software implementation of a Universal Asynchronous Receiver/Transmitter.

- *Hardware Interfacing*

Hardware and software interface solutions that include SoundPort® interfaces, a DRAM interface, loading a program through the serial port (SPORT), and a memory interface for the ADSP-2105.

Acknowledgments

The software and hardware implementations and the accompanying text for this book were provided by the Applications Engineering group of Analog Devices SPD Division. They designed, developed, and tested the solutions presented here and reviewed the final publication. Besides Jerry McGuire, who leads the group, contributors included: Dan Ash, Chris Cavigioli, Gordon Cooper, Ron Coughlin, Jeff Cuthbert, Colin Duggin, Kapriel Karagozyan, Noam Levine, Ann Mascarin, and Bruce Wolfeld.

Bob Fine and Jerry McGuire helped compile the index and provided editorial feedback on every chapter; Adele Hastings produced the layout, assisted with the illustrations, and assembled the finished book; Christine Hulme collected, compiled, and tested the code to produce the final diskette.

Norwood, Massachusetts   *Jere Babst*

Introduction ■ 1

1.1 OVERVIEW

This book is the second volume of digital signal processing applications based on the ADSP-2100 DSP microprocessor family. It contains a compilation of routines for a variety of common digital signal processing applications. As in the first volume, you may use these routines without modification or you can use them as a starting point for the development of routines tailored to your particular needs.

Besides showing the specific applications, these routines demonstrate a variety of programming techniques for getting the most performance out of the ADSP-2100 family processors. For example, several routines show you how to use address pointers efficiently to address circular buffers. We believe that you will benefit from reading every chapter, even if your present application uses only a single topic.

Some material in this book was originally published in an applications handbook that featured modem routines. The information in that volume was updated and integrated into this book, which supersedes the earlier publication.

1.2 ADSP-2100 FAMILY PROCESSORS

This section briefly describes the ADSP-2100 family of processors. For complete information, refer to the *ADSP-2100 Family User's Manual*, (ISBN 0-13-006958-2) available from Prentice Hall and Analog Devices. For the applications in this book, "ADSP-2100" refers to *any* processor in the ADSP-2100 family unless otherwise noted. At the time of publication, the ADSP-2100 Family consisted of the following members:

- ADSP-2100A—DSP microprocessor with off-chip Harvard architecture

- ADSP-2101—DSP microcomputer with on-chip program and data memory

- ADSP-2103—Low-voltage microcomputer, 3.3-volt version of ADSP-2101

1   Introduction

- ADSP-2105—Low-cost DSP microcomputer

- ADSP-2111—DSP microcomputer with Host Interface Port

- ADSP-2115—High-performance, Low-cost DSP microcomputer

- ADSP-2161/62/63/64—Custom ROM-programmed DSP microcomputers

- ADSP-2165/66—Custom ROM-programmed DSP microcomputers with larger on-chip memories and powerdown

- ADSP-21msp5x—Mixed-Signal DSP microcomputers with integrated, on-chip analog interface and powerdown

- ADSP-2171—Enhanced ADSP-2100 Family processor offering 33 MIPS performance, host interface port, powerdown, and instruction set extensions for bit manipulation, multiplication, biased rounding, and global interrupt masking

Since Analog Devices strives to provide products that exploit the latest technology, new family members will be added to this list periodically. Please contact your local Analog Devices sales office or distributor for a complete list of available products.

The ADSP-2100A is a programmable single-chip *microprocessor* optimized for digital signal processing and other high-speed numeric processing applications. The ADSP-2100A contains an ALU, a multiplier/accumulator (MAC), a barrel shifter, two data address generators and a program sequencer. It features an off-chip Harvard architecture, where data and program buses are available to external memories and devices.

The ADSP-2101 is a programmable single-chip *microcomputer* based on the ADSP-2100A. Like the ADSP-2100A, the ADSP-2101 contains computational units, as well as a program sequencer and dual address generators; these elements, combined with internal data and address busses, comprise the base architecture of the ADSP-2100 Family microcomputers. Additionally, all family members have the following core features:

- on-chip data memory, program memory, and boot memory
- one or two serial ports
- a programmable timer
- and enhanced interrupt capabilities.

Introduction   1

To expand the usefulness of the ADSP-2100 Family, the base architecture
has been enhanced with a variety of memory configurations, peripheral
devices, and features for improved performance. Table 1.1 is a matrix that
identifies the functional differences between members of the ADSP-2100
Family.

Model	Instruction Cycle Time ns	Internal Program Memory	Internal Data Memory	Host Interface Port	Program Memory Boot	Serial Ports	Programmable Timer	On-chip A/D & D/A	External Interrupts	Low Power Modes	Pin Count
ADSP-2100A	80	CACHE	—	—	—	—	—	—	4	—	100
ADSP-2101	50	2k x 24	1K X 16	—	√	2	√	—	3	1	80/68
ADSP-2102	50	2k x 24 RAM/ROM	1K X 16	—	√	2	√	—	3	1	80/68
ADSP-2103(3 V)	77	2k x 24	1K X 16	—	√	2	√	—	3	1	80/68
ADSP-2105	100	1k x 24	0.5K X 16	—	√	1	√	—	3	1	68
ADSP-2111	50	2k x 24	1K X 16	√	√	2	√	—	3	1	100
ADSP-2115	50	1k x 24	0.5K X 16	—	√	2	√	—	3	1	80/68
ADSP-21msp50A	77	2k x 24	1K X 16	√	√	2	√	√	3	2	144
ADSP-21msp55A	77	2k x 24	1K X 16	√	√	2	√	√	3	2	100
ADSP-21msp56A	77	2k x 24 & 2k x 24 ROM	1K X 16	√	√	2	√	√	3	2	100
ADSP-2161	60	8k x 24 ROM	0.5K X 16	—	√	2	√	—	3	1	80/68
ADSP-2162(3 V)	100	8k x 24 ROM	0.5K X 16	—	√	2	√	—	3	1	80/68
ADSP-2163	60	2k x 24 ROM	0.5K X 16	—	√	2	√	—	3	1	80/68
ADSP-2164(3 V)	100	2k x 24 ROM	0.5K X 16	—	√	2	√	—	3	1	80/68
ADSP-2165	60	1K X 24 RAM 12k x 24 ROM	4K X 16	—	√	2	√	—	3	1	80
ADSP-2166(3 V)	100	2k x 24 ROM	0.5K X 16	—	√	2	√	—	3	1	80
ADSP-2171	30	2k x 24 RAM 8k x 24 ROM	2K X 16	√	√	2	√	—	3	1	128

Table 1.1 ADSP-2100 Family Functional Differences

1 Introduction

This chapter includes overviews of the ADSP-2100 family base architecture and the three family members that exhibit the most distinct features. These overviews include the following ADSP-2100 Family members:

- ADSP-2100 Family Base Architecture–contains the computational units, address generators, and program sequencer

- ADSP-2101–contains the base architecture, plus on-chip memory (program, data, and boot memory), a programmable timer, and enhanced interrupts

- ADSP-2111–contains the features of the ADSP-2101, plus a host interface port (HIP)

- ADSP-21msp50–contains the features of the ADSP-2101, plus a host interface port (HIP) and a voice-band analog front end

Other family members are variations on these DSPs. For example, the ADSP-2105 is based on the ADSP-2101, but it has less on-chip memory and only one serial port; the ADSP-2171 is similar to the ADSP-2111, except it has functional enhancements (low-power operation and expanded instruction set).

Because all the microcomputers of the ADSP-2100 family are code-compatible, the programs in this book can be executed on any DSP in the family, although some modifications for interrupt vectors, peripherals, and control registers may be necessary. All the programs in this book, however, are not designed to use the extra features and functions of some family members.

1.2.1 ADSP-2100 Family Base Architecture

This section gives a broad overview of the ADSP-2100 family base architecture (shown in Figure 1.1). Refer to the *ADSP-2100 Family User's Manual* for additional details.

The base architecture contains three full-function and independent computational units: an arithmetic/logic unit, a multiplier/accumulator and a barrel shifter. The computational units process 16-bit data directly and provide for multiprecision computation.

4

Introduction   1

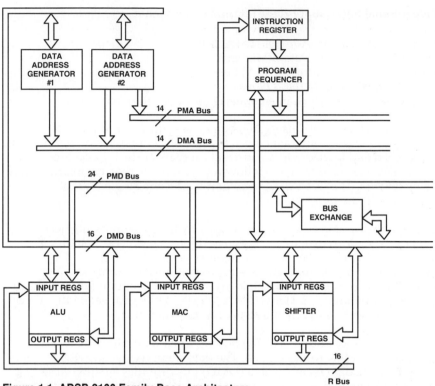

Figure 1.1 ADSP-2100 Family Base Architecture

A program sequencer and two dedicated data address generators (used to simultaneously access data in two locations) supply addresses to memory. The sequencer supports single-cycle conditional branching and executes program loops with zero overhead. Dual address generators allow the processor to output simultaneous addresses for dual operand fetches. Together the sequencer and data address generators allow computational operations to execute with maximum efficiency. The ADSP-2100 family uses an enhanced Harvard architecture in which data memory stores data, and program memory stores instructions and data. This feature lets ADSP-2100 family processors fetch two operands on the same instruction cycle.

1 Introduction

Five internal buses support the internal components.

- Program Memory Address (PMA) bus
- Program Memory Data (PMD) bus
- Data Memory Address (DMA) bus
- Data Memory Data (DMD) bus
- Result (R) bus (which interconnects the computational units)

The program memory data (PMD) bus serves primarily to transfer instructions from program memory to the instruction register. Instructions are fetched and loaded into the instruction register during one processor cycle; they execute during the following cycle while the next instruction is being fetched. The instruction register introduces a single level of pipelining in the program flow.

The next instruction address is generated by the program sequencer depending on the current instruction and internal processor status. This address is placed on the program memory address (PMA) bus. The program sequencer uses features such as conditional branching, loop counters and zero-overhead looping to minimize program flow overhead. The program memory address (PMA) bus is 14 bits wide, allowing direct access to up to 16K words of instruction code and data.

The data memory address (DMA) bus is 14 bits wide allowing direct access of up to 16K words of data. The data memory data (DMD) bus is 16 bits wide. The data memory data (DMD) bus provides a path for the contents of any register in the processor to be transferred to any other register, or to any data memory location, *in a single cycle*. The data memory address can come from two sources: an absolute value specified in the instruction code (direct addressing) or the output of a data address generator (indirect addressing). Only indirect addressing is supported for data fetches through the program memory bus.

The program memory data (PMD) bus can also be used to transfer data to and from the computational units through direct paths or through the PMD-DMD bus exchange unit. The PMD-DMD bus exchange unit permits data to be passed from one bus to the other. It contains hardware to overcome the 8-bit width discrepancy between the two buses when necessary.

Each computational unit contains a set of dedicated input and output registers. Computational operations generally take their operands from input registers and load the result into an output register. The

Introduction 1

computational units are arranged in parallel rather than cascaded. To avoid excessive delays when a series of different operations is performed, the internal result (R) bus allows any of the output registers to be used directly (without delay) as the input to another computation.

There are two independent data address generators (DAGs). As a pair, they allow the simultaneous fetch of data stored in program and in data memory for executing dual-operand instructions in a single cycle. One data address generator (DAG1) can supply addresses to the data memory only; the other (DAG2) can supply addresses to either the data memory or the program memory. Each DAG can handle linear addressing as well as modulo addressing for circular buffers.

With its multiple bus structure, the ADSP-2100 family architecture supports a high degree of operational parallelism. In a single cycle, a family processor can fetch an instruction, compute the next instruction address, perform one or two data transfers, update one or two data address pointers and perform a computation. Every instruction can be executed in a single cycle.

1.2.2 ADSP-2101 Architecture

Figure 1.2 shows the architecture of the ADSP-2101 processor. In addition to the base architecture, the ADSP-2101 has two serial ports, a programmable timer, enhance interrupts, and internal program, data and boot memory.

The ADSP-2101 has 1K words of 16-bit data memory on-chip and 2K words of 24-bit program memory on-chip. The processor can fetch an operand from on-chip data memory, an operand from on-chip program memory and the next instruction from on-chip program memory in a single cycle.

This scheme is extended off-chip through a single external memory address bus and data bus that may be used for either program or data memory access and for booting. Consequently, the processor can access external memory once in any cycle.

Boot circuitry provides for loading on-chip program memory automatically after reset. Wait states are generated automatically for interfacing to a single low-cost EPROM. Multiple programs can be selected and loaded from the EPROM with no additional hardware.

1 Introduction

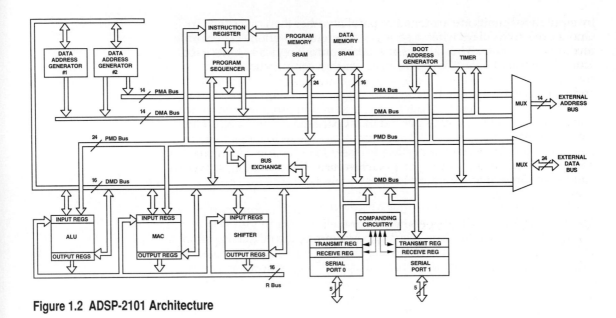

Figure 1.2 ADSP-2101 Architecture

The memory interface supports memory-mapped peripherals with programmable wait-state generation. External devices can gain control of buses with bus request and grant signals ($\overline{BR}$ and $\overline{BG}$). An optional execution mode allows the ADSP-2101 to continue running from internal memory while the buses are granted to another master as long as an external memory operation is not required.

The ADSP-2101 can respond to six user interrupts. There can be up to three external interrupts, configured as edge- or level-sensitive. Internal interrupts can be generated from the timer and the serial ports. There is also a master $\overline{RESET}$ signal.

The two serial ports ("SPORTs") provide a synchronous serial interface; they interface easily and directly to a wide variety of popular serial devices. They have hardware companding (data compression and expansion) with both μ-law and A-law available. Each port can generate an internal programmable clock or accept an external clock.

The SPORTs are synchronous and use framing signals to control data flow. Each SPORT can generate its serial clock internally or use an external clock. The framing synchronization signals may be generated internally or by an external device. Word lengths may vary from three to sixteen bits. One SPORT (SPORT0) has a multichannel capability that allows the receiving or transmitting of arbitrary data words from a 24-word or 32-word bitstream. The SPORT1 pins have alternate functions and can be configured as two additional external interrupt pins and Flag Out (FO) and Flag In (FI).

The programmable interval timer provides periodic interrupt generation. An 8-bit prescaler register allows the timer to decrement a 16-bit count register over a range from each cycle to every 256 cycles. An interrupt is generated when this count register reaches zero. The count register is automatically reloaded from a 16-bit period register, and the count resumes immediately.

1.2.3   ADSP-2111 Architecture

Figure 1.3 shows the architecture of the ADSP-2111 processor. The ADSP-2111 contains the same architecture of the ADSP-2101—plus a host interface port (HIP). This section only contains a brief overview; for detailed descriptions of the HIP and its operation, refer to the ADSP-2111 data sheet and *ADSP-2100 Family User's Manual*.

The host interface port is a parallel I/O port that lets the DSP act as a memory mapped peripheral (slave DSP) to a host computer or processor. You can think of the host interface port as a collection of dual-ported memory, or mailbox registers, that let the host processor communicate with the DSP's processor core. The host computer addresses the HIP as a section of 8-bit or 16-bit words of memory. To the processor core, the HIP is a group of eight data mapped registers.

The host interface port is completely asynchronous. This means that the host computer can write data into the HIP while the ADSP-2111 is operating at full speed.

The ADSP-2111 supports two types of booting operations. One method boots the DSP from external memory (usually an EPROM) through the boot memory interface. The *ADSP-2100 Family User's Manual* describes the boot memory interface in detail. In the second method, a boot program is loaded from the host computer through the HIP. Chapter 12, *Hardware Interface* includes a sample of code to load a program through the HIP.

1 Introduction

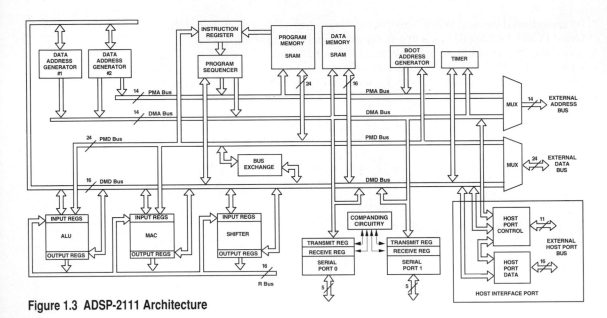

Figure 1.3 ADSP-2111 Architecture

1.2.4   ADSP-21msp50 Architecture

Figure 1.4 shows the architecture of the ADSP-21msp50 processor The ADSP-21msp50 contains the same core architecture of the ADSP-2101— plus a host interface port (described in the previous section) and an analog interface. This section only contains a brief overview; for detailed descriptions of the analog interface and its operation, refer to the ADSP-21msp50 data sheet and *ADSP-2100 Family User's Manual*.

The ADSP-21msp50 has an analog interface that provides the following features:

- linear-coded 16-bit sigma-delta ADC
- linear-coded 16-bit sigma-delta DAC
- on-chip anti-aliasing and anti-imaging filters
- individual interrupts for the ADC and DAC
- 8 kHz sampling frequency
- programmable gain for DAC and ADC
- on-chip voltage reference

Introduction   1

The analog interface is configured and operated through several memory mapped control and data registers. The ADC and DAC I/O can be transmitted and received through individual memory mapped registers, or the data can be autobuffered directly into the processor's data memory.

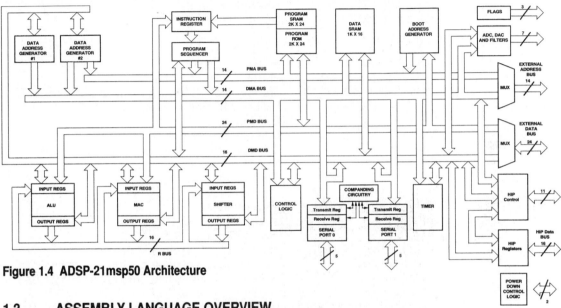

Figure 1.4 ADSP-21msp50 Architecture

1.3   ASSEMBLY LANGUAGE OVERVIEW

The ADSP-2100 family's assembly language uses an algebraic syntax for ease of coding and readability. The sources and destinations of computations and data movements are written explicitly in each assembly statement, eliminating cryptic assembler mnemonics. Each assembly statement, however, corresponds to a single 24-bit instruction, executable in one cycle. Register mnemonics, listed below, are concise and easy to remember.

1 Introduction

Mnemonic	Definition
AX0, AX1, AY0, AY1	ALU inputs
AR	ALU result
AF	ALU feedback
MX0, MX1, MY0, MY1	Multiplier inputs
MR0, MR1, MR2	Multiplier result (3 parts)
MF	Multiplier feedback
SI	Shifter input
SE	Shifter exponent
SR0, SR1	Shifter result (2 parts)
SB	Shifter block (for block floating-point format)
PX	PMD-DMD bus exchange
I0 - I7	DAG index registers
M0 - M7	DAG modify registers
L0 - L7	DAG length registers (for circular buffers)
PC	Program counter
CNTR	Counter for loops
ASTAT	Arithmetic status
MSTAT	Mode status
SSTAT	Stack status
IMASK	Interrupt mask
ICNTL	Interrupt control modes
RX0, RX1	Receive data registers (not on ADSP-2100A)
TX0, TX1	Transmit data registers (not on ADSP-2100A)

Instruction sets for other family members are upward-compatible supersets of the ADSP-2100A instruction set; thus, programs written for the ADSP-2100A can be executed on any family member with minimal changes.

Here are some examples of the ADSP-2100 family assembly language. The statement

```
MR = MR + MX1*MY1;
```

performs a multiply/accumulate operation. It multiplies the input values in registers MX1 and MY1, adds that product to the current value of the MR register (the result of the previous multiplication) and then writes the new result to MR.

Introduction   1

The statement

```
DM(buffer1) = AX0;
```

writes the value of register AX0 to data memory at the location that is the value of the variable *buffer1*.

1.4   DEVELOPMENT SYSTEM

The ADSP-2100 family is supported with a complete set of software and hardware development tools. The ADSP-2100 Family Development System consists of Development Software, to aid in software design, and in-circuit emulators, like the EZ-ICE®, to facilitate the debug cycle. Development tools, like the EZ-LAB® Development Board, are also available to provide a hardware platform for experiments and to evaluate processor functions. Additional development tool capabilities continue to be added as new members of the processor family are introduced.

The Development Software includes:

• System Builder

This module allows the designer to specify the amount of RAM and ROM available, the allocation of program and data memory and any memory-mapped I/O ports for the target hardware environment. It uses high-level constructs to simplify this task. This specification is used by the linker, simulators, and emulators.

• Assembler

This module assembles a user's source code and data modules. It supports the high-level syntax of the instruction set. To support modular code development, the Assembler provides flexible macro processing and "include" files. It provides a full range of diagnostics.

• Linker

The Linker links separately assembled modules. It maps the linked code and data output to the target system hardware, as specified by the System Builder output.

1 Introduction

- Simulator

This module performs an instruction-level simulated execution of ADSP-2100 family assembly code. The interactive user interface supports full symbolic assembly and disassembly of simulated instructions. The Simulator fully simulates the hardware configuration described by the System Builder module. It flags illegal operations and provides several displays of the internal operations of the processor.

- PROM Splitter

This module reads the Linker output and generates PROM-programmer-compatible files.

- C Compiler

The C Compiler reads ANSI C source and outputs source code ready to be assembled. It also supports inline assembler code.

In-circuit emulators provide stand-alone, real-time, in-circuit emulation. The emulators provide program execution with little or no degradation in processor performance. The emulators duplicate the simulators' interactive and symbolic user interface.

Complete information on development tools is available from your local authorized distributor or Analog Devices sales office.

1.5 CONVENTIONS OF NOTATION

The following conventions are used throughout this book:

- Many listings begin with a comment block that summarizes the calling parameters, the return values, the registers that are altered, and the computation time of the routine (in terms of the routine's parameters, in some cases).

- In listings, all keywords are uppercase; user-defined names (such as labels, variables, and data buffers) are lowercase. In text, keywords are uppercase and user-defined names are lowercase italics. Note that this convention is for readability only.

Introduction   1

- In comments, register values are indicated by "=" if the register contains the value or by "—>" if the register points to the value in memory.

- All numbers are decimal unless otherwise specified. In listings, constant values are specified in binary, octal, decimal, or hexadecimal by the prefixes B#, O#, D#, and H#, respectively.

1.6   PROGRAMS ON DISK
This book includes an IBM PC 3½ inch high-density diskette containing the routines that appear in this book. As with the printed routines, we cannot guarantee suitability for your application.

1.7   FOR FURTHER SUPPORT
If you need applications engineering assistance with the applications in this book, please contact:

Applications Engineering from your local Analog Devices distributor

Analog Devices, Inc.
DSP Applications Engineering
One Technology Way
Norwood, MA 02062-9106
Tel: (617) 461-3672
Fax: (617) 461-3010
e_mail: dsp_applications@analog.com

Or log into the DSP Bulletin Board System:
Tel: (617) 461-4258
300, 1200, 2400, 9600, 14400 baud, no parity, 8 bits data, 1 stop bit

Modems ■ 2

2.1 OVERVIEW

The International Telegraph and Telephone Consultative Committee (CCITT), which determines protocols and standards for telephone and telegraph equipment, has authored a number of recommendations describing modem operation. This chapter surveys the fundamental algorithms of the V.32 modem recommendation, which describes the operation of a high-speed modem. Implementations of the algorithms on the ADSP-2100 family of DSP microprocessors are shown.

A modem is an electronic device that incorporates both a **mo**dulator and a **dem**odulator into a single piece of signal conversion equipment. Interfacing directly to the communication channel, modems establish communication links between various computer systems and terminal equipment. In most cases the communications channel is the general switched telephone network (GSTN) or a two- or four-wire leased circuit. The GSTN is, for the most part, a copper wire network. The bandwidth of this channel is limited to 200 Hz to 3400 Hz.

Traditionally, a modem was implemented using analog discrete components. Today, digital circuits centered around a high performance digital signal processor can meet the demands of modem algorithms without the difficulties associated with analog circuitry. A digital modem implementation offers programmability, temperature insensitivity, ease of design and often reduced cost when compared with analog implementations.

2.2 V.32 MODEM DEFINITION

The V.32 recommendation describes a full duplex synchronous modem that operates on the general switched telephone network (GSTN) as well as point-to-point leased circuits. The V.32 modem communicates at a rate of 9600 bits per second (with a 4800 bit per second slow down mode) utilizing quadrature amplitude modulation (QAM). Four-bit symbols (bauds) modulate a carrier frequency of 1800 Hz with a modulation rate of 2400 bauds per second. *The modulation of 4-bit symbols at a rate of 2400 symbols per second yields the 9600 bit per second specification.*

17

2 Modems

There are three signal coding modes to choose from in the V.32 recommendation.

- 9600 bit/second 16-point QAM. Four bits per symbol are transmitted.
- 9600 bit/second 32-point trellis-coded QAM. Transmitted symbols contain four information bits and an additional trellis encoded bit for error correction.
- 4800 bit/second 4-point QAM.

The second method, which produces a redundant bit for error correction, is the method used in the implementation described in this chapter.

Channel separation is achieved through echo cancellation. Echo cancellers are subject to CCITT specification G.165. An ADSP-2100 family implementation of an echo canceller is described in this chapter.

The V.32 modem transmits with a carrier frequency of 1800 ±1 Hz and must be able to operate with received carrier frequency offsets of ±7 Hz. The V.32 recommendation also specifies the transmitted spectrum.

2.2.1   Transmitter Algorithms

A block diagram of the transmitter section of the V.32 modem implemented in this chapter is shown in Figure 2.1. The input serial bit stream is subject to a number of algorithms prior to modulation and transmission. Each step is described briefly below and in greater detail in the following sections.

Scrambler. The input serial bit stream is first scrambled by a self-synchronizing (requires no clock signal) scrambler. Scrambling takes the input serial bit stream and produces a pseudo-random sequence. The purpose of the scrambler is to whiten the spectrum of the transmitted data. Without the scrambler, a long series of identical symbols could cause the receiver to lose carrier lock. Scrambling makes the transmitted spectrum resemble white noise, to utilize the bandwidth of the channel more efficiently, makes carrier recovery and timing synchronization easy and makes adaptive equalization and echo cancellation possible.

Encoders. The scrambled bit stream is divided into groups of four bits. The first two bits of each 4-bit group are first differentially encoded and then convolutionally encoded. This produces a 5-bit symbol in which the first bit is a redundantly coded bit.

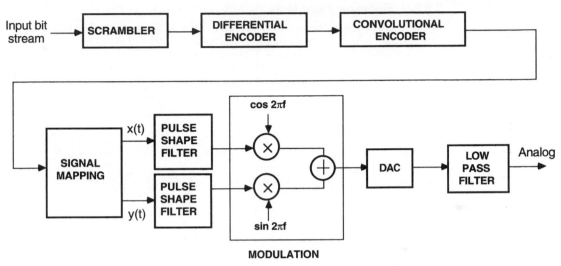

Figure 2.1 Transmitter Block Diagram

Signal Mapping. The 5-bit symbols are mapped into the signal space (defined in the V.32 recommendation) for modulation. The signal space mapping produces two coordinates, one for the real part of the QAM modulator and one for the imaginary part.

Pulse Shape Filters. The pulse shape filter is based on the impulse response of a raised cosine function. Used prior to modulation, these filters attenuate frequencies above the Nyquist frequency that are generated in the signal mapping process. The filters are designed to have zero crossings at the appropriate frequencies to cancel intersymbol interference.

Modulation. The modulation for all coding schemes in the V.32 modem recommendation is quadrature amplitude modulation (QAM). The carrier frequency is 1800 Hz and the modulation rate is 2400 symbols/second.

After modulation, the samples are converted to an analog signal. The analog output is filtered through a smoothing filter.

2   Modems

2.2.2   Receiver Algorithms

A block diagram of the receiver section of the V.32 modem described in this chapter is shown in Figure 2.2. Each step is described briefly below and in greater detail in the following sections.

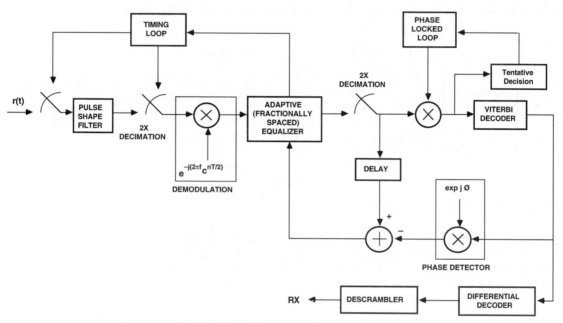

Figure 2.2 Receiver Block Diagram

Input Filter. The received analog signal is oversampled by a factor of 4 at 9600 samples per second. The sampled input is filtered with a raised cosine pulse shape filter. The output is then decimated by a factor of 2.

Demodulation. Multiplication by $e^{-j(2\pi fCnT/2)}$ demodulates the signal. QAM demodulation techniques are described in this chapter.

Adaptive Equalizer. An adaptive equalizer compensates for distortions introduced in the communications channel. A 64-tap fractionally spaced equalizer provides the performance necessary for V.32 applications. The equalizer also feeds a timing loop which adjusts the 4X sampling input and the 2X sampling output of the input filter. An ADSP-2100 family implementation of an adaptive equalizer is described in this chapter.

Viterbi Decoder. The decoder takes as input a demodulated, pulse shaped, equalized signal. The Viterbi algorithm is employed as a decoder in order to determine the appropriate signal constellation point received. This algorithm is a soft-decision maximum likelihood sequence decoder. By keeping a past history of 20 or so baud, the decoder can determine the signal point received in noisy conditions. The phase detector and delay adjust the feedback from the Viterbi decoder to the equalizer, which is constantly adapting in response to the received data.

Differential Decoder and Descrambler. Once the amplitude and phase of the signal point received is known, the corresponding symbol must be back-mapped to decode the encoded bits. The decoded 4-bit symbol is then descrambled utilizing the same generating polynomials as the scrambler.

2.2.3   Scrambler

The V.32 modem recommendation calls for the use of a scrambler in the transmit section of the modem and descrambler in the receive section of the modem. The scrambler and descrambler are based on simple polynomials. Each transmission direction uses a different scrambler, i.e., a different generating polynomial, as specified in the V.32 specification. The calling or call mode modem uses the following generating polynomial (GPC):

$$GPC = 1 + x^{-18} + x^{-23}$$

where x is the input sample and the exponent on x indicates a time delay, e. g., x^{-23} is the twenty-third previous sample. The answering or answer mode modem uses a similar scrambler with the following generating polynomial (GPA):

$$GPA = 1 + x^{-5} + x^{-23}$$

The additions are modulus 2 additions, that is, the bitwise exclusive-OR of the data values. The transmitting modem scrambles the input data sequence by dividing the message sequence by the generating polynomial. The receiving modem multiplies the scrambled sequence by the same polynomial to descramble and recover the original message sequence.

2   Modems

These polynomials can be thought of as digital filters. The scrambler has an all pole transfer function and the descrambler has an all zero transfer function.

The scrambler output is pseudo-random. For a repetitive input signal, the scrambler output is also repetitive with a maximum period of 2^k-1 samples, where k is the order of the generating polynomial (23 in the case of the V.32 scrambler). In order to maximize the period of the pseudo-random output patterns, the specified GPC and GPA are irreducible and primitive.

A block diagram of the call mode scrambler is shown in Figure 2.3; x_{in} is the serial bit input stream and D_S is the scrambled data bit stream. Each delay block corresponds to a serial port cycle and each addition block is an exclusive OR operation.

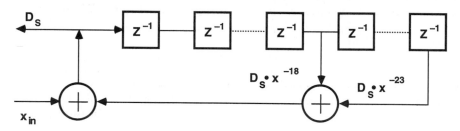

Figure 2.3 Call Mode Scrambler

The answer mode scrambler block diagram (Figure 2.4) is similar. The fifth delay line sample, x^5, is used in the answer mode scrambler rather than the eighteenth delay line value as in the call mode scrambler.

2.2.4   Descrambling

The descrambler is implemented using a delay line, similar to the scrambler. The descrambler is the last functional block that the data passes through in the receiver. The data that is input to the descrambler is in effect multiplied by the appropriate generating polynomial. This multiplication performs the inverse operation of the scrambler.

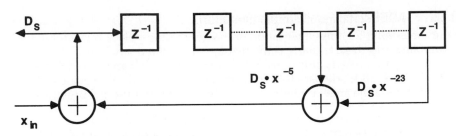

Figure 2.4 Answer Mode Scrambler

There are two versions of the descrambler, one for call mode and one for answer mode. Block diagrams for the call mode and answer mode descramblers are shown in Figures 2.5 and 2.6.

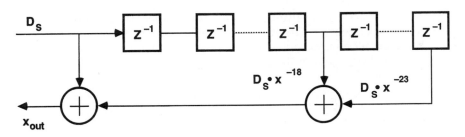

Figure 2.5 Call Mode Descrambler

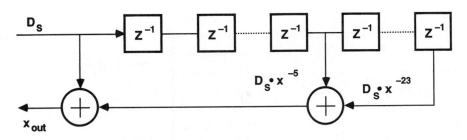

Figure 2.6 Answer Mode Descrambler

2 Modems

2.2.5 ADSP-2100 Family Implementation

Fundamentally, the implementation of the generating polynomials for scrambling and descrambling is the management of a delay line. The scrambler generates its output from the current input bit and two delayed outputs. The call mode uses the eighteenth and twenty-third previous outputs, while the answer mode uses the fifth and twenty-third previous outputs.

The ADSP-2100 family processors have two key features to facilitate efficient delay line management. First, each of two independent data address generators (DAGs) has four independent data pointers. An index register pointer can be programmed to handle each of the delay values and can be separately updated. Second, the DAGs support circular buffers into which delay lines are easily mapped.

In either scrambler, the twenty-third value is the oldest value, and once used is no longer needed. Thus the newest value can be written over it, so the circular buffer always contains only the 23 most recent values. Figure 2.7 illustrates the circular buffer implementation and shows the appropriate pointers.

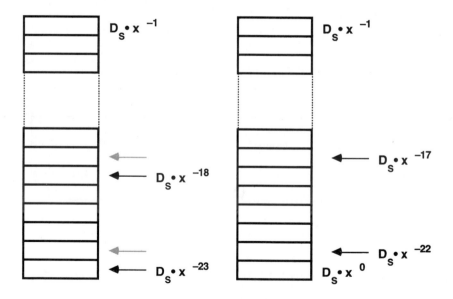

Figure 2.7 Circular Buffer Implementation For Scrambler

The value x^0 is the current input value. This value is put into an ALU register. The delayed value, $D_s \cdot x^{-18}$, is read from the circular buffer using the address supplied by a pointer (represented in the above diagram with an arrow). Once the location is read, the pointer is decremented to the next location in the buffer, shown with the light arrow. The oldest value is then written to an ALU register; the pointer's address is not yet modified. The necessary XOR operations are performed and the result is output, as well as written to the last buffer location. This pointer is now decremented to the next value, now the oldest.

This process is repeated with each new input bit. When a pointer comes to the first location in the circular buffer and is decremented, it wraps around to the last location in the circular buffer. Eighteen and twenty-three unit delays are maintained in the circular buffer, with no need to move data values, just pointer addresses.

The answer mode scrambler works similarly, except with a delay of five units instead of eighteen units. The descrambler, for both call and answer modes, also uses the same basic structure, but with a different flow of data to accomplish the inverse operation.

2.2.6   Scrambler/Descrambler Programs

The code in Listings 2.1 and 2.2 implements the V.32 scrambler (call mode) on the ADSP-2100 family processors. There are two modules, a main module and a scrambler module. The main module sets up interrupts, initializes the appropriate registers for interrupt control, initializes index registers for maintenance of the circular buffer, clears the circular buffer to zero and waits in an infinite loop for an interrupt. The only interrupt active in this program is IRQ3. This is the highest priority interrupt, and in this case it corresponds to a sampling interrupt. When a sample is ready to be scrambled, this interrupt is asserted.

The second program module is the actual scrambling routine. Included as part of this module is the *bits* subroutine, which takes 16-bit data values and strips off bits one at a time. The output of this subroutine is a string of simulated serial data values in the most significant bit position of 16-bit words. That is, a 16-bit word is input and 16 words (each of whose value is either H#8000 or H#0000) are output. These simulated serial bits are then passed to the scrambler. The scrambler output is in the AR register at the end of each pass and is written to the data memory location *dac*.

The descrambler program, in Listing 2.3, has the same fundamental structure as the scrambler program, performing the inverse operation of the scrambler.

2 Modems

```
.MODULE/RAM/ABS=0 cms_main_routine;

{      This module initializes registers, clears a buffer}
{      of length 23 for the call mode scrambler, sets IMASK}
{      and waits in a loop for sampling interrupt}
{      CALLS:      initial, clear_buffer}
{      INTERRUPTS: only interrupt 3 active}

.CONST              no_bits_per_word=16;
.VAR/DM/RAM/CIRC    buffer[23], input_buffer [no_bits_per_word];
.GLOBAL             input_buffer;
.PORT               cntl_port;
.EXTERNAL           start_scramble;

{interrupt jump table}
            RTI;                    {only INT3 is used}
            RTI;
            RTI;
            JUMP start_scramble; {INT3 8 kHz from codec}

{main routine}
            CALL initial;
            CALL clear_buffer;
            IMASK=H#8;              {enable interrupt 3}
mainloop:   JUMP mainloop;         {loop until interrupted}

{————————INIT SUBROUTINE————————}
{One time initialization subroutine, sets up registers}

initial:    IMASK=B#0000;          {disable interrupts}
            ICNTL=H#F;             {edge sensitive interrupts}
            SI=0;
            DM(cntl_port)=SI;      {load codec control register}

            L0=%buffer;            {length registers}
            L1=%buffer;            {circular buffer length 23}
            L2=%buffer;
            L3=0;                  {no other index circ buffer}
            L4=0;
            L5=0;
            L6=0;
            L7=0;
                                   {index registers}
            I0=^buffer;            {ds(n-5)}
            I1=^buffer + 17;       {ds(n-18)}
            I2=^buffer + 22;       {ds(n-23)}
```

```
              I3=0000;
              I4=^input_buffer + 15;

              M0=0;                    {modify registers}
              M1=-1;
              M2=1;
              M4=-1;
              M5=1;
              SE=4;                    {SE for nibble pack}
              RTS;

{————————CLEAR BUFFER SUBROUTINE————————}
{initialize scramble buffer to zero}

clear_buffer:  CNTR=%buffer;
               DO clear UNTIL CE;
clear:            DM(I0,M1)=0;
               RTS;

.ENDMOD;
```

Listing 2.1 Call Mode Scrambler Main Routine

2 Modems

```
.MODULE          call_mode_scrambler;

{     This module performs V.32 call mode scrambling}
{     The generating polynomial is: xin+y(n-18)+y(n-23)}
{     CALLS: bits}

.EXTERNAL        input_buffer;
.CONST           no_bits_per_word=16;
.PORT            codec;
.PORT            dac;
.ENTRY           start_scramble;

start_scramble:  AY0=DM(codec);      {read from port}
                 CALL bits;          {show as serial stream}
                 CNTR=no_bits_per_word;  {scramble 16 times}
                                     {once for every bit of input}
                 DO scrambl UNTIL CE;
                    AY0=DM(I4,M5);
                    AX0=DM(I1,M1);   {d(n-18)}
                    AY1=DM(I2,M0);   {d(n-23)}
                    AR=AX0 XOR AY1;  {d(n-18) + d(n-23)}
                    AR=AR XOR AY0;   {d(n) + d(n-18) + d(n-23)}
                    DM(I2,M1)=AR;    {store scramble in buffer}
                                     {write new value over oldest}
                    DM(dac)=AR;      {out to dac}
                    MODIFY(I4,M4);   {reset pointer to last buffer}
                                     {value for next input word}
scrambl:         NOP;

                 RTI;


{────────────BITS SUBROUTINE────────────}
{ takes output from u_expand (16-bit word) and separates out }
{ the bits; stores as MSB in a 16-word buffer 'input_buffer'}
{ The most significant bit of the input word is at the top of }
{ the buffer}

bits:            AX0=AY0;                 {expanded output into ALU}
                 SE=15;
                 CNTR=no_bits_per_word;
                 AY0=H#8000;
                 DO bit_loop UNTIL CE;
                    AR=AX0;
                    SR=LSHIFT AR (LO);  {shift so next bit is}
                                        {MSB in reg SR0}
```

```
                AR=AR0 AND AY0;    {mask out all except MS}
                DM(I4,M4)=AR;
                AY1=SE;            {decrement SE for next}
                AR=AY1-1;
bit_loop:       SE=AR;
            I4=^input_buffer;
            SE=4;
            RTS;

.ENDMOD;
```

Listing 2.2 Call Mode Scrambler Scrambling Routine

2 Modems

```
.MODULE/RAM/ABS=0 main_routine;

{       Descrambling Routine }
{       Call Mode Functions implemented:}
{            d(n)=di(n)+d(n-18)+d(n-23)}

{       System file:       fullpm.sys}
{       CALLS:       initial, clear_buffer, output}

.VAR/DM/RAM/CIRC  buffer[23];
.PORT             codec;
.PORT             dac;
.PORT             cntl_port;

                  RTI; RTI; RTI;          {int0-2 not used}
                  JUMP start_descramble;  {INT3 8 kHz from codec}
                  CALL initial;
                  CALL clear_buffer;

                  IMASK=h#8;          {enable interrupts}
mainloop:         JUMP mainloop;      {loop until interrupted}

{————— descramble subroutine —————————}
{addressing circular buffer with 2 pointers for modem scrambler}

start_descramble: AY0=DM(codec);     {read from port}
                  AX0=DM(I1,M1);     {d(n-18)}
                  AY1=DM(I2,M0);     {d(n-23)}
                  AR=AX0 XOR AY1;    {d(n-18)+d(n-23)}
                  AR=AR XOR AY0;     {d(n)+d(n-18)+d(n-23)}
                  DM(I2,M1)=AY0;     {store scramble in buffer}
                                     {input stored... not output}
                  CALL output;
                  AR=0;              {clear AR for next time}
                  RTI;

{————— initialize subroutine —————————}
{initialize registers}

initial:          IMASK=B#0000;      {disable interrupts}
                  ICNTL=H#F;         {edge level interrupts}
                  SI=0;
                  DM(cntl_port)=SI;  {load codec control reg}
                  L0=%buffer;        {circular buffer length 23}
                  L1=%buffer;
```

```
                L2=%buffer;
                L3=0;
                L4=0;
                L5=0;
                L6=0;
                L7=0;
                I0=^buffer;
                I1=^buffer + 17;
                I2=^buffer + 22;
                M0=0;
                M1=-1;
                SR0=0;
                SR1=0;
                SE=16;
                RTS;

{———— clear buffer subroutine ————————}
{initialize buffer to zero}

clear_buffer:   CNTR=%buffer;
                DO clear UNTIL CE;
clear:             DM(I0,M1)=0;
                RTS;

{—— output routine packs serial into 16 bit words ——}
output:         SR=SR OR LSHIFT AR(LO);
                AY0=SE;
                AR=AY0 -1;
                SE=AR;
                IF EQ CALL out;
                RTS;

out:            DM(dac)=SR1;
                SR0=0;
                SR1=0;
                SE=16;
                RTS;

.ENDMOD;
```

Listing 2.3 Call Mode Descrambler Routine

2 Modems

2.2.7 Raised Cosine Filter

For the V.32 modem recommendation, 5-bit symbols are modulated by a carrier of 1800 Hz. This modulation is performed digitally. Coupled with the modulator and the demodulator are pulse shaping low pass filters. These digital filters eliminate intersymbol interference (ISI) on the bandlimited GSTN.

A brief development of the theory of pulse shaping filters follows. For a more complete theoretical discussion of pulse shaping filters, see "References" at the end of this chapter: Bingham, Lee and Messerschmitt, Proakis.

Low pass transmitted signals can be shown to have the form

$$\sum_{n=0}^{\infty} I_n g(t-nT)$$

where I_n is the discrete code word and $g(t)$ is a pulse. For the bandlimited channel, we desire a transmitted pulse $g(t)$ that produces no ISI. If the channel is ideally bandlimited, then an ideally bandlimited pulse can be used. In the frequency domain, this ideally bandlimited pulse can be described as:

$$G(f) = \begin{array}{l} T \text{ for } f < 1/2T \\ 0 \text{ for } f \geq 1/2T \end{array}$$

This spectrum has an ideal rectangular shape.

In the time domain, this ideal spectrum shape is the sinc function:

$$g(t) = \sin(\pi t/T)/(\pi t/T)$$

The nulls (zero values of the pulse function) occur at multiples of T, the baud rate. Because of the placement of the nulls, there is no additive interference due to previous symbols; there is no ISI.

The ideal pulse shaping filter is not practical to implement. The ideally bandlimited frequency response has a corresponding infinite impulse response. Although the impulse response has a zero value at all multiples of T, any mistiming in the modem produces an infinite series of ISI terms.

A pulse shaping filter that is practical and widely used in digital communications is the raised cosine pulse shaping filter. The raised cosine pulse shaping filter is realizable, unlike the ideal pulse shaping filter. The raised cosine function has tails that decay proportional to $1/t^3$, whereas the ideal pulse tails off proportional to $1/t$. Mistiming errors in sampling in the modem therefore have a much less dramatic effect on the amount of ISI in the raised cosine pulse filter.

A generic formula for the impulse response of the raised cosine filter, p(t), is shown below. T is the symbol rate in Hz, t is the sampling rate in Hz, and α is the rolloff factor.

$$p(t) = \frac{\sin{(\pi t/T)} \bullet \cos{(\alpha \pi t/T)}}{(\pi t/T) \bullet (1 - (2\alpha \pi t/T)^2)}$$

The rolloff factor, α, represents the amount of excess bandwidth required. A raised cosine with a rolloff factor of 0 needs the least excess bandwidth. As α varies from 0 to 1, the amount of excess bandwidth required increases from 0 to 100%. For purposes of this implementation, a common rolloff factor of 0.25 is used. For the V.32 modem, the symbol rate, T, is specified at 2400 symbols per second. The sampling rate, t, is usually 9600 Hz. The frequency response of the raised cosine pulse shaping filter with these parameter values is shown in Figure 2.8.

The pulse shaping filter usually spans four baud intervals. For a sampling rate of 9600 Hz and a symbol rate of 2400 Hz, a 17-tap FIR filter can be used.

2.2.8   ADSP-2100 Family Implementation

The raised cosine pulse shaping filter can be implemented in the modem as a simple FIR filter. Implementation of FIR filters on the ADSP-2100 family is straightforward. The dual DAGs with circular buffering and the on-chip Harvard architecture allows for efficient realization of FIR filter structures. A complete description of FIR filters as well as other fixed-coefficient filters can be found in *Digital Signal Processing Applications Using the ADSP-2100 Family*, Chapter 5 (see "Literature" at the beginning of this book).

Filter coefficients are arrived at using the formula above, generated with a C program. The coefficients are scaled to provide a filter with 0 dB gain.

2   Modems

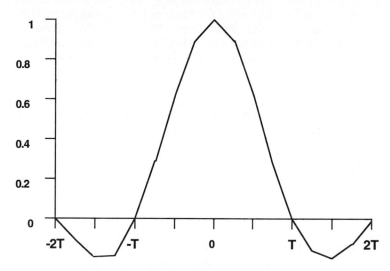

Impulse response

Figure 2.8 Raised Cosine Pulse Shaping Filter, α=0.25

The coefficients represent a rolloff factor of 0.25, and the generated impulse response spans four baud intervals.

For the V.32 modem, the filter input is a digitally modulated value (1800 Hz carrier). Samples are processed at the baud rate (2400 baud) and are interpolated, zero-filled, to provide filter input at a rate of 9600 Hz. Samples are processed in quadrature. Figure 2.9 shows the relationship of the filter to the digital modulator and the data rates.

Listing 2.4 contains the ADSP-2100 family code for implementation of the raised cosine filter. The coefficients can be found in the data file *coef.dat*.

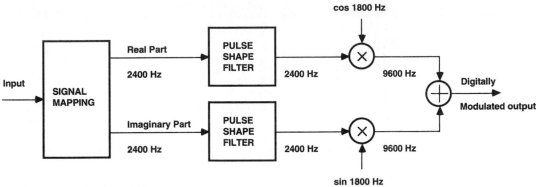

Figure 2.9 Modem Transmitter

```
.MODULE/boot=0          fir_sub;
{──────────────────────────────
        Pulse Shape filter routine for V.32
        ICASSP DEMO

        Rev History     2/8/90 take APP VOL I FIR routine
                        adapt for V.32

}

.ENTRY              pulse_shape;

.CONST              PSF_length=89;
.EXTERNAL           Real_PSF_delay_line, Imag_PSF_delay_line, Pulse_Shape_Coeff;
.EXTERNAL           real_PSF_i0, imag_PSF_i0;

.VAR/DM             psf_save_I0;
.VAR/DM             psf_save_L0;
.VAR/DM             psf_save_I4;
.VAR/DM             psf_save_L4;
.VAR/DM             test_psf1;
.VAR/DM             test_psf2;
```

(listing continues on next page)

2 Modems

```
pulse_shape:        DM(psf_save_I0)=I0; DM(psf_save_L0)=L0;  {save I0,L0,I4,L4}
                    DM(psf_save_I4)=I4; DM(psf_save_L4)=L4;

                    I0=DM(real_PSF_i0);
                    I4=^Pulse_Shape_Coeff;
                    L0=psf_length; L4=psf_length;

{—— Do real part of the filter.  ax0 contains the x value
        from the signal map module.}

                    DM(I0,M2)=AX0;          {dump new vals into delay line}
                    CNTR=PSF_Length-1;
                    MR=0,  MX0=DM(I0,M2), MY0=PM(I4,M5);
sop:                MR=MR+MX0*MY0(SS), MX0=DM(I0,M2), MY0=PM(I4,M5);
                    IF NOT CE JUMP sop;
                    MR=MR+MX0*MY0(RND);
                    IF MV SAT MR;
                    AX0=MR1;                {filtered X in ax0}
                    DM(real_PSF_i0)=I0;

{—— Do the imaginary part of the Pulse Shape filter.  ax1 contains
        the imaginary part of the point from the signal map module. }

                    I0=DM(imag_PSF_i0);
                    DM(I0,M2)=AX1;          {dump new vals into delay line}
                    CNTR=PSF_Length-1;
                    MR=0,  MX0=DM(I0,M2), MY0=PM(I4,M5);
imag_sop:           MR=MR+MX0*MY0(SS), MX0=DM(I0,M2), MY0=PM(I4,M5);
                    IF NOT CE JUMP imag_sop;
                    MR=MR+MX0*MY0(RND);
                    IF MV SAT MR;
                    AX1=MR1;                {filtered Y in ax1}
                    DM(imag_PSF_i0) = I0;

                    I0=DM(psf_save_I0); L0=DM(psf_save_L0);
                    I4=DM(psf_save_I4); L4=DM(psf_save_L4);

            RTS;
```

Listing 2.4 Raised Cosine Filter

2.2.9   Trellis Encoding

The GSTN was intended for voiceband transmission and is bandlimited 200 Hz to 3400 Hz. Data rates in excess of the upper band limit can be realized only by the transmission of multiple bits per symbol interval. Data rates of 9.6 Kbits per second can be achieved on unconditioned circuits and data rates of up to 16.8 Kbits per second can be realized on conditioned leased lines using the technique known as trellis coded modulation (TCM).

The V.32 modem recommendation specifies trellis encoding as an option. Four-bit symbols are encoded into 5-bit symbols that are made up of four information bits and a redundant bit. These 5-bit symbols are used with a 32 carrier state QAM modulator. A 2400 baud rate is used and 9600 information bits per second are transmitted. A trellis encoded scheme offers much better performance than a non-encoded scheme. It results in a much higher immunity to noise for a given error rate and can reduce the block error rate by three orders of magnitude for a given signal-to-noise ratio.

There are two fundamental types of codes used in channel encoding. Linear block codes include Hamming codes, BCH (Bose-Chadhuri-Hocquenghem) codes, Reed-Solomon codes, Galay codes and many others. The convolutional code, which is specified for V.32 modems can be implemented using a shift register and can be described using a diagram called a trellis diagram.

Suppose we can achieve a certain P_e (probability of error) in an uncoded system operating on a bandlimited channel. We can attempt to improve system performance by coding. If we add a single redundant bit to a binary symbol with k bits, we increase the number of waveforms that the modulator must produce from 2^k to 2^{k+1}. An increase in alphabet size on the same bandwidth requires a 3 dB increase in the signal to noise ratio to achieve the same P_e. That is, coding alone decreases the performance of the system.

Trellis coded modulation employs signal set partitioning in addition to redundant coding in order to increase the system performance. In the case of the V.32 modem, there are 32 modulator states. Of the four input bits to the encoder, only two are encoded. Two bits pass through uncoded and two bits are encoded into three output bits. The three bits provide a mechanism for dividing the 32 modulator states into 8 subsets of 4 modulator carrier states. The coded bits identify the subset of the 32

2   Modems

modulator states and the uncoded bits select a point within the subset. Figure 2.10 shows the input and output bits of the trellis encoder. Bits Q1 through Q4 are the input bits. Bits Q3 and Q4 pass through the encoder unchanged. Bits Q1 and Q2 are encoded to give Y1, Y2 and the redundant error correcting bit Y0. Bits Y0, Y1, Y2 identify the subset while the bits Q3 and Q4 identify the point within the subset.

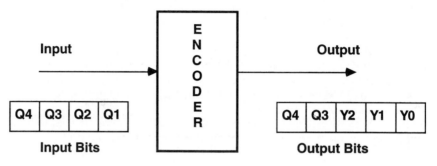

Figure 2.10 Encoder Block Diagram

The signal set for the V.32 modem (and other TCM schemes) has been designed so that there is a large distance between the members of each subset. The 32-state signal constellation for the V.32 modem is shown in Figure 2.11. Bits are ordered on this diagram left to right, most significant to least significant: Y0 Y1 Y2 Q3 Q4. The signal space mapping for the redundant coding is from Figure 3/V.32 of the V.32 recommendation.

The signal set is located on a quadratic grid known as a Z_2 lattice and the signal set type is known as 32 CROSS. In order to transmit m bits per signalling interval, 2^{m+1} signals are needed. The coding gain (performance of the coded signals versus uncoded signals) is approximately 4 dB for any m. The closest distance between any two points on the signal set is Δ_0. The closest distance between any two points in a subset (i.e., points that have the same Y0, Y1 and Y2 bits) is $\sqrt{8}\,\Delta_0$ for the 32 CROSS signal set.

All bit patterns that begin with the same three bits are spread out on the signal constellation. This signal set partitioning along with the redundant coding are the fundamentals of TCM.

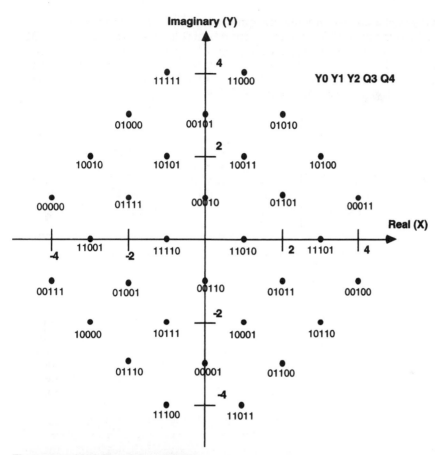

Figure 2.11 V.32 Signal Constellation

2.2.10   ADSP-2100 Family Implementation

Trellis encoding for the V.32 modem consists of two encoding operations: a differential encoder, implemented as a lookup table and a convolutional encoder, performed using a shift register and Boolean logic. Together, these two encoders generate a 5-bit symbol from a 4-bit input word.

The serial input bits to the encoder are Q1, Q2, Q3 and Q4 (Q1 first, Q4 last). Three of the output bits are Y0, Y1 and Y2, and the other two output

2   Modems

bits are Q3 and Q4, unchanged from the input. Y1 and Y2 are generated in the differential encoder. Y0, the redundant bit for error correction, is generated in the convolutional encoder.

The differential encoder takes as input the first two bits, Q1 and Q2, and produces two output bits, Y1 and Y2. Previous output bits, Y1(n–1) and Y2(n–1) are also used in the differential encoder. The encoder is easily implemented on the ADSP-2100 family as a lookup table. The input bits and the previous output bits are combined to a 4-bit value that serves as a pointer into the lookup table. For example, assume that the current input bits are Q1=1, Q2=0, Y1(n–1)=0 and Y2(n–1)=1, for a 4-bit value of 1001. This corresponds to the 1001 (ninth) entry in the lookup table, from which the current Y1 and Y2 outputs are read. Table 2.1 shows the lookup table for differential encoding.

Inputs		Previous Outputs		Outputs	
Q1	Q2	Y1(n-1)	Y2(n-1)	Y1	Y2
0	0	0	0	0	0
0	0	0	1	0	1
0	0	1	0	1	0
0	0	1	1	1	1
0	1	0	0	0	1
0	1	0	1	0	0
0	1	1	0	1	1
0	1	1	1	1	0
1	0	0	0	1	0
1	0	0	1	1	1
1	0	1	0	0	1
1	0	1	1	0	0
1	1	0	0	1	1
1	1	0	1	1	0
1	1	1	0	0	0
1	1	1	1	0	1

Table 2.1 Differential Encoder Lookup Table

Modems 2

The convolutional encoder (Figure 2.12) uses a shift register structure to examine the four incoming bits (the output of the differential encoder) and build a 5-bit symbol. The five output bits of the convolutional encoder consist of the four input bits plus an additional redundantly coded fifth bit. This additional bit increases the complexity of the signal set, but limits the number of possible transitions between bit patterns. For any given 5-bit convolutionally encoded word, only half of the signal states can follow. In other words, the process of convolutional encoding prohibits transitions from any particular signal state to only half of the possibilities. This property is exploited in the Viterbi decoder in the receiver.

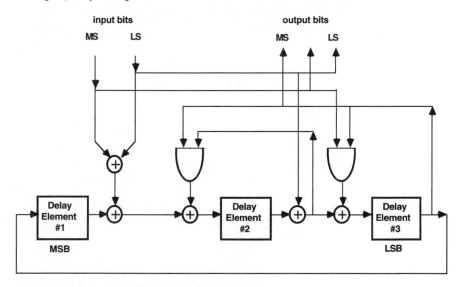

Figure 2.12 Convolutional Encoder Block Diagram

Listing 2.5 contains a ADSP-2100 family subroutine that provides both the differential encoder and the convolutional encoder. The input is assumed to be a single bit residing in the most significant bit position of a 16-bit word. Listing 2.6 shows the convolutional encoder routine that is called by the program in Listing 2.5, and Listing 2.7 contains the routine that performs signal mapping on the encoded data.

2 Modems

```
.MODULE/RAM         trellis;

.VAR/DM/RAM         t_table[16];
.VAR/DM/RAM         last_ys;
.VAR/DM/RAM         bit_count;
.VAR/DM/RAM         diff_out;
.VAR/DM/RAM         delay_val_1;
.VAR/DM/RAM         delay_val_2;
.VAR/DM/RAM         delay_val_3;
.VAR/DM/RAM         Y1;
.VAR/DM/RAM         Y2;
.INIT               t_table: 0,1,2,3,1,0,3,2,2,3,1,0,3,2,0,1;
.ENTRY              trellis_encode;
.PORT               dac;
.PORT               adc;
.GLOBAL             t_table, bit_count, last_ys;


{—bit count is intially 4—}
trellis_encode:  SE=DM(bit_count);
                 SI=DM(adc);          {take in new 8000 or 0000}


Q1Q2_pack:       SR=SR OR LSHIFT SI (LO); {count up 4 bits,}
                 AY0=SE;              {shift into SR register}
                 AR=AY0 -1;
                 SE=AR;
                 DM(bit_count)=SE;    {store decremented count}
                 IF EQ JUMP packed;
                 RTI;


packed:          AX0=SR1;             {stored as 4 bits}
                 AX1=4;               {Q1 Q2 Q3 Q4}
                 DM(bit_count)=AX1;
                 SR0=0;
                 SR1=0;
                 CALL d_encode;
                 RTI;
```

```
{————————————ENCODE————————————}
{input:  AX0 -> 0 0 0 X where X -> bits 0 0 0 0 Q1Q2Q3Q4}

d_encode:        I3=^t_table;
                 AY0=h#000C;          {mask to keep Q1 Q2}
                 AR=AX0 AND AY0;
                 AY1=DM(last_ys);     {last output Y1 Y2}
                 AR=AR XOR AY1;       {AR is Q1 Q2 Y1 Y2}

                 M3=AR;               {address in lookup}
                 MODIFY(I3,M3);       {for new Y1 Y2}

                 SI=DM(I3,M0);
                 DM(last_ys)=SI;      {AY0 ->encoded Y1 Y2}

                 AY1=3;
                 AF=AX0 AND AY1;      {keep Q3 Q4}
                 SR=LSHIFT SI BY 2(LO);
                 AR=SR0+AF;           {AR ->Y1 Y2 Q3 Q4}
                 DM(diff_out)=AR;     {store output of diff encode}
                 DM(dac)=AR;
                 CALL c_encode;       {call convolutional encode}
                 RTS;

.ENDMOD;
```

Listing 2.5 Trellis Encoder Program

2 Modems

```
.MODULE/RAM          conv_encode;

{  Trellis Encoder for V.32 Modem
   Implements convolutional encoder

Input:   Four bit symbols, output of the differential encoder

Output:  Five bit symbol in the LSB positions}

.VAR/DM/RAM        diff_out;           {differential encode output}
.VAR/DM/RAM        conv_out;           {convolutional encode output}
.VAR/DM/RAM        packed_4_bits;      {Q1Q2Q3Q4 as 4 LSBs}
.VAR/DM/RAM        delay_val_1;        {conv. enc delay element}
.VAR/DM/RAM        delay_val_2;        {conv. enc delay element}
.VAR/DM/RAM        delay_val_3;        {conv. enc delay element}
.VAR/DM/RAM        intermed_1;
.VAR/DM/RAM        intermed_2;
.VAR/DM/RAM        Y0;                 {output bit Y0}
.VAR/DM/RAM        Y1;                 {output bit Y1}
.VAR/DM/RAM        Y2;                 {output bit Y2}

.GLOBAL            conv_out;
.GLOBAL            delay_val_1, delay_val_2, delay_val_3;
.GLOBAL            intermed_1, intermed_2, packed_4_bits;

.ENTRY            c_encode;
.EXTERNAL         sig_map, dac;

{————————— CONVOLUTIONAL ENCODE—————————}
{Input is Y1Y2Q3Q4 located in "diff_out" 4 LSBs}
{Output is 3 encoded bits in data mem locations Y0 Y1 Y2}
{calls "pack_up_5_bits" for output to dac}

c_encode:   SR0=0;                     {clear shift result}
            SR1=0;
            SI=DM(diff_out);           {get input from diff encoder}
            SE=-3;
            SR=LSHIFT SI BY -3(HI);  {put Y1 in LSB position}
            AY0=1;
            AR=SR1 AND AY0;            {separate Y1}
            DM(Y1)=AR;
            AX0=AR;
            SR=LSHIFT SI BY -2(HI);
            AR=SR1 AND AY0;            {separate Y2 and store}
            DM(Y2)=AR;
            AY0=AR;
```

44

```
              AR=AX0 XOR AY0;           {op #1}
              AY1=DM(delay_val_3);
              AR=AR XOR AY1;            {op #2}
              DM(intermed_1)=AR;

              AX0=DM(delay_val_1);
              AR=AX0 XOR AY0;           {delay val 1 XOR Y2 op #5}
              DM(intermed_2)=AR;

              AY0=DM(delay_val_2);
              DM(delay_val_3)=AY0;      {update delay val 3}
              AR=AR AND AY0;            {and_1}

              AY1=DM(intermed_1);
              AR=AR XOR AY1;
              DM(delay_val_1)=AR;       {update delay_val_1}

              AX1=DM(Y1);
              AR=AX1 AND AY0;           {and_2}
              AY0=DM(intermed_2);
              AR=AR XOR AY0;

              DM(delay_val_2)=AR;       {update delay val 2}
              DM(Y0)=AR;

              CALL pack_up_5_bits;
              RTS;

{─────────── OUTPUT FORMATTER ───────────}
{Packs up convolutional bits as 5 LSBs Y0 Y1 Y2 Q3 Q4}
{Outputs to DAC}

pack_up_5_bits:   SR0=0;               {pack up bits as Y0Y1Y2Q3Q4}
                  SR1=0;               {clear SR}

                  SR1=DM(diff_out);
                  SI=DM(Y0);

                  SR=SR OR LSHIFT SI BY 4 (HI);
                  DM(conv_out)=SR1;
                  DM(dac)=SR1;

                  SR0=0;
                  SR1=0;

                  CALL sig_map;
                  RTS;

.ENDMOD;
```

Listing 2.6 Convolutional Encoder Routine

2 Modems

```
.MODULE      signal_map;

{  This module takes the output of the convolutional encoder,
   that is, a five bit code residing in the LSBs of the data
   memory location "conv_out", and looks up the x and y coordinates
   as defined by the CCITT spec for the V.32 modem.

   The coordinates are given in the CCITT spec as whole integers.
   They are represented in a 16-bit fixed format as follows:

           integer      hexadecimal
           0            0000
           1            2000
           2            4000
           3            6000
           4            7FFF
           -1           E000
           -2           C000
           -3           A000
           -4           8000

   Registers used:
}

.VAR/DM      x_table[32];
.VAR/DM      y_table[32];

.INIT        x_table: H#8000, H#0000, H#0000, H#7FFF, H#7FFF,
                      H#0000, H#0000, H#8000, H#C000, H#C000, H#4000,
                      H#4000, H#4000, H#4000, H#C000, H#C000, H#A000,
                      H#2000, H#A000, H#2000, H#6000, H#E000, H#6000,
                      H#E000, H#2000, H#A000, H#2000, H#2000, H#E000,
                      H#6000, H#E000, H#E000;

.INIT        y_table: H#2000, H#A000, H#2000, H#2000, H#E000,
                      H#6000, H#E000, H#E000, H#6000, H#E000, H#6000,
                      H#E000, H#A000, H#2000, H#A000, H#2000, H#C000,
                      H#C000, H#4000, H#4000, H#4000, H#4000, H#C000,
                      H#C000, H#7FFF, H#0000, H#0000, H#8000, H#8000,
                      H#0000, H#0000, H#7FFF;
```

```
.EXTERNAL    conv_out, dac;
.ENTRY       sig_map;

sig_map:     I1=^x_table;
             I2=^y_table;

             M0=0;

             M1=DM(conv_out);
             MODIFY(I1,M1);
             MODIFY(I2,M1);

             AX0=DM(I1,M0);      {x value in ax0}
             AX1=DM(I2,M0);      {y value in ax1}

             DM(dac)=ax0;
             DM(dac)=ax1;

             RTS;
.ENDMOD;
```

Listing 2.7 Signal Mapping Routine

2.2.11   Viterbi Decoding

The V.32 recommendation specifies a trellis or convolutional encoding of data before transmission. The most common technique used for decoding received data is Viterbi decoding. The Viterbi algorithm is a general purpose technique for making an error-corrected decision. Viterbi decoding provides a certain degree of error correction by determining from the received bit pattern the value that was the most likely to have been transmitted. The Viterbi algorithm can be used for many applications where error correcting is required. Its application in the V.32 modem is similar to that used in other digital data communication schemes, such as digital telephones.

In order for the Viterbi algorithm to decode received data properly, the model for encoding the transmitted data must be known. In trellis encoding, it is assumed that the three delay elements of the encoder contain zeros initially. At each time period, a new 2-bit input is presented. The contents of the delay elements are changed accordingly and a 3-bit output is produced. If the three delay elements are treated as a 3-bit word, where delay element 1 is the most significant bit and delay element 3 is

2 Modems

the least significant bit, then the state of the delay elements collectively can be represented by that 3-bit value.

It is possible to derive a state diagram or table from this specification. The three delay elements in the encoder are labelled from left to right as element 1, 2 and 3, respectively, in Figure 2.12. At any moment, each delay element has stored in it a 1 or a 0. The possible combinations of bits in the three delay elements or the possible states is eight. The state table shows the eight possible states of these three storage elements. It also shows that for any 2-bit input to the encoder, the three delay elements go to some new state and the encoder also produces an output. The state table showing the state transitions with the encoder inputs and outputs is shown in Table 2.2.

Beginning State	Input	Output	End State	Beginning State	Input	Output	End State
000	00	000	000	100	00	000	010
000	01	101	011	100	01	101	001
000	10	010	010	100	10	010	000
000	11	111	001	100	11	111	011
001	00	000	100	101	00	000	110
001	01	101	101	101	01	101	111
001	10	110	111	101	10	110	101
001	11	011	110	101	11	011	100
010	00	100	001	110	00	100	011
010	01	001	010	110	01	001	000
010	10	110	011	110	10	110	001
010	11	011	000	110	11	011	010
011	00	100	111	111	00	100	101
011	01	001	110	111	01	001	100
011	10	010	100	111	10	010	110
011	11	111	101	111	11	111	111

Table 2.2 State Table For Convolutional Encoder

Table 2.2 can also be used to derive a trellis diagram. The trellis diagram and the state diagram convey equivalent information. The trellis diagram for the convolutional encoder of the V.32 modem is shown in Figure 2.13.

Each node of the trellis represents a state and each node is labelled with the three-bit value of that particular state out of the eight possible states. A line is drawn from a state in one time window to a state of the next time window and represents the transition from one state to another for any given 2-bit input. Figure 2.13 shows some of the trellis paths labelled with the 3-bit output that was produced as the delay elements went from one state to another.

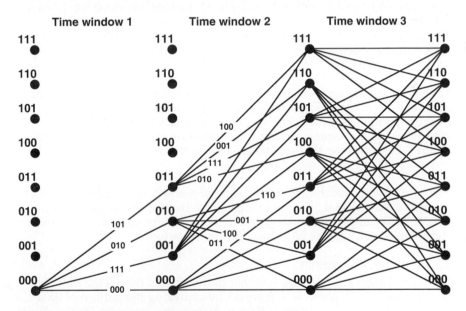

Figure 2.13 Trellis Diagram For Convolutional Encoding

It is assumed that at time t=0, the contents of each delay element is 0. Therefore the starting point for the trellis is at state 000. There are four possible combinations of 2-bit inputs and therefore, four lines that come out of state 000 and connect to the corresponding states at time window 2 as specified by the state table. For example, an input of 01 results in a

2 Modems

change in the state of the delay elements from 000 to 011 with an output of 101. This information is conveyed in the trellis diagram by a line from state 000 to 011 labelled 101. The trellis diagram in Figure 2.13 has some of the branches labelled with the output value that is produced for a specific state transition; the rest can be determined from the state table.

2.2.12   Data Constellation

A 2-bit input to the convolutional encoder produces a 3-bit output containing a redundant bit. Because of redundancy, this 3-bit data value can be corrected for errors that occur during transmission.

In the transmission of information in a V.32 modem, the three bits from the output of the convolutional encoder are combined with two bits coming directly from the data bit stream. In essence, four bits from the data stream are being encoded to five bits (one redundant bit is added to the four original bits).

To modulate a carrier with this information, a constellation is created that maps any 5-bit data value to an X and Y coordinate or a real and imaginary term associated. The real and imaginary terms are used to modulate sine and cosine carriers for quadrature amplitude modulation. Figure 2.14 shows the V.32 constellation with the 3-bit output of the convolutional encoder underlined.

The demodulated carrier yields the original X and Y coordinates which determine the original 5-bit data value. Since the transmission medium for the carrier is noisy, the demodulated data may not be correct. The Viterbi algorithm corrects errors introduced in transmission.

2.2.13   Viterbi Algorithm

The Viterbi algorithm decides whether demodulated data is the data that was sent and if not, corrects it. It works by analyzing the pattern of data values received over a period of time to deduce the data value that is most likely to have occurred at the beginning of the period.

The received carrier is demodulated to produce X and Y coordinates of a point on the signal constellation. The distances from that point on the constellation to the nearest eight points that all have different leading three bits are calculated. These Euclidean distances are then used to label the branches of the trellis diagram. After a number of samples have been received and mapped to the trellis diagram in this fashion, the diagram can be read to determine the shortest path back to the original state, which determines the data value that has the highest probability of having been transmitted at that time.

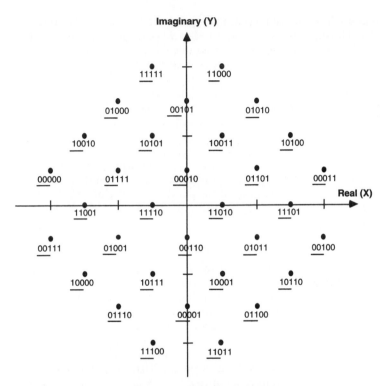

Figure 2.14 Signal Constellation Showing Convolutional Encoder Output

For example, assume that the received signal at time window 1 is mapped into the constellation at coordinate 2, 2 (x, y). This does not correspond to a five-bit code on the constellation. The Euclidean distances from this point to the nearest eight points are calculated. Because of the way the signal map is configured, each of these points has a different value for its first three bits (underlined in Figure 2.14).

In the trellis diagram, the line connecting state 000 to state 011 in time window 1 is labelled 101. The point in the signal constellation that is nearest to 2, 2 and has the value 101 as its first three bits is 10100, at coordinate 3, 2. The Euclidean distance between coordinate 2, 2 and 3, 2 is:

$$[(2–3)^2 + (2–2)^2]^{1/2} = 1$$

2 Modems

Therefore, the branch of the trellis diagram going from state 000 to state 011 is labelled 1. This process is repeated to label the other branches on the trellis diagram. As a new sample is received in each time window, the trellis branches are labelled with the corresponding Euclidean distances.

After a given number of time windows have elapsed, the shortest path back to the start of the first time window is calculated. The branch of the shortest path in the first time window represents the original data value that was transmitted.

Since the data point is determined only after a given number of time windows has elapsed, a delay of (number of time window multiplied by the symbol rate) is incurred. The more time windows that elapse before a decision is made, the more accurate the decision. Thus there is a tradeoff between accuracy and execution time.

2.2.14 ADSP-2100 Family Implementation

The first task of the program is to determine which eight points in the data constellation are the nearest to the X and Y coordinates produced by the demodulator. This is done using a lookup table. Each group in the lookup table contains the X and Y coordinates of the four points in the constellation that have the same 3-bit leading sequences. There are 32 points in the constellation, and therefore eight groups. Because the ADSP-2100 is a 16-bit machine, the X and Y values are normalized for 16-bit data. A negative full scale value of H#8000 and a positive full scale value of H#7FFF are used for both the X and Y values.

For example, 00000, 00001, 00010 and 00011 are in group 0. The Euclidean distance between the received point and the points in the group 0 are calculated. The shortest distance is then written into another table called *min_dist* in which the first location holds the shortest distance of the first group, the second location holds the shortest distance of the second group, etc. Table 2.3 shows the X and Y coordinates in each of the eight groups.

Group	X	Y		Group	X	Y
000	4	1		100	1	2
	0	1			–3	2
	–4	1			1	–2
	0	–3			–3	–2
001	4	–1		101	3	2
	0	–1			–1	2
	–4	–1			3	–2
	0	3			1	0
010	2	3		110	1	0
	–2	3			1	4
	2	–1			–3	0
	–2	–1			1	–4
011	2	1		111	3	0
	–2	1			–1	0
	2	–3			–1	4
	–2	–3			–1	–4

Table 2.3 Lookup Table Of X & Y Coordinates

2.2.15   Shortest Path Through Trellis Diagram

After the distance from the received point for the current time window to the closest point in each group is known, the total distance back to the beginning of the trellis diagram can be calculated. Each time, only the incremental distance for the time window, not the total distance, is calculated.

An 8-location table *acc_dist* stores the accumulated distance through the trellis diagram. Because the trellis diagram starts at state 000, the first location of the table is initialized with a 0 and all other locations with the positive full scale value. This ensures that, for the first time window, all paths converge back to state 000, since this state starts with the shortest accumulated distance.

At each time window, the surviving path to each state is determined and the accumulated distance table is updated with the accumulated distance of each of the eight surviving paths. The surviving path is determined by taking the length of all of the possible paths going into a state and adding that distance to the accumulated distance of the state at the other end of the path.

2 Modems

For example, Figure 2.15 shows the four paths that lead into state 001. The length of each path is added to the accumulated distance of the state from where the path emanates. The length of path 111 is added to the accumulated distance of state 000, the length of path 100 to is added the accumulated distance of state 010, the length of path 101 to is added the accumulated distance of state 100, and the length of path 110 is added to the accumulated distance of state 110. The lengths of these paths are read from the *min_dist* table.

The minimum of these four distances becomes the new accumulated distance to state 001 and is written into the appropriate location of the accumulated distance table (*acc_dist*). As each surviving path leg is determined, a table is filled with the distance of the path and the state from which it came, to allow the program to trace back along the surviving path to the beginning of the trellis diagram.

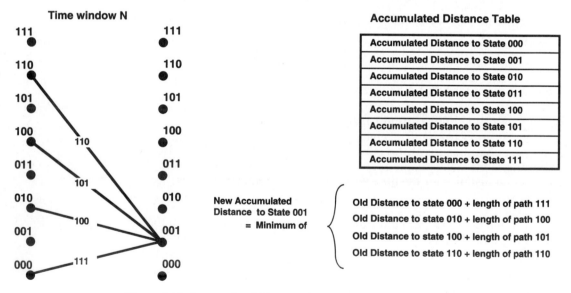

Figure 2.15 Accumulated Distance Table Update Example

After all eight accumulated distances are updated, the shortest of the eight accumulated distances is determined. This path is traced back the given number of time windows. The distance of the branch in the first time window determines the data value most likely to have been transmitted. The point in the data constellation that is this distance from the received point represents the error-corrected symbol.

2.2.16   Viterbi Program

The example program uses N=20 time windows. In general, a value of N which is greater than or equal to three times the constraint length gives good results. In this case, the constraint length is 3, the number of bits needed to describe the possible states at each time window. The larger the value of N, the better the performance of the Viterbi algorithm, but the longer the execution time and the larger the table sizes.

2.2.16.1 Initialization

The first part of the program declares buffers and initializes variables. A buffer to store input data, eight tables holding the coordinates of the eight data groups, eight tables holding the 5-bit codes for the eight data groups, the accumulated distance buffer, eight state-tracing tables, eight buffers to hold surviving path distances and some pointer tables are all declared in the initialization section.

2.2.16.2 Data Input & Euclidean Distance

Data values are placed in registers AX0 and AX1 as X and Y coordinates, respectively, for input to the Viterbi program. The code starting at *find_dist* calculates the distances by calling the subroutine *dist* (which calculates the Euclidean distance squared) followed by the subroutine *sqrt*. This subroutine is repeated for each data group. The table *min_dist* is filled with the shortest distance for each group.

2.2.16.3 Shortest Path

The code starting at *short_path* determines the shortest surviving path to each state for the current time window. It also fills the eight state tables with the distance of the surviving branch and the state from which the branch came. The subroutine *min_calc* compares the four possible surviving paths and determine the shortest.

2.2.16.4 Last Surviving Path

After the accumulated distances to all eight states are calculated, the shortest is determined. The code starting at *search* determines the shortest path and traces this path back to the start of the trellis diagram.

2 Modems

2.2.16.5 Determination Of Error Corrected Data

When the surviving branch of the first time window is determined, the closest point of the data constellation in that data group is found. This 5-bit code is put into the SR1 register.

```
.MODULE/RAM        viterbi;

{Viterbi decoder program for convolutional encoded data for a V.32 modem. This
program decodes information using N=20 levels or time windows of Viterbi decoding.

Demodulated data is stored as input to this routine in registers AX0 and AX1 as
follows;

        AX0=X coordinate
        AX1=Y coordinate

This data is used as input.

The 5-bit data word output by this routine is placed in register SR1.}

.CONST             N=20;
.CONST             base=h#0D49, sqrt2=h#5A82;      {required for square root}
.VAR/PM/RAM        sqrt_coeff[5];
.INIT              sqrt_coeff: h#5D1D00, h#A9ED00, h#46D600,
                               h#DDAA00, h#072D00;


{table for storing last N inputs, as X and Y coordinate
table will contain alternating X, Y for each time window}

.VAR/DM/RAM/CIRC   inputs[N+N];

{variables to hold new X and Y inputs}
.VAR/DM/RAM        x_input;
.VAR/DM/RAM        y_input;
```

56

{tables for X and Y coordinates of data constellation points. Coordinates of both axes are -4, -3, -2 ,-1, 0, 1, 2, 3, 4. They are represented in binary as:

```
        -4      H#8000
        -3      H#A000
        -2      H#C000
        -1      H#E000
         0      H#0000
         1      H#2000
         2      H#4000
         3      H#6000
         4      H#7FFF
}
```

```
.VAR/PM/RAM        group0[8];
.VAR/PM/RAM        group1[8];
.VAR/PM/RAM        group2[8];
.VAR/PM/RAM        group3[8];
.VAR/PM/RAM        group4[8];
.VAR/PM/RAM        group5[8];
.VAR/PM/RAM        group6[8];
.VAR/PM/RAM        group7[8];

.INIT group0:   H#7FFF00, H#200000, H#000000, H#200000,
                H#800000, H#200000, H#000000, H#A00000;
.INIT group1:   H#7FFF00, H#E00000, H#000000, H#E00000,
                H#800000, H#E00000, H#000000, H#600000;
.INIT group2:   H#400000, H#600000, H#C00000, H#600000,
                H#400000, H#E00000, H#C00000, H#E00000;
.INIT group3:   H#400000, H#200000, H#C00000, H#200000,
                H#400000, H#A00000, H#C00000, H#A00000;
.INIT group4:   H#200000, H#400000, H#A00000, H#400000,
                H#200000, H#C00000, H#A00000, H#C00000;
.INIT group5:   H#600000, H#400000, H#E00000, H#400000,
                H#600000, H#C00000, H#E00000, H#C00000;
.INIT group6:   H#200000, H#000000, H#200000, H#7FFF00,
                H#A00000, H#000000, H#200000, H#800000;
.INIT group7:   H#600000, H#000000, H#E00000, H#000000,
                H#E00000, H#7FFF00, H#E00000, H#800000;
```

{lookup table to get proper group}
```
.VAR/DM/RAM     group_table[8];

.INIT group_table:      ^group0, ^group1, ^group2, ^group3,
                        ^group4, ^group5, ^group6, ^group7;
```

(listing continues on next page)

2 Modems

```
{eight tables which show the 5-bit codes that correspond to the X and Y
coordinates in the 8 group tables}
.VAR/DM/RAM          codes0[4];
.VAR/DM/RAM          codes1[4];
.VAR/DM/RAM          codes2[4];
.VAR/DM/RAM          codes3[4];
.VAR/DM/RAM          codes4[4];
.VAR/DM/RAM          codes5[4];
.VAR/DM/RAM          codes6[4];
.VAR/DM/RAM          codes7[4];

.INIT codes0:   h#0003, h#0002, h#0000, h#0001;
.INIT codes1:   h#0004, h#0006, h#0007, h#0005;
.INIT codes2:   h#000A, h#0008, h#000B, h#0009;
.INIT codes3:   h#000D, h#000F, h#000C, h#000E;
.INIT codes4:   h#0013, h#0012, h#0011, h#0010;
.INIT codes5:   h#0014, h#0015, h#0016, h#0017;
.INIT codes6:   h#001A, h#0018, h#0019, h#001B;
.INIT codes7:   h#001D, h#001E, h#001F, h#001C;

.VAR/DM/RAM      codes_table[8];

.INIT codes_table:      ^codes0, ^codes1, ^codes2, ^codes3,
                        ^codes4, ^codes5, ^codes6, ^codes7;


{table for accumulated distances at each state}
.VAR/DM/RAM/CIRC  acc_dist[8];
.VAR/DM/RAM       temp_dist[8];

{eight tables where each table contains the possible states from where a
path could come for each of the eight states}

.VAR/DM/RAM          to_state0[4];
.VAR/DM/RAM          to_state1[4];
.VAR/DM/RAM          to_state2[4];
.VAR/DM/RAM          to_state3[4];
.VAR/DM/RAM          to_state4[4];
.VAR/DM/RAM          to_state5[4];
.VAR/DM/RAM          to_state6[4];
.VAR/DM/RAM          to_state7[4];
```

```
{table is stored with state numbers in backwards order}
.INIT to_state0:  2,4,6,0;
.INIT to_state1:  0,6,4,2;
.INIT to_state2:  6,0,2,4;
.INIT to_state3:  4,2,0,6;
.INIT to_state4:  5,3,7,1;
.INIT to_state5:  3,5,1,7;
.INIT to_state6:  1,7,3,5;
.INIT to_state7:  7,1,5,3;

{eight tables, each with N entries, where each entry contains the label of
the leg of the surviving path for a given time window}

.VAR/DM/RAM/CIRC  state0[N];
.VAR/DM/RAM/CIRC  state1[N];
.VAR/DM/RAM/CIRC  state2[N];
.VAR/DM/RAM/CIRC  state3[N];
.VAR/DM/RAM/CIRC  state4[N];
.VAR/DM/RAM/CIRC  state5[N];
.VAR/DM/RAM/CIRC  state6[N];
.VAR/DM/RAM/CIRC  state7[N];

{eight variables to hold the most recent pointer into the eight state
tables above}

.VAR/DM/RAM        pointer0;
.VAR/DM/RAM        pointer1;
.VAR/DM/RAM        pointer2;
.VAR/DM/RAM        pointer3;
.VAR/DM/RAM        pointer4;
.VAR/DM/RAM        pointer5;
.VAR/DM/RAM        pointer6;
.VAR/DM/RAM        pointer7;

.INIT pointer0:^state0;
.INIT pointer1:^state1;
.INIT pointer2:^state2;
.INIT pointer3:^state3;
.INIT pointer4:^state4;
.INIT pointer5:^state5;
.INIT pointer6:^state6;
.INIT pointer7:^state7;

{table used to look up pointers declared above}
.VAR/DM/RAM        point_table[8];
```

(listing continues on next page)

2 Modems

```
{initialize table with the addresses of the pointers}
.INIT point_table:      ^pointer0, ^pointer1, ^pointer2,
                        ^pointer3, ^pointer4, ^pointer5,
                        ^pointer6, ^pointer7;

{table to hold the eight possible distances, minimum of each group}
.VAR/DM/RAM    min_dist[8];


{interrupt vectors}
            RTI;
            RTI;
            RTI;
            JUMP decode;

            IMASK=0;        {disable all interrupts}
            ICNTL=8;        {interrupts edge sensitive, non-nested}
            ENA AR_SAT;

            I0=^inputs;     {init. I0 to start of input buffer}
            L0=%inputs;     {init. L0 to size of input buffer}
            M0=1;
            M1=0;
            M3=-1;

            L3=N;
            L5=0;

{initialize input buffer to all 0s}
            CNTR=%inputs;    {load counter with size of buffer}
            SI=0;            {put a 0 into register si}
            DO clear_buf UNTIL CE;
clear_buf:      DM(I0,M0)=SI; {transfer 0 into buffer location}

{initialize accumulated distance table}
            I1=^acc_dist;
            L1=%acc_dist;
            DM(I1,M0)=0;
            CNTR=%acc_dist-1;
            DO clear_acc UNTIL CE;
clear_acc:      DM(I1,M0)=h#7FFF;
```

```
{initialize eight tables with 0}
                I2=^state0;
                L2=%state0;
                CNTR=N;
                DO init_table0 UNTIL CE;
init_table0:      DM(I2,M0)=SI;

                I2=^state1;
                L2=%state1;
                CNTR=N;
                DO init_table1 UNTIL CE;
init_table1:      DM(I2,M0)=SI;

                I2=^state2;
                L2=%state2;
                CNTR=N;
                DO init_table2 UNTIL CE;
init_table2:      DM(I2,M0)=SI;

                I2=^state3;
                L2=%state3;
                CNTR=N;
                DO init_table3 UNTIL CE;
init_table3:      DM(I2,M0)=SI;

                I2=^state4;
                L2=%state4;
                CNTR=N;
                DO init_table4 UNTIL CE;
init_table4:      DM(I2,M0)=SI;

                I2=^state5;
                L2=%state5;
                CNTR=N;
                DO init_table5 UNTIL CE;
init_table5:      DM(I2,M0)=SI;

                I2=^state6;
                L2=%state6;
                CNTR=N;
                DO init_table6 UNTIL CE;
init_table6:      DM(I2,M0)=SI;

                I2=^state7;
                L2=%state7;
                CNTR=N;
                DO init_table7 UNTIL CE;
```

(listing continues on next page)

2 Modems

```
init_table7:      DM(I2,M0)=SI;

                  L2=0;
                  IMASK=8;                {enable interrupt 3}
waitlp:           JUMP waitlp;

{————————————————————————————————}

decode:           AX0=DM(codec);
                  AX1=DM(codec);
                  DM(I0,M0)=AX0;          {store X input in input buffer}
                  DM(I0,M0)=AX1;          {store Y input in input buffer}
                  DM(x_input)=AX0;
                  DM(y_input)=AX1;
```

{Calculate Euclidean distances from received point to 32 points of data
constellation. The shortest distance in each data group is saved and will
represent the distance for the trellis branch for the current time window}

```
find_dist:        M4=1;
                  L4=0;
                  I4=^group0;
                  CALL dist;
                  AR=PASS AF;             {put distance squared into AR}
                  MR0=0;
                  MR1=AR;
                  CALL sqrt;
                  DM(min_dist)=SR1;       {store shortest dist in table}


                  I4=^group1;
                  CALL dist;
                  AR=PASS AF;             {put distance squared into AR}
                  MR0=0;
                  MR1=AR;
                  CALL sqrt;
                  DM(min_dist+1)=SR1;     {store shortest dist in table}

                  I4=^group2;
                  CALL dist;
                  AR=PASS AF;             {put distance squared into AR}
                  MR0=0;
                  MR1=AR;
                  CALL sqrt;
                  DM(min_dist+2)=SR1;     {store shortest dist in table}
```

62

```
        I4=^group3;
        CALL dist;
        AR=PASS AF;              {put distance squared into AR}
        MR0=0;
        MR1=AR;
        CALL sqrt;
        DM(min_dist+3)=SR1;   {store shortest dist in table}

        I4=^group4;
        CALL dist;
        AR=PASS AF;              {put distance squared into AR}
        MR0=0;
        MR1=AR;
        CALL sqrt;
        DM(min_dist+4)=SR1;   {store shortest dist in table}

        I4=^group5;
        CALL dist;
        AR=PASS AF;              {put distance squared into AR}
        MR0=0;
        MR1=AR;
        CALL sqrt;
        DM(min_dist+5)=SR1;   {store shortest dist in table}

        I4=^group6;
        CALL dist;
        AR=PASS AF;              {put distance squared into AR}
        MR0=0;
        MR1=AR;
        CALL sqrt;
        DM(min_dist+6)=SR1;   {store shortest dist in table}

        I4=^group7;
        CALL dist;
        AR=PASS AF;              {put distance squared into AR}
        MR0=0;
        MR1=AR;
        CALL sqrt;
        DM(min_dist+7)=SR1;   {store shortest dist in table}

        SR1=H#7fff;
        DM(min_dist+8)=SR1;
```

{Add each path distance to accumulated distance to yield 4 accumulated distances for each state. The shortest accumulated distance becomes the new accumulated distance to that state.}

(listing continues on next page)

2   Modems

{Find shortest path into state 0. Choose from 0, 1, 2, 3 of min_dist table; these correspond to paths back to states 0, 6, 4, 2 respectively. The accumulated distances to these states are added with the paths of the current time window to determine the shortest accumulated path to this point.}

```
short_path:      I2=^min_dist;
                 I3=^to_state0+3;
                 CNTR=4;
                 CALL min_calc;
                 DM(temp_dist)=AR;       {store temporarily}

                 AX0=4;
                 AY0=SI;
                 AR=AX0-AY0;      {calc. label from index of survivor}
                 SR1=AR;          {store label into SR1, pack later}
```

{find the state from which the shortest path came}
```
                 I2=^to_state0-1;
                                  {point to 1 before start of table}
                 M2=SI;           {get index into table}
                 MODIFY(I2,M2);   {point into table}
                 SI=DM(I2,M1);    {get state at end of surviving path}
```

{now that state at end of path is known, store for later along with the 3-bit output label of the suriving path; pack both into 1 word; state in high byte, label low byte}

```
                 SR=SR OR LSHIFT SI BY 8 (HI);
                 I3=DM(pointer0);    {get pointer for state path}
                 DM(I3,M0)=SR1;      {store state for current time window}
                 DM(pointer0)=I3;    {store new pointer}
```

{find shortest path into state 1, choose from 4, 5, 6, 7 of min_dist table these correspond to paths back to states 2, 4, 6, 0 respectively}
```
                 I2=^min_dist+4;
                 I3=^to_state1+3;
                 CNTR=4;
                 CALL min_calc;
                 DM(temp_dist+1)=AR;  {store temporarily}

                 AX0=8;
                 AY0=SI;
                 AR=AX0-AY0;     {calc. label from index of survivor}
                 SR1=AR;         {store label into SR1, pack later}
```

{find the state from which the shortest path came.}

```
          I2=^to_state1-1;        {point to start of table}
          M2=SI;                  {get index into table}
          MODIFY(I2,M2);          {point into table}
          SI=DM(I2,M1);           {get state at end of surviving path}
```

{now that state at end of path is known, store for later use along with the 3-bit output label of the suriving path pack both into 1 word state is in high byte, label lo byte.}

```
          SR=SR or LSHIFT SI BY 8 (HI);
          I3=DM(pointer1);        {get pointer for state path}
          DM(I3,M0)=SR1;          {store state for current time window}
          DM(pointer1)=I3;        {store new pointer}
```

{find shortest path into state 2, choose from 0, 1, 2, 3 of min_dist table these correspond to paths back to states 4, 2, 0, 6 respectively}

```
          I2=^min_dist;
          I3=^to_state2+3;
          CNTR=4;
          CALL min_calc;
          DM(temp_dist+2)=AR;   {store temporarily}

          AX0=4;
          AY0=SI;
          AR=AX0-AY0;     {calc. label from index of survivor}
          SR1=AR;          {store label into SR1, pack later}
```

{find the state from which the shortest path came.}

```
          I2=^to_state2-1;     {point to start of table}
          M2=SI;               {get index into table}
          MODIFY(I2,M2);       {point into table}
          SI=DM(I2,I1);        {get state at end of surviving path}
```

{now that state at end of path is known, store for later use along with the 3-bit output label of the suriving path pack both into 1 word state is in high byte, label lo byte.}

```
          SR=SR or LSHIFT SI BY 8 (HI);
          I3=DM(pointer2);        {get pointer for state path}
          DM(I3,M0)=SR1;          {store state for current time window}
          DM(pointer2)=i3;        {store new pointer}
```

(listing continues on next page)

2 Modems

{find shortest path into state 3, choose from 4, 5, 6, 7 of min_dist table
these correspond to paths back to states 6, 0, 2, 4 respectively}

```
                I2=^min_dist+4;
                I3=^to_state3+3;
                CNTR=4;
                CALL min_calc;
                DM(temp_dist+3)=AR;   {store temporarily}

                AX0=8;
                AY0=SI;
                AR=AX0-AY0;     {calc. label from index of survivor}
                SR1=AR;         {store label into SR1, pack later}
```

{find the state from which the shortest path came.}
```
                I2=^to_state3-1;    {point to start of table}
                M2=SI;              {get index into table}
                MODIFY(I2,M2);      {point into table}
                SI=DM(I2,M1);       {get state at end of surviving path}
```

{now that state at end of path is known, store for later use along with the
3-bit output label of the suriving path pack both into 1 word state is in
high byte, label lo byte.}

```
                SR=SR OR LSHIFT SI BY 8 (HI);
                I3=DM(pointer3);    {get pointer for state path}
                DM(I3,M0)=SR1;      {store state for current time window}
                DM(pointer3)=I3;    {store new pointer}
```

{find shortest path into state 4, choose from 0, 1, 2, 3 of min_dist table
these correspond to paths back to states 1, 7, 3, 5 respectively}

```
                I2=^min_dist;
                I3=^to_state4+3;
                CNTR=4;
                CALL min_calc;
                DM(temp_dist+4)=AR;   {store temporarily}

                AX0=4;
                AY0=SI;
                AR=AX0-AY0;     {calc. label from index of survivor}
                SR1=AR;         {store label into SR1, pack later}
```

66

{find the state from which the shortest path came.}

```
            I2=^to_state4-1;      {point to start of table}
            M2=SI;                {get index into table}
            MODIFY(I2,M2);        {point into table}
            SI=DM(I2,M1);         {get state at end of surviving path}
```

{now that state at end of path is known, store for later use along with the
3-bit output label of the suriving path pack both into 1 word state is in
high byte, label lo byte.}

```
            SR=SR OR LSHIFT SI BY 8 (HI);
            I3=DM(pointer4);      {get pointer for state path}
            DM(I3,M0)=SR1;        {store state for current time window}
            DM(pointer4)=I3;      {store new pointer}
```

{find shortest path into state 5, choose from 4, 5, 6, 7 of min_dist table
these correspond to paths back to states 7, 1, 5, 3 respectively}

```
            I2=^min_dist+4;
            I3=^to_state5+3;
            CNTR=4;
            CALL min_calc;
            DM(temp_dist+5)=AR;   {store temporarily}

            AX0=8;
            AY0=SI;
            AR=AX0-AY0;    {calc. label from index of survivor}
            SR1=AR;        {store label into SR1, will pack later}
```

{find the state from which the shortest path came.}

```
            I2=^to_state5-1;      {point to start of table}
            M2=SI;                {get index into table}
            MODIFY(I2,M2);        {point into table}
            SI=DM(I2,M1);         {get state at end of surviving path}
```

{now that state at end of path is known, store for later use along with the
3-bit output label of the suriving path pack both into 1 word state is in
high byte, label lo byte.}

```
            SR=SR OR LSHIFT SI BY 8 (HI);
            I3=DM(pointer5);      {get pointer for state path}
            DM(I3,M0)=SR1;        {store state for current time window}
            DM(pointer5)=I3;      {store new pointer}
```

(listing continues on next page)

2 Modems

{find shortest path into state 6, choose from 0, 1, 2, 3 of min_dist table these correspond to paths back to states 5, 3, 7, 1 respectively}

```
            I2=^min_dist;
            I3=^to_state6+3;
            CNTR=4;
            CALL min_calc;
            DM(temp_dist+6)=AR;   {store temporarily}

            AX0=4;
            AY0=SI;
            AR=AX0-AY0;      {calc. label from index of survivor}
            SR1=AR;          {store label into SR1, pack later}
```

{find the state from which the shortest path came.}

```
            I2=^to_state6-1;    {point to start of table}
            I2=SI;              {get index into table}
            MODIFY(I2,M2);      {point into table}
            SI=DM(I2,I1);       {get state at end of surviving path}
```

{now that state at end of path is known, store for later use along with the 3-bit output label of the suriving path pack both into 1 word state is in high byte, label lo byte}

```
            SR=SR or LSHIFT SI BY 8 (HI);
            I3=DM(pointer6);    {get pointer for state path}
            DM(I3,M0)=SR1;      {store state for current time window}
            DM(pointer6)=I3;    {store new pointer}
```

{find shortest path into state 7, choose from 4, 5, 6, 7 of min_dist table these correspond to paths back to states 3, 5, 1, 7 respectively}

```
            I2=^min_dist+4;
            I3=^to_state7+3;
            CNTR=4;
            CALL min_calc;
            DM(temp_dist+7)=AR;   {store temporarily}

            AX0=8;
            AY0=SI;
            AR=AX0-AY0;      {calc. label from index of survivor}
            SR1=AR;          {store label into SR1, pack later}
```

```
{find the state from which the shortest path came.}
                I2=^to_state7-1;        {point to start of table}
                M2=SI;                  {get index into table}
                MODIFY(I2,M2);          {point into table}
                SI=DM(I2,M1);           {get state at end of surviving path}
```

{now that state at end of path is known, store for later use along with the 3-bit output label of the suriving path pack both into 1 word state is in high byte, label lo byte.}

```
                SR=SR OR LSHIFT SI BY 8 (HI);
                I3=DM(pointer7);        {get pointer for state path}
                DM(I3,M0)=SR1;          {store state for current time window}
                DM(pointer7)=I3;        {store new pointer}
```

{Put data from temp_dist back into acc_dist as new accumulated distance up to this point.}

```
replace:        CNTR=8;
                I2=^acc_dist;
                I1=^temp_dist;
                I1=0;
                DO move_buf UNTIL CE;
                    SI=DM(I1,M0);       {read data from temp_dist}
move_buf:           DM(I2,M0)=SI;       {put back as new acc_dist}
```

{Search through the acc_dist table for the shortest distance. This will indicate the end point of the surviving path.}

```
search:         I2=^acc_dist;
                CNTR=8;

                SI=CNTR;
                AY0=h#7FFF;             {initialize with largest number}
                AF=PASS AY0;
                AX0=DM(I2,M0);
                DO short_dst UNTIL CE;
                    AR=AF-AX0;
                    IF LE JUMP short_dst;
                    SI=CNTR;                {save index of smallest}
                    IF GE AF=PASS AX0;      {if smaller, update}
short_dst:          AX0=DM(I2,M0);

                AX0=8;
                AY0=SI;
                AR=AX0-AY0;             {calc. which state is at end of surviving path}
```

(listing continues on next page)

2 Modems

{Now that the end of surviving path is known (in AR), trace back N time
windows to find starting path or path of survivor in first time window.}

```
trace:          CNTR=N;                 {trace back N time windows}
                DO search_back UNTIL CE;
```

{read entry from proper state table to find from which state path came}
```
                I2=^point_table;        {point to start of table}
                M2=AR;                  {get offset into table}
                MODIFY(I2,M2);          {modify pointer to point into table}
                AX0=DM(I2,M1);          {read pointer address from table}

                I2=AX0;                 {put pointer address into I2}
                AY1=DM(I2,M2);          {get pntr value, add. into state table}
                I2=AY1;
                AY0=N+1;                {calculate index into state table}
                AX0=CNTR;
                AR=AX0-AY0;
                M2=AR;
                L2=N;
                MODIFY(I2,M2);          {point into state table using circ}
                L2=0;
                SI=DM(I2,M1);           {read contents of state table}
                AX0=SI;
                AY0=h#FF;               {set up mask to isolate path label}
                AF=AX0 AND AY0;             {extract path label}

                SR=LSHIFT SI BY -8 (HI);    {extract state info}
search_back:    AR=SR1;
```

{At this point the surviving leg label is in AF and the state number in AR
find the 5-bit code in the group specified by value in AF that is closest
to the data recieved N time windows ago.}

```
final_stage:    AR=PASS AF;             {put leg label into AR}
                MX1=AR;                 {store leg label in MX1,for later}
                I2=^group_table;        {point to start of group table}
                M2=AR;                  {get displacement into table}
                MODIFY(I2,M2);          {update pointer}
                AX0=DM(I2,M1);          {get address of proper table}
                I4=AX0;                 {load i4 with start of group table}

                AX0=DM(I0,M0);          {get X coord. of input N windows ago}
                M2=-1;
                AX1=DM(I0,M2);          {get Y coord. of input N windows ago}
```

```
                AY0=32767;                          {init with max distance}
                AF=PASS AY0, AY0=PM(I4,M4);         {get X value from table}
                CNTR=4;                             {4 points in group}
                DO ptloop2 UNTIL CE;
                   AR=AX0-AY0, AY1=PM(I4,M4);       {do X-X' and get Y}
                   IF AV JUMP ptloop2;              {if overflow, go on}
                   MY0=AR, AR=AX1-AY1;              {copy X-X', do Y-Y'}
                   IF AV JUMP ptloop2;              {if overflow, go on}
                   MY1=AR;                          {copy Y-Y'}
                   MR=AR*MY1(SS), MX0=MY0;          {square Y-Y', copy X-X'}
                   MR=MR+MX0*MY0(RND);              {add square of X-X'}
                   AR=MR1-AF;                       {compare with previous}
                   IF GE JUMP ptloop2;              {if larger, no update}
                   AF=PASS MR1;                     {if smaller, update}
                   SI=CNTR;                         {save index of closest point}
ptloop2:           AY0=PM(I4,M4);                   {get next X value}

                AX0=4;
                AY0=SI;
                AR=AX0-AY0;              {calculate index from min pointer}
                I2=^codes_table;        {point to start of codes_table}
                M2=MX1;                  {leg label is offset into table}
                MODIFY(I2,M2);
                SI=DM(I2,M1);           {get address of which codes buf}
                I2=SI;
                M2=AR;                  {get index into codes table}
                MODIFY(I2,M2);
                SR1=DM(I2,M1);          {get 5-bit code from table}

{SR1 now contains the answer}
answer:         DM(dac)=SR1;

        RTI;

{————————— SUBROUTINES —————————}
```

{Calculate the Euclidean distance squared between the point specified by the x and y coordinates found data memory locations x_input and y_input and the points specified by the x and y coordinates found in the table pointed to by index register i4. The index denoting the table entry which is closest to the input point is left in register SI and the shortest distance squared is left in register AF.}

```
dist:           AY0=32767;                       {init min distance to max num}
                AX0=DM(x_input);
                AX1=DM(y_input);
                AF=PASS AY0, AY0=PM(I4,M4);       {get X value from table}
                CNTR=4;                           {4 points in group}
```

(listing continues on next page)

2 Modems

```
                DO ptloop UNTIL CE;
                    AR=AX0-AY0, AY1=PM(I4,M4);  {do X-X' and get Y}
                    IF AV JUMP ptloop;          {if overflow, go on}
                    MY0=AR, AR=AX1-AY1;         {copy X-X', do Y-Y'}
                    IF AV JUMP ptloop;          {if overflow, go on}
                    MY1=AR;                     {copy Y-Y'}
                    MR=AR*MY1(SS), MX0=MY0;     {square Y-Y', copy X-X'}
                    IF MV SAT MR;
                    MR=MR+MX0*MY0(RND);         {add square of X-X'}
                    IF VM SAT MR;
                    AR=MR1-AF;                  {compare with previous}
                    IF GE JUMP ptloop;          {if larger, no update}
                    AF=PASS MR1;                {if smaller, update}
                    SI=CNTR;                    {save index of closest point}
ptloop:             AY0=PM(I4,M4);              {get next X value}
                RTS;
```

{————————————————————}

{Take a 32-bit number whose most significant portion is in register MR1 and least significant portion in register MR0 and calculate the 16-bit square root. If the input is interpreted as a 16.16 unsigned number, the output in register SR1 is in 8.8 signed format.}

```
sqrt:           I7=^sqrt_coeff;              {pointer to coeff. buffer}
                M4=1;
                L7=0;
                SE=EXP MR1(HI);              {check for redundant bits}
                SE=EXP MR0(LO);
                AX0=SE, SR=NORM MR1(HI);     {remove redundant bits}
                SR=SR OR NORM MR0(LO);
                MY0=SR1, AR=PASS SR1;
                IF EQ RTS;
                MR=0;
                MR1=base;                    {load constant value}
                MF=AR*MY0(RND), MX0=PM(I7,M4);    {MF = x squared}
                MR=MR+MX0*MY0(SS), MX0=PM(I7,M4); {MR = base + CX}
                CNTR=4;
                DO approx UNTIL CE;
                    MR=MR+MX0*MF(SS), MX0=PM(I7,M4);
approx:             MF=AR*MF(RND);
                AY0=15;
                MY0=MR1, AR=AX0+AY0;                 {SE + 15 = 0?}
                IF NE JUMP scale;                    {no, compute square-root}
                SR=ASHIFT MR1 BY -7 (HI);
                RTS;
```

```
scale:          MR=0;
                MR1=sqrt2;              {load 1 over square rt of 2}
                MY1=MR1, AR=ABS AR;
                AY0=AR;
                AR=AY0-1;
                IF EQ JUMP pwr_ok;
                CNTR=AR;                {compute (1/sqr-rt 2)^(SE+15)}
                DO compute UNTIL CE;
compute:            MR=MR1*MY1(RND);
pwr_ok:         IF NEG JUMP frac;
                AY1=h#0080;             {load a 1 in 9.23 format}
                AY0=0;
                DIVS AY1, MR1;          {compute reciprocal MR}
                DIVQ MR1;
                DIVQ MR1;
                DIVQ MR1;
                DIVQ MR1;
                DIVQ MR1;
                DIVQ MR1;
                DIVQ MR1;
                DIVQ MR1;
                DIVQ MR1;
                DIVQ MR1;
                DIVQ MR1;
                DIVQ MR1;
                DIVQ MR1;
                DIVQ MR1;
                MX0=AY0;
                MR=0;
                MR0=h#2000;
                MR=MR+MX0*MY0(US);
                SR=ASHIFT MR1 BY 1(HI);
                SR=SR OR LSHIFT MR0 BY 1(LO);
                RTS;
frac:           MR=MR1*MY0(RND);
                SR=ASHIFT MR1 BY -7(HI);
                RTS;
```

(listing continues on next page)

2 Modems

{────────────────────────────}

{Take the distances found in the table pointed to by register I2, add them
to the accumulated distance to the state specified in the state table
pointed to by register I3, and determine the shortest of these total
distances. The shortest distance is placed in register AR and the index of
the shortest distance is placed in register SI.}

```
min_calc:       L3=0;
                SI=CNTR;
                AY0=h#7FFF;                 {initialize with largest number}
                AF=PASS AY0;

                MR1=DM(I2,M0);
                SR=ASHIFT MR1 BY -1(HI);     {half scale}
                AX0=SR1;

                DO short_dist UNTIL CE;

                    AY1=DM(I3,M3);           {read state number}
                    I5=^acc_dist;
                    M5=AY1;
                    MODIFY(I5,M5);           {point to proper acc_dist val}
                    MR1=DM(I5,M4);           {get acc_dist value}
                    AR=MR1-AY0;              {check for max value of acc_dist}
                    IF EQ JUMP read_nxt;         {if max go to next}
                    SR=ASHIFT MR1 BY -1(HI);     {half scale}
                    AY1=SR1;

                    AR=AX0+AY1;              {add new path to acc_dist}

                    AX0=AR;

                    AR=AF-AX0;
                    IF LE JUMP read_nxt;
                    SI=CNTR;                 {save index of smallest}
                    IF GE AF=PASS AX0;       {if smaller, update}
read_nxt:           MR1=DM(I2,M0);
                    SR=ASHIFT MR1 BY -1(HI);       {half scale}
short_dist:         AX0=SR1;
                AX0=DM(I2,M0);
                AR=PASS AF;
                L3=N;
                RTS;

.ENDMOD;
```

Listing 2.8 Viterbi Decoder

74

Modems 2

2.3 QUADRATURE AMPLITUDE MODULATION

The CCITT V.32 modem recommendation calls for the use of quadrature amplitude modulation (QAM) in the transmit section and quadrature amplitude demodulation in the receive section of the modem. The encoded digital sequence to be transmitted is amplitude modulated in the digital domain and then converted to analog form (via a D/A converter) for transmission over the telephone wires. At the receiving end of the V.32 system, the received analog signal is digitized (via an A/D converter) and demodulated in the digital domain in order to recover the information that was sent.

This section describes the implementation of quadrature amplitude modulation and demodulation on the ADSP-2100 family of processors.

2.3.1 QAM Methodology

Double-sideband quadrature amplitude modulation (QAM) is a very efficient modulation technique in terms of bandwidth usage. In QAM, two quadrature (90° phase-shifted) carriers, $\cos \omega_c k$ and $\sin \omega_c k$, are amplitude-modulated by two separate information-bearing signals, as shown in Figure 2.16.

The synthesized digital sequence can be expressed as:

$$x(k) = m_1(k) \cos \omega_c k + m_2(k) \sin \omega_c k$$

where $m_1(k)$ and $m_2(k)$ are the two separate information-bearing signals. The QAM signal sequence $x(k)$ has the spectrum:

$$X(2\pi F) = 1/2\ [M_1(\omega - \omega_c) + M_1(\omega + \omega_c)] - j\ 1/2\ [M_2(\omega - \omega_c) - M_2(\omega + \omega_c)]$$

2 Modems

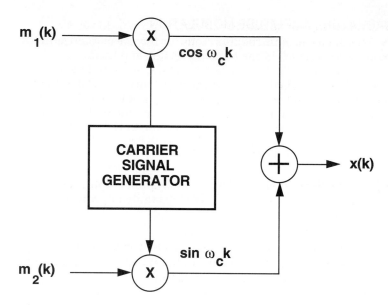

Figure 2.16 QAM Modulator Block Diagram

The spectrum components of the information-bearing signals overlap. However, the quadrature phase relationship in the carrier components $\cos \omega_c k$ and $\sin \omega_c k$ allows the receiving end of the V.32 system to separate the two signals.

The demodulation is performed as shown in Figure 2.17. A digital phase-locked loop is used to obtain the carrier component $\cos \omega_c k$ and to generate $\sin \omega_c k$.

Subsequently, the received sequence is multiplied by the two quadrature carriers. This multiplication results in two signal sequences:

$$x(k) \cos \omega_c k = 1/2\, m_1(k) + 1/2\, m_1(k) \cos 2\omega_c k + 1/2\, m_2(k) \sin 2\omega_c k$$

$$x(k) \sin \omega_c k = 1/2\, m_2(k) + 1/2\, m_2(k) \cos 2\omega_c k + 1/2\, m_1(k) \sin 2\omega_c k$$

The information-bearing signal components $m_1(k)$ and $m_2(k)$ can be recovered by passing each of the sequences through a filter that rejects the double-frequency terms centered at 2ω.

In this particular V.32 implementation, the carrier frequency (F_c) is 1800 Hz, the symbol rate is 2400 Hz and the sample rate of the modulator is 9600 Hz. Thus, the desired cosine carrier is:

$$\cos \omega_c k = \cos 2\pi F_c k T_s = \cos 2\pi(1800)(1/9600)\, k = \cos 3\pi/8\, k$$

and similarly the sine carrier is:

$$\sin \omega_c k = \sin 3\pi/8\, k$$

Again, in this particular V.32 implementation, the sequences $m_1(k)$ and $m_2(k)$ correspond to $i(k)$ and $q(k)$ respectively. These input streams are the filtered versions of quadrature and in-phase portions of the encoded symbols to be transmitted.

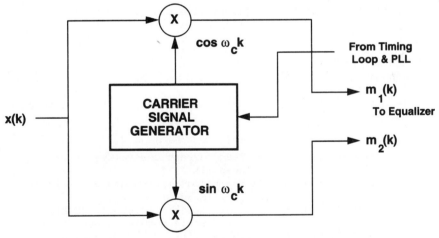

Figure 2.17 QAM Demodulator Block Diagram

2 Modems

2.3.2 ADSP-2100 Family Implementation

There are two ADSP-21XX assembly modules that handle the modulation and demodulation tasks separately. These modules are arranged as interrupt service routines that can be called from a main program which is presumably managing the V.32 modem.

Modulation is performed by the *modulator* routine shown on Listing 2.9. The first section of the code contains the necessary variable, constant and buffer declarations. The *cosine* table contains 16 discrete values of a cosine wave between 0 and 2π, in increments of $\pi/8$. This table is used to generate the $\cos 3\pi/8k$ and $\sin 3\pi/8k$ values for the modulation process. The variable *mod_ptr* stores a pointer into the *cosine* table between interrupts. The *mod_ptr* points to the cosine value to be modulated with the next arriving data sample.

```
.MODULE/RAM        modulator;
.VAR/PM/CIRC       cosine[16];                {Declare cosine table}
.VAR               cos_ptr;
.PORT              mod_out;

.INIT              cosine:<cosval.dat>;       {Initialize the cosine table}
.INIT              cos_ptr:^cosine;           {and the pointer}

.EXTERNAL          q_in, i_in;                {Input ports for i(k) and q(k)}
.GLOBAL            cosine, mod_out;
.ENTRY             modulate;

modulate:          I4=DM(cos_ptr);            {Read current pointer to cosine table}
                   M4=-4;
                   M5=7;
                   L4=16;
                   MX0=PM(I4,M4);             {Read current cos value}
                   MY0=DM(i_in);              {Read I(k)}
                   MR=MX0*MY0(SS),MX0=PM(I4,M5); {cos(k)*I(k) and get -sin value}
                   MY0=DM(q_in);              {Read Q(k)}
                   MR=MR+MX0*MY0(RND);        {cos(k)*I(k)-sin(k)*Q(k)}
                   SR=ASHIFT MR2 BY -1(HI);      {Scale modulated output by 1/2}
                   SR=SR OR LSHIFT MR1 BY -1(LO);
                   DM(mod_out)=SR0;           {Send scaled output}
                   DM(cos_ptr)=I4;            {Save the cosine table pointer}
                   RTI;

.ENDMOD;
```

Listing 2.9 Modulator Code

The main body of the modulator code starts at the label *modulate*. The current cosine pointer is read and used to fetch the proper cosine value from the table. This fetch is done using M4=–4, which modifies the I4 register to point to the proper sine value on the following program memory (PM) fetch. Next, the i(k) input is read and multiplied with the cosine value. Subsequently, the proper sine value is fetched, multiplied with the q(k) input and added to the previous multiplication result. The sine value is fetched using M5=7 which modifies the I4 register to point to the proper cosine value on the following PM fetch. At this point, the MR register contains the output of the QAM modulator. Next, the contents of MR are scaled down by 1/2 using the shifter. This is necessary to keep the output of the modulator within a 16-bit field without causing overflows or underflows. Finally the current I4 value is saved as *mod_ptr* and the output is sent to the D/A converter.

The demodulation is handled by the *demodulator* routine shown in Listing 2.10. The first section of the code contains the necessary variable, constant and buffer declarations. This module also uses the *cosine* table that is declared and initialized in the modulator program. The variable *demod_ptr* points to the next cosine value for the demodulator, just as *mod_ptr* does for the modulator.

The main body of the demodulator code starts at the label *demodulate*. First, the current cosine pointer is read into I4. Next, the variable *phase_shift* is read in order to determine whether the phase-locked loop requires a phase shift in the cosine values to be used in demodulation. If a shift is required, the subroutine *cos_gen* is called to compute new values for the cosine table. Once this is completed, the appropriate cosine value is read from program memory using M4=–4. This value is multiplied with the input from the A/D converter and sent out to the memory location *xcos* which represents $x(k) \cos \omega_c k$. Subsequently, the proper sine value is fetched from program memory using M5=7 and multiplied with the A/D input. This result is sent to the memory location *xsin* which represents $x(k) \sin \omega_c k$. Finally, the current I4 value is saved as *demod_ptr*.

2 Modems

```
.MODULE/RAM       demodulator;
.VAR              cos_ptr;
.PORT             xsin;                    {Sine demodulated received signal}
.PORT             xcos;                    {Cosine demodulated received signal}
.PORT             ad_in;                   {Input port from the A/D}

.INIT             cos_ptr:^cosine;         {Initialize cosine table pointer}

.EXTERNAL         ph_shift_flag, cosine;
.GLOBAL           xsin, xcos;
.ENTRY            demodulate;

demodulate:       I4=DM(cos_ptr);          {Read current ptr to cosine table}
                  AY0=DM(ph_shift_flag);   {Read phase shift flag from the}
                                           {carrier recovery routine}
                  AR=PASS AY0;
                  IF NE CALL phase_shift;  {Call if phase shift desired}
                  M4=-4;
                  M5=7;
                  L4=16;
                  MX0=PM(I4,M4);           {Read the current cosine value}
                  MY0=DM(ad_in);                  {Read the A/D input}
                  MR=MX0*MY0(RND),MX0=PM(I4,M5);     {cos(k)*x(k), get sine value}
                  DM(xcos)=MR1;            {Output cosine demodulated sample}
                  MR=MX0*MY0(RND);         {sin(k)*x(k)}
                  DM(xsin)=MR1;            {Output sine demodulated sample}
                  DM(cos_ptr)=I4;          {Save the cosine table pointer}
                  RTI;

phase_shift:      MODIFY(I4,M4);
                  MODIFY(I4,M5);
                  RTS;

.ENDMOD;
```

Listing 2.10 Demodulator Code

Modems 2

2.4   ECHO CANCELLATION

Most voiceband telephone connections involve several connections through the telephone network. The 2-wire subscriber line available at most sites is generally converted to a 4-wire signal at the telephone central office. The signal must be converted back to a 2-wire signal at the far-end subscriber line. The 2-to-4-wire interface is implemented with a circuit called a *hybrid*. The hybrid intentionally inserts impedance mismatches to prevent oscillations on the 4-wire trunk line. The mismatch forces a portion of the transmitted signal to be reflected or echoed back to the transmitter. This echo can corrupt data the transmitter receives from the far-end modem.

The telephone system and sources of echo are shown in Figure 2.18. There are two types of echo in a typical voiceband telephone connection. The first echo is the reflection from the near-end hybrid, and the second echo is from the far-end hybrid.

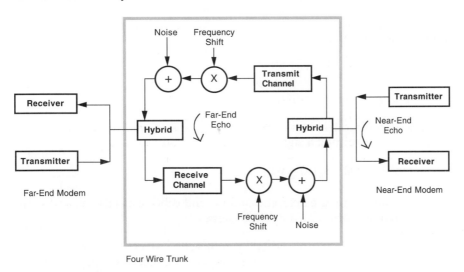

Figure 2.18 Telephone Channel Block Diagram

2   Modems

In long distance telephone transmissions, the transmitted signal is heterodyned to and from a carrier frequency. Since local oscillators in the network are not exactly matched, the carrier frequency of the far-end echo is offset from the frequency of the transmitted carrier signal. In modem applications this shift can affect the degree to which the echo signal can be cancelled. It is therefore desirable for the echo canceller to compensate for this frequency offset.

2.4.1   Echo Cancellation Algorithm

A data signal produced by a modem with a two-dimensional signal constellation has the form

$$s(t) = RE \left[\sum b_m g(t-mT) \, e^{\, j2\pi ft} \right]$$

where b_m is the complex data symbol and $g(t)$ is the baseband pulse shape. The frequency f is the carrier frequency. The echo signal is the transmitted signal convolved with the channel transfer function, $H(f)$. This transfer function usually involves a linear delay and some dispersive filtering. The echo signal has the form

$$s_e(t) = RE \left[\sum b_m h(t-mT) \, e^{\, j2\pi(f+f')t} \right]$$

where f' is the frequency offset (Weinstein, 1977).

If the near-end modem is transmitting a signal $s(n)$ and the far-end modem is transmitting a signal $y(n)$, the near-end received signal is:

$$r(n) = y(n) + s_{ne}(n) + s_{fe}(n) + w(n)$$

where s_{ne} and s_{fe} are the near-end and far-end echo respectively, and $w(n)$ is random noise introduced by the system.

Echo cancellation is accomplished by subtracting an estimate of the echo return signal from the actual received signal. The received signal after echo cancellation is

$$r'(n) = y(n) + (s_{ne}(n) - {}^{\wedge}s_{ne}(n)) + (s_{fe}(n) - {}^{\wedge}s_{fe}(n)) + w(n)$$

where ${}^{\wedge}s_{fe}(n)$ is the estimate of the far-end echo and ${}^{\wedge}s_{ne}(n)$ is the estimate of the near-end echo. Ideally, the estimates are equal to the echo signals and the echo terms drop out (Quatieri and O'Leary, 1989).

The estimated echo is generated by feeding the transmitted signal into an adaptive filter whose transfer function tries to model the telephone channel's (see Figure 2.19). The filter coefficients are determined using the stochastic gradient (Least Mean Squared, or LMS) algorithm (Kamilo and Messerschmitt, 1987) during a training sequence prior to full duplex communications. The LMS algorithm attempts to minimize the mean squared error $|E(n)^2|$. A more detailed description of the LMS algorithm can be found later in this chapter.

In the training sequence, because the far-end modem is not transmitting, the received signal consists of echo:

$r(n) = s_{ne}(n) + s_{fe}(n)$

The output of the filter is an estimate of the received signal,

$r^{\wedge}(n) = {}^{\wedge}s_{ne}(n) + {}^{\wedge}s_{fe}(n)$

and the difference is the error term that the LMS algorithm operates on.

$E(n) = r(n) - r^{\wedge}(n)$

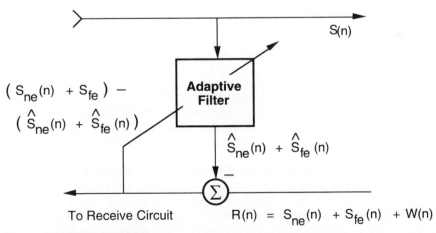

Figure 2.19 Echo Canceller

2 Modems

The adaptive filter is commonly implemented with a transverse FIR filter. The structure of this filter is shown in Figure 2.20. The LMS update equation for tap C at sample time n is

$$C(n)_{k+1} = C(n)_k + \beta A(n)E(n)$$

where $A(n)$ is the sample transmitted at sample time n, $E(n)$ is the residual error and β is an adaptation constant related to the rate of convergence.

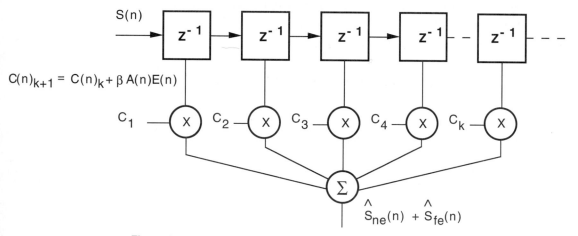

Figure 2.20 LMS Adaptive Filter

In a modem application, the filter taps are only updated during the training periods. The tap update algorithm is either disabled or the adaptation constant β is greatly reduced during full duplex operation. In the second case, reducing β allows the echo canceller to track a slowly changing telephone channel without retraining the modem.

2.4.2   ADSP-2100 Family Implementation Of LMS Algorithm

Figure 2.21 shows a flowchart for implementing the LMS stochastic gradient algorithm on the ADSP-2100 family of processors. The LMS algorithm is implemented in an interrupt service routine so that the arrival of a new sample forces one iteration of the algorithm. In this example, the FIR filter and the tap update are implemented as subroutine calls from the interrupt service routine.

In applications such as V.32 modems, the tap update algorithm gets disabled during full duplex operation.

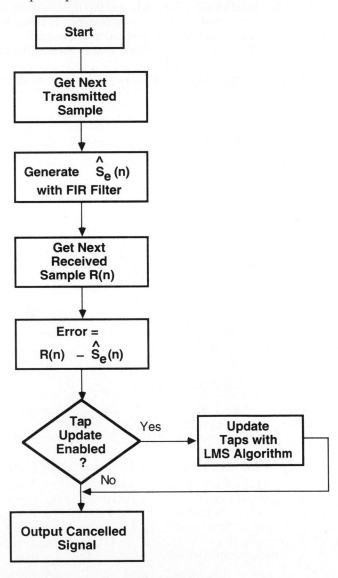

Figure 2.21 Flowchart For LMS Stochastic Gradient Algorithm

2 Modems

Listing 2.11 contains the LMS filter code. The ADSP-2100 family can execute a multiply/accumulate operation and fetch two operands in a single cycle. The FIR filter loop and the tap update loop are executed without any additional cycles for loop overhead. These features allow the FIR filter to execute in one cycle per tap and the coefficient update to execute in two cycles per tap. Table 2.4 summarizes the execution speeds.

Some applications require the echo canceller to operate on complex data. A complex data implementation of the LMS algorithm is described later in this chapter.

```
.MODULE/RAM/ABS=0      adaptive;

{ Near and Far End Echo Canceller
                    INPUT: Received Data from Channel
                    Transmitted Data
                    OUTPUT: To Rest of Modem
}

.PORT      received_data;            {Received sample from channel}
.PORT      transmitted_data;         {Transmitted sample from modem}
.PORT      out;                      {Output to rest of modem}

.CONST              A=154;                      {Adaptive filter length}
.CONST              beta=H#CC;                  {Adaptation constant}
.VAR/DM/RAM/CIRC    enable;                     {Update enabled bit}
.VAR/DM/RAM/CIRC    afilt_data[A];              {Filter delay line}
.VAR/PM/RAM/CIRC    afilt_coeff[A];             {Filter coefficients}

{ Each new sample asserts interrupt 3}
start:    RTI;
          RTI;
          RTI;
          JUMP sample;
```

```
{ Initialize Routine: This is executed during system startup}
.ENTRY     setup;

setup:     ICNTL=B#01111;              {Initialize Interrupts}
           M0=0;                       {Initialize DAGS}
           M1=1;
           M3=-1;
           M4=1;
           M5=1;
           M6=-1;
           M7=2;
           I0=^afilt_data;
           I4=^afilt_coeff;
           L0=%afilt_data;
           L4=%afilt_coeff;
           AX0=H#0000;
           AY1=H#0000;                 {Initialize filter to 0}
           CNTR=%afilt_data;
           DO foo3 UNTIL CE;
foo3:          PM(I4,M4)=AY1,DM(I0,M1)=AX0;
           IMASK=B#1000;               {Enable IRQ2}
fevr:      JUMP fevr;                  {Wait for Interrupt}

{ Interrupt Routine: This code processes one data sample}
sample:    AY0=DM(received_data);      {Received data: r(n)}
           SR0=DM(transmitted_data);   {Transmitted data: A(n)}
           CALL fir;                   {Calculate r^(n)}
           AR=AY0-MR1;                 {AR=error=r-r^}
           DM(out)=AR;                 {Output cancelled data}
           AX0=DM(enable);             {Update taps if enabled}
           AF=PASS AX0;
           IF EQ CALL update;
done:      RTI;

{ FIR Filter
           INPUTS:
               I0=Start of data buffer in DM
               I4=Start of coeff buffer in PM
               SR0=Newest input value
               M1,M4=1
           OUTPUTS:
               MR=Output value
           ALTERS:
               MR, MY0, MX0
}
```

(listing continues on next page)

2 Modems

```
.ENTRY     fir;

fir:       DM(I0,M1)=SR0;
           MR=0, MX0=DM(I0,M1), MY0=PM(I4,M4);
           CNTR=A-1;
           DO floop UNTIL CE;
floop:         MR=MR+MX0*MY0(SS), MX0=DM(I0,M1), MY0=PM(I4,M4);
           MR=MR+MX0*MY0(RND);
           RTS;

{ Adaptive Filter Coefficient Update
           INPUTS:
               I0=Start of data buffer in DM
               I4=Start of coeff buffer in PM
               M1,M4=1
               M6=-1
               M7=+2
               AR=error of last iteration

  Executes the coeff update algorithm as follows:
           Ck+1=Ck+Beta*Error*A(n)
}
.ENTRY     update;

update:    MY1=beta;                            {Load Beta}
                                                {MF=Beta*Error, Load Ck, A(n)}
           MF=AR*MY1(RND), AY0=PM(I4,M4), MX0=DM(I0,M1);
           MR=MX0*MF(RND);
           CNTR=A;                              {Tap update loop}
           DO uloop UNTIL CE;
               AR=MR1+AY0, AY0=PM(I4,M6), MX0=DM(I0,M1);
uloop:         PM(I4,M7)=AR, MR=MX0*MF(RND);
           MODIFY(I0,M3);
           MODIFY(I4,M6);
           RTS;
.ENDMOD;
```

Listing 2.11 LMS Stochastic Gradient Implementation

2.4.3 Frequency Offset Compensation

Frequency offset in the far-end echo can limit convergence of the adaptive filter. In order to compensate for shifts in the carrier frequency, it is necessary to shift the received signal back to the original carrier frequency. Figure 2.22 shows a block diagram for performing this operation. The frequency shifter is a first-order digital phase locked loop (DPLL). The magnitude of the frequency shift is defined as

$$\emptyset^\wedge(n+1) = \emptyset^\wedge(n) + \beta\, A(n)\, (\emptyset(n) - \emptyset^\wedge(n))\, r(n)$$

where β is the adaptation constant, $\emptyset(n)$ is the frequency offset of sample n, $\emptyset^\wedge(n)$ is the estimate of the frequency offset, $A(n)$ is the transmitted sample, and $r(n)$ is the received sample from the echo channel (Wang and Werner, 1988).

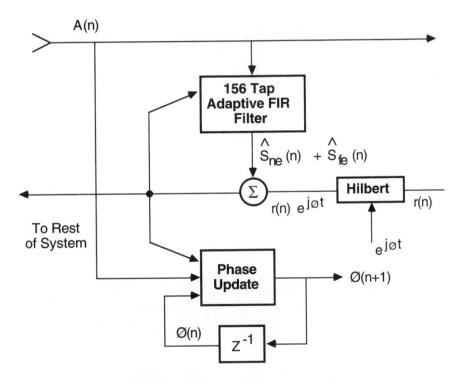

Figure 2.22 Block Diagram Of Echo Canceller With Frequency Shift

2 Modems

When compensating for frequency offset, the received sample must be rotated before the error term is calculated. The new error equation is

$$E(n) = r(n)\ e^{j\emptyset t} - r(n)^\wedge$$

In a real system, the frequency shift is implemented in the time domain with a Hilbert transform algorithm. Figure 2.23 shows the general structure of this algorithm.

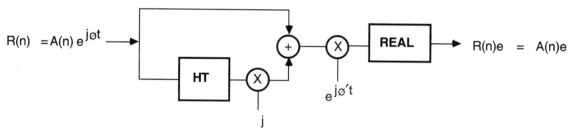

Figure 2.23 Block Diagram Of Hilbert Transform

The Hilbert algorithm is best understood in the frequency domain. Consider the real, bandlimited signal shown in Figure 2.24a. The Hilbert transfer function is

$$
\begin{aligned}
H(\omega) \quad &= \quad -j \quad \omega > 0 \\
&= \quad +j \quad \omega < 0
\end{aligned}
$$

The output of the Hilbert transform is multiplied by +j so that the frequency magnitude is real. The sum of the Hilbert transform and the original sample is complex in the time domain and contains only positive frequencies in the frequency domain. The magnitude in the frequency domain is equal to twice the magnitude of the original sample (Figure 2.24d).

The frequency shift is accomplished by convolving (in the frequency domain) the signal in Figure 2.24d with the desired frequency. This convolution is equivalent to multiplying the time domain signal by $e^{-j\omega_o t}$, where ω_o is the desired frequency shift. The sample is converted back to a real signal by taking the real part of the complex waveform.

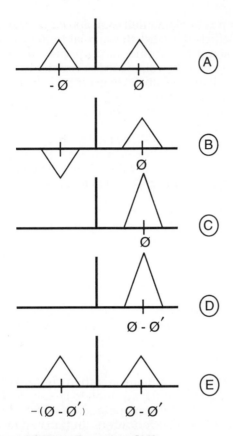

Figure 2.24 Spectrum Of Hilbert Frequency Shift

2.4.4   ADSP-2100 Family Implementation Of Hilbert Transform

Code implementing a Hilbert transform is shown in Listing 2.12. The received signal must be rotated before E_n, the error signal for the adaptive filter, can be calculated. The Hilbert transform is thus performed in a subroutine called from the LMS interrupt service routine.

2　Modems

The Hilbert transform is implemented with a 31-tap transverse FIR filter. Since every other coefficient is zero, the circular buffers in the ADSP-2100 are programmed to access every other data sample. This is possible using multiple modify registers with a single index register in the data address generators. The 31-tap Hilbert transform executes in 20 cycles.

To compensate for the group delay in the Hilbert transform, a 15-cycle linear delay is required for the real-valued input signal. Again, the circular buffering capabilities of the ADSP-2100 family allow for a simple implementation. Once the delay line is initialized, the index registers automatically increment to the next value, even when the end of the buffer is reached. The 15-tap delay line executes in just 3 cycles per sample.

The addition operation described shown in Figure 2.23 is actually summing of a real and a complex number. Since a real and imaginary number cannot be added, this operation is not implemented in the code. Instead, the real and imaginary parts are used in the complex multiplication.

The complex multiply by $e^{-j\omega_o t}$ would normally require four multiplications and two additions. In practice, the desired output is contained entirely in the real part of the product. Therefore, only two multiplications and one addition are required. The values for $\sin(\omega_o t)$ and $\cos(\omega_o t)$ must be calculated for each successive sample.

The single cycle multiply/accumulate operation on the ADSP-2100 family allows both multiplications and the addition to be executed in two cycles. Execution time is also reduced when operands are fetched from data memory in parallel with the multiplications. In transmit mode, the entire Hilbert frequency shift requires about 100 cycles to execute.

Modems   2

```
.MODULE/RAM/ABS=0          hilbert_rotator;

{ Hilbert Rotator
        INPUT: Received Sample
        OUTPUT: To Adaptive Filter
}

.CONST                     H=31;                      {Length of Hilbert xform filter}
.PORT                      received_data;             {Received sample from channel}
.PORT                      out;                       {Output to rest of modem}
.VAR/DM/RAM/CIRC           hdelay[H];                 {Delay line for phase matching}
.VAR/DM/RAM/CIRC           hil_dat[H];                {filter data values}
.VAR/PM/RAM/CIRC           hilbert_coeff[16];         {Hilbert filter coefficients}
.VAR/DM/RAM                time;
.VAR/DM/RAM                delta_time;                {Delta for frequency shift}
.VAR/DM/RAM                high;
.VAR/DM/RAM                low;
.VAR/DM/RAM                ovr;

.INIT                      hilbert_coeff: <hilb.dat>;
                                                      {Hilbert filter coefficients}

{ Initialize Routine: This is executed during system startup}
.ENTRY    setup;

setup:    AX0=H#00;
          DM(time)=AX0;
          AX0=H#02;
          DM(delta_time)=AX0;
          CNTR=^HIL_DAT;               {Init Delay line, Hilbert data}
          DO iloop UNTIL CE;
              DM(I0,M1)=H#0000;
iloop:        DM(I1,M1)=H#0000;
          IMASK=B#1000;                {Enable IRQ2}
fevr:     JUMP fevr;                   {Wait for Interrupt}

{ Interrupt Routine: This code processes one data sample}
sample:   AY0=DM(received_data);       {Received data: r(n)}
          CALL delay;                  {Insert r(n) into delay line}
          CALL hilb;                   {Execute Hilbert transform}
          CALL rotate2;
          AR=MR1;
          DM(out)=AR;
          RTI;
```

(listing continues on next page)

2 Modems

```
{ 31 Tap Linear Delay Line
        INPUTS:         AY0=Newest Input Value
                        I0=Oldest value in delay
                        M0=0
                        M1=1
        OUTPUTS:        AX1=Delay line output
}
.ENTRY  delay;

delay:  AX1=DM(I0,M0);
        DM(I0,M1)=AY0;
        RTS;

{ 31 Tap Fir Hilbert Filter
        INPUTS:         AY0=Newest Input Data
                        I1=Oldest data value
                        I4=First Coeff value
                        M0=0
                        M1=1
                        M4=1
        OUTPUTS:        AY0=Hilbert output
}
.ENTRY  hilb;

hilb:   MR=0, MX0=DM(I1,M2), MY0=PM(I4,M4);
        CNTR=16;
        DO hil_loop UNTIL CE;
hil_loop:       MR=MR+MX0*MY0(SS), MX0=DM(I1,M2), MY0=PM(I4,M4);
        MR=MR+MX0*MY0(RND);
        DM(I1,M1)=AY0;
        AY0=MR1;
        RTS;

{ Hilbert Rotator
        Perform the calculation:
            Y(t)=RE[(Xr(t)+jXi(t)*(exp(-jWt))]

        INPUTS:         AY0=Xi(t)
                        AX1=Xr(t)
                        AY1=W in degrees-q15 format
                        W*t=DM(time)=time in q15
        OUTPUTS:        MR=Y(t)
}
```

```
.ENTRY    rotate2;

rotate2: AX0=DM(time);                  {Get and update rotate time}
         AY1=DM(delta_time);            {on unit circle}
         AR=AX0+AY1, MY0=AY0;           {MY0=im(x)}
         IF AC AR=PASS 0;
         DM(time)=AR;
         CALL sin;                      {Xi(t)*IM[exp(-jwt)]}
         MR=AR*MY0(SS), MY0=AX1;
         DM(ovr)=MR2;
         DM(high)=MR1;
         DM(low)=MR0;
         AY0=H#4000;                    {Xr(t)*sin(wt+90)}
         AR=AX0+AY0;
         AX0=AR;
         CALL sin;
         MR0=DM(low);
         MR1=DM(high);
         MR2=DM(ovr);
         MR=MR+AR*MY0(RND);
         RTS;

{ Sine Calculation
         Sine Approximation:  Y=Sin(x)

         INPUTS:        AX0=x in scaled 1.15 format
                        M3=1
                        L3=0
         OUTPUTS:       AR=y in 2.14 format

         Computation Time: 25 cycles
}
```

(listing continues on next page)

2 Modems

```
.VAR/DM   sin_coeff[5];
.INIT     sin_coeff: H#3240, H#0053, H#AACC, H#08B7, H#1CCE;
.ENTRY    sin;

sin:    I3=^sin_coeff;                   {Pointer to coeff. buffer}
        AY0=H#4000;
        AR=AX0, AF=AX0 AND AY0;      {Check 2nd or 4th quad}
        IF NE AR=-AX0;               {If yes, negate input}
        AY0=H#7FFF;
        AR=AR AND AY0;                   {Remove sign bit}
        MY1=AR;
        MF=AR*MY1(RND), MX1=DM(I3,M3);           {MF=x2}
        MR=MX1*MY1(SS), MX1=DM(I3,M3);           {MR=C1x}
        CNTR=3;
        DO approx UNTIL CE;
           MR=MR+MX1*MF(SS);
approx:    MF=AR*MF(RND), MX1=DM(I3,M3);
        MR=MR+MX1*MF(SS);
        SR=ASHIFT MR1 BY 2(HI);
        SR=SR OR LSHIFT MR0 BY 2(LO);  {Convert to 2.14 format}
        AR=PASS SR1;
        IF LT AR=PASS AY0;          {Saturate if needed}
        AF=PASS AX0;
        IF LT AR=-AR;                  {Negate output if needed}
        RTS;
.ENDMOD;
```

Listing 2.12 Hilbert Transform Implementation

2.4.5   V.32 Modem Implementation

V.32 modems operate in full duplex mode; both the near-end and far-end modem are transmitting data at the same time. The echo canceller is responsible for channel separation as well as cancelling the near-end and far-end echos.

The echo canceller can be implemented in the passband or the baseband. The advantage of passband cancellation is reduced computation. A baseband echo canceller must execute all algorithms on complex data. In addition, compensating for frequency shift in the baseband is difficult. The disadvantage of passband echo canceller is a longer convergence time for the adaptive filter and the digital phase locked loop. Figure 2.25 shows a block diagram of a V.32 modem with a passband echo canceller.

Modems   2

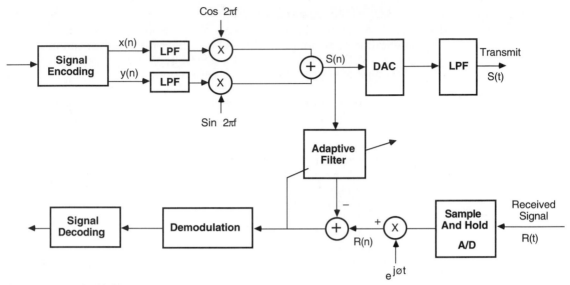

Figure 2.25 V.32 Modem Block Diagram

The CCITT specification for V.32 modems recommends a carrier frequency of 1800±7 Hz. The echo canceller must be able to cancel 16 ms of echo. At 9600 samples/second, a 154-tap FIR filter is required to cancel the echo. It is recommended that the echo canceller be implemented with a minimum number of taps.

Assuming that the canceller and frequency shifter have converged during the training period, about 200 cycles are required to cancel a V.32 signal. Benchmarks are summarized in Table 2.4.

Operation	Cycles	@12.5 MHz
Real FIR Filter	N + 6	80 ns per tap
Complex FIR Filter	4 (N–1) + 21	240 ns per tap
Real LMS Update (Stochastic)	2N + 9	160 ns per tap
Complex LMS Update (Stochastic)	6N + 10	480 ns per tap
154-Tap LMS Filter With Update	935	74.8 μs

N = Number of Taps

Table 2.4 ADSP-2100 Family Benchmarks For Echo Cancellation

2 Modems

2.5   ADAPTIVE EQUALIZATION

This section presents subroutines for an ADSP-2100 family implementation of an adaptive channel equalizer for a high speed modem. The CCITT's V.32 recommendation for a 9600 bps modem specifies the use of this type of equalizer in the receiver section.

The architecture used in this equalizer is a fractionally-spaced tapped delay line with a least-mean-squared (LMS) algorithm for adapting the tap weights.

The topics discussed in this section are:

- Historical perspective of adaptive filters
- Applications of adaptive filters
- Channel equalization in a modem
- Equalizer structures
- Least Mean Square (LMS) Algorithm
- Program Structure
- Practical considerations

2.5.1   History Of Adaptive Filters

Until the mid-1960s, telephone-channel equalizers were either fixed equalizers that caused performance degradation or manually adjustable equalizers that were cumbersome to adjust.

In 1965, Lucky (see "References" at the end of this chapter) introduced the zero-forcing algorithm for automatic adjustment of the equalizer tap weights. This algorithm minimizes a certain distortion, which has the effect of forcing the intersymbol interference (ISI) to zero. This breakthrough by Lucky inspired other researchers to investigate different aspects of the adaptive equalization problem, leading to new improved solutions.

Modems   2

Proakis and Miller (1969) reformulated the adaptive equalizer problem using a new criterion known as the mean squared error (MSE). This formulation requires a relatively modest amount of computation and remains the most popular approach for data rates up to 9600 bits/s.

Three years later, Ungerboeck (1972) improved on this work by presenting a detailed mathematical analysis of the convergence properties of an adaptive transversal equalizer using the least-mean-squared (LMS) algorithm. This algorithm is described later in this chapter.

A more powerful algorithm for adjusting the tap weights based on Kalman filtering theory was developed soon afterward (Godard, 1974). This algorithm is computationally demanding, but it was later modified by Falcomer and Ljing (1978) to simplify its computational complexity.

All of these adaptive equalizer implementations are synchronous, that is, the spacing between taps is equal to the reciprocal of the symbol interval. Other possible structures include the fractionally spaced equalizer (FSE) and the decision feedback equalizer (DFE).

The FSE has the ability to better compensate for channel distortion by spacing the tap weights more closely than in the conventional synchronous equalizer. Brady (1970) did some early work on this class of equalizers and was followed by Ungerboeck (1976). The DFE, on the other hand, uses a more elaborate structure and can yield good performance in the presence of severe ISI as experienced in fading radio channels.

2.5.2   Applications Of Adaptive Filters

Adaptive filters offer a significant improvement in performance over fixed-tap-weight digital filters because of their ability to detect signals in environments of unknown characteristics. They are successfully used in several areas including:

System Identification And Modeling
An adaptive transversal filter can be forced to converge to the same impulse response as an unknown linear system and then can be used to model the unknown system. To determine the taps for this filter, an excitation input drives both the unknown system and the adaptive filter. The outputs of these two systems are compared, and the error signal generated is used to adjust the tap weights of the adaptive filter to reduce the error size. After a sufficiently large number of iterations, the error is reduced to some small value (in a statistical sense) and the tap weights converge to model the real system.

2 Modems

If the unknown system is dynamic and time-variant, the adaptive filter can track these variations provided they are sufficiently slow compared to the convergence time of the filter.

Echo Cancellation
In telephone systems that include both 2-wire and 4-wire loops, hybrid circuits couple these lines. These hybrid circuits create impedance mismatches which in turn create signal reflections, heard at both ends of the line as echo. This echo is tolerable to some degree over long distance voice connections, but can be catastrophic in high-speed data transmission over cross-Atlantic links.

Echo cancellers, in the form of adaptive filters, model the impulse response of the echo path. Cancellation is achieved by making an estimate of the echo and subtracting it from the return signal.

Linear Predictive Coding
In the past 20 years, digital coding of speech waveforms has become a popular technique for reducing speech degradation due to transmission. Of the speech coding techniques, linear predictive coding (LPC) stands out for its ability to produce low data rates. Basic speech parameters (e.g. pitch, vocal tract, formants) are estimated, transmitted and then used at the receiver to resynthesize the speech through a speech production model. Adaptive filters can be used to estimate speech parameters in model-based speech coding systems.

The speech quality of LPC is synthetic when compared to other coding techniques such as PCM or ADPCM; however, its significantly lower data rates make it attractive. The GSM standard for the Pan-European cellular digital mobile radio network specifies an LPC-based coding scheme.

Adaptive Beamforming
A spatial form of adaptive signal processing finds applications in radar and sonar. By combining signals from an array of sensors, it is possible to change the directivity pattern of the array. Independent sensors (e.g. antennas or hydrophones) placed at various locations in space or water detect incoming waveforms. The collection of sensor outputs at a particular instant is analogous to the set of consecutive tap inputs in a transversal filter. The sensitivity and directivity of the sensor array can be adaptively adjusted. Beamforming is discussed in Chapter 15 of *Digital Signal Processing Using the ADSP-2100 Family*.

Modems 2

Adaptive Channel Equalization For Data Transmission
Adaptive filters used in digital communication systems as channel equalizers minimize transmission distortion and maximize the use of channel bandwidth. A typical bandlimited telephone channel or radio link suffers from intersymbol interference (ISI) and additive noise. To improve system performance in additive-noise channels, transmission power can be increased. However, increased power has no effect on ISI since it amplifies both the intended symbol sample as well as interfering ones.

The traditional technique for alleviating ISI is an equalizing filter at the receiver. The receiver equalizer filter combines the channel characteristics and the transmitter filter to minimize ISI distortions. Channel characteristics, however, vary over time. An adaptive equalizer is needed to ensure a constant transmission quality.

Since the channel conditions are unknown, a training sequence is transmitted to bring up the equalizer from its initial (usually zero) state. This sequence is known at the receiver and therefore the deviation error of received samples from the expected sequence is used to adjust the equalizer tap weights. Once the training period is completed, the weights can still be continually updated in a decision directed mode. In this mode, a minimum distance detector at the receiver decides which symbol was transmitted. In normal operation these decisions have a high probability of being correct, and thus are good enough to allow the equalizer to maintain proper adjustment.

2.5.3   Channel Equalization In A Modem

The International Telegraph and Telephone Consultative Committee (CCITT) sets standards and protocols for telephone and telegraph equipment. Its V.32 modem recommendation specifies a fractionally spaced transversal filter as the channel equalizer in the receiver. This equalizer, along with trellis coding and quadrature amplitude modulation (QAM), maximizes data rates over the bandlimited telephone channel.

2 Modems

A telephone channel can suffer from a variety of limitations as a communications medium:

- As a bandlimited channel, it creates an environment for ISI.
- Channel additive noise requires increased transmitted power to improve signal-to-noise ratio.
- Radio links create fading channels and echo in cross-Atlantic connections
- When several connections are frequency multiplexed, baseband speech signals are modulated into the passband using different carrier frequencies for transmission. Demodulating these passband signals can create frequency offsets as well as amplitude and phase distortion.
- Phase jitter (poor timing recovery).
- Envelope delay or harmonic distortion is another limitation.

These channel limitations combined with the dense symbol constellation of the V.32 modem necessitate adaptive equalization for acceptable error rates at 9600 bits/s.

2.5.3.1 Equalization

The basic function of the equalizer is to create an ideal transmission medium from a real channel. An example channel's short impulse response {h1, h2, h3, h4} is shown in Figure 2.26. The ideal medium is characterized as a pure delay, shown in Figure 2.27.

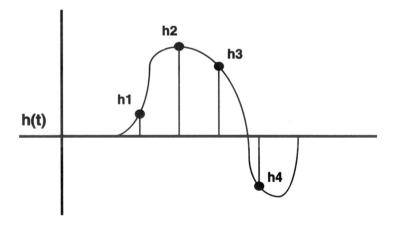

Figure 2.26 Example Short Impulse Response

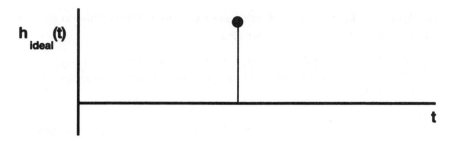

Figure 2.27 Pure Delay Impulse Response

Take for example the equalizer shown in Figure 2.28 which has three taps
{c1, c2, c3}. Convolving this response with the channel's impulse response
from Figure 2.26 yields

$$
\begin{vmatrix} y_1 \\ y_2 \\ y_3 \\ y_4 \\ y_5 \\ y_6 \end{vmatrix} = \begin{vmatrix} c_1 & 0 & 0 & 0 \\ c_2 & c_1 & 0 & 0 \\ c_3 & c_2 & c_1 & 0 \\ 0 & c_3 & c_2 & c_1 \\ 0 & 0 & c_3 & c_2 \\ 0 & 0 & 0 & c_3 \end{vmatrix} \times \begin{vmatrix} h_1 \\ h_2 \\ h_3 \\ h_4 \end{vmatrix}
$$

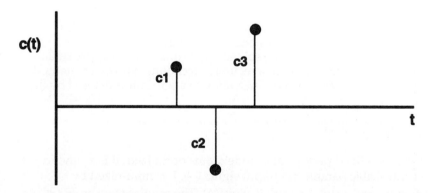

Figure 2.28 Equalizer Impulse Response

2 Modems

The outputs $\{y_1, y_2, y_3, y_4, y_5, y_6\}$ represent samples of the impulse response of the combined channel/equalizer system.

If the equalizer is to create ideal conditions for transmission, all the y's should be zeros except for one main sample. Rewriting the equation for ideal equalization yields:

$$
\begin{vmatrix} 0 \\ 0 \\ 1 \\ 0 \\ 0 \\ 0 \end{vmatrix} = \begin{vmatrix} c_1 & 0 & 0 & 0 \\ c_2 & c_1 & 0 & 0 \\ c_3 & c_2 & c_1 & 0 \\ 0 & c_3 & c_2 & c_1 \\ 0 & 0 & c_3 & c_2 \\ 0 & 0 & 0 & c_3 \end{vmatrix} \times \begin{vmatrix} h_1 \\ h_2 \\ h_3 \\ h_4 \end{vmatrix}
$$

or

$$0 = c_1 h_1$$
$$0 = c_1 h_2 + c_2 h_1$$
$$1 = c_1 h_3 + c_2 h_2 + c_3 h_1$$
$$0 = c_1 h_4 + c_2 h_3 + c_3 h_2$$
$$0 = c_2 h_4 + c_3 h_3$$
$$0 = c_3 h_4$$

The system of equations above has only three controllable variables (unknowns) but six simultaneous equations. The system is overdetermined and can only be solved approximately. To approximate this solution, a reformulation of a recursive technique known as method of steepest descent can be used. This iterative algorithm is defined by the equation:

(1) $C_{k+1} = C_k - \Delta \, \partial E / \partial C_k$

where E is a defined performance index to be optimized. It is a function of some controllable parameters (tap weights C_k). E is minimized by adjusting the tap weights in small steps (Δ). The gradient vector $\partial E / \partial C_k$ indicates the direction of the adjustment required to minimize E. This method converges to an optimum solution when $\partial E / \partial C_k$ is zero.

2.5.3.2 Performance Index

It is important to choose a meaningful performance index that is a linear function of the tap weights and that defines a smooth error surface (bowl) in the space spanned by the tap weight vector. This ensures the convergence of the algorithm to the lowest point (minimum) of the error surface.

In some cases, a desirable performance index is a nonlinear function of the adjustable parameters and the solution is unrealizable. As an example, consider the probability of error in a digital communication system. Even though this is a meaningful measure of system performance, it is a highly nonlinear function of the equalizer tap weights. Using the method of steepest descent, it cannot be determined whether the adaptive equalizer has converged to the optimum solution or to one of the relative minima of the surface. For this reason some desirable performance indices must be rejected.

A practical and popular index for performance is the mean squared error (MSE). The error is measured as the difference between the received signal and the ideal signal value. The MSE index is a measure of the energy in this error signal averaged over a signaling interval. It results in a quadratic performance surface as a function of the filter coefficients and thus has a single minimum (optimal solution). An implementation of an MSE-based iterative adaptation algorithm is developed for the ADSP-2100 processor family in this chapter; it is discussed in a later section.

2.5.4   Equalizer Architectures

The preferred form of a linear equalizer is a tapped delay line. The delay line consists of delay elements in a feedforward path and possibly a feedback path.

If the delay line has feedforward delays only, its transfer function can be expressed as a single polynomial in Z^{-1} and therefore the equalizer has a finite impulse response (FIR). This type of equalizer is often called a nonrecursive or transversal equalizer (Figure 2.29).

If the delay line also has feedback delay elements, its transfer function is a rational function of Z^{-1} and the equalizer has an infinite impulse response (IIR) due to its nonzero poles (Figure 2.30).

The V.32 modem equalizer has no feedback delay elements and is therefore an FIR equalizer.

2 Modems

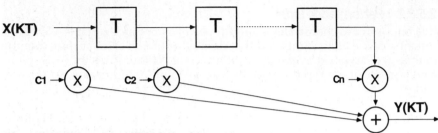

Figure 2.29 Transversal (FIR) Delay Line

2.5.4.1   *Real Or Complex*

In a one-dimensional communication system (e.g. pulse amplitude modulation or PAM), the signal is real and the equalizer has real coefficients. The V.32 modem, which uses quadrature amplitude modulation (QAM), transmits complex data by modulating two

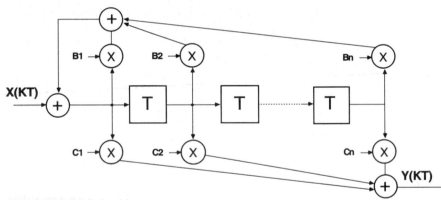

Figure 2.30 IIR Delay Line

orthogonal carrier signals. Because of cross-distortion between the in-phase and quadrature channels in this two-dimensional communication system, an equalizer with complex tap coefficients is required.

Algorithms for the complex equalizer are essentially the same as for the real equalizer with the added burden of complex arithmetic. A complex equalizer typically requires four times as many multiplications and introduces the complex conjugation operator in recursive algorithms such as LMS adaptation.

2.5.4.2 Sampling Rates

It is often advantageous to space the delay elements in an equalizer more closely than the symbol rate, as shown in Figure 2.31. This has the effect of oversampling the input to the filter and thus increasing the effective bandwidth of the equalizer. The input is pushed onto the delay line twice for every one output computed. Fractionally spaced equalizers have superior performance because of wider bandwidth, and they simplify the problem of phase synchronization between transmitter and receiver. They do, however, suffer from stability problems in low noise conditions and are more computationally demanding (Ungerboeck, 1976).

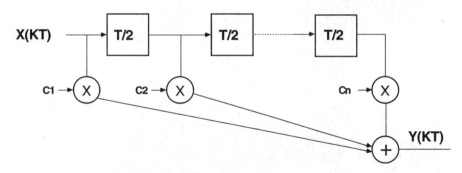

Figure 2.31 Fractionally Spaced Delay Line (FSE)

A fractionally spaced filter can be designed the same way as a T-spaced delay line filter. The basic delay line structure is the same for both. For a T/2 FSE filter, the samples are shifted in at $2f_s$ (twice the sampling frequency) but the output is only computed at f_s, i.e. every other input time.

The ADSP-2100 routine to implement the delay line with complex tap weights is in Listing 2.13.

2   Modems

{   Fractionally Spaced Filter (FSE) Subroutine

This Complex Fractionally Spaced Filter (FSE) Subroutine is used in the V32 equalizer. The basic structure for the delay line is the same as that of a T-Spaced Filter (TSE). In the FSE case, however, samples are shifted in at 2Fs (Fs=Sampling Frequency) and the output is computed at Fs, i.e. at alternate times. This subroutine will therefore be called after 2 new input samples have been pushed onto the delay line.

```
        Calling Parameters
           I0-->Oldest data value in real delay line (Xr's)
           L0=filter length (N)
           I1-->Oldest data value in imag. delay line (Xi's)
           L1=filter length (N)
           I4-->Beginning of real coefficient table (Cr's)
           L4=filter length (N)
           I5-->Beginning of imaginary coefficient table (Ci's)
           L5=filter length (N)
           M0,M6=1
           AX0=filter length minus one (N-1)
           CNTR=filter length minus one (N-1)

        Return Values
           I0-->Oldest data value in real delay line
           I1-->Oldest data value in imaginary delay line
           I4-->Beginning of real coefficient table
           I5-->Beginning of imaginary coefficient table
           SR1=real output (rounded, cond. saturated)
           MR1=imaginary output (rounded, cond. saturated)

        Altered Registers
           MX0,MY0,MR,SR1

        Computation Time
           2*(N-1)+2*(N-1)+13+8 cycles
```

All coefficients and data values are assumed to be in 1.15 format.
}

```
fir:      MR=0, MX0=DM(I1,M0), MY0=PM(I5,M6);
          DO realloop UNTIL CE;
              MR=MR-MX0*MY0(SS), MX0=DM(I0,M0), MY0=PM(I4,M6);    {Xi*Ci}
realloop:     MR=MR+MX0*MY0(SS), MX0=DM(I1,M0), MY0=PM(I5,M6);    {Xr*Cr}
          MR=MR-MX0*MY0(SS), MX0=DM(I0,M0), MY0=PM(I4,M6);        {last Xi*Ci}
          MR=MR+MX0*MY0(RND);                                     {last Xr*Cr}
          IF MV SAT MR;
          SR1=MR1;                                                {Store Yr}
          MR=0, MX0=DM(I0,M0), MY0=PM(I5,M6);
          CNTR=AX0;
          DO imagloop UNTIL CE;
              MR=MR+MX0*MY0(SS), MX0=DM(I1,M0), MY0=PM(I4,M6);    {Xr*Ci}
imagloop:     MR=MR+MX0*MY0(SS), MX0=DM(I0,M0), MY0=PM(I5,M6);    {Xi*Cr}
          MR=MR+MX0*MY0(SS), MX0=DM(I1,M0), MY0=PM(I4,M6);        {last Xr*Ci}
          MR=MR+MX0*MY0(RND);                                     {last Xi*Cr}
          IF MV SAT MR;                                           {MR1=Yi}
          RTS;
```

Listing 2.13 Delay Line Routine, Complex Tap Weights

2.5.5   Least Mean Squared (LMS) Algorithm

Since the mean squared error (MSE) performance index is a convex function of the tap weights (has a bowl-shaped surface), the optimum tap weights can be obtained by the steepest descent algorithm. In this algorithm, tap weights are assumed to have an arbitrary initial setup and are moved in the direction of optimum value when MSE is minimized. The direction is determined by the gradient of the objective function of performance,

(2) $E = \mid e(kt) \mid^2$

where $e(kt)$ is the error between the estimated symbol and the received sample and the bar above the expression denotes time averaging. Optimum tap weights are determined when the derivative of the MSE surface with respect to all the tap weights is zero.

(3) $\partial E / \partial C_k = 0$ $1 \leq k \leq N$, for an N-tap filter

The error function E is a complex quadratic function because of the 2-dimensional modulation scheme (QAM). The derivative expression is:

(4) $\partial E / \partial C_n(k) = -2\ \overline{e(kt)\ y(kT_{sym} - nT_{taps})}$

109

2 Modems

where $\quad$ T_{taps} is the spacing between the taps
$\quad\quad\quad\quad$ T_{sym} is the spacing between symbols

Combining with equation (1) yields:

$$(5) \qquad C_n(k+1) = C_n(k) + \beta e(kt)\, \overline{y^*(kT_{sym} - nT_{taps})}$$

The implementation of the steepest descent algorithm requires the evaluation of the cross-correlation of error signal e(kt) and received signal y(t). Cross-correlation requires time-averaging, which is not a viable option considering the real time requirements of the equalizer. To alleviate this problem, the approximation:

$$(6) \qquad \overline{e(kt)\, y^*(kT_{sym} - nT_{taps})} \approx e(kt)\, y^*(kT_{sym} - nT_{taps})$$

is used instead of time-averaging. This simplification of the steepest descent algorithm greatly reduces the amount of computation. It is very popular and is generally referred to as the least mean square (LMS) algorithm.

An LMS algorithm updates the equalizer tap weights according to

$$(7) \qquad C_n(k+1) = C_n(k) + \beta e(kt)\, y^*(kT_{sym} - nT_{taps})$$

Listing 2.14 shows an LMS algorithm implemented on the ADSP-2100 family.

```
{          Complex SG Update LMS Subroutine.

This routine updates the complex taps according to the relation:

          Cn(k+1)=Cn(k)-Beta.E(k).Y*(n-K)

     where:        <Beta>=Adaptation step size
          <E(k)>=estimation error at time k
          <Y*(n-k)>=Received signal complex
                    conjugated & sampled at time (n-k)

     Calling Parameters
          I0-->Oldest data value in real delay line    L0=N
          I1-->Oldest data value in imag. delay line   L1=N
          I4-->Beginning of real coefficient table     L4=N
          I5-->Beginning of imag coefficient table     L5=N
          MX0=real part of Beta*Error
          MX1=imag part of Beta*Error
          M0,M5=1
          M1=-1
          M6=0
          CNTR=Filter length (N)

     Return Values
          Coefficients updated
          I0-->Oldest data value in real delay line
          I1-->Oldest data value in imag delay line
          I4-->Beginning of real coefficient table
          I5-->Beginning of imag coefficient table

     Altered Registers
          MY0,MY1,MR,SR,AY0,AY1,AR

     Computation Time
          6*N+10 cycles

     All coefficients and data values are assumed to be in
     1.15 format.
}

upd_taps:   MY0=DM(I0,M0);                            {Get Xr}
            MR=MX0*MY0(SS), MY1=DM(I1,M0);            {Er*Xr, get Xi}
            DO adaptc UNTIL CE;
               MR=MR+MX1*MY1(RND), AY0=PM(I4,M6);     {Ei*Xi, get Cr}
               AR=AY0-MR1, AY1=PM(I5,M6);             {Cr-(Er*Xr+Ei*Xi), get Ci}
               PM(I4,M5)=AR, MR=MX1*MY0(SS);          {Store new Cr, Ei*Xr}
               MR=MR-MX0*MY1(RND), MY0=DM(I0,M0);     {Er*Xi, get Xr}
               AR=AY1-MR1,MY1=DM(I1,M0);              {Ci-(Ei*Xr-Er*Xi), get Xi}
adaptc:     PM(I5,M5)=AR, MR=MX0*MY0(SS);             {Store new Ci, Er*Xr}
            MODIFY (I0,M1);                           {point back to start}
            MODIFY (I1,M1);                           {of complex delay line}
            RTS;
```

Listing 2.14 LMS Routine

2 Modems

2.5.6 Program Structure

The flowchart shown in Figure 2.32 depicts the sequence of operations of an equalizer program. Each program section is discussed below.

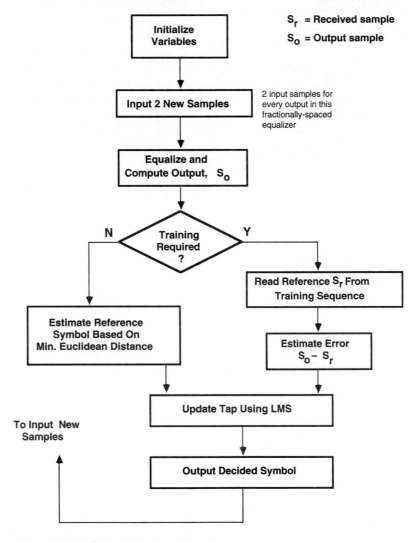

Figure 2.32 Adaptive Equalizer Flowchart

Modems   2

2.5.6.1   Input New Sample

The equalizer program is interrupt-driven. The arrival of a new complex sample causes the equalizer to start executing. The *sample_in* port in Listing 2.15 holds the new sample (real, then imaginary). Index registers I0 and I1 point to the complex delay line.

The V.32 modem recommendation specifies a fractionally spaced equalizer. The delay line therefore consists of delays that are spaced at one-half the symbol rate. This means that the output (at 2400 symbols/s) is only computed for every two input samples (at 4800 symbols/s). The variable *decimator_flag* is used to decide whether to get another sample or to start computing the output.

```
{        input_new_sample routine

This part will read a new sample from the port 'sample_in' and
place it on the delay line. This new complex sample will overwrite
the oldest value on the delay line (complex also).
}

start:    AR=DM(sample_in);           {read in real & imag. values}
          DM(I0,M0)=AR;               {of new sample and store them}
          AR=DM(sample_in);           {in delay line}
          DM(I1,M0)=AR;
          AR=DM(decimator_flag);    {check flag to see if filtering}
                                      {is required this time through.}
          AR=NOT AR;                  {Then toggle the flag}
          DM(decimator_flag)=AR;      {to ensure that we filter}
                                      {every other sample}
          IF EQ RTS;                  {as required in an FSE}
```

Listing 2.15 Input Routine

2.5.6.2   Filtering (Equalizing)

The actual filtering is performed in the subroutine in Listing 2.16. The calling parameters for the filter are initialized, and after the subroutine is called the return values are stored in data memory.

2 Modems

```
{          do the fir filtering (equalization)

Performs the actual fir filtering. Takes the input sample
from the receiver front end & produces an output value
(fir_out_real & fir_out_imag)
}
          AX0=no_of_taps-1;
          CNTR=no_of_taps-1;
          CALL fir;
          DM(fir_out_real)=SR1;      {save return values of}
                                     {subroutine in}
          DM(fir_out_imag)=MR1;      {their designated var names:}
                                     {fir_out_real & fir_out_imag}
```

Listing 2.16 Filter Routine

2.5.6.3 Training Sequence

Initially the tap weights of the equalizer are at some arbitrary state (possibly zero) that is typically far from the optimum state. The receiver decisions based on the output of the equalizer are therefore incorrect with a high probability. Decision-directed adaptation is not guaranteed to work because of the initial high error rate. The equalizer might be unable to move into the error-free region and the adaptation would diverge or stop (MSE neither increasing nor decreasing significantly).

To train the equalizer through this blind stage, a data sequence that is known at the receiver is used for initial transmission. If the locally generated reference is properly synchronized to the received signal, this training brings the equalizer to its optimum state. After training, slow channel variations are tracked using decision-directed adaptation.

The stored training sequence at the receiver is read at the *training_list* port (real, then imaginary) in Listing 2.17. The received signal is read in from the filter outputs *fir_out_real* and *fir_out_imag*. A complex error value which is equal to the Euclidean distance between the two samples is generated. The estimation error is stored in data memory (*error_real* and *error_imag*).

114

```
{         estimate the transmitted symbol ( training )

Given fir_out_real & fir_out_imag, we compute the error value
(real and complex) using the training sequence as a reference.
This estimate for error is used only initially to train the
equalizer. Following the training, decision directed adaptation
would take over.
}

        AX0=DM(fir_out_real);         {inputs are fed in directly}
        AX1=DM(fir_out_imag);         {from output of fir}
        AY0=DM(training_list);
        AY1=DM(training_list);
        CALL est_error_train;

{────────────────────────────}
{
Est_error_train subroutine: Returns the equalizer output minus the
ideal value available from the training sequence.

        AX0=fir_out_real
        AX1=fir_out_imag
        AY0=ideal_symbol_real
        AY1=ideal_symbol_imag

        Returns:
           error_real
           error_imag

}

est_error_train:  AR=AX0-AY0;
                  DM(error_real)=AR;
                  AR=AX1-AY1;
                  DM(error_imag)=AR;
                  RTS;
```

Listing 2.17 Training Sequence Routine

2.5.6.4 *Decision-Directed Adaptation*

Once the equalizer is trained, decision-directed adaptation is possible. In this mode, symbols estimated at the receiver are used as the reference from which to measure the deviation error and subsequently adjust the taps. With the equalizer trained, low decision-error rates make it possible to continue to adapt to small changes in channel conditions.

2 Modems

In Listing 2.18, the estimated symbol is chosen as the symbol geometrically closest to the received coordinates. A 15-instruction loop (worst case) computes the distance to each of the 32 symbols in the symbol table and determines the nearest one. The routine returns a pointer to the estimated symbol in the table as well as the real and imaginary values of the error.

```
{                 estimate the transmitted symbol     ( no training )

Given fir_out_real & fir_out_imag, we compute the error value (real and
complex) using a Euclidean distance routine (decision directed adaptation).
In this mode the estimated symbol is the geometrically closest to the
received coordinates.  This routine also returns the complex error value.
}
        AX0=DM(fir_out_real);   {these inputs are fed in directly}
        AX1=DM(fir_out_imag);   {from the output of the fir}
        CALL est_error_eucl;
{──────────────────────────────}
{Estimate_error_euclidean Symbol Subroutine
(normal mode, i.e. no training):

Maps input sample onto an ideal symbol in the constellation table This
routine also returns the value of the error measured as the Euclidean
distance between received signal and its ideal value.

        Calling Parameters
           AX0 contains Xr
           AX1 contains Xi
           M0=1
           M1=-1

        Return Values
           SI=decision index j
           (position with respect to end of table)
           AF=minimum distance (squared)
           I2—>Beginning of constellation table

        Altered Registers
           AY0,AY1,AF,AR,MX0,MY0,MY1,MR,SI
           AR_SAT mode enabled

        Computation Time
           15*N+5 (maximum)
}
```

```
est_error_eucl:   I2=^constellation_table;
                  L2=3;                                 {number of symbols in}
                                                        {constellation table}
                  AY0=32767;                            {Initialize minimum distance to}
                                                        {largest possible value}
                  ENA AR_SAT;                           {put ALU in saturation mode to}
                                                        {prevent overflow}
                  AF=PASS AY0, AY0=DM(I2,M0);       {Get Cr}
                  CNTR=32;
                  DO ptloop UNTIL CE;
                      AR=AX0-AY0, AY1=DM(I2,M0);        {Xr-Cr, Get Ci}
                      MY0=AR, AR=AX1-AY1;               {Copy Xr-Cr, Xi-Ci}
                      MY1=AR;                           {Copy Xi-Ci}
                      MR=AR*MY1(SS), MX0=MY0;           {(Xi-Ci)^2,}
                                                        {Copy Xr-Cr}
                      MR=MR+MX0*MY0(SS);                {(Xr-Cr)^2}
                      IF MV SAT MR;                 {clip result to max value}
                      AR=MR1-AF;                    {Compare with previous minimum}
                      IF GE JUMP ptloop;
                      AF=PASS MR1;                  {New minimum if MR1<AF}
                      AR=AX0-AY0;                   {error is euclidean distance}
                      DM(error_real)=AR;            {between actual received}
                      AR=AX1-AY1;                   {signal and ideal symbol}
                      DM(error_imag)=AR;            {coordinates}
                      SI=CNTR;                      {Record constellation index}
ptloop:               AY0=DM(I2,M0);
                  MODIFY(I2,M1);                    {Point to beginning of table}
                  RTS;
```

Listing 2.18 Decision-Directed Adaptation Routine

2.5.6.5   Tap Update (LMS Algorithm)

Once an estimate error is computed, it is possible to adapt the equalizer coefficients to a new set of values closer to the optimum vector. The LMS routine in Listing 2.19 performs the computation. The estimation error is first scaled down by the adaptation step size (β). This constant provides a mechanism to trade off convergence speed against the amount of jitter in the steady state value of the tap vector.

2 Modems

```
{        update the taps

Takes the estimation error values previously computed multiply
them by the step size (beta). The upd_taps routine is then called
to update coefficients of the equalizer.

}


        MY0=DM(error_real);  {MX0=beta x error_real}
        MX0=beta;
        MR=MX0*MY0(SS);
        MX0=MR1;

        MY1=DM(error_imag);
        MX1=beta;            {MX1=beta x error_imag}
        MR=MX1*MY1(SS);
        MX1=MR1;
        CNTR=No_of_taps;
        CALL upd_taps;
```

Listing 2.19 Tap Update Routine

2.5.6.6   Output
These equalizer routines can be integrated into other modules to form the
receiver block of a V.32 modem. As specified in the V.32 recommendation,
the equalized sample is decoded using the Viterbi algorithm. The
equalizer output (real and the imaginary) is therefore written to an I/O
port *sample_out*.


```
{        output the resulting sample of the equalizer}

        AR=DM(fir_out_real); {output the equalizer output}
        DM(sample_out)=AR;   {to the outport port}
        AR=DM(fir_out_imag);
        DM(sample_out)=AR;
        RTS;                 {return from equalizer routine and}
                             {wait for a new sample interrupt}
```

Listing 2.20 Output Routine

2.5.7 Practical Considerations

This section describes considerations for using and modifying the routines in this chapter.

2.5.7.1 Viterbi Decoder

In the implementation of decision-directed adaptation, the received sample is matched to the nearest symbol and the error is used to adjust the taps. A few wrong decisions could cause the equalizer to wander off temporarily, but because right decisions have a proportionately larger effect, convergence is ensured.

If a sophisticated algorithm such as Viterbi decoding is used to improve the decision, the signal sample and error are not available until several symbol intervals after the input time. This Viterbi delay requires a modified LMS updating routing with delayed coefficient adaptation (DLMS). It can be shown that the DLMS adaptation has the same steady state behavior as the LMS adaptation, provided the adaptation constant is within a certain range (Long et al, 1989).

2.5.7.2 Pseudo-Random Training Sequence

The routines in this chapter have been validated with a pseudo-random training sequence. This training sequence consists of a set of symbol values with a repetition period that is much longer than the convergence time of the equalizer. The benefit of using such a sequence is that the approximation of the gradient vector $\partial E/\partial C_k$ is less noisy. Noisy estimates of the gradient vector can cause the tap coefficients to wander a long way from the path of the steepest descent (Bingham, 1988).

2.5.7.3 Delay Line Length

If the exact source of the channel's distortion is known and the impulse response can be modeled precisely, it is possible to calculate the minimum order of the equalizer transfer function needed to reduce the MSE to an acceptable level. In general, the only practical method of deciding the length of the delay line is to derive a theoretical length based on several worst-case channel characteristics. The equalizer is then designed slightly longer than the theoretical minimum to compensate for the cumulative effects of finite precision arithmetic in the ADSP-2100 family processor. For a discussion of quantization effects in the LMS algorithm, see Bershad, 1989.

2 Modems

2.6 CONTINUOUS PHASE FREQUENCY-SHIFT KEYED MODULATION

Constant phase modulation (CPM) techniques find applications in satellite communications. Because of power amplifier considerations, satellite communications require a modulation technique with a constant or nearly constant envelope versus time (no amplitude modulation). Technological and regulatory limitations also require low error probability for a given signal-to-noise ratio and high bits per second of transmitted information for a given bandwidth. The technique of multi-h CPM, which combines encoding and modulation, achieves all of these goals.

This chapter describes an implementation of continuous phase frequency-shift keying (CPFSK), a sub-class of multi-h CPM, on the ADSP-2100 family of processors. Only modulation is described here; demodulation is usually performed with the Viterbi algorithm.

Fast frequency-shift keying (FFSK) is a special case of CPFSK with h=1/2.

2.6.1 CPFSK Methodology

The general form for a multi-h CPM signal is:

$$s(t; \alpha) = \sqrt{(2E_s/T_s)} \cos [2\pi f_0 t + \varphi(t; \alpha) + \varphi_0]$$

E_s = symbol energy
T_s = symbol duration
f_0 = carrier frequency (Hz)
φ_0 = carrier phase (arbitrary)
$\varphi(t; \alpha)$ = information-carrying phase function, expressed as:

$$2\pi \int_{-\infty}^{t} \sum_{i=-\infty}^{\infty} h_i a_i \, g(\tau - iT_s) \, d\tau \qquad -\infty < t < \infty$$

120

where:

α $= (\ldots, a_{-2}, a_{-1}, a_0, a_1, a_2, \ldots)$, representing the data sequence

h_i = set of K modulation indices cycled through periodically,
     i.e., $h_{i+K} = h_i$

$g(t)$ = frequency pulse-shape function

For CPFSK, all h_i are equal and the pulse-shape function is:

$$g(t) = T_s/2 \qquad\qquad \text{for } 0 \le t \le T_s, \text{ otherwise } 0$$

2.6.2   ADSP-2100 Family Implementation

Figure 2.33 shows a flowchart of the CPFSK program implemented on the ADSP-2100 family of processors. This particular implementation uses the ADSP-2101 to take advantage of its on-chip serial ports and timer. The timer generates a clock at the symbol rate (2400 baud) for reading input data. The ADSP-2101 outputs CPFSK modulated data to a digital-to-analog converter (DAC) at the rate of 8 kHz.

The CPFSK program is shown in Listing 2.21. This program sets up a buffer of dummy data for demonstrations; in actual use, the data would come from an input device and could be read from the FI (Flag In) input of the ADSP-2101.

The CPFSK routine calls two external routines not shown here. The *cntlreg_inits* routine initializes the ADSP-2101's control registers. The *boot_sin* routine computes the sine of the input in AX0, returning the output in AR.

After setup (initializing variables, etc.) the processor waits for one of two interrupts. The SPORT0 interrupt causes the processor to calculate the next output sample by adding the current phase increment to the phase accumulator and computing the sine of the result. The output samples are transmitted from SPORT0 and are also sent to a DAC for display (for demonstration).

The timer interrupt causes the processor to select a new phase increment based on the value of the input data. Because the data is binary (1 or 0) it could be input through the flag input (FI) pin instead of data memory as shown. The code would have to be modified to use the state of the input flag as a condition for selecting the phase increment.

2 Modems

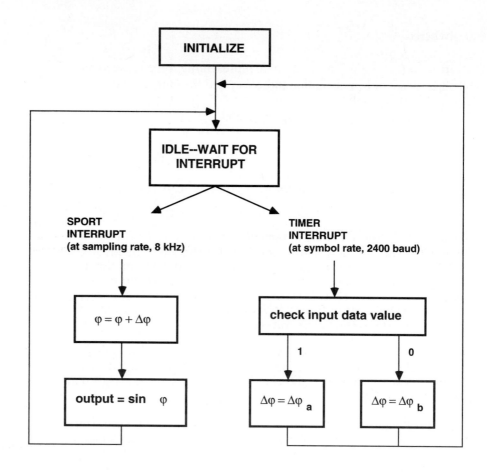

φ = current phase value (stored in "phase accumulator")

$\Delta\varphi$ = current phase increment

$\Delta\varphi_a$ = phase increment for tone a

$\Delta\varphi_b$ = phase increment for tone b

Figure 2.33 CPFSK Flow Diagram

122

Modems   2

```
.MODULE/BOOT=0/ABS=0     cpfsk_modulator;

{ CPFSK - Continuous Phase Frequency Shift Keying modulator

    input:              data stream stored in DM circ buffer (for demo)
                        in actual use, data could be state of FLAG_IN pin
    output:             dac0 - CPFSK output waveform
                        dac1 - input data stream (echoed for demo display)
                        spkr - CPFSK "sound"
}

.EXTERNAL               boot_sin;
.EXTERNAL               cntlreg_inits;
.PORT                   write_dac0;
.PORT                   write_dac1;
.PORT                   load_dac;

.CONST                  lo_tone=220;{Hertz}
.CONST                  hi_tone=880;{Hertz}
.CONST                  logic_one=H#7F00;
.CONST                  logic_zero=0;

.VAR/CIRC               demo_input_data[7];
.VAR                    hertz0, phase_incr_0;
.VAR                    hertz1, phase_incr_1;
.VAR                    phase_accumulator, phase_increment;

{————————————————————————}
    JUMP start; RTI; RTI; RTI;       {Reset Vector}
    RTI; RTI; RTI; RTI;              {irq2}
    RTI; RTI; RTI; RTI;              {sport0 TX}
    JUMP sample; RTI; RTI; RTI;      {sport0 RX} {at 8 kHz rate}
    RTI; RTI; RTI; RTI;              {irq0}
    RTI; RTI; RTI; RTI;              {irq1}
    CALL symbol; RTI; RTI; RTI;      {timer}      {at 2400 baud}

{————————————————————————}
start:  CALL cntlreg_inits;          {set up SPORTS, TIMER, etc}
        M7=1; L7=0;                  {used by bootsin routine}

{————————————————————————}
baud_clock:     L0=0;
                M0=1;
                I0=H#3FFB;           {point to DM-mapped TIMER ctrl regs}
                                     {2400 baud=5120 cycles @ 12.288 MHz}
{H#3FFB}        DM(I0,M0)=0;              {TIMER - TSCALE}
{H#3FFC}        DM(I0,M0)=5119;          {TIMER - TPERIOD}
{H#3FFD}        DM(I0,M0)=5119;          {TIMER - TCOUNT}
```

(listing continues on next page)

123

2 Modems

```
{─────────────────────────}
make_demo_data:    SI=lo_tone;              DM(hertz0)=SI;
                   SI=hi_tone;              DM(hertz1)=SI;
                   SI=logic_one;            DM(demo_input_data)=SI;
                   SI=logic_zero;           DM(demo_input_data+1)=SI;
                                            DM(demo_input_data+2)=SI;
                                            DM(demo_input_data+3)=SI;
                                            DM(demo_input_data+4)=SI;
                                            DM(demo_input_data+5)=SI;
                                            DM(demo_input_data+6)=SI;
                   I0=^demo_input_data;
                   L0=%demo_input_data;

{─────────────────────────}
{These segments convert "Hertz" to 8 kHz Phase_Increment}

load_tone1:        SI=DM(hertz1);
                   SR=ASHIFT SI BY 3(HI);
                   MY0=H#4189;              {mult Hz by .512*2}
                   MR=SR1*MY0(RND);         {i.e. mult by 1.024}
                   SR=ASHIFT MR1 BY 1(HI);
                   DM(phase_incr_1)=SR1;

load_tone0:        SI=DM(hertz0);
                   SR=ASHIFT SI BY 3(HI);
                   MY0=H#4189;              {mult Hz by .512*2}
                   MR=SR1*MY0(RND);         {i.e. mult by 1.024}
                   SR=ASHIFT MR1 BY 1(HI);
                   DM(phase_incr_0)=SR1;

{─────────────────────────}
        SI=0;
        DM(phase_accumulator)=SI;   {clear phase accumulator on startup}
        CALL symbol;                {start with first symbol}
        ICNTL=B#01111;
        IMASK=B#001001;             {enable SPORT0_RX, TIMER now}
        ENA TIMER;                  {start baud_clock now}

{─────────────────────────}
here:   JUMP here;      {wait for symbol and sample interrupts}
```

Modems 2

```
{================================================================}
{========== P R O C E S S   A   N E W   S A M P L E ==========}
{================================================================}
sample: AX0=DM(phase_accumulator);
        AY0=DM(phase_increment);
        AR=AX0+AY0;
        DM(phase_accumulator)=AR;
        AX0=AR;
        CALL boot_sin;
sound:  DM(write_dac0)=AR;              {"display" CPFSK on oscilloscope}
        DM(load_dac)=AR;
        SR=ASHIFT AR BY -2(HI);
        TX0=SR1;                        {"hear" CPFSK from speaker (PCM out)}
        RTI;


{================================================================}
{========== P R O C E S S   A   N E W   S Y M B O L ==========}
{================================================================}
symbol: AX1=DM(I0,M0);                  {get input data (could be FLAG_IN)}
        DM(write_dac1)=AX1;             {echo input data stream for demo}
        DM(load_dac)=AR;
        AF=PASS AX1;
        IF EQ JUMP zero;
one:    SI=DM(phase_incr_1);           DM(phase_increment)=SI; RTS;
zero:   SI=DM(phase_incr_0);           DM(phase_increment)=SI; RTS;

.ENDMOD;
```

Listing 2.21 CPFSK Program (ADSP-2101)

2 Modems

2.7 V.27 *ter* & V.29 MODEM TRANSMITTERS

V.27 *ter* and V.29 modem transmitters are often used in facsimile transmission systems. This section contains example programs for implementing these transmitters. The subroutines for each transmitter are listed at the end of each subsection.

These subroutines include a V.27 *ter* transmitter and a V.29 transmitter with two V.29 fallback modes. The code includes the scrambler, the IQ encoder, pulse-shaped filter and modulator.

The V.27 *ter* scrambler does not implement the extra functions required to check for repeating patterns. Both transmitters use the same pulse-shaped filter code and random number generator code. These listing are included only in section 2.7.1.

The code is contained in two subdirectories, V.27 and V.29. Each directory contains the code and ancillary files necessary for that demonstration. The file MAINXX.DSP calls a random number generator that creates data to be transmitted. The file BUILD.BAT assembles and links the files.

The demonstration code is configured for an ADSP-2101 EZ-LAB® Evaluation Board. To observe the encoder constellation, attach an analog oscilloscope in X-Y configuration to the DAC0 and DAC1 pins on the EZ-Lab board. To observe the eye pattern, attach an oscilloscope probe (in sweep mode) to the analog output pin of the codec. A synchronization pulse is available on DAC2 of the EZ-LAB board. The V.29 demonstration enters the fallback modes when the IRQ2 button is pressed.

2.7.1 V.27 *ter* Transmitter

The CCITT Recommendation is a 4800 bits/s modem standard for data transmission over the general switched telephone network. The recommendation defines the following characteristics:

- Data rate of 4800 bits/s with 8-phase differentially encoded modulation

- Fallback mode of 2400 bits/s with a 4-phase differentially encoded modulation signal

- Provision for backward channel with a modulation rate of 75 bauds

Modems 2

- An adaptive equalizer in the receiver

- The carrier frequency is 1800 Hz ± 1 Hz

CCITT also specifies a raised cosine filter on the transmitter and the receiver, a data scrambler, the phase encoding and the turn on sequence.

Figure 2.34 is a block diagram of a modem transmitter. The data stream to be transmitted is first scrambled to randomize the data. The data scrambler has a generating polynomial of the form:

$$1 + x^{-6} + x^{-7}$$

with additional logic to guard against repeating patterns.

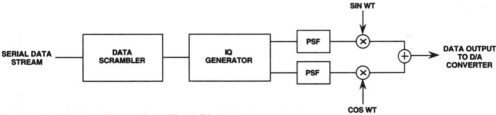

Figure 2.34 Modem Transmitter Block Diagram

The scrambled data stream is divided into groups of three bits (tribits) and is encoded as a phase change relative to the phase of the preceding word transmitted. The phase change for each tribit combination is shown in Table 2.5.

Tribit Values			*Phase Change*
0	0	1	0°
0	0	0	45°
0	1	0	90°
0	1	1	135°
1	1	1	180°
1	1	0	225°
1	0	0	270°
1	0	1	315°

Table 2.5 8-Point V.27 *ter* Phase Changes

2 Modems

The output of the encoder is then mapped to one point in an 8-point signal space, or constellation. The signal space mapping produces two coordinates, one for the real part and one for the imaginary part of a quadrature amplitude modulator. Figure 2.35 shows the 8-point constellation.

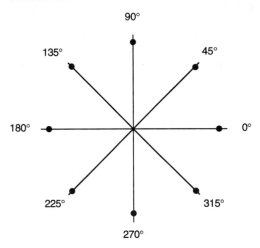

Figure 2.35 8-Point V.27 *ter* Constellation

The output of the signal mapping is interpolated to a 9.6 kHz sample rate and pulse-shaped filtered to assure zero inter-symbol interference. The signal is then modulated onto an 1800 Hz carrier and then sent to the DAC for transmission.

The 2400 bits per second fallback mode is similar to the 4800 bits per second mode. The scrambled data stream is divided into groups of two bits (dibits) before they are encoded into phase information. Table 2.6 and Figure 2.36 show the phase change table and constellation of the 4-point signal space used in the mode.

Dibit Values		Phase Change
0	0	0°
0	1	90°
1	1	180°
1	0	270°

Table 2.6 4-Point V.27 *ter* Phase Changes

128

Modems   2

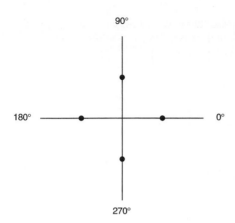

90°

180° ———————————— 0°

270°

Figure 2.36 4-Point V.27 *ter* Constellation

```
{Main routine for V.27 ter Modem                                            }
{Analog Devices                                                             }
{DSP Applications                                                           }
{                                                                           }
.module/ram/boot=0/abs=0        V27_MOD;
{——————————External Function Declaration——————————————————————}
.external get1;
.external scramble;
.external iq;
.external psf;
.external rand;
{——————————Global Variable Declaration——————————————————————}
                                {For Baud/Sample Count                       }
.var/dm              sample_count;   {Number of samples until next          }
     .                               {baud. Default is ? between baud       }
                                {For Scrambler                               }
.var/dm              seed_hi;        {Seeds for random number generator     }
.var/dm              seed_lo;
.var/dm              tri_bit;        {Next bit to transmit                  }
.var/dm              bit_count;      {Number of bits left in current word   }
.var/dm/ram/circ  buffer[12];       {Delay Buffer                           }
.var/dm              buf_ptr1;       {Pointer to y(n-6) buffer value        }
.var/dm              buf_ptr2;       {Pointer to y(n-7) buffer value        }
                                {For IQ Generator                            }
.var/pm/ram        delta_phi[8];    {Tribit phase change table             }
.var/dm/ram        phi_value;       {Current phi value                     }
.var/pm/ram        I_table[8];      {Tables for IQ Output Values           }
.var/pm/ram        Q_table[8];
```

(listing continues on next page)

129

2 Modems

```
                                          {For Pulse Shaped FIlters              }
.var/dm/ram/circ   i_delay_line[5];       {Data Delay line for I channel PSF     }
.var/dm/ram        i_delay_ptr;           {Oldest value in delay line            }
.var/pm/ram/circ   i_coef_line[17];       {Coefficients for I challen PSF        }
.init i_coef_line : <17.dat>;
.var/dm/ram        i_coef_ptr;            {Last Coeff read by PSF                }
.var/dm/ram        i_psf_output;          {Last output value of PSF              }
                                          {For DAC Outputs
.var/dm/ram        sync;                  {Sync value for DAC                    }
.var/dm/ram        i_output;              {Pulse shaped filter output value      }
.var/dm/ram        i_value;               {IQ section I output                   }
.var/dm/ram        q_value;               {IQ section Q output                   }

.init delta_phi:   0x200, 0x000, 0x400, 0x600,
                   0xC00, 0xE00, 0xA00, 0x800;
.init I_table:     0x7fff00, 0x5b0000, 0x000000, 0xa40000,
                   0x800000, 0xa40000, 0x000000, 0x5b0000;
.init Q_table:     0x000000, 0x5b0000, 0x7fff00, 0x5b0000,
                   0x000000, 0xa40000, 0x800000, 0xa40000;
.port    write_dac0;
.port    write_dac1;
.port    write_dac2;
.port    write_dac3;
.port    load_dac;
.global  delta_phi;
.global  phi_value;
.global  I_table;
.global  Q_table;
.global  buffer;
.global  buf_ptr1;
.global  buf_ptr2;
.global  tri_bit;
.global  bit_count;
.global  i_delay_line;
.global  i_delay_ptr;
.global  i_coef_line;
.global  i_coef_ptr;
.global  i_psf_output;
.global  sync;
.global  i_output;
.global  i_value;
.global  q_value;
.global  seed_hi;
.global  seed_lo;
```

Modems 2

```
{————————Interrupt Vectors————————————————}
RESETV:   jump start;          {Reset Vector      }
          rti; rti; rti;
IRQ2V:    rti; rti; rti; rti;  {IRQ2 Vector                        }
HIPWV:    rti; rti; rti; rti;  {Hip Write Interrupt Vector         }
HIPRV:    rti; rti; rti; rti;  {Hip Read Interrupt Vector          }
SPRT0T:   rti; rti; rti; rti;  {Sport0 Transmit Interrupt Vector   }
SPRT0R:   jump next_output;    {Sport0 Receive Interrupt Vector    }
          rti; rti; rti;
IRQ1V:    rti; rti; rti; rti;  {IRQ1 Interrupt Vector              }
IRQ0V:    rti; rti; rti; rti;  {IRQ0 Interrupt Vector              }
TIMERV:   rti; rti; rti; rti;  {Timer Interrupt Vector             }
{————————Main Code Starts Here————————————————}
start:    call V27_INIT;       {Initialize ADSP-2101 Dags and Sports }
          imask = 0x08;        {Enable Sports and Interrupts         }
          ax0 = 0x101f;
          dm(0x3fff) = ax0;
fevr:     idle;                {Wait for Interrupt                 }
          jump fevr;
{————————MAIN ROUTINE————————————————}
next_output:
          {This interrupt routine is executed every output sample rate }
          {One output sample is calculated and a new baud is calulated }
          {if interrupt occurs at baud rate                            }
                               {Output values to the DAC             }
          ax0 = dm(i_value);
          dm(write_dac0) = ax0;
          ax0 = dm(q_value);
          dm(write_dac1) = ax0;
          ax0 = dm(sync);
          dm(write_dac2) = ax0;
          si = dm(i_output);
          sr = ashift si by -2 (lo);
          dm(write_dac3) = sr0;
          dm(load_dac) = sr0;
          tx0 = sr0;

          {Check if new baud should be calculated               }
          ax0 = dm(sample_count);
          ay0 = 1;
          my1 = 0x7fff;        {Set sync value for new sample        }
          dm(sync) = my1;
          ar = ax0 - ay0;
          dm(sample_count) = ar;
          if gt jump next_sample;
```

(listing continues on next page)

131

2 Modems

```
next_baud:
        i4 = ^i_coef_line;      {Reset Coefficient Pointer            }
        dm(i_coef_ptr) = i4;
        ax0 = 4;                {Reset sample counter                 }
        dm(sample_count) = ax0;
        si = 0;                 {Zero SI Register }
        cntr = 3;
        do scram_bit until ce;
            call get1;          {Get next 3 data bits to be transmitted }
            call scramble;
scram_bit:
        my1 = 0x8000;           {Set sync value for new baud          }
        dm(sync) = my1;
        call IQ;                {Calculate delta phi,phase sum,I and Q }
        jump next_sample;

        {Generate Next output sample to DAC      }
next_sample:
        call psf;               {Low Pass Filter I and Q components   }
done:   rti;

{————————Initialization for V.27————————————————————————}
V27_INIT:
                        {————————SPORT INIT————————————————————}
ax0 = 0x02;             {SCLKDIV = 2.048 MHz                          }
        dm(0x3ff5) = ax0;
        ax0 = 213;      {RFSDIV = 213 for 9600                        }
        dm(0x3ff4) = ax0;
        ax0 = 0x6b27;   {Internal SCLK, Frame Syncs                   }
        dm(0x3ff6) = ax0;
                        {————————Wait States————————————————————}
        ax0 = 0xffff;
        dm(0x3ffe) = ax0;


                        {————————Data Buffer Inits————————————————}
        I0 = ^buffer;
        m0 = 1;
        L0 = %buffer;
        cntr = %buffer;
        ax0 = 0;
        do zloop until ce;
 zloop:     dm(i0,m0) = ax0;

                        {————————Variable Initializations——————————}
        ax0 = 0;
        dm(tri_bit) = ax0;
        dm(bit_count) = ax0;
        dm(phi_value) = ax0;
        dm(sample_count) = ax0;
```

132

```
        ax0 = ^buffer+5;
        dm(buf_ptr1) = ax0;
        ax0 = ^buffer+6;
        dm(buf_ptr2) = ax0;
        ax0 = 0xa5a5;
        dm(seed_hi) = ax0;
        ax0 = 0x1234;
        dm(seed_lo) = ax0;

                        {————Pulse Shaped Filter Inits————————}
        i0 = ^i_coef_line;
        dm(i_coef_ptr) = i0;
        i0 = ^i_delay_line;
        dm(i_delay_ptr) = i0;
        l0 = %i_delay_line;
        m0 = 1;
        ax0 = 0;
        cntr = %i_delay_line;
        do dloop until ce;
dloop:
        dm(i0,m0) = ax0;
        ax0 = 0x7fff;
        dm(sync) = ax0;

                        {————————DAG INIT————————————————————}
        m0 = 0;         {Init M0,M1,M5,M5                     }
        m1 = 1;
        m4 = 0;
        m5 = 1;
        l0 = 0;         {Set All L registers to zero          }
        l1 = 0;
        l2 = 0;
        l3 = 0;
        l4 = 0;
        l5 = 0;
        l6 = 0;
        l7 = 0;
        rts;
{————————————————END INITIALIZATION——————————————————————————}

.endmod;
```

Listing 2.22 Main V.27 *ter* Routine (MAIN27.DSP)

2 Modems

```
{GET2 routine for v.27 Modem                                                     }
{Analog Devices                                                                  }
{DSP Applications                                                                }
{                                                                                }
.module/ram/boot=0        get1mod;
{Get 1 data bit to be transmitted                                                }
{                                                                                }
{   INPUTS:                                                                       }
{         dm(data_word)   16 bit words for transmission                           }
{         dm(bit_count)   number of bits left in word                             }
{         m0 = 0                                                                  }
{         m1 = 1                                                                  }
{         m4 = 0                                                                  }
{         m5 = 1                                                                  }
{                                                                                }
{   OUTPUTS:                                                                      }
{         dm(tri_bit):   bit to transmit                                          }
{                                                                                }
{   USAGE:                                                                        }
{         AX0, AY0, AR, SI, SR                                                    }
{                                                                                }
{Global Variable Declaration                                                      }
.external tri_bit;    {bit to transmit                                            }
.external bit_count; {number of bits left in word                                 }

{Local Variable Declaration                                                       }
.var/dm/ram        data_word;              {Data word to be transmitted           }
.external          rand;
{────────────Code Start───────────────────────────────────────────────────────}
.entry   GET1;
GET1:
          ay0 = dm(bit_count);        {If bit_count is zero                       }
          ar = pass ay0;              {load new word                              }
          if ne jump dec_count;
          call rand;                  {Load new data word here                    }
          dm(data_word) = ay0;
          ay0 = 16;
dec_count:
          ar = ay0 -1;                {Decrement counter                          }
          dm(bit_count) = ar;
get_bit:
          sr0 = dm(data_word);        {Get next bit                               }
          ay0 = 0x01;
          ar = sr0 and ay0;
          dm(tri_bit) = ar;
          sr = lshift sr0 by -1 (lo);
          dm(data_word) = sr0;        {Write data_word for next time              }
          rts;
.endmod;
```

Listing 2.23 Data Acquisition Routine (GET27.DSP)

```
{DSP Applications                                                          }
{                                                                          }
{ 2/20/91          Start Date                                             }
{This module performs v.27 ter scrambling on one input bit                 }
{                                                                          }
{The scrambler generating polynomial is xin + y(n-6) + y(n-7)              }
{V.27 ter specifies additional checking for certain pattarns.             }
{This has not been implemented.                                            }
{                                                                          }
{  INPUTS:                                                                 }
{       dm(tri_bit) = bit to be scrambled                                  }
{       si = partial tri bit                                               }
{                                                                          }
{  OUTPUTS:                                                                }
{       si = scrambled output bit                                         }
{            si is left shifted by 1 and the new bit is put in b0          }
{                                                                          }
{  USAGE:                                                                  }
{       I0 = ^buffer                                                       }
{       I1 = Y(n-6)                                                        }
{       I2 = y(n-7)                                                        }
{                                                                          }
.module/ram/boot=0          SCRAMBLE_MOD;
{Local variable Declarations                                               }
.external   buffer;                        {Delay Buffer                   }
.external   buf_ptr1;                       {Pointer to y(n-6) buffer value }
.external   buf_ptr2;                       {Pointer to y(n-7) buffer value }
.external   tri_bit;
.entry      SCRAMBLE;
SCRAMBLE:
   i1 = dm(buf_ptr1);              {Get Pointer values                 }
   i2 = dm(buf_ptr2);
   m3 = -1;
   L1 = %buffer;                   {Set circular buffer registers      }
   L2 = %buffer;

   ax0 = dm(tri_bit);              {Get next bit to scramble           }
   ay0 = dm(i1,m3);                {y(n-6)                             }
   ar = ax0 xor ay0, ay0=dm(i2,m0); {y(n-7)                            }
   ar = ar xor ay0;                { = y(n)                            }
   dm(i2,m3) = ax0;                {Store Oldest Value                 }

   sr = lshift si by 1 (lo);       {Store output in sr0                }
   sr = sr or lshift ar by 0 (lo);
   si = sr0;

   dm(buf_ptr1) = i1;              {Save buffer pointers               }
   dm(buf_ptr2) = i2;
   L1 = 0;                         {Clear L registers                  }
   L2 = 0;
   rts;
.endmod;
```

Listing 2.24 Data Scrambler Routine (SCRAM27.DSP)

2 Modems

```
{Analog Devices                                                          }
{DSP Applications                                                        }
{ 2/20/91        Start Date                                              }
.module/ram/boot=0        PHI_MOD;
{Calculate delta phi, phase sum, I and Q{Analog Devices                 }
{                                                                        }
{   INPUTS:                                                              }
{           SI = TriBit Value                                           }
{           dm[PHI] = current Phi Value                                  }
{                                                                        }
{   OUTPUTS:                                                             }
{           MX0 = I Value                                               }
{           MX1 = Q Value                                               }
{                                                                        }
{   USAGE:                                                               }
{           I3, M3, AR, AX0, AY0                                         }
{                                                                        }
.external   Q_table;
.external   I_table;
.external   delta_phi;
.external   phi_value;
.external   i_coef_line;
.external   i_coef_ptr;
.external   i_delay_line;
.external   i_delay_ptr;
.external   i_value;
.external   q_value;
.entry      IQ;
IQ:
{Look up Delta_phi value and add to phi for new phase                    }
            I7 = ^delta_phi;
            m7 = si;
            modify(i7, m7);
            ax0 = dm(phi_value);
            ay0 = pm(i7,m4);
            ar = ax0 + ay0;
            ay1 = 0x0f;
            ar = ar and ay1;
            dm(phi_value) = ar;
{Look up I and Q components                                              }
{To find I/Q values, use phi/2 as offset into table                     }
            sr = lshift ar by -1 (lo);
            I6 = ^I_table;
            I7 = ^Q_table;
            m7 = sr0;
            modify(I6,M7);
            modify(I7,M7);
            mx0 = pm(i6,m4);
            mx1 = pm(i7,m4);
```

```
{Write new values for DAC                                    }
      dm(i_value) = mx0;
      dm(q_value) = mx1;
{Write new I value to PSF Delay Line                         }
      I6 = dm(i_delay_ptr);
      L6 = %i_delay_line;
      M6 = -1;
      dm(i6,m6) = mx0;
      dm(i_delay_ptr) = i6;
      L6 = 0;
      rts;
.endmod;
```

Listing 2.25 IQ Generator Routine (IQ27.DSP)

2 Modems

```
{Low Pass Filter I component only                                   }
{Analog Devices                                                     }
{DSP Applications                                                   }
{                                                                   }
{        INPUTS:                                                    }
{            MX0 = I Value                                          }
{                                                                   }
{        OUTPUTS:                                                   }
{            dm(I_LPF) = I filter output                            }
{                                                                   }
{        USAGE:                                                     }
{            MX0, MX1, MY0, MR, I3, I7, L3, L7                      }
{                                                                   }
.module/boot=0/ram       PSF_MOD;
.external         i_delay_line;
.external         i_delay_ptr;
.external         i_coef_line;
.external         i_coef_ptr;
.external         i_output;
.entry            PSF;
PSF:
                                        {Initialize DAGs           }
        L0 = %i_delay_line;
        L4 = %i_coef_line;
        I0 = dm(i_delay_ptr);
        I4 = dm(i_coef_ptr);
        modify(i0,m1);                  {Point to newest data value }
        m6 = 4;                         {Interpolate by factor of 4 }
                                        {I filter                   }
        mr = 0, mx0 = dm(i0,m1);        {load first data word       }
        my0 = pm(i4,m6);                {load first coefficient     }
        cntr = 3;
        do i_loop until ce;
    i_loop:
        mr = mr+mx0*my0(SS), mx0 = dm(i0,m1), my0 = pm(i4,m6);
        mr = mr+mx0*my0(RND);
        if mv sat mr;
        dm(I_output) = mr1;
        I4 = dm(i_coef_ptr);            {Point to next set of coefficients }
        modify(i4,m5);
        dm(i_coef_ptr) = I4;
        L0 = 0;
        L4 = 0;
        rts;
.endmod;
```

Listing 2.26 Pulse Shape Filter Routine (PSF.DSP)

```
.MODULE/ram/boot=0          rand_sub;
{
            Linear Congruence Uniform Random Number Generator

    INPUTS:
            dm(Seed_hi) = MSW of seed value
            dm(Seed_lo) = LSW of seed value

    OUTPUTS:
            ay0 = random number;
            dm(Seed_hi) = MSW of updated seed value
            dm(Seed_lo) = LSW of updated seed value
}

.external   seed_hi;
.external   seed_lo;

.ENTRY      rand;
rand:       MY1=25;                             {Upper half of a}
            MY0=26125;                          {Lower half of a}
            sr1 = dm(seed_hi);
            sr0 = dm(seed_lo);
            mr=sr0*my1(uu);                     {A(HI)*X(LO)}
            mr=mr+sr1*my0(uu);                  {A(HI)*X(LO) + A(LO)*X(HI)}
            mr2=mr1;
            mr1=mr0;
            {mr2=si;}
            mr0=h#fffe;                         {C=32767, LEFT-SHIFTED BY 1}
            mr=mr+sr0*my0(uu);                  {(ABOVE) + A(LO)*X(LO) + C}
            sr=ashift mr2 by 15 (hi);
            sr=sr or lshift mr1 by -1 (hi);     {RIGHT-SHIFT BY 1}
            sr=sr or lshift mr0 by -1 (lo);
            dm(seed_hi) = sr1;
            dm(seed_lo) = sr0;
            ay0 = sr1;
        rts;
.endmod;
```

Listing 2.27 Random Number Generator Routine (RAND.DSP)

2   Modems

```
.module/ram/boot=0                  MODULATE_MOD;
{Modulate and sum signals                                          }
{                                                                  }
{   INPUTS:                                                        }
{       dm(i_lpf) = I LPF output                                   }
{       dm(q_lpf) = Q LPF output                                   }
{                                                                  }
{   OUTPUTS:                                                       }
{       dm(result) = output value                                 }
{                                                                  }
{   USAGE:                                                         }
{       I4, M6, M7, MX0, MY0, MR, SR                              }
{                                                                  }
.var/pm/ram/circ  cosine[16];          {Cosine Table              }
.var/dm/ram       cos_ptr;             {Current Pointer to Cosine Table }
.var/dm/ram       output;              {Output Value              }
.external         i_lpf;
.external         q_lpf;
.init             cosine: <cosval.dat>;
.init             cos_ptr: ^cosine;
.entry            modulator;
MODULATOR:
   cntr = 6;
   do mod_loop until ce;
       i4 = dm(cos_ptr);                    {Initialize DAGs           }
       m6 = -4;
       m7 = 7;
       L4 = %cosine;
       mx0 = pm(i4,m6);                     {Read Cosine, point to sine }
       my0 = dm(i_lpf);                     {Read I value              }
       mr = mx0*my0(ss), mx0=pm(i4,m7);     {cos(k)*I(k), get -sin     }
       my0 = dm(q_lpf);                     {Read Q Value              }
       mr = mr + mx0*my0(rnd);              {-Q * Sine                 }
       dm(cos_ptr) = i4;                    {Save cosine Pntr          }
       sr = ashift mr2 by -1 (HI);          { Scale output by 1/2      }
       sr = sr or lshift mr1 by -1 (lo);
mod_loop:
       dm(output) = sr0;
       L4 = 0;
   rts;
.endmod;
```

Listing 2.28 Signal Modulation Routine (MODULATE.DSP)

2.7.2 V.29 Transmitter

The CCITT Recommendation V.29 is a 9600 bits per second modem standard for data transmission over the general switched telephone network. The specification defines the following characteristics:

- Data rate of 9600 bits/s with 8-phase differentially encoded modulation

- Fallback rates of 7200 and 4800 bits/s

- Provision for backward channel with a modulation rate of 75 bauds

- An adaptive equalizer in the receiver

- The carrier frequency is 1700 Hz ± 1 Hz

CCITT also specifies a raised cosine filter on the transmitter and the receiver, a data scrambler, the phase encoding and the turn on sequence.

The data stream to be transmitted is first scrambled to randomize the data. The data scrambler has a generating polynomial of the form:

$$1 + x^{-18} + x^{-23}$$

The scrambled data is divided into for bits (quadbits). The first bit (Q1) is used to determine the signal amplitude and the remaining three bits (Q2, Q3, and Q4) are encoded as a phase change relative to the phase of the preceding word transmitted. Table 2.7 and Figure 2.37 show the phase change table and the V.29 8-point constellation.

Q2	Q3	Q4	Phase Change
0	0	1	0°
0	0	0	45°
0	1	0	90°
0	1	1	135°
1	1	1	180°
1	1	0	225°
1	0	0	270°
1	0	1	315°

Table 2.7 8-Point V.29 Phase Changes

2 Modems

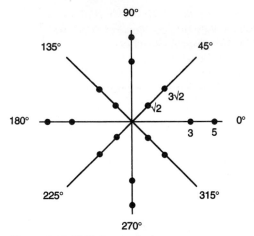

Figure 2.37 V.29 Constellation

The output of the signal mapping is interpolated to a 9.6 kHz sample rate and pulse-shaped filtered to assure zero inter-symbol interference. The signal is then modulated onto a 1700 Hz carrier and then sent to the DAC for transmission.

The 7200 bits per second fallback mode is similar to the 9600 bits per second mode. The scrambled data is divided into groups of three bits (tribits). Table 2.7 can be used to determine the phase change. The first

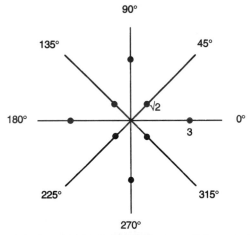

Figure 2.38 V.29 Constellation For 7200 bits/s Fallback Mode

142

tribit represents Q2 in the Table 2.7 and the next two tribits represent Q3 and Q4 respectively. Q1 is assumed to be zero. Figure 2.38 shows the 8-point V.29 constellation for the 7200 bits/s fallback mode.

The 4800 bits per second fallback mode is similar to the 4800 bits per second mode. The scrambled data stream is divided into groups of two bits (dibits). Table 2.7 can be used to determine the phase change. The tribits represent bit Q2 and Q3 in the figure. Bit Q4 in Table 2.7 is the inversion of the mod 2 sum of the two data bits and Q1 is assumed to be zero. Figure 2.39 shows the constellation for this mode.

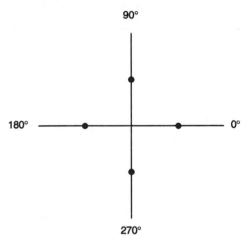

Figure 2.39 V.29 Constellation For 4800 bits/s Fallback Mode

2 Modems

```
{Main routine for V.29 ter Modem                                            }
{Analog Devices                                                             }
{DSP Applications                                                           }
{                                                                           }
{ 2/20/91   Start Date                                                      }
.module/boot=0/ram/abs=0        29_MOD;
{———————————External Function Declaration———————————————}
.external get1;
.external scramble;
.external IQ;
.external psf;
.external rand;
{———————————Global Variable Declaration———————————————————}
                                {For baud/sample count                      }
.var/dm/ram       sample_count;  {Number of samples until next baud         }
                                {Default is 3 samples between each baud     }
                                {For Scrambler                             }
.var/dm/ram       seed_hi;       {Seeds for random number generator         }
.var/dm/ram       seed_lo;
.var/dm/ram       quad_bit;      {next bit to transmit                      }
.var/dm/ram       bit_count;     {Number of bits left in current word       }
.var/dm/ram/circ  buffer[23];    {Scrambler Delay Buffer                    }
.var/dm           buf_ptr2;      {Pointer to y(n-18) buffer value           }
.var/dm           buf_ptr3;      {Pointer to y(n-23) buffer value           }
                                {For IQ Generator                          }
.var/pm/ram       delta_phi[8];  {Quadbit phase change table                }
.var/dm/ram       phi_value;     {Current phi value                         }
.var/pm/ram       I_table_0[8];  {Tables for IQ Output Values               }
.var/pm/ram       Q_table_0[8];
.var/pm/ram       I_table_1[8];
.var/pm/ram       Q_table_1[8];
.var/dm/ram       fallback_mode;  {0=9600bps, 3=7200bps, 6=4800bps          }
.var/dm/ram       fallback_count; {Number of bits to scramble               }
                                {For Pulse Shaped FIlters                  }
.var/dm/ram/circ  i_delay_line[5]; {Data Delay line for I channel PSF       }
.var/dm/ram       i_delay_ptr;    {Oldest value in delay line               }
.var/pm/ram/circ  i_coef_line[17]; {Coefficients for I challen PSF          }
.init             i_coef_line: <17.dat>;
.var/dm/ram       i_coef_ptr;     {Last Coeff read by PSF                   }
.var/dm/ram       i_psf_output;   {Last output value of PSF                 }
                                {For DAC Outputs                           }
.var/dm/ram       sync;          {Sync value for DAC                        }
.var/dm/ram       i_output;      {Pulse shaped filter output value          }
.var/dm/ram       i_value;       {IQ section I output                       }
.var/dm/ram       q_value;       {IQ section Q output                       }
```

144

Modems   2

```
.init delta_phi:   0x000200, 0x000000, 0x000400, 0x000600,
                   0x000C00, 0x000E00, 0x000A00, 0x000800;

.init I_table_1:   0x7fff00, 0x5bff00, 0x000000, 0xa40000,
                   0x800000, 0xa40000, 0x000000, 0x5bff00;
.init Q_table_1:   0x000000, 0x5bff00, 0x7fff00, 0x5bff00,
                   0x000000, 0xa40000, 0x800000, 0xa40000;

.init I_table_0:   0x4c0000, 0x1e0000, 0x000000, 0xe10000,
                   0xb30000, 0xe10000, 0x000000, 0x1e0000;
.init Q_table_0:   0x000000, 0x1e0000, 0x4c0000, 0x1e0000,
                   0x000000, 0xe10000, 0xb30000, 0xe10000;

{───────────Ports and Global ──────────────────────────────────}
.port    write_dac0;
.port    write_dac1;
.port    write_dac2;
.port    write_dac3;
.port    load_dac;
.global  delta_phi;
.global  phi_value;
.global  I_table_1;
.global  Q_table_1;
.global  I_table_0;
.global  Q_table_0;
.global  buffer;
.global  buf_ptr2;
.global  buf_ptr3;
.global  quad_bit;
.global  bit_count;
.global  fallback_mode;
.global  i_delay_line;
.global  i_delay_ptr;
.global  i_coef_line;
.global  i_coef_ptr;
.global  i_psf_output;
.global  sync;
.global  i_output;
.global  i_value;
.global  q_value;
.global  seed_hi;
.global  seed_lo;
```

(listing continues on next page)

2 Modems

```
{──────────Interrupt Vectors──────────────────────────────}
RESETV:   jump start;                 {Reset Vector                        }
          rti; rti; rti;
IRQ2V:    jump set_fall_back;         {IRQ2 Vector                         }
          rti; rti; rti;
HIPWV:    rti; rti; rti; rti;         {Hip Write Interrupt Vector          }
HIPRV:    rti; rti; rti; rti;         {Hip Read Interrupt Vector           }
SPRT0T:   rti; rti; rti; rti;         {Sport0 Transmit Interrupt Vector    }
SPRT0R:   jump next_output;           {Generate Next Sample                }
          rti; rti; rti;
IRQ1V:    rti; rti; rti; rti;         {IRQ1 Interrupt Vector               }
IRQ0V:    rti; rti; rti; rti;         {IRQ0 Interrupt Vector               }
TIMERV:   jump set_fall_back;         {Timer Interrupt Vector              }
          rti; rti; rti;
{──────────Main Code Starts Here──────────────────────────}
start:    call V29_INIT;             {Initialize ADSP-2101 Dags and Sports}
          ena timer;
          ax0 = 0x101f;              {Enable SPORT0                       }
          dm(0x3fff) = ax0;
          imask = 0x88;              {Enable Timer,Sport0,IRQ2 Interrupts }
fevr:     idle;                      {Wait for Interrupt                  }
          jump fevr;
{──────────MAIN ROUTINE───────────────────────────────────}
next_output:
   {This interrupt routine is executed every 8 kHz.                       }
   {One sample is output to codec and a new Baud is calculate if          }
   {interrupt occurs at baud rate                                         }

                           {Output values to the DAC                      }
          ax0 = dm(i_value);
          dm(write_dac0) = ax0;
          ax0 = dm(q_value);
          dm(write_dac1) = ax0;
          ax0 = dm(sync);
          dm(write_dac2) = ax0;
          si = dm(i_output);
          sr = ashift si by -2 (lo);
          dm(write_dac3) = sr0;
          dm(load_dac) = sr0;
          tx0 = sr0;

          {Check if new baud should be calculated                        }
          ax0 = dm(sample_count);
          ay0 = 1;
          my1 = 0x7fff;              {Set sync value for new sample       }
          dm(sync) = my1;
          ar = ax0 - ay0;
          dm(sample_count) = ar;
          if gt jump next_sample;
```

146

```
next_baud:
        i4 = ^i_coef_line;     {Reset Coefficient Pointer           }
        dm(i_coef_ptr) = i4;
        ax0 = 4;               {Reset sample counter                }
        dm(sample_count) = ax0;
        si = 0;                {Zero SI Register                    }
        cntr = dm(fallback_count);
        do scram_bit until ce;
            call get1;         {Get next N data bits to be transmitted }
            call scramble;
scram_bit:
        ax0 = dm(fallback_mode);   {Test if in 4800 bps mode        }
        ay0 = 0x06;
        ar = ax0 - ay0;
        if ne jump not48;

mode48:                            {Calculate Q4 = inv(Q3 + Q2)     }
        sr = lshift si by -1(lo);  {sr0 = Q2                        }
        ay0 = si;
        ar = sr0 xor ay0;
        ar = not ar;
        ay1 = 1;
        ar = ar and ay1;           {AY1 must be preset to 1         }
        sr = lshift si by 1 (lo);  {Store output in si              }
        sr = sr or lshift ar by 0 (lo);
        si = sr0;
not48:
        call IQ;               {Calculate delta phi,phase sum,I and Q }
        my1 = 0x8000;          {Set sync value for new baud         }
        dm(sync) = my1;
        jump next_sample;
next_sample:
        call psf;              {Low Pass Filter I and Q components  }
                               {Output values to the DAC            }
{       call modulator;        Modulate and sum signals            }
done:   rti;

{─────────────Change Fallback Mode──────────────────────────────}
set_fall_back:
        ena sec_reg;
        ax0 = dm(fallback_mode);
        mx0 = i4;              {store i4 in temporary location      }
        mx1 = m4;              {store m7 in temporary location      }
        my0 = L4;              {store L4 in temporary location      }
        l4 = 0;
        m4 = ^jump_table;      {CASE Statement                     }
        i4 = ax0;
        modify(i4,m4);
        jump (i4);
```

(listing continues on next page)

2   Modems

```
jump_table:
   {Current fallback = 9600 bps, change to 7200bps                              }
         ay0 = 0x03;
         ax0 = 0x03;
         jump case_end;

   {Current fallback = 7200 bps, change to 4800bps                              }
         ay0 = 0x06;
         ax0 = 0x02;
         jump case_end;

   {Current fallback = 4800 bps, change to 7200bps                              }
         i4 = ^buffer;          {Reset Delay Buffer and Data Pointers }
         l4 = %buffer;
         m4 = 1;
         ax0 = 0;
         dm(bit_count) = ax0;
         cntr = %buffer;
         do z2 until ce;
z2:      dm(i4,m4) = ax0;
         ay0 = 0x0;
         ax0 = 0x04;
         jump case_end;
case_end:
         dm(fallback_count) = ax0;
         dm(fallback_mode) = ay0;
         ax0 = 0;                              {Zero Phi Value         }
         dm(phi_value) = ax0;
         i4 = mx0;                             {Restore i4             }
         m4 = mx1;                             {Restore m7             }
         L4 = my0;                             {Restore L4             }
         rti;
{————————Initialization for V.29————————————————————}
V29_INIT:
                        {————————SPORT INIT————————————————————}
         ax0 = 0x02;          {SCLKDIV = 2.048 MHz                    }
         dm(0x3ff5) = ax0;
         ax0 = 213;           {RFSDIV = 639 for 9600                  }
         dm(0x3ff4) = ax0;
         ax0 = 0x6b27;        {Internal SCLK, Frame Syncs             }
         dm(0x3ff6) = ax0;
                        {————————Wait States————————————————————}
         ax0 = 0xffff;
         dm(0x3ffe) = ax0;
```

148

Modems 2

```
                        {————————Data Buffer Inits————————————}
        i0 = ^buffer;
        l0 = %buffer;
        m1 = 1;
        ax0 = 0;
        cntr = %buffer;
        do z1 until ce;
z1:     dm(i0,m1) = ax0;
                        {————————Variable Initializations————————————}
ax0 = 0;
        dm(quad_bit) = ax0;
        dm(bit_count) = ax0;
        dm(phi_value) = ax0;
        dm(sample_count) = ax0;
        ax0 = 6;
        dm(fallback_mode) = ax0;
        ax0 = 2;
        dm(fallback_count) = ax0;
        ax0 = ^buffer+17;
        dm(buf_ptr2) = ax0;
        ax0 = ^buffer+22;
        dm(buf_ptr3) = ax0;
        ax0 = 0xa5a5;
        dm(seed_hi) = ax0;
        ax0 = 0x1234;
        dm(seed_lo) = ax0;


                        {————————Pulse Shaped Filter Inits————————————}
        i0 = ^i_coef_line;
        dm(i_coef_ptr) = i0;
        i0 = ^i_delay_line;
        dm(i_delay_ptr) = i0;
        l0 = %i_delay_line;
        m0 = 1;
        ax0 = 0;
        cntr = %i_delay_line;
        do dloop until ce;
   dloop:
        dm(i0,m0) = ax0;
        ax0 = 0x7fff;
        dm(sync) = ax0;
                        {————————Timer Init————————————}
ax0 = 0xffff;
        dm(0x3ffd) = ax0;
        dm(0x3ffc) = ax0;
        dm(0x3ffb) = ax0;
```

(listing continues on next page)

2 Modems

```
                        {————————DAG INIT————————————————————————————}
        m0 = 0;         {Init M0,M1,M5,M5                             }
        m1 = 1;
        m4 = 0;
        m5 = 1;
        l0 = 0;         {Set All L registers to zero                 }
        l1 = 0;
        l2 = 0;
        l3 = 0;
        l4 = 0;
        l5 = 0;
        l6 = 0;
        l7 = 0;
        rts;
{————————————END INITIALIZATION—————————————————————————————————————}
.endmod;
```

Listing 2.29 Main V.29 Routine (MAIN29.DSP)

```
{GET2 routine for v.29   Modem                                              }
{Analog Devices                                                             }
{DSP Applications                                                           }
{                                                                           }
.module/boot=0/ram            get1mod;
{Get 1 data bit to be transmitted                                           }
{                                                                           }
{     INPUTS:                                                               }
{          dm(data_word)   16 bit words for transmission                    }
{          dm(bit_count)   number of bits left in word                      }
{          m0 = 0                                                           }
{          m1 = 1                                                           }
{          m4 = 0                                                           }
{          m5 = 1                                                           }
{                                                                           }
{   OUTPUTS:                                                                }
{          dm(quad_bit):  bit to transmit                                   }
{                                                                           }
{     USAGE:                                                                }
{          AX0, AY0, AR, SI, SR                                             }
{                                                                           }

{Global Variable Declaration                                               }
.external    quad_bit;                        {bit to transmit             }
.external    bit_count;                       {number of bits left in word }
.port        in_port;

{Local Variable Declaration                                                }
.var/dm/ram        data_word;                 {Data word to be transmitted }
.external          rand;
{———————————Code Start———————————————————————}
.entry GET1;
GET1:
          ay0 = dm(bit_count);        {If bit_count is zero           }
          ar = pass ay0;              {load new word                  }
          if ne jump dec_count;
          ay0 = 0xc123;               {Load new data word here        }
          call rand;
          dm(data_word) = ay0;
          ay0 = 16;
          dec_count: ar = ay0 -1;     {Decrement counter              }
          dm(bit_count) = ar;
get_bit:
          sr0 = dm(data_word);        {Get next bit                   }
          ay0 = 0x01;
          ar = sr0 and ay0;
          dm(quad_bit) = ar;
          sr = lshift sr0 by -1 (lo);
          dm(data_word) = sr0;        {Write data_word for next time  }
          rts;
.endmod;
```

Listing 2.30 Data Acquisition Routine (GET29.DSP)

2 Modems

```
.module/boot=0/ram        SCRAMBLE_MOD;
{Analog Devices                                                          }
{DSP Applications                                                        }
{                                                                        }
{ 2/20/91   Start Date                                                   }
{This module performs v.29 scrambling on one input bit                   }
{                                                                        }
{The scrambler generating polynomial is xin + y(n-18) + y(n-23)          }
{v.29 specifies additional data pattarns on startup.                     }
{This has not been implemented.                                          }
{                                                                        }
{  INPUTS:                                                               }
{        dm(quad_bit) = bit to be scrambled                              }
{        si = partial quad bit                                           }
{                                                                        }
{  OUTPUTS:                                                              }
{        si = scrambled output bit                                       }
{             si  is left shifted by 1 and the new bit is put in b0      }
{                                                                        }
{  USAGE:                                                                }
{        I0 = ^buffer                                                    }
{        I1 = Y(n-18)                                                    }
{        I2 = y(n-23)                                                    }
{Local variable Declarations                                            }
.external   buffer;                    {Delay Buffer                    }
.external   buf_ptr2;                  {Pointer to y(n-18) buffer value }
.external   buf_ptr3;                  {Pointer to y(n-23) buffer value }
.external   quad_bit;
.entry      SCRAMBLE;
SCRAMBLE:
        i1 = dm(buf_ptr2);             {Get Pointer values              }
        i2 = dm(buf_ptr3);
        m3 = -1;
        L1 = %buffer;                  {Set circular buffer registers   }
        L2 = %buffer;

        ax0 = dm(quad_bit);            {Get next bit to scramble        }
        ay0 = dm(i1,m3);               {y(n-18)                         }
        ar = ax0 xor ay0, ay0=dm(i2,m0); {y(n-23)                       }
        ar = ar xor ay0;               { = y(n)                         }
        dm(i2,m3) = ax0;               {Store Oldest Value              }

        sr = lshift si by 1 (lo);      {Store output in sr0             }
        sr = sr or lshift ar by 0 (lo);
        si = sr0;

        dm(buf_ptr2) = i1;             {Save buffer pointers            }
        dm(buf_ptr3) = i2;
        L1 = 0;                        {Clear L registers               }
        L2 = 0;
        rts;
.endmod;
```

152   **Listing 2.31 Data Scrambler Routine (SCRAM29.DSP)**

Modems   2

```
{Analog Devices                                                      }
{DSP Applications                                                    }
{                                                                    }
{ 2/20/91        Start Date                                          }
.module/boot=0/ram      PHI_MOD;
{Calculate delta phi, phase sum, I and Q                             }
{                                                                    }
{        INPUTS:                                                     }
{            SI = QuadBit Value in four LSBs                         }
{            Q1 (Amplitude Bit) is in b3                             }
{            Q2-4 (Phase Change) is b b 2-0                          }
{            dm[PHI] = current Phi Value                             }
{                                                                    }
{        OUTPUTS:                                                    }
{            MX0 = I Value                                           }
{            MX1 = Q Value                                           }
{                                                                    }
{        USAGE:                                                      }
{            I3, M3, AR, AX0, AY0                                    }
{                                                                    }
.external   Q_table_0;
.external   I_table_0;
.external   Q_table_1;
.external   I_table_1;
.external   delta_phi;
.external   phi_value;
.external   i_coef_line;
.external   i_coef_ptr;
.external   i_delay_line;
.external   i_delay_ptr;
.external   i_value;
.external   _value;
.entry      IQ;
IQ:
    {Look up Delta_phi value and add to phi for new phase      }
            ax1 = si;
            ay0 = 0x07;
            ar = ax1 and ay0;
            I7 = ^delta_phi;
            m7 = ar;
            modify(i7, m7);
            ax0 = dm(phi_value);
            ay0 = pm(i7,m4);
            ar = ax0 + ay0;
            ay1 = 0x0f;
            ar = ar and ay1;
            dm(phi_value) = ar;
```

(listing continues on next page)

2 Modems

```
{Look up I and Q components                                          }
{To find I/Q values, use phi/2 as offset into table                  }
        sr = lshift ar by -1 (lo);
{Look at Amplitude bit and chose lookup table                        }
        ay0 = 0x08;
        ar = ax1 and ay0;
        if eq jump zamp;              {if zero amplitude bit is zero }
nzamp:  I6 = ^I_table_1;              {amplitude bit is equal to one }
        I7 = ^Q_table_1;
        jump lookup;
zamp:      I6 = ^I_table_0;           {amplitude bit is equal to zero}
        I7 = ^Q_table_0;
lookup:
        m7 = sr0;
        modify(I6,M7);                {I6 = I Value                  }
        modify(I7,M7);                {I7 = Q Value                  }
        mx0 = pm(i6,m4);
        mx1 = pm(i7,m4);
   {Write new I value to PSF Delay Line                              }
        I6 = dm(i_delay_ptr);
        L6 = %i_delay_line;
        M6 = -1;
        dm(i6,m6) = mx0;
        dm(i_delay_ptr) = i6;
        L6 = 0;

        {Write IQ values for DAC                                     }
        dm(i_value) = mx1;
        dm(q_value) = mx0;
        rts;
.endmod;
```

Listing 2.32 IQ Generator Routine (IQ29.DSP)

2.8   REFERENCES

Bershad, J. N. 1989. "Nonlinear Quantization Effects in the LMS and Block LMS Adaptive Algorithms," *IEEE Trans. ASSP*, vol. 37, No. 10, pp.1504-1512.

Bingham A. J. 1988. *The Theory and Practice of Modem Design*. New York, NY: John Wiley & Sons.

Brady, D. M. 1970. "An Adaptive Coherent Diversity Receiver for Data Transmission through Dispersive Media," *IC Conference Record*, ICC 70 pp. 21-40.

CCITT, Eighth Plenary Assembly. 1985. *Red Book, Volume VIII, Fascicle VIII.1: Data Communication Over the Telephone Network*. Geneva: International Telecommunication Union.

Falconer, D. D. and L. Ljung. 1978. "Application of Fast Kalman Estimation to Adaptive Equalization," *IEEE Trans. Commun.*, vol. COM-26, pp. 1439-1446.

Godard, D. 1974. "Channel Equalization Using a Kalman Filter for Fast Data Transmission," *IBM. J. Res. Develop.*, vol. 18, pp. 267-273.

Kamilo, F. and D. Messerschmitt. 1987. *Advanced Digital Communications*. Englewood Cliffs, NJ: Prentice Hall.

Lee, Edward A. and David G. Messerschmitt. 1988. *Digital Communication*. Boston, MA: Kluwer Academic Publishers.

Lin, Shu and Daniel J. Costello, Jr. 1983. *Error Control Coding: Fundamentals and Applications*. Englewood Cliffs, N.J.: Prentice-Hall.

Long, G., Fuyun, L. and J. G. Proakis. 1989. "The LMS Algorithm with Delayed Coefficient Adaptation," *IEEE Trans. ASSP*, vol. 37, No. 9, pp. 1397-1405.

Lucky, R.W. 1965. "Automatic Equalization for Digital Communication," *Bell Syst. Tech. J.*, vol 44. pp. 547-588.

Proakis, J.G. and J. H. Miller. 1969. "An Adaptive Receiver for Digital Signaling through Channels with Intersymbol Interference," *IEEE Trans. Information Theory*, vol IT-15, pp. 484-497.

2   Modems

Proakis, John G. 1989. *Digital Communications*. Second Edition. New York, N.Y.: McGraw-Hill.

Proakis, John, G. and D.G. Manolakis. 1988. *Introduction to Digital Signal Processing*. New York, N.Y.: McMillan Publishing Co.

Quatieri, T. and G. O'Leary. 1989. "Far-Echo Cancellation in the Presence of Frequency Offset", *IEEE Transactions on Communications*. Volume 37, No. 6, pp. 635-634.

Satotius, E. H. and J. D. Pack. 1981. "Application of Least Squares Lattice Algorithms to Adaptive Equalization," *IEEE Trans. Commun.*, vol. COM-29, pp. 136-142.

Sklar, Bernard. 1988. *Digital Communications - Fundamentals and Applications*. Englewood Cliffs, N.J.: Prentice-Hall.

Ungerboeck, G. 1972. "Theory on the Speed of Convergence in Adaptive Equalizers for Digital Communication," *IBM. J. Res. Develop.*, vol. 16, pp. 546-555.

Ungerboeck, G. 1976. "Fractional Tap-spacing Equalizer and Consequences for Clock Recovery in Data Modems," *IEEE Trans. Commun.*, vol. COM-24, pp. 856-864.

Wang, J. D. and J. J. Werner. 1988. "Performance Analysis of an Echo Cancellation Arrangement that Compensates for Frequency Offset in the Far Echo", *IEEE Transactions on Communications*. Volume COM-36, No. 3, pp. 364-372.

Weinstein, S. 1977. "A Passband Data-Driven Echo Canceller for Full-Duplex Transmission on Two-Wire Circuits", *IEEE Transactions on Communications*. Volume COM-25, pp. 654-665.

Ziemer and Peterson. 1985. *Digital Communications and Spread Spectrum Systems*. New York: MacMillan Publications.

Linear Predictive Coding 3

3.1 OVERVIEW

Digital Signal Processing Applications Using the ADSP-2100 Family, Volume 1, contains a chapter about Linear Predictive Coding. That chapter (Chapter 10) discusses the following topics:

- LPC theory
- Correlation functions (auto-correlation and cross-correlation)
- Levinson-Durbin Recursion
- Pitch Detection

Also, the chapter includes program listings and subroutines for the following applications:

- Correlation subroutine
- LPC coefficient calculation
- Pitch Detection
- LPC synthesis

After Volume 1 was published, additional application programs for 7.8 kbits/s and 2.4 kbits/s LPCwere developed to build on the usefulness of the information included in that text. This chapter introduces those applications.

3.2 LINEAR PREDICTION

Linear Predictive Coding is a speech coding technique that models the human vocal tract. According to the model, the human body produces two basic types of sounds: voiced and unvoiced. If the vocal folds vibrate when air from the lungs is forced through them, voiced sound is produced. Unvoiced sound is produced by the tongue, lips, teeth, and mouth.

For voiced and unvoiced sounds, the vocal tract can be modeled as a series of cylinders with different radii and different amounts of energy at the boundaries between the cylinders. Mathematically, you can represent this model as a linear filter excited by a fundamental frequency (voiced sound) or random noise (unvoiced sound).

3   Linear Predictive Coding

The goal of linear predictive analysis is to derive the necessary parameters for reconstructing the sound: voiced/unvoiced decision, fundamental frequency, system gain, and the coefficients that describe the filter. The objective of Linear Predictive Coding (LPC) is to predict the next output of the system based on previous outputs and inputs. This is an effective coding technique because speech is a highly correlated signal when considered during a short interval of time (frame length). That is, given a sequence of speech samples, subsequent speech samples can be predicted with a minimum of error over a short period of time.

LPC relies on a technique that uses a linear combination of previous outputs, past inputs, and an input excitation. The filter equation or speech output, s(n), can be represented as:

$$s(n) = -\sum_{k=1} a_k s(n-k) + G\sum_{j=0} b_j u(n-j), b0 = 1$$

where s() is the speech sequence and u() is the excitation sequence.

The resulting synthesis filter has the following domain representation:

$$H(z) = \frac{\omega(z)}{u(z)} = G * \frac{1-z}{\left(1 + a_j z^{-i}\right)}$$

To synthesize sound with the model described above, you must specify the complete parameter set for the synthesis filter. Table 3.1 defines these parameters.

Parameter	Symbol
Filter Coefficients	a_k, b_k
Gain	G
Excitation	
Voiced/Unvoiced	1/0
Pitch Period	P

Table 3.1 Parameter Set For The Sound Synthesis Model

Linear Predictive Coding   3

3.3   7.8 kbits/s LPC

The 7.8 kbits/s LPC (which is of higher speech quality) is derived by using a data frame size of 180 samples sampled at 8 kSa/s. For each frame, the analysis section generates ten 16-bit coefficients and an additional 16-bit word containing the gain and pitch period. Therefore, the bit rate is derived by:

$$\text{Bit Rate} = \frac{(11 \text{ words})*(16 \text{ bits / word})}{(180 \text{ samples})}*(125 \text{ }\mu s \text{ / sample}) = 7822 \text{ bits / s } (7.8 \text{ kbits / s})$$

Listing 3.1 shows the 7.8 kbits/s LPC Routine. This routine calls several subroutines listed in Section 3.4, *"LPC Subroutines."*

```
.module/boot=3/abs=0        lpc7k8_through;
{   LPC7k8.DSP - talk through, encoding and decoding using LPC.
Input: Speech samples from microphone (using autobuffering) via sport0
Ouput: Speech samples to speeaker (using autobuffering) via sport0
Modules used:
     - pre_emphasize_speech          (PREEMP.DSP)
     - gain_calculation              (GAIN.DSP)
     - autocorrelation_of_speech     (AUTOCOR.DSP)
     - durbin_double/single          (DURBIN2.DSP/DURBIN.DSP)
     - pitch_detection               (PITCH.DSP)
     - lpc_sync_synth                (SSYNTH.DSP)
     - de_emphasize_speech           (DEEMP.DSP)
     - constant header               (LPC.H)
Description:
   This program implements a shell to demonstrate the LPC algorithm on an EZ-LAB
   board. Speech is autobuffered in from the codec, compressed, decompressed and
   autobuffered back out to the codec, providing a "talk-through" program.
NOTE: The framesyncs of sport0 (TFS0 & RFS0) SHOULD NOT be tied together EXTERNALY!!
   (On the EZ-LAB's serial connector it is pins 4 & 5)
}

{include constant definitions}
#include "lpc.h";

{Buffers used by autobuffering, input swaps between analys_buf & receive_buf,
   whereas output swaps between synth_buf & transit_buf}
.var/dm/ram/circ  analys_buf[FRAME_LENGTH];
.var/dm/ram/circ  receive_buf[FRAME_LENGTH];

{The LPC-parameters are stored in trans_line to "simulate" transmission}
.var/dm/ram       trans_line[WORDS_PR_LPCFRAME];
```

(listing continues on next page)

3   Linear Predictive Coding

```
{Pointers to the buffers that are NOT currently being used by autobuffering}
.var/dm/ram        p2_analysis,p2_synth;

{Intermediate variables}
.var/pm/ram        autocor_speech[FRAME_LENGTH];
.var/dm/ram/circ   k[N];
.var/dm/ram        pitch, gain;
.var/dm/ram        lpc_flag;
.external          pre_emph;
.external          calc_gain;
.external          a_correlate;
.external          levinson;
.external          detect_pitch;
.external          clear_filter;
.external          synthesis;
.external          de_emph;

{load interrupt vectors}
jump start_test; nop; nop; nop;                {reset interrupt}
    rti; nop; nop; nop;
    rti; rti; nop; nop;                        {sport0 transmit}
    call rcv_ir; rti; nop; nop;                {sport0 receive}
    rti; nop; nop; nop;
    rti; nop; nop; nop;
    rti; nop; nop; nop;

change_demo:
    gd: if not flag_in jump gd;
    ay0 = 0x0338;                              {set bootforce bit}
    dm(0x3fff) = ay0;
    nop;
    rts;

start_test:
    {configure sports etc.}
    ax0 = 0x0000;
    dm(0x3ffe) = ax0;                          {dm waits = 0}
    ax0 = 0x6327;
    dm(0x3ff6) = ax0;                          {set sport0 control reg}
    ax0 = 2;
    dm(0x3ff5) = ax0;                          {set sclkfreq to 2.048 Mhz}
    ax0 = 255;
    dm(0x3ff4) = ax0;                          {set rclkfreq to 64 khz}

{Default register values, these values can always be asumed, and must
    be reset if altered}
    l0 = 0; l1 = 0; l2 = 0; l3 = 0;
    l4 = 0; l5 = 0; l6 = 0; l7 = 0;
```

Linear Predictive Coding   3

```
{DEDICATED REGISTERS. These registers must NOT be altered by any routine at any
   time! (used by autobuffering)}
   m0 = 0;   m4 = 0;
   m1 = 1;   m5 = 1;
   i3 = ^receive_buf;
   l3 = %receive_buf;

{Setup and clear intermediate buffers}
   i6 = ^analys_buf;
   l6 = 0; dm(p2_analysis) = i6;

   ena sec_reg;
   ax0 = FRAME_LENGTH;
   af  = pass ax0;
   dis sec_reg;

   ax0 = 0;
   dm(lpc_flag) = ax0;

{clear the synthesis filter}
   call clear_filter;

{enable sport0}
   ax0 = 0x1038;
   dm(0x3fff) = ax0;

{enable sport0}
   icntl = b#00111;
   imask = b#001000;                                {Enable receive}

   wait: idle;


   if not flag_in call change_demo;
   ax0 = dm(lpc_flag);
   ar  = pass ax0;
   if eq jump wait;
   ax0 = 0;
   dm(lpc_flag) = ax0;

{Parameters: i0 = p2_analysis (-> speech)
 Returns:    filtered speech}
   i0 = dm(p2_analysis);
   l0 = 0;
   call pre_emph;
```

(listing continues on next page)

3   Linear Predictive Coding

```
{Parameters: i0  = p2_analysis (-> speech)
 Returns:    sr1 = gain}
    i0 = dm(p2_analysis);
    l0 = 0;
    call calc_gain;
    dm(gain) = sr1;

{Parameters: i0 = p2_analysis (-> speech)
 Returns:    autocor_speech[]}
    i0 = dm(p2_analysis); l0 = 0;
    i6 = ^autocor_speech; l6 = 0;
    call a_correlate;

{Parameters: i4 -> autocor_speech[])
 Returns:    i0 -> k[]}
    i4 = ^autocor_speech; l4 = 0;
    i0 = ^k; l0 = 0;
    call levinson;

{Parameters: i0 -> k[], i6 -> autocor_speech[]
 Returns:    si = pitch}
    i0 = ^k; l0 = 0;
    i6 = ^autocor_speech; l6 = 0;
    call detect_pitch;
    dm(pitch) = si;

{TRANSMISSION LINE}

{Parameters: ax1 =  pitch, mx1 = gain, i0  -> k[]
 Returns:    i2  -> speech}
    i0 = ^k; l0 = 0;
    i1 = ^k + N - 1; l1 = 0;               {store k's in revers order - }
    cntr = 5;                              {N/2}{required by lattice routine}
    m2 = -1;
    do reverse_ks until ce;
         ay0 = dm(i0,m0);
         ay1 = dm(i1,m0);
         dm(i0,m1) = ay1;
    reverse_ks: dm(i1,m2) = ay0;
    ax1 = dm(pitch);
    mx1 = dm(gain);
    i1 = ^k; l1 = N;
    i2 = dm(p2_analysis); l2 = 0;
    call synthesis;

{Parameters: i0 = p2_analysis (-> speech)
 Returns:    filtered speech}
    i0 = dm(p2_analysis); l0 = 0;
    call de_emph;

jump wait;
```

Linear Predictive Coding   3

```
{End of main routine}

{Autobuffering interrupt routines}
trns_ir:
rts;
rcv_ir:
   ena sec_reg;
   ax0 = dm(i3,m0);
   tx0 = ax0;
   ax0 = rx0;
   dm(i3,m1) = ax0;
   af = af - 1;
   if gt jump no_lpc;                          {switch pointers}
   ay0 = i3;
   ay1 = dm(p2_analysis);
   i3 = ay1; l3 = %receive_buf;
   dm(p2_analysis) = ay0;
   ax0 = FRAME_LENGTH;
   af  = pass ax0;
   ax0 = 1;
   dm(lpc_flag) = ax0;
   no_lpc:
   dis sec_reg;
   rts;

{END of main code}

.endmod;
```

Listing 3.1 7.8 kbits/s LPC Routine

3.4   2.4 kbits/s LPC

The speech quality of 2.4 kbits/s LPC is somewhat deminished from the quality of 7.8 kbits/s LPC, but the compression ration is much higher. Although the bit rate is more than 3 times slower than 7.8 kbits/s LPC, the quality is considered acceptable for most applications. The lower bit rate is achieved by reducing the number of bits per frame from 176 (11 words X 16 bits/word) unquantized bits to 54 quantized bits. Therefore, the bit rate for the 2.4 kbits/s LPC is derived by:

$$\text{Bit Rate} = \frac{(54 \text{ bits / frame})}{(180 \text{ samples})} * (125 \text{ μs / sample}) = 2400 \text{ bits / s } (2.4 \text{ kbits / s})$$

The quantization routines are called only by the 2.4 kbits/s version of the code (main module is LPC2K4.DSP); they are called encode.dsp and decode.dsp.

3   Linear Predictive Coding

Listing 3.2 shows the 2.4 kbits/s LPC Routine. This routine calls the subroutines listed in Section 3.4, *"LPC Subroutines."*

```
.module/boot=4/abs=0        lpc2k4_through;
{   LPC2k4.DSP - talk through, encoding and decoding using LPC.
Input: Speech samples from microphone (using autobuffering) via sport0
Ouput:Speech samples to speeaker (using autobuffering) via sport0
Modules used:
   - pre_emphasize_speech        (PREEMP.DSP)
   - gain_calculation            (GAIN.DSP)
   - autocorrelation_of_speech   (AUTOCOR.DSP)
   - durbin_double/single        (DURBIN2.DSP/DURBIN.DSP)
   - pitch_detection             (PITCH.DSP)
   - lpc_sync_synth              (SSYNTH.DSP)
   - de_emphasize_speech         (DEEMP.DSP)
   - constant header             (LPC.H)
Description:
   This program implements a shell to demonstrate the LPC algorithm on an EZ-LAB
   board. Speech is autobuffered in from the codec, compressed, decompressed and
   autobuffered back out to the codec, providing a "talk-through" program.
NOTE: The framesyncs of sport0 (TFS0 & RFS0) SHOULD NOT be tied together EXTERNALY!!
   (On the EZ-LAB's seriel connector it is pins 4 & 5)
}

{include constant definitions}
#include "lpc.h";

{Buffers used by autobuffering, input swaps between analys_buf & receive_buf,
   whereas output swaps between synth_buf & transit_buf}
.var/dm/ram/circ  analys_buf[FRAME_LENGTH];
.var/dm/ram/circ  receive_buf[FRAME_LENGTH];

{The LPC-parameters are stored in trans_line to "simulate" transmission}
.var/dm/ram       trans_line[WORDS_PR_LPCFRAME];

{Pointers to the buffers that are NOT currently being used by autobuffering}
 .var/dm/ram      p2_analysis,p2_synth;

{Intermediate variables}
.var/pm/ram       autocor_speech[FRAME_LENGTH];
.var/dm/ram/circ  k[N];
.var/dm/ram       pitch, gain;
.var/dm/ram       lpc_flag;


.external         pre_emph;
.external         calc_gain;
.external         a_correlate;
.external         levinson;
```

164

Linear Predictive Coding   3

```
.external        detect_pitch;
.external        encode;
.external        decode;
.external        clear_filter;
.external        synthesis;
.external        de_emph;

{load interrupt vectors}
jump start_test; nop; nop; nop;                    {reset interrupt}
   rti; nop; nop; nop;
   rti; rti; nop; nop;                             {sport0 transmit}
   call rcv_ir; rti; nop; nop;                     {sport0 receive}
   rti; nop; nop; nop;
   rti; nop; nop; nop;
   rti; nop; nop; nop;
change_demo:
   gd: if not flag_in jump gd;
   ay0 = 0x0238;                                   {set bootforce bit}
   dm(0x3fff) = ay0;
   nop;
   rts;

start_test:
   {configure sports etc.}
   ax0 = 0x0000;
   dm(0x3ffe) = ax0;                               {dm waits = 0}
   ax0 = 0x6327;
   dm(0x3ff6) = ax0;                               {set sport0 control reg}
   ax0 = 2;
   dm(0x3ff5) = ax0;                               {set sclkfreq to 2.048 Mhz}
   ax0 = 255;
   dm(0x3ff4) = ax0;                               {set rclkfreq to 64 khz}

{Default register values, these values can always be asumed, and must
   be reset if altered}
   l0 = 0; l1 = 0; l2 = 0; l3 = 0;
   l4 = 0; l5 = 0; l6 = 0; l7 = 0;

{DEDICATED REGISTERS. These registers must NOT be altered by any routine at
any time! (used by autobuffering)}
   m0 = 0;   m4 = 0;
   m1 = 1;   m5 = 1;
   i3 = ^receive_buf;
   l3 = %receive_buf;
```

(listing continues on next page)

3 Linear Predictive Coding

```
{Setup and clear intermediate buffers}
    i6 = ^analys_buf;
    l6 = 0; dm(p2_analysis) = i6;

    ena sec_reg;

    ax0 = FRAME_LENGTH;
    af  = pass ax0;
    dis sec_reg;

    ax0 = 0;
    dm(lpc_flag) = ax0;

{clear the synthesis filter}
    call clear_filter;

{enable sport0}
    ax0 = 0x1038;
    dm(0x3fff) = ax0;

{enable sport0}
    icntl = b#00111;
    imask = b#001000;                      {Enable receive}


    wait: idle;


    if not flag_in call change_demo;


    ax0 = dm(lpc_flag);
    ar  = pass ax0;
    if eq jump wait;


    ax0 = 0;
    dm(lpc_flag) = ax0;

{Parameters: i0 = p2_analysis (-> speech)
 Returns:    filtered speech}
    i0 = dm(p2_analysis);
    l0 = 0;
    call pre_emph;

{Parameters: i0  = p2_analysis (-> speech)
 Returns:    sr1 = gain}
    i0 = dm(p2_analysis);
    l0 = 0;
    call calc_gain;
    dm(gain) = sr1;
```

Linear Predictive Coding   3

```
{Parameters: i0 = p2_analysis (-> speech)
 Returns:    autocor_speech[]}
   i0 = dm(p2_analysis); 10 = 0;
   i6 = ^autocor_speech; 16 = 0;
   call a_correlate;

{Parameters: i4 -> autocor_speech[])
 Returns:    i0 -> k[]}
   i4 = ^autocor_speech;
   14 = 0;
   i0 = ^k;
   10 = 0;
   call levinson;

{Parameters: i0 -> k[],
         i6 -> autocor_speech[]
 Returns:    si =  pitch}
   i0 = ^k;
   10 = 0;
   i6 = ^autocor_speech; 16 = 0;
   call detect_pitch;
   dm(pitch) = si;

{Parameters: i1 -> k[],
         ar = pitch,
         si = gain
Returns:  parameters encoded ar = pitch,
         si = gain}
   i1 = ^k; 11 = 0;
   ar = dm(pitch);
   si = dm(gain);
   call encode;
   dm(pitch) = ar;
   dm(gain)  = si;

{TRANSMISSION LINE}


{Parameters: i1 -> k[],
         si  = pitch,
         ax0 = gain
Returns:  k's decoded
         si  = pitch,
         ax0 = gain}
   i1  = ^k;   11 = 0;
   si  = dm(pitch);
   ax0 = dm(gain);
   call decode;
   dm(pitch) = si;
   dm(gain)  = ax0;
```

(listing continues on next page)

3   Linear Predictive Coding

```
{Parameters: ax1 =  pitch,
         mx1 =  gain,
         i0  -> k[]
 Returns:  i2  -> speech}
   i0 = ^k;
   l0 = 0;
   i1 = ^k + N - 1;
   l1 = 0;                                    {store k's in revers order - }
   cntr = 5;                                  {N/2} {required by lattice routine}
   m2 = -1;
   do reverse_ks until ce;
         ay0 = dm(i0,m0);
         ay1 = dm(i1,m0);
         dm(i0,m1) = ay1;
   reverse_ks: dm(i1,m2) = ay0;
   ax1 = dm(pitch);
   mx1 = dm(gain);
   i1 = ^k;
   l1 = N;
   i2 = dm(p2_analysis); l2 = 0;
   call synthesis;

{Parameters: i0 = p2_analysis (-> speech)
 Returns:    filtered speech}
   i0 = dm(p2_analysis);  l0 = 0;
   call de_emph;
jump wait;


{End of main routine}

{Autobuffering interrupt routines}
trns_ir:
rts;

rcv_ir:
   ena sec_reg;
   ax0 = dm(i3,m0);
   tx0 = ax0;
   ax0 = rx0;
   dm(i3,m1) = ax0;
   af = af - 1;
   if gt jump no_lpc;
   {switch pointers}
   ay0 = i3;
   ay1 = dm(p2_analysis);
   i3 = ay1;
   l3 = %receive_buf;
   dm(p2_analysis) = ay0;
```

Linear Predictive Coding   3

```
   ax0 = FRAME_LENGTH;
   af  = pass ax0;


   ax0 = 1;
   dm(lpc_flag) = ax0;
   no_lpc:


   dis sec_reg;
rts;

{END of main code}

.endmod;
```

Listing 3.2 2.4 kbits/s LPC Routine

3.5   LPC SUBROUTINES

This section contains the subroutines called by the 7.8 and 2.4 kbits/s LPC routines.

```
.module/boot=3/boot=4        autocorrelation_of_speech;
{   AUTOCOR.DSP - perform autocorrelation on input speech frame.
INPUT: i0 -> frame of speech (dm)
    l0 =  0
    i6 -> buffer for autocorrelation (pm)
    l6 =  0

OUTPUT: autocorrelation buffer filled

FUNCTIONS CALLED: None
DESCRIPTION:
    First the speech is autoscaled (IN PLACE!!), to avoid overflow.
    The autocorrelation is calculated, and normalized so r[0] = 1 (0x7fff). }

{include constant and macro definitions}
#include "lpc.h";
.entry     a_correlate;
.external  overflow;

{.var/pm/ram/circ  copy_of_speech[FRAME_LENGTH];}
.var/dm     p2_speech;
.var/dm     p2_autocor_speech;
```

(listing continues on next page)

3 Linear Predictive Coding

```
a_correlate:

{store pointers for later use}
    dm(p2_speech) = i0;
    dm(p2_autocor_speech) = i6;

{auto scale input before correlating}
{first: detect largest exp in speech}
    {i0 -> speech)}
    cntr = FRAME_LENGTH;
    sb = -16;
    do max_speech until ce;
          si = dm(i0,m1);
    max_speech: sb = expadj si;

{adjust speech input: normalize to largest and then right shift to |AC_SHIFT|.
    (16-|AC_SHIFT|) format (done in one shift). At the same time copy to pm for
    correlation}
    i0 = dm(p2_speech);            l0 = 0;
    i5 = dm(p2_autocor_speech);    l5 = 0;
    cntr = FRAME_LENGTH;
    ax0 = sb;
    ay0 = AC_SHIFT;                        {scale down to avoid overflow (worst case)}
    ar = ay0 - ax0;                        {effective scale value}
    se = ar;
    do adj_speech until ce;
          si = dm(i0,m0);
          sr = ashift si (hi);
          pm(i5,m5) = sr1;
    adj_speech: dm(i0,m1) = sr1;

{do autocorrelation, R[i] = sum_of s[j]*s[i+j]}
{NOTE: the counter updating scheme, might cause a "acces to non-existing memory"
    in the simulator/emulator}
    i5 = dm(p2_autocor_speech); l5 = 0;              {s[i+j]}
    {i6 -> autocor_speech}                           {->R[i]}
    i2 = FRAME_LENGTH; l2 = 0;
    m2 = -1;
    cntr = FRAME_LENGTH;
    do corr_loop until ce;                           {i loop}
    i0 = dm(p2_speech); l0 = 0;                      {->s[j]}
    i4 = i5; l4 = 0;                                 {->s[i+j]} cntr=i2;
        mr=0, my0=pm(i4,m5), mx0=dm(i0,m1);
        do cor_data_loop until ce;   {j loop}
    cor_data_loop: mr=mr+mx0*my0(ss),my0=pm(i4,m5),mx0=dm(i0,m1);
        if mv call overflow;
    mx0 = dm(i2,m2), my0 = pm(i5,m5);                {update counters: - }
                                                     {(innerloop cnt'er)-, i++)}
    corr_loop: pm(i6,m5) = mr1;                      {store R[i]}
```

170

Linear Predictive Coding   3

```
{Normalize autocorrelation sequence}
{shift sequnece for maximum precision before division}
    i5 = dm(p2_autocor_speech); 15 = 0;
    cntr = FRAME_LENGTH - 1;
    si = pm(i5,m4);                          {R(0)}
    se = exp si (hi);
    sr = norm si (hi);
    pm(i5,m5) = sr1;                         {new R(0)}
    do sh_cor until ce;                      {shift remaining sequence accordingly}
          si = pm(i5,m4);
          sr = norm si (hi);
    sh_cor: pm(i5,m5) = sr1;

{calculate R(i)/R(0)}
    i5 = dm(p2_autocor_speech); 15 = 0;
    cntr = FRAME_LENGTH - 1;
    ax0 = pm(i5,m4);                         {ax0 = divisor = R(0)}
    ay0 = 0x7fff;
    pm(i5,m5) = ay0;                         {new R(0) = 1}
    do nrm_cor until ce;
    ay1 = pm(i5,m4);                         {ay1 = MSW of dividend }
    ay0 = 0x0000;                            {ay0 = LSW of dividend}
    divide(ax0,ay1);
    nrm_cor: pm(i5,m5) = ay0;
rts;

.endmod;
```

Listing 3.3 AUTOCOR.DSP Subroutine

3   Linear Predictive Coding

```
.module/boot=4          decode_parameters;
{   DECODE.DSP - decompresses the lpc parameters.
INPUT:
   i1  -> k  (reflection coeffs) l1 = 0
   si  = pitch
   ax0 = gain

OUTPUT:
   k's decoded inplace
   si  = pitch
   ax0 = gain

The log coded parameters are decompressed using:
     k[i] = (10^(g[i]*4)+1)/(10^(g[i]*4)-1)
}

#include "lpc.h"
.const    DELOG_ORDER = 8;
.var/pm   delog_coeffs[2*DELOG_ORDER];              {Ci lsb, Ci msb, Ci-1 lsb .....}
.init         delog_coeffs: <delog.cff>;            {scaled down by 512 = 2^9}
.const    LAR_ORDER = 16;
.var/pm   lar_coeffs[2*LAR_ORDER];                  {Ci lsb, Ci msb, Ci-1 lsb .....}
.init         lar_coeffs: <dec.cff>;                {scaled down by 1024 = 2^10}
.var/dm   temp_pitch;
.var/dm   temp_gain;
.entry    decode;
.external   poly_approx;

decode:
{decode}
   {si = pitch}
   se = -9;
   sr = lshift si (lo);
   dm(temp_pitch) = sr0;

        {ax0 = gain}
        sr1 = ax0;
        ar  = pass ax0;
        if eq jump zero_gain;
           my0 = 0x0000;                  {gain lsb}
           my1 = ax0;                     {gain msb}
           ax0 = DELOG_ORDER - 1;
           i6  = ^delog_coeffs; l6 = 0;
           call poly_approx;              {log10 function}
           si = mx0;
        sr = lshift si by 9 (lo);         {scale up by 512, comes with the coeff's}
        sr = sr or ashift ar by 9 (hi);
        zero_gain:
        dm(temp_gain) = sr1;
```

```
            cntr = 2 {N};
            do dec_k until ce;
                my0 = 0x0000;                  {k lsb}
                my1 = dm(i1,m0);               {k msb}
                ax0 = LAR_ORDER - 1;
                i6  = ^lar_coeffs; l6 = 0;
                call poly_approx;             {log area ratio function}
                si = mx0;
            sr = lshift si by 10 (lo);        {scale up by 1024}
            sr = sr or ashift ar by 10 (hi);
            dec_k: dm(i1,m1) = sr1;
    {setup return parameters}
    si  = dm(temp_pitch);
    ax0 = dm(temp_gain);
rts;

.endmod;
```

Listing 3.4 DECODE.DSP Subroutine

3 Linear Predictive Coding

```
.module/boot=3/boot=4          de_emphasize_speech;
{   DEEMP.DSP - deemphasizes a speech frame. (filters it)
INPUT:
    i0 -> frame of speech
    l0 =  0
OUTPUT: speech deemphasized
FUNCTIONS CALLED:
    None
DESCRIPTION:
    Filters speech using: H(z) = 1/(1 - 0.75)
}

{Include constant definitions}
#include "lpc.h"
.entry    de_emph;
.external    overflow;
.var/dm/ram delay;

de_emph:
    {deemphasize}
    mx0 = 0x6000;                          {a1 = 0.75}
    cntr = FRAME_LENGTH;
    do filt_speech until ce;
         mr  = 0;
         mr1 = dm(i0,m0);                  {x(n)}
         my0 = dm(delay);                  {y(n-1)}
         mr = mr + mx0*my0 (ss);
              if mv call overflow;
    dm(delay) = mr1;                       {update delay with y(n)}
    filt_speech: dm(i0,m1) = mr1;
rts;

.endmod;
```

Listing 3.5 DEEMP.DSP Subroutine

Linear Predictive Coding   3

```
.module/boot=3/boot=4         durbin_single;
{  DURBIN.DSP - single precision Levinso-Durbin routine
INPUT:
   i4 -> buffer with autocorrelated speech (pm)
   14 =  0
   i0 -> buffer for reflection coeffs
   10 =  0
OUTPUT: reflection coeffs calculated
     mr1 = Ep (minimum total squared prediction error
FUNCTIONS CALLED:
     None
DESCRIPTON:
   The routine implements Durbins recursion method of solving a set of linear
   equations forming a Toeplitz matrix. The algorithm in C is as follows:

   Where R[] is the autocorrelation, and k[] the reflection coeffs. e[] is the
   total squared error, and a[][] is the predictor coeff matrix (since only the
   i'th and the i+1'th column is used at any one time, the matrix is implemented
   as two (a_old and a_new) columns swapping place after each iteration.


        e[0] = R[0]
        k[1] = R[1] / e[0]
        alpha[1][1] = k[1]
        e[1] = (1 - k[1]*k[1]) * e[0]

        for (i=2; i<=N; i++)
        begin
        k[i] = 0
        for (j=1; j<=i-1; j++)
        k[i] = k[i] + R[i-j] * alpha[i-1][j]
        k[i] = R[i] - k[i]
        k[i] = k[i] / e[i-1]
        alpha[i][i] = k[i]
        for (j=i-1; j>0; j++)
        alpha[i][j] = alpha[i-1][j] - k[i]*alpha[i-1][i-j]
        e[i] = (1 - k[i]*k[i]) * e[i-1]
        end
}

{Include constant definitions}
#include "lpc.h"
.entry    levinson;
.external    overflow;
.global e;
```

(listing continues on next page)

3 Linear Predictive Coding

```
.var/dm/ram i_1;
.var/dm/ram e[N+1];                     {error values}
.var/dm/ram a_new[N],a_old[N];
.var/dm/ram ap_new,ap_old;              {pointers to a_*}
.var/dm/ram p2_k_i;                     {pointer to k[i]}
.var/dm  p2_autocor_speech;

{determines the format that a-values are stored in format: (SBITS+1).(16-SBITS-1)}
.const   SBITS  =  3;
.const   NSBITS = -SBITS;

levinson:
    i1 = ^a_new; l1 = 0;
    dm(ap_new) = i1;
    i2 = ^a_old; l2 = 0;
    dm(ap_old) = i2;
    dm(p2_autocor_speech) = i4;
    i5 = ^e;        l5 = 0;
    m2 = -1;
    m6 = -1;

se = NSBITS;

    {e[0] = R[0]}
    ax0 = pm(i4,m5);
    dm(i5,m5) = ax0;

    {k[1] = R[1]/e[0]}
    {ax0 = e[0] = divisor}
    ay1 = pm(i4,m4);                    {MSW of dividend}
    ay0 = 0000;                         {LSW of dividend}
    divide(ax0,ay1);
    ar = -ay0;                          {reverse sign of k before storing}
    dm(i0,m1)   = ar;
    dm(p2_k_i) = i0;

    {a_old[1] = k[1]}
    si = ay0;
    sr = ashift si (hi);               {store in (SBITS+1).(16-SBITS-1) format}
    dm(i2,m0) = sr1;

    {e[1] = (1 - k[1]*k[1])*e[0]}
                                        {ay0 = k[1]}
    mx0 = ay0;
    my0 = ay0;
    mr0 = 0xffff;                       {mr = 1 in 1.31 format}
    mr1 = 0x7fff;
    mr = mr - mx0*my0 (ss);
```

Linear Predictive Coding   3

```
                                        {ax0 = e(0)}
   my0 = ax0;
   mr = mr1 * my0 (ss);
   dm(i5,m4) = mr1;

   {for(i = 2; i <= N; i++)}
   cntr = N-1;
   ax0 = 1;                             {i-1}
   dm(i_1) = ax0;
   do pass_two until ce;

   {k[i] = 0}
         mr = 0;
{for(j = 1; j <= i-1; j++)}
         ay0 = dm(i_1);
         cntr = ay0;
         m3 = ay0;                      {i-1}
         m7 = ay0;                      {i-1}

{prepare: k[i] = k[i] + R[i-j]*a_old[j]}
         i2 = dm(ap_old);
         l2 = 0;
         i4 = dm(p2_autocor_speech);
         l4 = 0;
         modify(i4,m7);   {->R[i-1]}

         {loop}
         do calc_ks until ce;
            mx0 = pm(i4,m6);
            my0 = dm(i2,m1);
         calc_ks: mr = mr + mx0*my0 (ss);
         if mv call overflow;

         {k[i] = R[i] - k[i]}
         i4 = dm(p2_autocor_speech); l4 = 0;
   modify(i4,m7);
   modify(i4,m5);                       {->R[i]}
   si = pm(i4,m4);                      {R[i]}
   sr = ashift si (hi);                 {shift to (SBITS+1).(16-SBITS-1) format}
   ay1 = mr1;      {k[i]}
   ar = sr1 - ay1;
   if av call overflow;

{k[i] = k[i]/e[i-1]}
   i5 = ^e; l5 = 0;
   modify(i5,m7);
   ax0 = dm(i5,m5);                     {e[i-1]}
   ay1 = ar;                            {MSW of k[i]}
   ay0 = 0000;                          {LSW of k[i]}
```

(listing continues on next page)

3   Linear Predictive Coding

```
{overflow check}
   si = ax0;
   sr = ashift si (hi);
   ar = sr1 - ay1; {e[i-1] - k[i]}
   if ge jump e_ok;

{call overflow;}
       si = 0x7fff;                        {sat k[i]}
       sr = ashift si (hi);

{ay1 = sr1;}
       e_ok:
       divide(ax0,ay1);
       si = ay0;
       sr = ashift si by SBITS(hi);        {shift to 1.15 format before
storing}
       i0 = dm(p2_k_i);  0 = 0;
       ay1 = sr1;
       ar  = -ay1;                         {reverse sign of k before
storing}
       dm(i0,m1) = ar;                     {k[i] store}
       dm(p2_k_i) = i0;

{a_new[i] = k[i]}
   i1 = dm(ap_new); l1 = 0;
   modify(i1,m3);                          {->a_new[i]}
   dm(i1,m2) = ay0;

{for(j = i-1; j>0; j—)}
   cntr = dm(i_1);

{prepare: a_new[j] = a_old[j] - k[i]*a_old[i-j]}
   i2 = dm(ap_old);
   l2 = 0;
   modify(i2,m3);                          {modify by j (= i-1)}
   modify(i2,m2);                          {-> a_old[j]}
   i0 = dm(ap_old);                        {-> a_old[i-j]}
   l0 = 0; mx0 = sr1;                      {k[i]}

{loop}
   do calc_as until ce;
       mr0 = 0;
       mr1 = dm(i2,m2);                    {a_old[j]}
       my0 = dm(i0,m1);                    {a_old[i-j]}
       mr  = mr - mx0*my0 (ss);
   if mv {sat mr} call overflow;

   calc_as: dm(i1,m2) = mr1;
```

```
{e[i] = (1 - k[i]*k[i]) * e[i-1]}          {ay0 = k[i]}
        mx0 = sr1;
        my0 = sr1;
        mr0 = 0xffff;                      {mr = 1 in 1.31 format}
        mr1 = 0x7fff;
        mr = mr - mx0*my0 (ss);
        if mv call overflow;

        {ax0 = e(i-1)}
        my0 = ax0;
        mr = mr1 * my0 (ss);
        dm(i5,m4) = mr1;

        {switch the a pointers}
        ax0 = dm(ap_old);
        ay0 = dm(ap_new);
        dm(ap_new) = ax0;
        dm(ap_old) = ay0;

        {i++ }
        ay0 = dm(i_1);
        ar  = ay0 + 1;
        pass_two: dm(i_1) = ar;
    rts;

.endmod;
```

Listing 3.6 DURBIN.DSP Subroutine

3   Linear Predictive Coding

```
.module/boot=3/boot=4          durbin_double;
{  DURBIN2.DSP - single precision Levinso-Durbin routine
INPUT:

   i4 -> buffer with autocorrelated speech (pm)
   l4 = 0
   i0 -> buffer for reflection coeffs
   l0 =  0

OUTPUT  reflection coeffs calculated
   mr1 = Vp minimum total squared prediction error (normalized)
FUNCTIONS CALLED:
   None
DESCRIPTON:
   The routine implements Durbins recursion method of solving a set of linear
   equations forming a Toeplitz matrix. The algorithm in C is as follows:

   Where R[] is the autocorrelation, and k[] the reflection coeffs. e[] is the
   total squared error, and a[][] is the predictor coeff matrix (Since only the
   i'th and the i+1'th column is used at any one time, the matrix is implemented
   as two (a_old and a_new) columns swapping place after each iteration.


   e[0] = R[0]
   k[1] = R[1] / e[0]
   alpha[1][1] = k[1]
   e[1] = (1 - k[1]*k[1]) * e[0]


   for (i=2; i<=N; i++)
   begin
   k[i] = 0
   for (j=1; j<=i-1; j++)
   k[i] = k[i] + R[i-j] * alpha[i-1][j] k[i] = R[i] - k[i]
   k[i] = k[i] / e[i-1]
   alpha[i][i] = k[i]
   for (j=i-1; j>0; j++)
   alpha[i][j] = alpha[i-1][j] - k[i]*alpha[i-1][i-j]
   e[i] = (1 - k[i]*k[i]) * e[i-1]
   end

In this version the alpha's (a's) are stored as 32 bit numbers.
}
```

Linear Predictive Coding   3

```
#include "lpc.h"
.entry    levinson;
.external   overflow;
.global e;
.var/dm/ram i_1;
.var/dm/ram e[N+1];                          {error values}
.var/dm/ram a_new[2*N],a_old[2*N];           {msw0, lsw0, msw1, lsw1,.....}
.var/dm/ram ap_new,ap_old;                   {pointers to a_*}
.var/dm/ram p2_k_i;                          {pointer to k[i]}
.var/dm  p2_autocor_speech;

{determines the format that a-values are stored in format: (SBITS+1).(32-SBITS-1)}
.const    SBITS = 4;
.const    NSBITS = -SBITS;

levinson:
    i1 = ^a_new; l1 = 0;
    dm(ap_new) = i1;
    i2 = ^a_old; l2 = 0;
    dm(ap_old) = i2; dm(p2_autocor_speech) = i4;
    i5 = ^e; l5 = 0;
    m2 = -1;
    m6 = -1;
    se = NSBITS;

{e[0] = R[0] }
    ax0 = pm(i4,m5);
    dm(i5,m5) = ax0;

{k[1] = R[1]/e[0]}
{ax0 = e[0] = divisor}
    ay1 = pm(i4,m4);                         {MSW of dividend}
    ay0 = 0000;                              {LSW of dividend}
    divide(ax0,ay1);
    ar = -ay0;                               {reverse sign of k before storing}
    dm(i0,m1) = ar;
    dm(p2_k_i) = i0;

{a_old[1] = k[1]}
    si = ay0;
    sr = ashift si (hi);                     {store in (SBITS+1).(32-SBITS-1) format}
    dm(i2,m1) = sr1;
    dm(i2,m1) = sr0;

{e[1] = (1 - k[1]*k[1])*e[0]}
{ay0 = k[1]}
    mx0 = ay0;
    my0 = ay0;
    mr0 = 0xffff;                            {mr = 1 in 1.31 format}
    mr1 = 0x7fff;
    mr = mr - mx0*my0 (ss);
```

(listing continues on next page)

181

3   Linear Predictive Coding

```
{ax0 = e(0)}
   my0 = ax0;
   mr = mr1 * my0 (ss);
   dm(i5,m4) = mr1;

{for(i = 2; i <= N; i++)}
   cntr = N-1;
   ax0 = 1;                          {i-1}
   dm(i_1) = ax0;
   do pass_two until ce;

        {k[i] = 0}
        ay0 = 0;                     {LSW}
        ay1 = 0;                     {MSW}

{for(j = 1; j <= i-1; j++)}
        ax0 = dm(i_1);
        cntr = ax0;
        m3 = ax0;                    {i-1}
        m7 = ax0;                    {i-1}

{prepare: k[i] = k[i] + R[i-j]*a_old[j]}
   i2 = dm(ap_old); l2 = 0;
   i4 = dm(p2_autocor_speech); l4 = 0;
   modify(i4,m7);                    {->R[i-1]}

   {loop}
   do calc_ks until ce;
        my1 = pm(i4,m6);            {R[i-j]}
        mx1 = dm(i2,m1);           {msw of a_old[j]}
        mx0 = dm(i2,m1);           {lsw of a_old[j]}
        mr  = mx0 * my1 (us);      {lsw * msw}
        mr0 = mr1;                  {shift down 16 bits}
        mr1 = mr2;
        mr  = mr + mx1*my1 (ss);    {msw * msw}
        if mv call overflow;
        ar  = mr0 + ay0;           {acum. lsw's}
        ay0 = ar;
        ar  = mr1 + ay1 + c;       {acum. msw's}
        if av call overflow;

   calc_ks: ay1 = ar;

   {k[i] = R[i] - k[i]}
   i4 = dm(p2_autocor_speech); l4 = 0;
   modify(i4,m7);
   modify(i4,m5);                   {->R[i]}
   si = pm(i4,m4);                  {R[i]}
   sr = ashift si (hi);            {shift to (SBITS+1).(32-SBITS-1) format}
                                    {ay0 = LSW of k[i]}
```

Linear Predictive Coding 3

```
    ar = sr0 - ay0;
    si = ar;                                {store for double precision upshift}
                                            {ay1 = MSW of k[i]}

    ar = sr1 - ay1 + c - 1;
    if av call overflow;
{       sr  = lshift si by SBITS (lo);
        si  = ar;
        se  = exp si (hi);
        ay1 = se;
        se  = NSBITS;
        ax1 = SBITS;
        ar  = ax1 + ay1;
        if gt call overflow;
        sr = sr or ashift si by SBITS (hi);
}

{k[i] = k[i]/e[i-1]}
    i5 = ^e; l5 = 0;
    modify(i5,m7);                          {->e[i-1]}
    ax0 = dm(i5,m5);                        {e[i-1]}
    ay1 = ar {sr1};                         {MSW of k[i]}

    {overflow check}
    ar  = abs ax0;
    ay0 = ar;
    ar  = pass ay1;
    ar  = abs ar;
    ar  = ar - ay0;                         {abs(k[i]) - abs(e[i-1])}
    if gt call overflow;
    ay0 = si {sr0};                         {LSW of k[i]}
    divide(ax0,ay1);


    si = ay0;
    sr = ashift si by SBITS (hi);
    ay0 = sr1;
    i0 = dm(p2_k_i); l0 = 0;
    ay1 = sr1;
    ar  = -ay1;                             {reverse sign of k before storing}
    dm(i0,m1) = ar;                         {k[i] store}
    dm(p2_k_i) = i0;
```

(listing continues on next page)

3   Linear Predictive Coding

```
{a_new[i] = k[i]}
   si = ay0;
   sr = ashift si (hi);                   {store in (SBITS+1).(32-SBITS-1) format}
   i1 = dm(ap_new); l1 = 0;
         modify(i1,m3);
         modify(i1,m3);                    {->a_new[i].msw}
         modify(i1,m1);                    {->a_new[i].lsw}
         dm(i1,m2) = sr0;                  {store lsw}
         dm(i1,m2) = sr1;                  {store msw}

         {for(j = i-1; j>0; j—)}
         cntr = dm(i_1);

{prepare: a_new[j] = a_old[j] - k[i]*a_old[i-j]}
         i2 = dm(ap_old);  l2 = 0;
         modify(i2,m3);
         modify(i2,m3);                    {-> a_old[j+1].msw}
         modify(i2,m2);                    {-> a_old[j].lsw}
         i0 = dm(ap_old);                  {-> a_old[i-j].msw}
         l0 = 0; my1 = ay0;               {k[i]}

         {loop}
         do calc_as until ce;
            ay0 = dm(i2,m2);              {a_old[j].lsw}
            ay1 = dm(i2,m2);              {a_old[j].msw}
            mx1 = dm(i0,m1);             {a_old[i-j].msw}
            mx0 = dm(i0,m1);             {a_old[i-j].lsw}
            mr  = mx0*my1 (us)           {lsw * msw}
            mr0 = mr1;                    {shift down by 16 bits}
            mr1 = mr2;
            mr  = mr + mx1*my1 (ss);      {msw * msw}
            if mv call overflow;
            ar  = ay0 - mr0;             {acum. lsw's}
            dm(i1,m2) = ar;
            ar  = ay1 - mr1 + c - 1;     {acum. msw's}
            if av call overflow;
         calc_as: dm(i1,m2) = ar;

{e[i] = (1 - k[i]*k[i]) * e[i-1]}         {my1 = k[i]}
         mx0 = my1;
         mr0 = 0xffff;                     {mr = 1 in 1.31 format}
         mr1 = 0x7fff;
         mr = mr - mx0*my1 (ss);
         if mv call overflow;

         {ax0 = e(i-1)}
         my0 = ax0;
         mr = mr1 * my0 (ss);
         dm(i5,m4) = mr1;
```

184

Linear Predictive Coding   3

```
        {switch the a pointers}
        ax0 = dm(ap_old);
        ay0 = dm(ap_new);
        dm(ap_new) = ax0;
        dm(ap_old) = ay0;

        {i++}
        ay0 = dm(i_1);
        ar  = ay0 + 1;
    pass_two: dm(i_1) = ar;
rts;

.endmod;
```

Listing 3.7 DURBIN2.DSP Subroutine

3   Linear Predictive Coding

```
.module/boot=4          encode_parameters;
{  ENCODE.DSP - truncates/compresses  the lpc parameters.

INPUT:
    i1 -> k (reflection coeffs) l1 = 0
    ar = pitch
    si = gain

OUTPUT: k's encoded inplace
     ar = pitch
     si = gain

The parameters are truncated into the required nr of bits, either by linearly (gain)
or logarithmic (k's) quantization.

Logarithmic: g[i] = log10((1+k[i])/(1-k[i]))/4
}

#include "lpc.h"
.const    LOG_ORDER = 8;
.var/pm   log_coeffs[2*LOG_ORDER];              {Ci lsb, Ci msb, Ci-1 lsb .....}
.init     log_coeffs: <log.cff>;                {scaled down by 512 = 2^9}
.const    LAR_ORDER = 16;
.var/pm   lar_coeffs[2*LAR_ORDER];              {Ci lsb, Ci msb, Ci-1 lsb .....}
.init      lar_coeffs: <enc.cff>;               {scaled down by 512 = 2^9}
.var/pm   round[WORDS_PR_LPCFRAME];
.init     round:
        0x000900, {pitch,    7 bit, (9=16-7) shifting NOT rounding}
        0x000600, {gain,     6 bit}
        0x000600, {k1, 6 bit}
        0x000600, {k2, 6 bit}
        0x000500, {k3, 5 bit}
        0x000500, {k4, 5 bit}
        0x000400, {k5, 4 bit}
        0x000400, {k6, 4 bit}
        0x000300, {k7, 3 bit}
        0x000300, {k8, 3 bit}
        0x000300, {k9, 3 bit}
        0x000200; {k10,      2 bit}
                {     ——
                   54 bit/frame}
.var/dm   temp_pitch;
.var/dm   temp_gain;
.entry    encode;
.external    poly_approx;
encode:
```

Linear Predictive Coding   3

```
{encode parameters}
   i4 = ^round; l4 = 0;

{ar = pitch}
   se = pm(i4,m5);                        {nr of bits to shift}
   sr = lshift ar (lo);
   dm(temp_pitch) = sr0;

        {si = gain}
        sr0 = si;
        ar  = pass sr0;
        if eq jump zero_gain;
           my0 = 0x0000;               {gain lsb}
           my1 = si;                   {gain msb}
           ax0 = LOG_ORDER - 1;
           i6  = ^log_coeffs; l6 = 0;
           call poly_approx;           {log10 function}
                si = mx0;
                sr = lshift si by 9 (lo);     {scale up by 512, comes with}
                                              {the coeff's}
                sr = sr or ashift ar by 9 (hi);
                si = sr1;
                se = pm(i4,m5);         {nr of bits to round to}
                call do_round;
        zero_gain:
        dm(temp_gain) = sr0;
        cntr = 2 {N};
        do enc2_k until ce;
                my0 = 0x0000;          {k lsb}
                my1 = dm(i1,m0);       {k msb}
                ax0 = LAR_ORDER - 1;
                i6  = ^lar_coeffs; l6 = 0;
                call poly_approx;      {log area ratio function}
                si = mx0;
                sr = lshift si by 9 (lo);     {scale up by 512, comes with}
                                              {the coeff's}
                sr = sr or ashift ar by 9 (hi);
                si = sr1;
                se = pm(i4,m5);
                call do_round;
        enc2_k: dm(i1,m1) = sr0;
        cntr = N-2;
        do enc_k until ce;
                si = dm(i1,m0);
                se = pm(i4,m5);
                call do_round;
        enc_k: dm(i1,m1) = sr0;
```

(listing continues on next page)

3 Linear Predictive Coding

```
            {setup return parameters}
            ar = dm(temp_pitch);
            si = dm(temp_gain);
              rts;

{ ROUNDING routine
   Input si = value to be rounded
   se = nr of bits to round to
   Output sr0 = rounded value}

   do_round:
            sr = ashift si (lo);
            mr0 = sr0;
            mr1 = sr1;
            mr  = mr (rnd), ay0 = se;
            ar  = -ay0;
            se  = ar;
            sr  = ashift mr1 (hi);
rts;

.endmod;
```

Listing 3.8 ENCODE.DSP Subroutine

Linear Predictive Coding   3

```
.module/boot=3/boot=4          gain_calculation;
{  GAIN.DSP - Calculates the gain factor for a speech frame.

INPUT:
    i0 -> speech frame
    l0 = 0

OUTPUT:
    sr1 = gain

FUNCTIONS CALLED:
    poly_approx - used to aproximate sqrt function
DESCRIPTION:
    The gain of a frame is calculated as:
    gain = sqrt(sum_over_frame(x(n)^2))

    A simple no-speech detection is implemented, if the gain is lower than
    NOISE_FLOOR the gain is set to zero. The result is scaled appropriately by
    GAIN_SCALE.
}

{Include constant definitions}
#include "lpc.h"
.const   NOISE_FLOOR = 0x0000;            {found as gain when no input is present}
.const   GAIN_SCALE  = 0;                 {appropriate scale value}
.entry   calc_gain;
.external   sqrt;

calc_gain:
{calulate energy of frame, R(0), as sum of input squared}
    mr=0;
    cntr = FRAME_LENGTH;
    do cor_data_loop until ce;
        si = dm(i0,m1);
        sr = ashift si by G_INP_SHIFT (hi);    {scale to avoid overflow}
        my0 = sr1;
    cor_data_loop: mr=mr+sr1*my0(ss);

{set gain = 0 if energy is under noise level}
    ay0 = NOISE_FLOOR;
    ar = mr1 - ay0;
    if gt jump speech;
        sr1 = 0;
        jump from_noise;
    speech:
```

(listing continues on next page)

```
        {calc the gain as the squareroot of R(0)}
    sr = lshift mr0 by -12 (lo);                    {shift to 16.16 format}
        sr = sr or ashift mr1 by -12 (hi);
        mr1 = sr1;                                  {msw of gain^2}
        mr0 = sr0;                                  {lsw of gain^2}
    call sqrt;                                      {result is in unsigned 8.8 format}
        sr  = lshift sr1 by 7 (hi);                 {shift back to 1.15 format}
        sr  = lshift sr1 by GAIN_SCALE (hi);
    from_noise:
rts;

.endmod;
```

Listing 3.9 GAIN.DSP Subroutine

```
.module/boot=3/boot=4    ovfl;
.entry                   overflow;

{used to break on overflows during debug}
    overflow:
rts;

.endmod;
```

Listing 3.10 OVERFLOW.DSP Subroutine

Linear Predictive Coding   3

```
.module/boot=3/boot=4      pitch_detection;
{  PITCH.DSP - extracts the pitch period, and makes a voiced/unvoiced decision.

INPUT:
    i0 -> k[N]
    l0 =  0
    i6 -> autocor_speech[FRAME_LENGTH]
    l6 =  0

OUTPUT:
    si = pitch (= 0 if unvoiced)
CONST:
PITCH_DETECT_LENGTH = part of frame used for pitch detection starting at 3 msec.
    mSEC_3 = sample to start pitch detection at
FUNCIOTNS CALLED:
    None
DESCRIPTION:
    The k's are autoscaled, and then autocorrelated. The new values are correlated
    with the autocorrelated speech R[]. The resulting sequence is searched for the
    largest peak in the interval mSEC_3 .....
    mSEC_3+PITCH_DETECT_LENGTH Depending on the relative size of the peak (to
    re[0]), a decision of voiced/unvoiced is made. In case of voiced the location
    is equal to the pitch period.
}

{Include constant definition}
#include "lpc.h"
.entry    detect_pitch;
.external    overflow;
.var/dm  rk[N];                        {autocorrelation of k[]}
{.var/pm re[FRAME_LENGTH];}            {cross correlation of R[] and rk[]}
.var/dm  k_dm[N];                      {scratch copy's of k}
.var/dm  p2_k;
.var/dm  p2_autocor_speech;
.var/dm  zero_crossings;               {in re[]}

detect_pitch:
{store pointers for later use}
    dm(p2_k) = i0;
    dm(p2_autocor_speech) = i6;

{autoscale before autocorrelation}
{detect largest value in k's}
{i0 = ^k;     l0 = 0;}
    cntr = N;
    sb = -16;
    do max_k until ce;
        si = dm(i0,m1);
    max_k: sb = expadj si;
```

(listing continues on next page)

191

3   Linear Predictive Coding

```
{adjust k input: normalize to largest and then right shift to |P_K_SHIFT|.
   (16-|P_K_SHIFT|) format (done in one shift). At the same time copy to
   pm for correlation}
   i0 = dm(p2_k); l0 = 0;
   i1 = ^k_dm;    l1 = 0;
   cntr = N;
   ax0 = sb;
   ay0 = P_K_SHIFT;                    {scale down to avoid overflow (worst case)}
   ar = ay0 - ax0;
   se = ar;
   do adj_k until ce;
        si = dm(i0,m1);
        sr = ashift si (hi);
   adj_k: dm(i1,m1) = sr1;

{calculate autocorrelation of k[], rk[i] = sum_of k[j]*k[i+j]}
   i5 = ^k_dm; l5 = 0;            {k[i+j]}
   i2 = N;      l2 = 0;           {innerloop counter 'refill'}
   i1 = ^rk;    l1 = 0;
   m2 = -1;
   cntr = N;
   do corr_loop until ce;
        i0 = ^k_dm; l0 = 0;       {k[j]}
        i4 = i5;     l4 = 0;
        cntr = i2;
        mr=0, my0 = dm(i4,m5);
        mx0=dm(i0,m1);
   do cor_data_loop until ce; mr=mr+mx0*my0(ss),my0=dm(i4,m5);
        cor_data_loop: mx0=dm(i0,m1);
        if mv call overflow;
        mx0 = dm(i2,m2);
        my0 = dm(i5,m5);              {(innerloop cnt'er)-, i++}
   corr_loop: dm(i1,m1) = mr1;

{shift down R[] (autocor_speech) to |P_R_SHIFT|.(16-|P_R_SHIFT|) format}
   cntr = FRAME_LENGTH;

{i6 = ^autocor_speech; l6 = 0;}
   do shft_ac until ce;
        si = pm(i6,m4);
        sr = ashift si by P_R_SHIFT (hi);
   shft_ac: pm(i6,m5) = sr1;

{Setup rk[] and R[] (autocor_speech) for correlation. Only calculate the necessary
   correlation coefficients, equivalent to the 0-15 ms portion of the frame
   (samples 0-120 (out of 160), re(0) is necesary for later voiced/unvoiced calc}
```

192

Linear Predictive Coding   3

```
{ re[i] = sum_of(rk[j]*R[i+j]) }
i5 = dm(p2_autocor_speech);        l5 = 0;        {R[i+j]}
i2 = N;                    l2 = 0;
i6 = dm(p2_autocor_speech);        l6 = 0;
ay0 = 0;                                {last 'sign' for zerocrossing count}
af = pass ay0;                          {zerocrossing counter}
cntr = PITCH_DETECT_LENGTH + mSEC_3;    {0-15 msec} {correlate rk's and R's}
        do cor_loop until ce;
                i0 = ^rk; l0 = 0;        {rk[j]}
                i4 = i5;  l4 = 0;
                cntr=i2;
                mr=0, my0=pm(i4,m5), mx0=dm(i0,m1);
                do cor_inner_loop until ce;
                cor_inner_loop: mr=mr+mx0*my0(ss),my0=pm(i4,m5),mx0=dm(i0,m1);
                ar = mr1 xor ay0;        {test for sign switch = zerocrossing}
                if ge jump no_crossing;
                        af  = af + 1;        {inc zerocrossing counter}
                        ay0 = mr1;           {store new sign}
                        no_crossing:
                        modify(i5,m5);        {i++}
        cor_loop: pm(i6,m5) = mr1;
        ar = pass af;
        i5 = ar;                        {zerocrossing count}
```

{find the largest peak in range 3-15 msec. The index of the largest peak is
equal to the pitch period. At the same time count the nr of zerocrossings in
re[]}

```
    si  = 0;                             {store for pitch of max peak}
    ay1 = 0;                             {store for value of max peak}
    i6 = dm(p2_autocor_speech); l6 = 0;
    ax0 = pm(i6,m4);                     {save re(0) for voiced/unvoiced check}
    m7  = mSEC_3;
    modify(i6,m7);                       {-> re(mSEC_3)}
    i2  = mSEC_3; l2 = 0;                {pitch counter}
    cntr = PITCH_DETECT_LENGTH;
    do find_max_peak until ce;
            ax1 = pm(i6,m5);             {re[j]}
            ar  = ax1 - ay1;             {re[j] > max?}
            if le jump not_bigger;
            ay1 = ax1;                   {new max value}
            si = i2;                     {corresponding pitch value}
            not_bigger:
            nop;
    find_max_peak: modify(i2,m1);        {(pitch period cnt'er)++}
```

(listing continues on next page)

3 Linear Predictive Coding

```
{Check for voiced/unvoiced excitation. If unvoiced set pitch = 0}
                                        {ax0 = re(0)}
                                        {ay1 = MSW of peakvalue}
                                        {LSB of peakvalue}
    ay0 = 0000;                         {ay0 = re[peak]/re[0]}
    divide(ax0,ay1);                    {nr of zerocrossings in re[]}
    ax0 = i5;
    ay1 = 70;
    ar = ax0 - ay1;                     {i5>treshold => unvoiced}
    if ge jump not_voiced;
    ax1 = 0x1999;                       {= 0.20}
    ar  = ay0 - ax1;                    {re(j)/re(0) - 0.20}
    if lt jump not_voiced;
    ax1 = 0x2666;                       {= 0.30}
    ar  = ay0 - ax1;                    {re(j)/re(0) - 0.30}
    if gt jump frame_voiced;
    ay1 = 60;
    ar = ax0 - ay1;                     {i5>treshold => unvoiced}
    if lt jump frame_voiced;
    not_voiced:
    si = 0;                        {unvoiced => pitch period = 0}
    i0 = dm(p2_k); l0 = 0;         {zero out k5 - k10 for unvoiced speech}
        m3 = 4;
        modify(i0,m3);
        ax0 = 0;
        cntr = 6;
        do zero_ks until ce;
        zero_ks:  dm(i0,m1) = ax0;
    frame_voiced:
rts;

.endmod;
```

Listing 3.11 PITCH.DSP Subroutine

194

Linear Predictive Coding   3

```
.module/boot=3/boot=4          approximate_func;
{  POLY.DSP - Calculates the polynomial approximation to a function given by the
   coefficients. Uses 32 bit; y = f(x); 

INPUT:
    my0 = x lsb
    my1 = x msb
    ax0 = POLY_ORDER - 1
    i6  = -> coeffs (in pm) (Ci lsb, Ci msb, Ci-1 lsb .....)

OUTPUT:
    mx0 = y msb
     ar = y lsb
    f(x) is approximated by a polynomial:
    f(x) = C[0] + C[1]*X^1 .... + C[(POLY_ORDER-1)]*X^(POLY_ORDER-1)
}

#include "lpc.h"
.entry   poly_approx;

poly_approx:
    mx0 = pm(i6,m5);                          {c lsb}
    ar  = pm(i6,m5);                          {c msb}
    cntr = ax0;
    do approx_loop until ce;
    mr = mx0 * my1 (us), ay0 = pm(i6,m5);     {c[i]lsb*xmsb, c[i-1] lsb}
    mr = mr + ar * my0 (su), ay1 = pm(i6,m5); {c[i]msb*xlsb, c[i-1] msb}
    mr0 = mr1;
    mr1 = mr2;                                {shift down by 16 bits}
    mr = mr + ar * my1 (ss);                  {c[i]msb*xmsb}
    ar = mr0 + ay0;                           {c[i]*x lsb + c[i-1] lsb}
    mx0 = ar;
    approx_loop: ar = mr1 + ay1 + c;          {c[i]*x msb + c[i-1] msb}
rts;

.endmod;
```

Listing 3.12 POLY.DSP Subroutine

3 Linear Predictive Coding

```
.module/boot=3/boot=4    pre_emphasize_speech;
{   PREEMP.DSP - pre-emphasizes a frame of speech. (filters it)

INPUT:
    i0 -> speech frame to be filtered
    l0 =  0
    Frame is altered!!

OUTPUT:
    frame of speech is emphasized
FUNCTIONS CALLED:
    None
DESCRIPTION:
    Filters the speech using   H(z) = 1 - 0.9375*z-1
}

{Include constant definitions}
#include "lpc.h"
.entry    pre_emph;
.external   overflow;
.var/dm/ram delay;

pre_emph:

    {preemphasize}
    mx0 = 0x8801;                     {u = -0.9375}
    cntr = FRAME_LENGTH;
    do filt_speech until ce;
         mr   = 0;
         my0 = dm(delay);             {x(n-1)}
         mr1 = dm(i0,m0);             {x(n)}
         dm(delay) = mr1;             {update delay with x(n)}
         mr   = mr  + mx0*my0 (ss);   {x(n) + u*x(n-1)}
                 if mv call overflow;
         filt_speech: dm(i0,m1) = mr1;    {store filtered sample}
rts;

.endmod;
```

Listing 3.13 PREEMP.DSP Subroutine

196

Linear Predictive Coding   3

```
.module/boot=3/boot=4              random;
{   RANDOM.DSP - Random number function.

INPUT :
    sr1 = msw of seed
    sr0 = lsw of seed

OUTPUT:
    for best result use ONLY sr1 as random number
    sr1 = msw of new seed between 0 and 2^32
    sr0 = lsw of new seed
FUNCTIONS CALLED:
    None
DESCRIPTION:
    The function (taken from the APPS handbook) implements
        x(n+1) = (a*x(n) + c) mod m
        m = 2^32 a = 1,664,525 c = 32767
}

.entry noise_rand;
noise_rand:
    my1 = 25;                          {upper half of a}
    my0 = 26125;                       {lower half of a}
    mr = sr0*my1 (uu);
    mr = mr + sr1*my0 (uu);            {a(hi)*x(lo)}
    si = mr1;                          {a(hi)*x(lo) + a(lo)*x(hi)}
    mr1 = mr0;
    mr2 = si;
    mr0 = 0xfffe;                      {c = 32767, leftshifted by 1}
    mr = mr + sr0*my0 (uu);            {(above) + a(lo)*x(lo) + c}
    sr = ashift mr2 by 15 (hi);
    sr = sr or lshift mr1 by -1 (hi);  {right shift by 1}
    sr = sr or lshift mr0 by -1 (lo);
rts;

.endmod;
```

Listing 3.14 RANDOM.DSP Subroutine

3   Linear Predictive Coding

```
.module/boot=3/boot=4    square_root;
{  SQRT.DSP - Calculate the squareroot

INPUT:
    mr1 = msw of x in 16.16 format
    mr0 = lsw of x
    m1  = 5

OUTPUT:
    sr1 = y in 8.8 unsigned format
CALLED FUNCTIONS:
    None
DESCRIPTION:
    Approximates the squareroot of x by a Taylor series y = sqrt(x)
COMPUTATION TIME:
    75 cycles (maximum)
}

.const   BASE=h#0d49, SQRT2=h#5a82;
.var/pm  sqrt_coeff[5];
.init       sqrt_coeff: h#5d1d00, h#a9ed00, h#46d600, h#ddaa00, h#072d00;
.entry   sqrt;

sqrt:
    i6=^sqrt_coeff; 16 = 0;             {pointer to coeff. buffer}
    se=exp mr1 (hi);                    {check for redundant bits}
    se=exp mr0 (lo);
    ax0=se, sr=norm mr1 (hi);          {remove redundant bits}
    sr=sr or norm mr0 (lo);
    my0=sr1, ar=pass sr1;
    if eq rts;
    mr=0;
    mr1=BASE;                           {load constant value}
    mf=ar*my0 (rnd), mx0=pm(i6,m5);     {mf = x**2}
    mr=mr+mx0*my0 (ss), mx0=pm(i6,m5);  {mr = BASE + c1*x}
    cntr=4;
        do approx until ce;
            mr=mr+mx0*mf (ss), mx0=pm(i6,m5);
        approx: mf=ar*mf (rnd);
            ay0=15;
            my0=mr1, ar=ax0+ay0;        {se + 15 = 0?}
            if ne jump scale;           {no, compute sqrt(s)}
            sr=ashift mr1 by -6 (hi);
            rts;
```

Linear Predictive Coding   3

```
scale:      mr=0;
            mr1=SQRT2;                          {load 1/sqrt(2)}
            my1=mr1, ar=abs ar;
            ay0=ar;
            ar=ay0-1;
            if eq jump pwr_ok;
            cntr=ar;                            {compute (1/sqrt(2))^(se+15)}
            do compute until ce;
compute: mr=mr1*my1 (rnd);
pwr_ok: if neg jump frac;
    ay1=h#0080;                                 {load a 1 in 9.23 format}
    ay0=0;                                      {compute reciprocal of mr}
    divs ay1, mr1;
    divq mr1; divq mr1; divq mr1;
    divq mr1; divq mr1; divq mr1;
    divq mr1; divq mr1; divq mr1;
    divq mr1; divq mr1; divq mr1;
    divq mr1; divq mr1; divq mr1;
    mx0=ay0;
    mr=0;
    mr0=h#2000;
    mr=mr+mx0*my0 (us);
    sr=ashift mr1 by 2 (hi);
    sr=sr or lshift mr0 by 2 (lo);
    rts;
frac:     mr=mr1*my0 (rnd);
    sr=ashift mr1 by -6 (hi);
rts;

.endmod;
```

Listing 3.15 SQRT.DSP Subroutine

3 Linear Predictive Coding

```
.module/boot=3/boot=4    lpc_sync_synth;
{  SSYNTH.DSP - synthesizes lpc speech on a pitch syncronious boundry.

INPUT:
    i1  -> (negative) reflection coefficient (k[]'s)
    l1  =  0
    i2  -> output (speech) buffer
    l2  =  0
    ax1 =  pitch period
    mx1 =  gain

OUTPUT:
    Ouput buffer filled
FUNCTIONS CALLED:
    noise_rand              (random number generator)
DESCRIPTION:
    clear_filter: clears delay line, initializes variables (no arguments required)


    synthesis:  updates the frame delay line -> new -> old -> then synthesises
                a frame of speech, based on interpolated parameters (cur_) from
                the last frame (old_) and the latest (new_). When a frame is
                considered voiced the filter is excited with an impuls on a
                pitchsyncronous boundry. When a frame is unvoiced, the filter is
                excited whith random noise. Before input to the filter the
                excitation is scaled to an appropriate value depending on
                voiced/unvoiced status and the gain. The lattice filter is
                taken from the apps handbook.

In order to interpolate the parameter a DELAY OF ONE FRAME is introduced!
}

{include constant definitions}
#include    "lpc.h";

.entry      synthesis, clear_filter;

.external            noise_rand;


.var/pm/ram/circ  e_back[N];                            {delay line}
.var/dm/ram       lo_noise_seed,hi_noise_seed;
.var/pm/ram       new_k[N];
.var/dm/ram       new_gain;
.var/dm/ram       new_pitch;
.var/pm/ram       old_k[N];
.var/dm/ram       old_gain;
.var/dm/ram       old_pitch;
.var/dm/ram/circ  cur_k[N];
```

Linear Predictive Coding   3

```
.var/dm/ram        cur_gain;
.var/dm/ram        cur_pitch;
.var/dm/ram        pif_cnt;                          {Place In Frame - cntr}
.var/dm/ram        pp_cnt;                           {pitch period - cntr}
.var/dm/ram first_time;

{clear filter and return}
clear_filter:
   {clear the filter}
   i4 = ^e_back; l4 = 0;
   ar = 0;
   cntr = N;
   do clear_loop until ce;
   clear_loop: pm(i4,m5) = ar;

{initialize seed value for random nr generation}
   ax0 = 0;
   dm(lo_noise_seed) = ax0;
   dm(hi_noise_seed) = ax0;
   ax0 = 1;
   dm(first_time) = ax0;                             {first_time = TRUE}
   rts;

{generate one frame of data and output to speech buffer}
synthesis:
   ax0 = dm(first_time);
   ar  = pass ax0;
   if eq jump not_first;

{copy parm's to new}                                 {first_time = TRUE}
   dm(new_pitch) = ax1;
   dm(new_gain) = mx1;
   i6 = ^new_k; l6 = 0;
   cntr = N;
   do move_to_new until ce;
           ax0 = dm(i1,m1);
   move_to_new: pm(i6,m5) = ax0;

   {start of by interpolating}
   ax0 = 1;
   dm(pp_cnt) = ax0;

       {don't do this anymore}
       ax0 = 0;
       dm(first_time) = ax0;
       jump done;
```

(listing continues on next page)

3 Linear Predictive Coding

```
not_first:                          {first_time = FALSE}
                                    {move parm's from old to new and update new}
    ax0 = dm(new_pitch);
    mx0 = dm(new_gain);
    dm(new_pitch) = ax1;
    dm(new_gain)  = mx1;
    dm(old_pitch) = ax0;
    dm(old_gain)  = mx0;
    i6 = ^new_k; l6 = 0;
    i5 = ^old_k; l5 = 0;
    cntr = N;
    do move_to_old until ce;
        ay0 = pm(i6,m4), ax0 = dm(i1,m1);
        pm(i5,m5) = ay0;
        move_to_old: pm(i6,m5) = ax0;
        {setup for lattice filter}
        i0 = ^cur_k;  l0 = N;
        i4 = ^e_back; l4 = N;
        m2 = -1;
        m6 = 3;
        m7 = -2;

{setup pitch_period cntr, in temp var (=af)}
        ax0 = dm(pp_cnt);
        af = pass ax0;

        {synthesize a whole frame}
        cntr = FRAME_LENGTH;
        ax0 = 0;
        dm(pif_cnt) = ax0;
        do frame_loop until ce;
            my0 = 0;                 {excitation default to 0}
            af = af - 1;
            if gt jump not_pitch_time;

        {time to interpolate}

{calculate interpolation factor = pif_cnt/(FRAME_LENGTH - 1)}
        mx0 = dm(pif_cnt);
        my0 = INTERP_FACTOR;         {= 1/(FRAME_LENGHT - 1) shift -1}
        mr  = mx0 * my0 (ss);        {product is in 17.15 format}
        my1 = mr0;                   {get 1.15 format}

        {calc interpolated gain}
        ax0 = dm(new_gain);
        ay0 = dm(old_gain);
        ar  = ax0 - ay0;             {new - old}
        mr  = ar * my1 (ss);         {(new-old)*int_factor}
        ar  = mr1 + ay0;             { + old}
        dm(cur_gain) = ar;
```

Linear Predictive Coding　3

```
{test for transition between voiced/unvoiced and unvoiced/voiced}
        ay0 = dm(old_pitch);
        ar  = pass ay0;
        if eq jump old_unv;
        ax0 = dm(new_pitch);
        ar  = pass ax0;
        if eq jump new_unv;
                                        {voiced - voiced}
        {calc interpolated pitch}
        ar  = ax0 - ay0;                {new - old}
        mr  = ar * my1 (ss);            {(new-old)*int_factor}
        ar  = mr1 + ay0;               { + old}
        dm(cur_pitch) = ar;

        {"interpolate" k's}
        i1 = ^cur_k;  l1 = 0;
        i5 = ^old_k;  l5 = 0;
        cntr = N;
        do interpolate_k until ce; ay0 = pm(i5,m5);
        interpolate_k: dm(i1,m1) = ay0; new_unv:

        {reinitialize pitch cntr}
        ar = dm(cur_pitch);
        af = pass ar;
{set my0 = excitation impuls, in case of voiced frame}
{multiply excitation by gain*(pitch/FRAME_LENGTH)}
    mx1 = ONE_OVER_FRAMEL;             { = 1/FRAME_LENGTH}
    my1 = dm(cur_pitch);
    mr = mx1 * my1 (ss);               {pitch/FRAME_LENGTH, result in 16.16 format}
my1 = dm(cur_gain);
    mr  = mr0 * my1 (ss);              {gain*pitch/FRAME_LENGTH}
    my1 = 0x7fff;                      {impuls}
    mr  = mr1 * my1 (ss);
    my0 = mr1;
    jump not_pitch_time;
    old_unv:            {unvoiced - *}
        ax0 = 1;        {set the pitch_period cntr to a apropriate spacing}
        af  = pass ax0;
        ax0 = 0;
        dm(cur_pitch) = ax0;           {set voiced state to unvoiced}

        {copy old_k to cur_k}
        i1 = ^cur_k;  l1 = 0;
        i5 = ^old_k;  l5 = 0;
        cntr = N;
        do move_to_cur until ce;
        ay0 = pm(i5,m5);
        move_to_cur:dm(i1,m1) = ay0;

        not_pitch_time:
```

(listing continues on next page)

3   Linear Predictive Coding

```
              {calculate driving sample if noised}
              ax0 = dm(cur_pitch);
              ar = pass ax0;                  {check for voiced/unvoiced}
              if ne jump voiced;
              sr1 = dm(hi_noise_seed);        {noised, old_pitch = 0}
              sr0 = dm(lo_noise_seed);
        call noise_rand;                      {random: 16 bit nr}
       dm(hi_noise_seed) = sr1;
              dm(lo_noise_seed) = sr0;
              ar  = abs sr1;
              sr  = ashift ar by -3 (hi);
              my1 = dm(cur_gain);             {multiply by gain}
              mr  = sr1 * my1 (ss);
              my0 = mr1;                       {my0 = excitation}
              jump do_filter;
              voiced:
              do_filter:

         {do allpole lattice filter (from apps book)}
              cntr = N - 1;
              mr = 0;
              mr1 = my0;
              mx0 = dm(i0,m1), my0 = pm(i4,m5);
              mr = mr - mx0*my0 (ss), mx0 = dm(i0,m1), my0 = pm(i4,m5);
              do dataloop until ce;
                    mr = mr - mx0*my0 (ss);
                    my1 = mr1, mr = 0;
                    mr1 = my0;
                    mr = mr + mx0*my1 (ss), mx0 = dm(i0,m1), my0 = pm(i4,m7);
                    pm(i4,m6) = mr1, mr = 0;
                    dataloop:  mr1 = my1;
                    my0 = pm(i4,m7), mx0 = dm(i0,m2);
                    dm(i2,m1) = my1;        {store synthesized sample}
                    pm(i4,m5) = mr1;        {store newest value in delay line}

                    {increment place_in_frame cntr}
                    ay0 = dm(pif_cnt);
                    ar        = ay0 + 1;
   frame_loop: dm(pif_cnt) = ar;
      {store pitch_period cntr for next iteration}
      ar = pass af;
      dm(pp_cnt) = ar;
      done:
      l4 = 0;
   rts;

   .endmod;
```

Listing 3.16 SSYNTH.DSP Subroutine

204

GSM Codec ◼ 4

4.1   OVERVIEW

This chapter describes the implementation of the Pan-European Digital Mobile Radio (DMR) Speech Codec Specification 06.10. This code was developed in accordance with the recommendation of the Conference of European Post and Telecommunications' (CEPT) Group Special Mobile (GSM). A copy of the recommendation can be obtained directly from this organization.

The recommendation describes how the software must perform, and provides a brief tutorial on the algorithm's operation. This chapter and the accompanying code were written to follow the structure of the recommendation.

For your reference, this chapter also includes subroutines for Voice Activity Detection (VAD, Specification 06.32) and Comfort Noise Insertion (CNI, Specification 06.12) . Together, these subroutines provide a more complete solution for GSM applications. For more information about these particular subjects, refer to the corresponding specifications.

4.1.1   Speech Codec

The speech codec for pan-European digital mobile radio is a modified version of a Linear Predictive Coder (LPC). The LPC algorithm uses a simplified model of the human vocal tract, which consists of a series of cylinders that vary in diameter. To produce voiced speech, you force air through these cylinders. You can represent this structure mathematically by a series of simultaneous equations that describe the cylinders.

Early LPC systems worked well enough for users to understand the coded speech, but often, not well enough to identify the speaker. The LPC system described in this chapter uses two techniques, Regular Pulse Excitation (RPE) and Long Term Prediction (LTP), to improve the quality of the coded speech. The improved speech quality is almost comparable to the speech quality produced by logarithmic Pulse Code Modulation (PCM).

4 GSM Codec

The input to the speech codec is a series of 13-bit speech data samples sampled at 8 kSa/s. The codec operates on a 20 ms window (160 samples) and reduces it to 76 coefficients (260 bits) that result in a coded data rate of 13 kbits/s.

4.1.2 Software Comments

This section includes several comments that apply to the program examples in this chapter.

4.1.2.1 Multiply With Rounding

The GSM recommendation requires a *multiply with rounding* operation that provides biased rounding. Although the ADSP-21xx family does have a multiply with rounding instruction, this implementation does not use it because the instruction performs unbiased rounding (see the *ADSP-2100 Family User's Manual*), and the RND mode of the multiplier introduced bit-errors during the codec testing.

To eliminate this problem, the code uses a pre-multiply that stores the value H#8000 in the MR register. Unbiased rounding is then completed by a multiply/accumulate that produces the desired result. The MF register is loaded with H#80, and, at various points, an X-register is also loaded with H#80. Multiplying these two registers places the H#0000008000 in MR.

4.1.2.2 Arithmetic Saturation Results

The GSM recommendation also requires that arithmetic results be saturated. The ALU's AR_SAT mode easily accomplishes this task. Whenever an ALU operation produces an overflow, the output is automatically saturated at the appropriate value.

An arithmetic overflow occurs when the arithmetic operation produces an output that does not fit completely in the proper word size. In other words, the MSB of the word is not the sign bit. Since only the Most Significant Word (MSW) of a multiprecision value contains a sign bit, it is appropriate to check for overflow only in the MSW. When an LSW result does not fit in the output word size, it produces a carry into the next word, not an overflow.

When the LSW of a double precision result is produced, the saturation mode must be disabled. When the MSW is produced, the entire word can be checked for overflow, and saturated as necessary. Throughout the code, the ALU saturation mode is turned on when producing MSWs, or single precision values, and turned off for LSWs.

4.1.2.3 Temporary Arrays

The GSM recommendation specifies the creation of temporary arrays during codec execution. You do not need to save the value of these arrays, and whenever possible, they are eliminated in this implementation to save memory space. For example, the code overwrites the input speech window array with the output of the short term filter (difference signal d() array) instead of creating a new array.

In many cases, the code uses a single array for several purposes. The code's in-line comments indicate what information is stored by a particular section of code.

4.1.2.4 Shared Subroutines

The encoder is designed to produce an estimated signal based on the same information that is available at the decoder. This structure allows both systems to operate in synchronization. The encoder uses only the decoded values of transmitted parameters, insuring that it acts on the same information available to the decoder.

This requires that the encoder uses many of the same subroutines used by the decoder. Routines that are used by both systems are placed at the end of the listing, and are described only in the encoder section of this chapter.

4.2   ENCODER

Listing 4.1, *GSM0610.DSP*, is a full-duplex codec program example that contains the encoder and decoder subroutines. The encoder has three main sections:

- The linear prediction coder (LPC)–The LPC computes a set of eight reflection coefficients that describe the entire window of data.

- The regular pulse excitation (RPE) grid selector–The RPE grid selector breaks the input window into 4 sub-windows and computes a different excitation signal for each. By using 4 separate excitation signals, the codec can process speech signals that may change within a given window.

- The long term prediction (LTP) system–The LTP system reduces the error of the signal over the entire window.

4 GSM Codec

4.2.1 Down Scaling & Offset Compensation Of The Input

The LPC encoder requires 160 samples of left-justified linear data as input. This window of data must be downshifted three bits, then upshifted two bits. The final result of this is to divide each value in half and set its two LSBs to zero. The first two instructions of the *offset_comp* loop perform this operation.

A double-precision high-pass filter is applied to the downshifted input to produce an offset-free signal. The code must execute a double-precision multiplication to maintain the necessary accuracy.

The rest of the *offset_comp* loop implements this filter. The shift instruction isolates the MSW of L_z2, which is held in the MR register. The AR register holds the LSW of L_z2. The LSW is multiplied by alpha (MY0) to produce the result *temp*. The new value of L_s2 is generated, shifted into position and added to *temp*. After the addition of these two values, the MSW is multiplied by alpha and added to L_s2 to produce L_z2.

The last steps of the loop compute the rounded value that is stored as output, and loads several registers for the next iteration. As in most of these operations, the compensation is performed in place, to conserve memory.

4.2.2 Pre-Emphasis Filtering

Before the LPC coefficients are determined, the input data is filtered by a first-order FIR filter. While filtering, the window is searched for the maximum value. This is necessary to ensure that the data can be properly scaled for the auto-correlation that follows. The *pre_emp* loop filters the input data.

This filter multiplies the delayed value and the filter coefficient, then adds the product to the current sample. The subroutine uses the SB register to check each sample for the number of redundant sign bits present. When the loop is completed, SB holds the negative number that corresponds to the number of growth bits in the maximum value of the window. The last step of the loop saves each output sample (written over the input), and prepares the MR register for the next multiply with round operation.

4.2.3   Auto-Correlation

The program uses the *auto_corr* loop for auto-correlation of the filtered input window to calculate the reflection coefficients for the entire window. To prevent an overflow during this procedure, the input data is scaled appropriately.

To compute the scale factor, the subroutine searches the input window for the maximum value, and determines the number of redundant sign bits (growth bits). The window is multiplied by a scale factor to insure that there are three redundant sign bits to handle any growth during the auto-correlation. The search operation is completed in the previous filtering section. The code loop labeled *scale* adjusts the data to ensure the necessary number of growth bits.

The *corr_loop* loop determines the first nine terms of the auto-correlation sequence. The auto-correlation is the sum of the products of the signal with itself offset for k = 0–8. The terms of the sequence are used to compute the reflection coefficients.

The auto-correlation code sets two pointers to the data areas (I1, I5), one pointer to the output array (I6), and uses another pointer as a down-counter for the inner loop (*data_loop*). Since the inner loop executes one less time for each successive value of the auto-correlation sequence, the CNTR is set to I2 for each new auto-correlation term.

After *data_loop* is completed, the next term of the sequence is in the MR register. This value is saved in the output array after incrementing the pointer to the data array, and decrementing the down-counter.

When *corr_loop* is completed, all nine terms of the auto-correlation sequence have been generated and stored in the double precision array L_ACF(). The input data is rescaled by the *rescale* loop before the reflection coefficients are computed.

4.2.4   The Schur Recursion

The theory behind any LPC voice coder is that the throat can be modeled as a series of concentric cylinders with varying diameters. An excitation signal is passed through these cylinders, and produces an output signal. In the human body, the excitation signal is air moving over the vocal cords. In a digital system, the excitation signal is a series of pulses input to a lattice filter with coefficients that represent the sizes of the cylinders.

4 GSM Codec

An LPC system is characterized by the number of cylinders it uses for the model. The DMR system uses eight cylinders, therefore, eight reflection coefficients must be generated. This system uses the *Schur recursion* to efficiently solve for each coefficient.

After a coefficient is determined, two equations are re-computed and used to solve for the next coefficient. The following equations are used:

$$r(n) = \frac{ABS[P(1)]}{P(0) \times SIGN[P(1)]} \qquad \text{for } n = 1 - 8$$

$$P(0) = P(0) + P(1) \times r(n)$$

$$P(m) = P(m+1) + r(n) \times K(9-m) \qquad \text{for } m = 1 - 8 - n$$

$$K(9-m) = K(9-m) + r(n) \times P(m+1)$$

The P() and K() arrays are initialized with values from the auto-correlation sequence determined earlier. If during the computation, the value of $ABS[P(1)] \div P(0)$ is greater than or equal to one, all r-values are set to zero, and the program proceeds with the transformation of the r-values to Logarithmic-Area-Ratios (LARs) described in the next section.

Before initializing the P() and K() arrays, the double precision auto-correlation sequence L_ACF() is normalized. The *set_acf* loop normalizes each of the nine values and places them in the array acf(). The SE register is initialized before entering the loop by the EXP instruction of the shifter. The first value of the auto-correlation sequence is always the largest value of the sequence. The normalization of the rest of the sequence is based on the number of redundant sign bits in the first value.

The *create_k* loop copies the values of the normalized auto-correlation sequence acf() into the appropriate locations in the P() and K() arrays.

The *compute_reflec* loop actually implements the Schur recursion. The I2 and I3 pointers are set to the beginning of the two arrays used to compute the r-values. The absolute values of P(1) and P(0) are compared. If the divide produces an invalid result (r > 1), the code executes a JUMP instruction to skip the remaining computations. Since this test is also performed after the exit from this loop (and since the P() array is not altered if the JUMP is executed) the program eventually jumps to the *zero_reflec* code block, and sets each r-value to zero.

If the divide is valid, it is computed with the ADSP-2100 family divide instructions. The DIVS command computes the sign bit of the quotient, and 15 DIVQs compute the remaining bits. These commands produce the 16-bit value in the AY0 register. After the division, another test is performed to see if the original dividend and divisor are equal (the division instruction does not saturate), if so, the quotient is saturated to 32767. The sign of the quotient is determined from the original sign of P(1), and the r-value is stored in the result array.

The new value for P(0) is computed according to the equation shown above. The two equations are re-computed in the *schur_recur* loop. The counter for this loop is set from the I6 register, which is used as a down-counter.

The *compute_reflec* loop generates the first seven reflection coefficients. The eighth r-value is computed outside of the loop. The code outside the loop is identical to the code inside, but it is not included in the loop since the K() and P() arrays do not need to be re-calculated after the final r-value is computed.

4.2.5 Transformation Of The Reflection Coefficients

The reflection coefficients generated by the Schur recursion are constrained to be in the range -1 < r() < 1. To produce a value that can be more easily quantized into a small number of bits, the following equation transforms the reflection coefficients to Logarithmic-Area-Ratios (LARs): This transformation process is similar to logarithmic companding used in

$$LAR(i) = Log_{10}\frac{1+r(i)}{1-r(i)}$$

log-PCM coding. Taking the logarithm of a number in a fixed precision n-bit machine allocates more bits for the smaller values, and tends to saturate for larger values.

In the implementation of the encoder, the logarithm is approximated with a linear segmentation (as in log-PCM) to simplify the computation. Instead of the divide and logarithm operations, the segmentation simplifies to multiplies, adds, and compares.

The code that transforms the reflection coefficients starts at label *real_rs*. The *compute_lar* loop executes once for each r-value, and produces one LAR-value for each iteration. The three values that *temp* can become are computed first, and stored in various registers. The final ELSE value is left in AR, which holds the result. The inner IF statement is checked, and if true, AR is set with the appropriate *temp* value.

4   GSM Codec

The first IF statement is checked last. This ensures that AR holds the correct value for *temp*. The last step of the loop generates the sign value for *temp*, and stores the LAR value.

4.2.6   Quantization & Coding Of The Logarithmic-Area-Ratios

The LARs produced in the last section of the program must be quantized and coded into a limited number of bits for transmission. The *quantize_lar* loop computes the following equation to generate the coded LARs or LAR$_c$s.

$$LAR_c(i) = N\operatorname{int}[A(i) \times LAR(i) + B(i)]$$

The function Nint defines the nearest integer value to its input. Since each LAR has a different dynamic range, they are coded into varying word sizes. Using a table, the values for A() and B() are defined to reflect these differences. In addition to A() and B(), the table defines the maximum and minimum values for each LAR$_c$. After each LAR$_c$() is computed, it is saturated at the appropriate value.

To implement this coding in the program, several Index (I) registers are set to data arrays representing a table. The AX0 register is set to 256 and is used for rounding the results within the loop.

The code is a straightforward implementation of the recommendation. The first multiply computes A() × LAR(), and the value for B() is added to the product. This sum, which is rounded by the addition of AX0, is downshifted nine bits for saturation. After limiting, the minimum value is subtracted from the final value to produce the LAR$_c$() that is transmitted.

The eight LAR$_c$s are copied from their array to the *xmit_buffer* that holds the entire window of 76 coefficients to be transmitted. A similar transfer is executed every time some of the code words are available for transmission.

4.2.7   Decoding Of The Logarithmic-Area-Ratios

The LARs that were just coded are now decoded (using the *decode_larc subroutine*), and used in the short term analysis section. The encoder uses the decoded LARs because that information matches the information that the receiving decoder uses. This lets the encoder and decoder produce results based on the same data.

The decoded LARs (or LAR$_{pp}$) are calculated from the coded LARs (LAR$_{cs}$) with the following equation:

$$LAR_{pp}(i) = \frac{LAR_c(i) - B(i)}{A(i)}$$

To simplify the implementation of this equation, a table in memory contains the reciprocal of A(i). The equation becomes a subtraction and a multiply, which is faster than a divide.

The same decoding subroutine is used in the encoder and decoder, so the code is written as a separate subroutine that can be called from either routine. The *decode_larc* subroutine is located near the end of the listing.

This subroutine is a straightforward implementation of the recommendation. The minimum value for the current LAR$_c$ (from the table) is added to the coded LAR$_c$. This value is upshifted ten bits, and B() (upshifted one bit) is subtracted. This remainder is multiplied by the reciprocal of A(). The final value is doubled before being stored in the LAR$_{pp}$() array.

4.2.8   Short Term Analysis Filtering

Once the LARs are decoded, they are transformed back into reflection coefficients and used in an 8-pole lattice filter. The short term analysis filter uses the input speech window and reflection coefficients as inputs, and produces a difference signal as output. The difference signal represents the difference between the actual input speech window, and the speech that would be generated based only on the reflection coefficients.

The difference signal is used by the long term predictor (LTP) section of the codec. The LTP is described in Section 4.2.9.

To avoid transients that could occur with a rapid change of filter coefficients, the LARs are linearly interpolated with the previous set of LARs. The input speech frame is broken into four sections (not at the same boundaries as sub-windows), and a different set of interpolated coefficients is used for each section. A table defines the coefficients that are used for each section of the speech frame.

4 GSM Codec

When the interpolated LAR value is generated for each section, it must be transformed from a Logarithmic-Area-Ratio back into a reflection coefficient. This sequence must also be performed in the decoder. To minimize code, the *st_filter* subroutine, called by the encoder and decoder, interpolates, transforms, and executes the short term filter for each section of the input frame.

This subroutine is similar for the encoder and decoder except that different 8-pole lattice filters are called for the encoder and decoder. This is easily coded as an indirect call through one of the index registers. Register I6 is set to the address of *st_analysis* (for the encoder) and the indirect *call (I6)* instruction jumps to that subroutine.

The LARs are interpolated at four points in the *st_filter* routine. The first section's coefficients are interpolated by the *k_end_12* loop. Every *k_end_xx* code loop uses the *old_larpp* array (pointed to by I4) and the *larpp* array (the current decoded LARs) to produce a weighted sum of the two, and stores the output in the array *larp*. The *larp* array is transformed into reflection coefficients that are used by the short term filter.

4.2.8.1 *Transformation Of The LARs Into Reflection Coefficients*

Before transmission, the computed reflection coefficients are transformed into LARs to provide favorable quantization characteristics. Although this transformation is useful for transmission, the LARs must be transformed back into reflection coefficients before they can be used as inputs to the synthesis filter.

The *make_rp* subroutine transforms the LARs back into reflection coefficients and stores them in the rp() array. This subroutine's implementation is similar to the subroutine that codes the LARs. The result for each IF-THEN-ELSE test is created first, with the final ELSE value stored in the AR register. The condition of each IF statement is tested from the inside out. The final test of the loop generates the sign of the output. The rp() array is stored in program memory for easy fetching during the filtering subroutine.

4.2.8.2   Short Term Analysis Filtering

The short term analysis filter implements a lattice structure by solving the following five equations:

1) $d_0(k) = s(k)$

2) $u_0(k) = s(k)$

3) $d_i(k) = d_{i-1}(k) - r'_i \times u_{i-1}(k-1)$ with $i = 1-8$

4) $u_i(k) = u_{i-1}(k-1) + r'_i \times d_{i-1}(k)$ with $i = 1-8$

5) $d(k) = d_8(k)$

The *st_analysis* subroutine computes the five equations shown above. Several registers are setup before calling this subroutine. The CNTR register is set with the number of output samples to be generated during this call. The *st_compute* loop executes once for each output sample created generated. Pointers to the rp() coefficient and u() delay line are setup, and the input sample is fetched.

The *st_loop* loop calculates the two iterative equations (3 and 4) shown above. The first multiply prepares the MR register and loads the coefficient and delay values. The second and third lines of the loop generate a new $u_i()$ value (equation 4). The fourth line saves the previous value of u() (for use in the next iteration) and prepares the MR register. The final two lines generate a new $d_i()$ value (equation 3) that is held in the AR register.

When the *st_loop* is exited, the value for $d_8(k)$ is in the AR register. This value is stored in the output array, and the loop re-executes as necessary.

4.2.9   Calculation Of The Long Term Parameters

The long term calculations of the LPC speech codec are performed four times for each window of data. The calculations are the same for each sub-window, so they are implemented as a set of subroutines that are called four times per frame.

Once the calculations are complete for a sub-window, the 17 coefficients (Nc, bc, mc, xmaxc, and xMc[0–12]), which are stored contiguously, are copied to the *xmit_buffer*. Since the previous sub-window's coefficients do not need to be saved, the same memory locations are used by the next sub-window.

4 GSM Codec

The code must set the I3 register to the input array before the first call to the subroutines. The I3 register is automatically incremented by the necessary number (40) during the *lt_analysis* section of code.

4.2.9.1 Long Term Analysis Filtering

The long term predictor (LTP) produces two coefficients to describe each sub-window. A long term correlation lag (Nc) represents the maximum cross-correlation between samples of the current sub-window and the previous two sub-windows. A gain parameter (bc) represents the quantized ratio of the power of delayed samples to the maximum cross-correlation value.

The value for Nc is determined by computing the cross-correlation between the short-term residual signal of the current sub-window and the signal of the previous sub-windows. The *cross_loop* loop computes each value of the cross-correlation and puts the maximum lag in AX1.

The transmitted value of Nc is not coded, but sent using a 7-bit word.

The coded value for bc is determined using the *table_dlb* lookup-table. This table holds values that indicate the ratio of the numbers. The coded value of bc is the index into a table that satisfies the relationship.

The *ltp_computation* subroutine searches the input sub-window for a maximum value. When the *find_dmax* code loop is exited, SB holds a negative number that corresponds to the number of redundant sign bits present in the maximum value of the sub-window.

The *init_wt* loop uses the value determined above, and shifts the data to ensure that there is at least six redundant sign bits for growth during the cross-correlation execution.

The execution of the cross-correlation is similar to the execution of the auto-correlation performed for the Schur recursion. The only difference is that the auto-correlation uses the same signal for both inputs, while the cross-correlation uses two different signals, dp() and wt(). Each term of the cross-correlation is checked, and if it exceeds the current maximum, the new value is taken as the maximum, and its index is saved as Nc. When the *cross_loop* loop is exited, the value in AX1 is the final value of Nc.

The *power* loop determines the power of the maximum cross-correlation and the gain (bc) value. The value for bc is the ratio of the power of the cross-correlation and the maximum value of the correlation. This ratio is expressed as one of the four values in *table_dlb*, which is stored in data memory. The transmitted value for bc is the index into the table that satisfies the relationship.

4.2.9.2 Long Term Synthesis Filtering

The short-term analysis filter computes a residual signal and stores it in the d() array. Using the LTP coefficients determined by this filter, an estimated short-term residual signal, stored in the dpp() array, is computed from the previously reconstructed short-term residual samples from the dp() array and the new Nc and bc parameters.

From the values of the dpp() array, the long-term residual signal is computed and stored in the e() array. The e() array will be applied to a FIR filter to generate the residual pulse excitation (RPE) signal.

4.2.10 Residual Pulse Excitation Encoding Section

After the long-term residual signal is produced, it is sent through a FIR filter to generate an excitation signal for the sub-window. After decimation, the maximum excitation sequence is determined and coded for transmission.

An Adaptive Pulse Code Modulation (APCM) technique codes the sequence. The maximum value in the sequence is determined and logarithmically coded into six bits. The sequence is normalized and uniformly coded into three bits.

4.2.10.1 Weighting Filter

The output of the long term analysis filtering section, e(), is applied as an input to an FIR filter. The filter's coefficients are stored in a table. This section of code uses a special "block" filter that produces the 40 central samples of a conventional filter. The x() output array is used in the RPE grid selector described in the following section.

The *compute_x_array* loop implements the FIR block filter. The e() input array is placed into the wt() temporary array with five zeros padded at each end. The zero padding is necessary because the block filter implementation tries to use values outside of the defined range of e().

4   GSM Codec

Pointers to the input and output arrays are initialized and the code enters the *compute_x_array* loop. The first two operands of the convolution are fetched, and the appropriate rounding value is placed in the MR register. An inner loop is executed to compute the convoluted output value.

The final double precision output value must be scaled by four before the MSW is stored. This is accomplished using two double-precision additions. After the first addition, the AV (overflow) flag is checked. If an overflow occurs, the output value is saturated and the second addition is skipped. The MS part of the second addition is performed with the saturation mode of the ALU enabled, which automatically causes saturation if an overflow occurs.

4.2.10.2 Adaptive Sample Rate Decimation By RPE Grid Selection
The output of the weighting filter, put in the x() array, is examined to determine the excitation sequence that is used. The x() array is decimated into four sub-sequences. The sub-sequence with the maximum energy is used as the excitation signal, and the value of m indicates the RPE grid selection. The following formula performs the decimation:

$$x_m(i) = x(m + 3 \times 1)$$
$$\text{where} \quad i = 0 - 12, \quad m = 0 - 3$$

The *find_mc* loop determines the sub-sequence with the maximum energy. The energy of each $X_m()$ array is determined by the *calculate_em* loop. This loop multiplies each element of the sequence (downshifted twice) by itself and computes the sum. The value of m that indicates the sub-sequence with the maximum energy is held in AX0.

Once the *find_mc* loop is completed, the value for mc is stored, and the appropriate sub-sequence is copied into the wt() array. The code then determines the maximum element of the $x_m()$ array and holds it in the AR register for quantizing.

4.2.10.3 APCM Quantization Of The Selected RPE Sequence
The maximum value of the sequence is coded logarithmically using six bits. The upper three bits of *xmaxc* hold the exponent of *xmax*, and the lower three bits hold the mantissa. Once *xmax* is coded, the array can be normalized without performing a division.

The $x_m()$ array is normalized by downshifting each element by the exponent of *xmaxc*, and multiplying it by the inverse of the *xmaxc*'s mantissa. The normalized array is uniformly quantized with three bits.

GSM Codec   4

The *quantize_xmax* loop performs the logarithmic quantization of *xmax* by determining the exponent and mantissa, and then positioning them appropriately. The call to *get_xmaxc_pts* decodes *xmaxc*, then returns to the calling routine with the exponent and mantissa of *xmax*.

The *compute_xm* loop performs the normalization of $x_m()$. The inverse of *xmax*'s mantissa is read from a table and stored in MY0, while the magnitude of the downshift is stored in SE. After normalization, the upper three bits of the result are biased by four, and stored in the $x_{mc}()$ array for transmission.

4.2.10.4 APCM Inverse Quantization & RPE Grid Positioning

The $x_{mc}()$ array must be decoded for use as the excitation signal. The subroutine *rpe_decoding* is used by the encoder and decoder. This subroutine assumes that the coded mantissa of *xmaxc* is available in MX0, and its exponent is in AY1.

The actual value for the mantissa is read from *table_fac* and stored in MY0, while the adjusted exponent is stored in SE and the value of *temp3* is placed in AY1. Various pointers are initialized before entering the *inverse_apcm* loop, which decodes the entire $x_{mc}()$ array. After decoding each element, it is stored in the $x_{mp}()$ array.

The ep() array is reconstructed from the decoded $x_{mc}()$ array. The ep() array is first set to zero over its entire length, then filled with the interpolated, decoded values of the $x_{mc}()$ array. The intermediate $x_{mp}()$ array is not used.

4.2.10.5 Update Of The Reconstructed Short Term Residual Signal

The final step of the encoder's sub-window computation is to update the short term residual signal, dp(). The process involves updating the array and computing the new short term residual signal based on the reconstructed long term residual signal and the long term analysis signal. Both of these steps are completed by the *update_dp_code* loop.

The *update_dp* loop updates the dp() array by delaying the data one sub-window. The *fill_dp* loop adds the dpp() array, generated by the long term analysis filter, and ep(), the reconstructed long term residual signal, then stores the result at the end of the dp() array.

4 GSM Codec

4.3 DECODER

Many of the sections in the decoder are also contained in the encoder, so they have already been described. The three sections unique to the decoder are the long term synthesis filter, the short term synthesis filter, and the post processing. Variables that are unique to the decoder and must be stored between calls have an "r" in their names, such as drp().

The decoder for the LPC speech codec creates an excitation signal for the short term synthesis filter. The excitation window is created using the 17 sub-window coefficients that were generated by the encoder. The excitation signal is used as input to a lattice filter with coefficients of the eight decoded LAR_cs. The output of this filter is a full window of speech data. The speech window is down-scaled and sent through a de-emphasis filter before returning.

The *dmr_decode* subroutine computes the output speech window from the 76 input coefficients. The *recv_data* subroutine copies coefficients from the input buffer to the appropriate location in memory. The transmitted LAR_cs are copied into their array and decoded using the *decode_larc* routine described in section 4.2.8. These values are used by the short term synthesis filter described below.

Computation of the sub-window data starts by copying the sub-window coefficients into their arrays. A call to *get_xmaxc_pts* breaks the coded value of xmaxc into its two parts for use by the *rpe_decode* routine (see section 4.2.10.4). The *lt_predictor* routine takes the reconstructed ep() array and computes the new values for the short term reconstructed residual signal drp(). Four calls to these subroutines are executed to compute the excitation signal for the short term synthesis filter.

The *post_process* loop completes the computation of the output window, then control is returned to the calling routine.

4.3.1 Short Term Synthesis Filtering

The decoder uses short term synthesis filtering that is almost identical to the encoder's short term synthesis filtering. The *st_filter* routine is called, but with different parameters. The I6 register is set to the address of *st_synthesis*, the lattice filter used by the decoder, and register I4 is set to the address of *old_larpp*, the array that holds the previous LARs for the decoder. Address register I0 points to a temporary array that holds the reconstructed short term residual signal that was generated for each sub-window.

Section 4.2.9.1 has a complete description of the *st_filter* routine. Section 4.2.9.2. describes the transformation of LARs into reflection coefficients.

4.3.1.1   Short Term Synthesis Filter

The short term synthesis filter is an implementation of an 8-pole lattice filter. It uses the reconstructed short term residual signal as an excitation, and computes the reconstructed speech signal as output. LARs that are averaged and transformed are used as the coefficients for the filter.

The lattice filter used in the decoder is different from the filter used in the encoder. It is defined by the following five equations.

1) $sr_0(k) = dr'(k)$

2) $sr_i(k) = sr_{i-1}(k) - rr'_{(9-i)} \times v_{8-i}(k-1)$ with $i = 1 - 8$

3) $v_{9-i}(k) = v_{8-i}(k-1) + rr'_{(9-i)} \times sr_i(k)$ with $i = 1 - 8$

4) $sr'(k) = sr_8(k)$

5) $v_0(k) = sr_8(k)$

The code that solves these equations is contained in the subroutine *st_synthesis*. The *st_synth_compute* loop generates one output value (sr) during each pass of the loop, while *st_synth_loop* recursively solves the two inner equations.

The first two instructions of the *st_synth_loop* loop generate a new value for sr(i). The next three instructions generate the new value for v(9-i). The address modification that points to the v() array uses a non-sequential modifier.

The first fetch to the v() array reads v(7) and points to v(6). The first fetch in the loop reads v(6) and modifies the pointer to v(8). The last instruction of the loop writes to the v() array, places the updated value in v(8), and modifies the pointer to v(5) for the next read. After the *st_synth_loop* is exited, the code must modify the pointer so the next write is to v(0).

4.3.2   Long Term Synthesis Filtering

The long term synthesis filtering used in the decoder takes the lag (Nc), gain (bc), and reconstructed long term residual signal in ep() and generates the reconstructed short term residual signal in drp(). This signal is used as an input to the short term filter.

4 GSM Codec

The received lag coefficient is checked to ensure that a transmission error did not cause an inappropriate value to be received. If the value falls outside its permissible range, it is set to the previous value. The decoded gain value is multiplied by the previous reconstructed short term residual signal (drp()) and subtracted from the reconstructed long term residual signal (ep()) to generate the reconstructed short term residual signal for the current sub-window. Also, the drp() array is updated by the subroutine.

The *compute_drp* loop generates the new set of reconstructed short term residual values, and *update_drp* updates (or delays) the values of the drp() array.

4.3.3 Post Processing

The final stage of the decoder involves the de-emphasis filtering and down scaling. These two operations are performed by the *post_process* loop. A first order IIR filter is applied to the output of the short term synthesis filter. The first two instructions of the loop accomplish this while the next two instructions double the value of the output.

The last two instructions mask the three LSBs of the output, and store the final value in the output array.

4.4 BENCHMARKS & MEMORY REQUIREMENTS

The following listings implement the entire set of GSM 06 series speech functions on the ADSP-2101. This code is validated to pass all available GSM test vectors. This code is also available on the diskette included with this book.

Table 4.1 presents benchmarks for the system that include encoding and decoding, voice activity detection, comfort noise insertion and generation, and discontinuous transmission functions. The ADSP-2100 family instruction set lets you code the entire set of GSM speech functions into 1988 words of program memory and 964 words of data memory. All the code fits in the internal memory of the ADSP-2101 or the ADSP-2171 microcomputer.

These benchmarks are for ADSP-2101 (13 MHz instruction rate) and ADSP-2171 (26 MHz instruction rate) GSM systems with a 20 ms frame. Most of the time in the frame is unused, leaving ample time and processing power to implement additional features, such as acoustic echo cancellation.

GSM Codec 4

	Cycle Count (maximum worst case)	Time Required (ms)	Processor Loading (%)
ADSP-2101 (13 MHz)			
RPE-LTP LPC Encoder	49300	3.8	19.0
RPE-LTP LPC Decoder	14400	1.1	05.5
Voice Activity Detector	02141	0.17	00.9
Total of 06 series functions	65841	5.07 ms	25.4 %
Free			74.6 %
ADSP-2171 (26 MHz)			
RPE-LTP LPC Encoder	49300	1.9	9.5
RPE-LTP LPC Decoder	14400	0.55	2.75
Voice Activity Detector	02141	0.09	0.45
Total of 06 series functions	65841	2.54 ms	12.7 %
Free			87.3 %

Table 4.1 GSM Implementation Benchmarks

4.5   LISTINGS
This section contains the listings for this chapter.

4   GSM Codec

   This routine performs all of the necessary initialization of variables
   in all of the various GSM speech processing routines. All of these
   variables are defined in RAM, in either Program or Data Memory.

   The subroutine "reset_codec" must be called following DAG initialization
   after system power-up or system reset, before any other subroutine is
   called and before the data acquisition routine is enabled.

   This program must also be called to set the initial state prior to
   validation with the GSM test vectors.

   ADSP-2101 Execution cycles: 894 maximum

Release History:
__Date___ _Ver_ _____Comments_____
01-Sep-89 58   Initial implementation
10-Jan-90 1.00 Second Release
01-Nov-90 2.00 Third release
_____}

```
.MODULE     software_reset;
.ENTRY      reset_codec;

{  from 06.10 (encoder/decoder) and 06.12 (comfort noise in encoder)
         and 06.31 (dtx in encoder) }

.EXTERNAL   u, dp, nrp;
.EXTERNAL   oldlar_buffer, oldxmax_buffer, cni_wait;
.EXTERNAL   speech_count, oldlar_pntr, oldxmax_pntr;
.EXTERNAL   old_LARrpp, old_LARpp;
.EXTERNAL   drp, mp, L_z2_l, L_z2_h;
.EXTERNAL   z1, msr, v;

{  from 06.32 (voice activity detection) }

.EXTERNAL   rvad, normrvad, L_sacf, L_sav0;
.EXTERNAL   pt_sacf, pt_sav0, L_lastdm;
.EXTERNAL   oldlagcount, veryoldlagcount;
.EXTERNAL   e_thvad, m_thvad, adaptcount;
.EXTERNAL   burstcount, hangcount, oldlag;

{  from 06.31 (dtx codeword decoding) and 06.11 (sub and mute) }
```

GSM Codec 4

```
.EXTERNAL   valid_sid_buffer, sub_n_mute, sid_inbuf, taf_count;

{  from 06.12 (comfort noise in decoder) }

.EXTERNAL   seed_lsw, seed_msw;

{  from shell }

.EXTERNAL   speech_1, speech_2, coeff_codeword;

reset_codec:AX0 = 0;

        I0  = ^L_sacf;
        CNTR = 54;
        CALL zero_dm;

        I0  = ^L_sav0;
        CNTR = 72;
        CALL zero_dm;

        I0  = ^speech_1;
        CNTR = 160;
        CALL zero_dm;

        I0  = ^speech_2;
        CNTR = 160;
        CALL zero_dm;

        I0  = ^drp;
        CNTR = 160;
        CALL zero_dm;

        I4  = ^dp;
        CNTR = 120;
        CALL zero_pm;

        I0  = ^msr;          { msr, old_LARrpp[8], v[9] }
        CNTR = 18;
        CALL zero_dm;

        I0  = ^u;            { u[8], oldLARpp[8], z1, L_z2_h, L_z2_l, mp }
        CNTR = 20;
        CALL zero_dm;

        I0  = ^L_lastdm;     { L_lastdm[2], oldlagcount, veryoldlagcount, }
        CNTR = 6;            { adaptcount, burstcount }
        CALL zero_dm;
```

(listing continues on next page)

4 GSM Codec

```
IO   = ^sub_n_mute;     { sub_n_mute, sid_inbuf }
CNTR = 2;
CALL zero_dm;

DM(coeff_codeword) = AX0;

AX0 = 40;
DM(oldlag) = AX0;
DM(nrp) = AX0;

AX0 = 15381;
DM(seed_lsw) = AX0;
AX0 = 7349;
DM(seed_msw) = AX0;
AX0 = 1;
DM(speech_count) = AX0;
AX0 = -4;
DM(cni_wait) = AX0;

AX0 = -1;
DM(hangcount) = AX0;
AX0 = 20;
DM(e_thvad) = AX0;
AX0 = 31250;
DM(m_thvad) = AX0;
AX0 = -7;
DM(normrvad) = AX0;

AX0 = -24;
DM(taf_count) = AX0;

AX0 = ^L_sacf;
DM(pt_sacf) = AX0;
AX0 = ^L_sav0;
DM(pt_sav0) = AX0;
AX0 = ^oldlar_buffer;
DM(oldlar_pntr) = AX0;
AX0 = ^oldxmax_buffer;
DM(oldxmax_pntr) = AX0;

IO   = ^rvad;
AX0 = 24576;
DM(IO,M1) = AX0;
AX0 = -16384;
DM(IO,M1) = AX0;
AX0 = 4096;
DM(IO,M1) = AX0;
AX0 = 0;
CNTR = 6;
CALL zero_dm;
```

```
        I0  = ^valid_sid_buffer;
        AX0 = 42;
        DM(I0,M1) = AX0;
        AX0 = 39;
        DM(I0,M1) = AX0;
        AX0 = 21;
        DM(I0,M1) = AX0;
        AX0 = 10;
        DM(I0,M1) = AX0;
        AX0 = 9;
        DM(I0,M1) = AX0;
        AX0 = 4;
        DM(I0,M1) = AX0;
        AX0 = 3;
        DM(I0,M1) = AX0;
        AX0 = 2;
        DM(I0,M1) = AX0;
        AX0 = 0;
        DM(I0,M1) = AX0;

        RTS;

zero_dm: DO dmloop UNTIL CE;
dmloop:     DM(I0,M1) = AX0;
        RTS;

zero_pm: DO pmloop UNTIL CE;
pmloop:     PM(I4,M5) = AX0;
        RTS;

.ENDMOD;
```

Listing 4.1 Initialization Routine (GSM_RSET.DSP)

4   GSM Codec

{ GSM0610.DSP

These subroutines: dmr_encode and dmr_decode, represent a full duplex codec for the Pan-European Digital Mobile Radio Network. The code implements a Linear Predicitive Coder (LPC) which incorporates a Long Term Predictor with Regular Pulse Excitation (LTP-RPE), as defined by the CEPT/GSM 06.10 specification. This code also includes support for the DTX functions of the GSM specification. Calls are made to Voice Activity Detection (06.32) and Comfort Noise Insertion (06.12) subroutines. This code has been verified and successfully transcodes the GSM 06.10 Test Sequence Version 3.0.0 dated April 15, 1988. The -Dnovad switch must be used at assembly to turn of Voice Activity Detection during validation. In-line comments refer to various sections of this recommendation. It is assumed that the reader is familiar with that document.

Release History:
    03-Feb-89 32 Initial release.
    20-Jun-89 56 Fix reflect coef sect to pass all 3.0.0 vectors.
    10-Jan-90 1.00 Second release.

Information furnished by Analog Devices is believed to be accurate and reliable. However, no responsibility is assumed by Analog Devices for its use; nor for any infringement of patents or other rights of third parties which may result from its use. Portions of the algorithms implemented in this code may have been patented; it is up to the user to determine the legality of their application.

Assembler Preprocessor Switches:
    -cp switch   must always be used when assembling
    -Dnovad switch   disables VAD for validation of 06.10
    -Dalias switch   aliases some variables to save RAM space
    -Ddemo switch   enables several functions necessary for
                          the eight-state demonstration

Calling Parameters:
    I0 -> Input Speech Buffer (for dmr_encode)
    I1 -> Coefficient Buffer (for both)
    I2 -> Output Speech Buffer (for dmr_decode)
    AX0 -> Silence Descriptor Frame flag (for dmr_decode)
    M0=0; M1=1;   M2=-1; M3=2;
    M4=0; M5=1;   M6=-1;
    L0=0; L1=0;   L2=0; L3=0;
    L4=0; L5=0;   L6=0; L7=0;

Return Values:
    I1 -> Coefficient Buffer (for dmr_encode)
    I2 -> Output Speech Buffer (for dmr_decode)

```
Altered Registers:
   AX0, AX1, AY0, AY1, AR, AF,
   MX0, MX1, MY0, MY1, MR, MF,
   SI, SE, SB, SR,
   I0, I1, I2, I3, I4, I5, I6
   M0, M7

ADSP-2101 Computation Time (without Voice Activity Detection):
Encoder   49300 cycles maximum
Decoder   14400 cycles maximum

State:                            Encoder    Decoder
   speech only                    46900      14000   cycles maximum
   comfort noise generation       47200      14400   cycles maximum
   speech hangover                49300      14000   cycles maximum
}

.MODULE/RAM/BOOT=0     Digital_Mobile_Radio_Codec;
.ENTRY      dmr_encode, dmr_decode, schur_routine, divide_routine;
.EXTERNAL   comfort_noise_generator;
.EXTERNAL   vad_routine, update_periodicity;
.EXTERNAL   vad, lags;

{_____Conditional Assembly_____}
{  Use (asm21 -cp -Dalias) to alias some variables to save RAM                }

#ifdef alias
.INCLUDE    <var0610.ram>;
   #define r dpp
   #define k dpp+25
   #define acf dpp+8
   #define p dpp+17
   #define LAR dpp+25
   #define rp wt
   #define LARp wt+8
   #define LARpp DPP
   #define LARc wt
   #define ep wt
   #define mean_larc dpp+17
#else
   .INCLUDE          <var0610.h>;
#endif
{_____}

.INCLUDE <init0610.h>;

{_____Global variable declarations_____}
   {variables used in the encoder }
.GLOBAL     u, dp, L_ACF, scaleauto;
.GLOBAL     old_LARpp, mp, L_z2_l, L_z2_h, z1;
```

(listing continues on next page)

4 GSM Codec

```
    {variables used in the decoder }
.GLOBAL     nrp, drp, old_LARrpp, msr, v;

    {variables used for comfort noise insertion in the encoder}
.GLOBAL     cni_wait, speech_count, oldlar_pntr, oldxmax_pntr;
.GLOBAL     oldlar_buffer, oldxmax_buffer, sp_flag;

    {variable used as a working buffer to alias VAD variables}
.GLOBAL     wt;
{_____}

{_____Encoder Subroutine_____}
dmr_encode: ENA AR_SAT;                     {Enable ALU saturation}
        DM(speech_in)=I0;                   {Save pointer to input window}
        DM(xmit_buffer)=I1;                 {Save pointer to coeff window}
        MX1=H#4000;                         {This multiply will place the}
        MY1=H#100;                          {vale of H#80 in MF that will}
        MF=MX1*MY1 (SS);                    {be used for unbiased rounding}

{  This section of code computes the downscaling and offset compensation
   of the input signal as described in sections 4.2.1 and 4.2.2 of the
recommendation}
        I0=DM(speech_in);                   {Get pointer to input data}
        I1=I0;                              {Set pointer for output data}
        SE=-15;                             {Commonly used shift value}
        MX1=H#80;                           {Used for unbaised rounding}
        AX1=16384;                          {Used to round result}
        MY0=32735;                          {Coefficient value}
        AY1=H#7FFF;                         {Used to mask lower L_z2}
        MY1=DM(z1);
        MR0=DM(L_z2_l);
        MR1=DM(L_z2_h);
        DIS AR_SAT;                         {Cannot do saturation}
        AR=MR0 AND AY1, SI=DM(I1,M1);       {Fill the pipeline}
        CNTR=window_length;
{       DO offset_comp UNTIL CE;}
gsm1:           SR=ASHIFT SI BY -3 (HI);{Shift input data to zero the}
        SR=LSHIFT SR1 BY 2 (HI);            {the LSB and half data}
        AX0=SR1, SR=ASHIFT MR1 (HI);        {Get upper part of L_z2 (msp)}
        SR=SR OR LSHIFT MR0 (LO);           {Get LSB of L_z2 (lsp)}
        MR=MX1*MF (SS), MX0=SR0;            {Prepare MR, MX0=msp}
        MR=MR+AR*MY0 (SS), AY0=MY1;         {Compute temp}
        AR=AX0-AY0, AY0=MR1;                {Compute new s1}
        SR=ASHIFT AR BY 15 (LO);            {Compute new L_s2}
        AR=SR0+AY0, MY1=AX0;                {MY1 holds z1, L_s2+temp is in}
        AF=SR1+C, AY0=AR;                   {SR in double precision}
        MR=MX0*MY0 (SS);                    {Compute msp*32735}
        SR=ASHIFT MR1 BY -1 (HI);           {Downshift by one bit }
        SR=SR OR LSHIFT MR0 BY -1 (LO);{before adding to L_s2}
```

```
            AR=SR0+AY0, AY0=AX1;              {Compute new L_z2 in }
            MR0=AR, AR=SR1+AF+C;              {double precision MR0=L_z2}
            MR1=AR, AR=MR0+AY0;              {MR1=L_z2, round result }
            SR=LSHIFT AR (LO);               {and downshift for output}
            AR=MR1+C, SI=DM(I1,M1);          {Get next input sample}
            SR=SR OR ASHIFT AR (HI);
offset_comp:      DM(I0,M1)=SR0, AR=MR0 AND AY1;{Store result, get next lsp}
{?} IF NOT CE JUMP gsm1;
            DM(L_z2_l)=MR0;                          {Save values for next call}
            DM(L_z2_h)=MR1;
            DM(z1)=MY1;
            ENA AR_SAT;                              {Re-enable ALU saturation}


{   This section of code computes the pre-emphasis filter and
    the autocorrelation as defined in sections 4.2.3 and 4.2.4 of
    the recommendation}
            MX0=DM(mp);                          {Get saved value for mp}
            MY0=-28180;                          {MY0 holds coefficient value}
            MX1=H#80;                            {These are used for biased}
            MR=MX1*MF (SS);                      {rounding}
            SB=-4;                               {Maximum scale value}
            I0=DM(speech_in);                    {In-place computation}
            CNTR=window_length;
{         DO pre_emp UNTIL CE;}
gsm2:           MR=MR+MX0*MY0 (SS), AY0=DM(I0,M0);
            AR=MR1+AY0, MX0=AY0;
            SB=EXPADJ AR;                        {Check for maximum value}
pre_emp:    DM(I0,M1)=AR, MR=MX1*MF (SS); {Save filtered data}
{?} IF NOT CE JUMP gsm2;
            DM(mp)=MX0;

            AY0=SB;                              {Get exponent of max value}
            AX0=4;                               {Add 4 to get scale value}
            AR=AX0+AY0;
            DM(scaleauto)=AR;                    {Save scale for later}
            IF LE JUMP auto_corr;                {If 0 scale, only copy data}
            AF=PASS 1;
            AR=AF-AR;
            SI=16384;
            SE=AR;
            I0=DM(speech_in);
            I1=I0;                               {Output writes over the input}
            SR=ASHIFT SI (HI);
            AF=PASS AR, AR=SR1;                  {SR1 holds temp for multiply}
            MX1=H#80;                            {Used for unbiased rounding}
            MR=MX1*MF (SS), MY0=DM(I0,M1);       {Fetch first value}
CNTR=window_length;
```

(listing continues on next page)

4 GSM Codec

```
{           DO scale UNTIL CE;}
gsm3:               MR=MR+SR1*MY0 (SS), MY0=DM(I0,M1);   {Compute scaled data}
scale:      DM(I1,M1)=MR1, MR=MX1*MF (SS);              {Save scaled data}
{?} IF NOT CE JUMP gsm3;

auto_corr:  I1=DM(speech_in);                {This section of code computes}
            I5=I1;                           {the autocorr section for LPC}
            I2=window_length;                {I2 used as down counter}
            I6=^L_ACF;                       {Set pointer to output array}
            CNTR=9;                          {Compute nine terms}
{           DO corr_loop UNTIL CE;}
gsm4:               I0=I1;                    {Reset pointers for mac loop}
            I4=I5;
            MR=0, MX0=DM(I0,M1);             {Get first sample}
            CNTR=I2;                         {I2 decrements once each loop}
{           DO data_loop UNTIL CE;}
gsm5:               MY0=DM(I4,M5);
data_loop:          MR=MR+MX0*MY0 (SS), MX0=DM(I0,M1);
{?} IF NOT CE JUMP gsm5;
            MODIFY(I2,M2);                   {Decrement I2, Increment I5}
            MY0=DM(I5,M5);
            DM(I6,M5)=MR1;                   {Save double precision result}
corr_loop:          DM(I6,M5)=MR0;           {MSW first}
{?} IF NOT CE JUMP gsm4;

            I0=DM(speech_in);                {This section of code rescales}
            SE=DM(scaleauto);                {the input data}
            I1=I0;                           {Output writes over input}
            SI=DM(I0,M1);
            CNTR=window_length;
{           DO rescale UNTIL CE;}
gsm6:               SR=ASHIFT SI (HI), SI=DM(I0,M1);
rescale:    DM(I1,M1)=SR1;
{?} IF NOT CE JUMP gsm6;

            call vad_routine;                {determine vad state}

{*****  This section of code sets the Voice Activity Flag (vad) and, if
    vad has been inactive four or more cycles (cni_wait), sets the
    Comfort Noise Insert Flag (cni_flag).  *****}
set_flags:  AX0 = DM(vad);          {AX0 holds vad}

{_____Conditional Assembly_____}
{  Use (asm21 -cp -Ddemo) to turn on the demonstration functions}
#ifdef demo
set_vad_demo:AY0 = 2;
            MR0 = M7;
            AF  = PASS 1;
            AR  = MR0 AND AF;                {extract force_vad_low}
            IF NE AF = PASS 0;
            AR  = AX0 AND AF;                {AR = vad AND /force_vad_low }
```

```
          AF  = MR0 AND AY0;                 {extract force_vad_high}
          AR  = AR OR AF;                    {AR = .. OR force_vad_high }
          DM(vad) = AR;
          AX0 = AR;
          M7  = 2;
#endif
{_____}

{_____Conditional Assembly_____}
{    Use (asm21 -cp -Dnovad) to turn VAD off for validation          }
#ifdef novad
          AX0 = 1;
          DM(vad) = AX0;
#endif
{_____}

          AY0 = DM(cni_wait);
          AY1 = DM(speech_count);

          MR0 = H#FFFF;                      {MR0 holds cni_flag}
          AR  = -4;                          {AR holds cni_wait}

          AF  = PASS AX0;
          IF NE MR = 0;                      {If vad<>0, set cni_flag=0}
          IF NE JUMP store_cni;

          AR  = AY0 + 1;                     {Increment cni_wait}
          IF LE MR = 0;                      {If cni_wait <= 0, cni_flag=0}
store_cni:  DM(cni_wait) = AR;
          DM(cni_flag)  = MR0;

          AY0 = -24;
          AF  = PASS MR0;
          IF NE AR = PASS AY0;
          IF NE JUMP store_spcnt;

          AF  = PASS AX0, AR = AY1;
          IF NE AR = AY1 + 1;
store_spcnt:DM(speech_count) = AR;

          AF  = PASS AX0, AY1 = AR; AR  = 0;
          IF NE AR = PASS 1;
          AF = PASS AY1;
          IF GE AR = PASS 1;
store_spflg:DM(sp_flag) = AR;
```

(listing continues on next page)

4   GSM Codec

```
{  Now begin section 4.2.5 of the recommendation}
set_up_schur:AY1 = ^L_ACF;              {in DM}
         MY1 = ^acf;
         M0  = ^r;
         CALL schur_routine;

{  This section of code transforms the r-values to log-area-ratios
   as defined in section 4.2.6 of the recommendation}

real_rs: I5=^r;                         {This section of code computes}
         I4=^LAR;                       {the log area ratio from r}
         CNTR=8;
{        DO compute_lar UNTIL CE;   }
gsm7:            AX0=DM(I5,M5);
            AR=ABS AX0;
            SR=ASHIFT AR BY -1 (HI);{Generate temp>>1}
            AX0=SR1;                {AX0 holds temp>>1}
            AY0=26112;
            AX1=AR, AR=AR-AY0;      {Generate temp-26112}
            SR=LSHIFT AR BY 2 (HI); {Generate (temp-26112)<<2}
            AY0=31130;
            AY1=11059;
            AR=SR1, AF=AX1-AY0;     {Default to AR=(temp-26112)<<2}
            IF LT AR=AX1-AY1;       {AR=temp-11059 (if necessary)}
            AY0=22118;
            AF=AX1-AY0;
            IF LT AR=PASS AX0;      {AR=temp>>1 (if necessary)}
            IF NEG AR=-AR;          {Compute sign of LAR[i]}
compute_lar:     DM(I4,M5)=AR;          {Save LAR[i]}
{?} IF NOT CE JUMP gsm7;

{*****  If necessary, the code will now average the LAR values, and write
   new values into oldlar_buffer.  The proper LAR values are then
   transmitted. *****}

         AX0 = DM(vad);
         AF  = PASS AX0;
         IF NE JUMP encode_lar;     {Voice Activity, skip the rest}
         AX0 = DM(cni_flag);
         AF  = PASS AX0;
         IF EQ JUMP write_oldlar;   {Not cni, so do not avg. oldlar}

{*****  The code will now average the four previous frames lar values as
   specified in GSM recommendation 06.12.  Note that the values were
   previously scaled. *****}

         I4  = ^oldlar_buffer;
         I5  = ^mean_lar;
         I6  = I4;
         M7  = 8;
         AX0 = DM(I6,M7);
         CNTR = 7;
```

```
{        DO average_lar UNTIL CE;}
gsm8:            MODIFY (I4,M5);
            AY0 = DM(I6,M7);
            AF  = AX0 + AY0, AX0 = DM(I6,M7);
            AF  = AX0 + AF, AX0 = DM(I6,M7);
            I6  = I4;
            AR  = AX0 + AF, AX0 = DM(I6,M7);
average_lar:    DM(I5,M5) = AR;          {store mean_lar[i]}
{?} IF NOT CE JUMP gsm8;
            AY0 = DM(I6,M7);
            AF  = AX0 + AY0, AX0 = DM(I6,M7);
            AF  = AX0 + AF, AX0 = DM(I6,M7);
            AR  = AX0 + AF;
            DM(I5,M5) = AR;              {store mean_lar[8]}
            M7  = 2;                     {restore M7}
```

```
{*****  This section of code will write the current lar values into one
    of four (eight location) buffers in the thirty-two location
    oldlar_buffer for use in the next frame.  The values are also
    scaled. *****}
```

```
write_oldlar:AX0 = ^oldlar_buffer;
            AY1 = ^oldlar_buffer + 32;
            AR  = DM(oldlar_pntr);
            AF  = AY1 - AR;
            IF LE AR = PASS AX0;
            I4  = AR;                    {Set the top of buffer}
            SE  = -2;                    {Roughly divide by four}
            I5  = ^LAR;
            SI  = DM(I5,M5);
            CNTR = 8;
{        DO write_buffer UNTIL CE;}
gsm9:            SR = ASHIFT SI (HI), SI = DM(I5,M5);   {last read will be junk}
write_buffer:    DM(I4,M5) = SR1;
{?} IF NOT CE JUMP gsm9;
            DM(oldlar_pntr) = I4;
```

```
{*****  This code will quantize the current LAR values and the mean_lar values, if
necessary.  One of these is then sent to the transmit buffer. *****}
```

```
encode_lar: I6  = ^LAR;
            I1  = ^LARc;
            CALL lar_encoding;

            AX0 = DM(sp_flag);
            AF = PASS AX0;
            IF NE JUMP transmit_lar;

            I6  = ^mean_lar;
            I1  = ^mean_larc;
            CALL lar_encoding;
```

(listing continues on next page)

235

4   GSM Codec

```
transmit_lar:     I1 = AX1;              {The quantized LAR values}
        CNTR=8;                          {can now be sent}
        CALL xmit_data;                  {Copy to the output buffer}

{  Now, continue with GSM recommendation 4.2.8.}

        CALL decode_larc;                {Decode the LARcs }
        I0=DM(speech_in);                {Input/output of the st filter}
        I6=^st_analysis;                 {Use the st analysis routine}
        I4=^old_larpp;                   {Use the previous LARpp}
        CALL st_filter;                  {Call st filter manager}

{  Compute sub-window information for each of the 4 sub-windows}

{*****  Check to see if Comfort Noise is being generated. *****}

        AX0 = DM(sp_flag);
        AF  = PASS AX0;
        IF NE JUMP speech_frame;

        AX0 = DM(cni_flag);
        AF  = PASS AX0;
        IF NE JUMP comp_mnxmax;

silence_frame:AR  = DM(mean_xmaxc);
        JUMP xmit_cmfrtnois;

{*****  This section will average the four xmax values from the previous
four frames as specified in GSM recommendation 06.12, section 2.1. Note
that the values have been pre-scaled. *****}

comp_mnxmax:I5  = ^oldxmax_buffer;
        AR  = DM(I5,M5);                 {AR holds mean_xmax.}
        AY0 = DM(I5,M5);
        CNTR = 15;
{       DO avg_xmax UNTIL CE;}
avg_xmax:         AR = AR + AY0, AY0 = DM(I5,M5); {Last read is junk.}
{?} IF NOT CE JUMP avg_xmax;


{*****  Now xmax must be quantized. *****}

        CALL quantize_xmax;              {mean_xmaxc returned in AR.}
        DM(mean_xmaxc) = AR;

{*****  The transmit buffer is filled next. *****}
xmit_cmfrtnois:CNTR = 4;
        AX0 = 0;
        I0  = DM(xmit_buffer);
{       DO xmit_sid UNTIL CE;}
```

236

```
gsm10:     DM(I0,M1) = AX0;
           DM(I0,M1) = AX0;
           DM(I0,M1) = AX0;
           DM(I0,M1) = AR;            {The fourth value is mean_xmaxc}
           CNTR = 12;
{          DO zero_rpe UNTIL CE;}
zero_rpe:              DM(I0,M1) = AX0;
{?} IF NOT CE JUMP zero_rpe;
xmit_sid:      DM(I0,M1) = AX0;
{?} IF NOT CE JUMP gsm10;

{*****  The Silence Descriptor (SID) frame has been sent to the transmit
    buffer. *****}

{*****  Must now compute the xmax values for the current frame. *****}

           I3  = DM(speech_in);
           I6=^lags;

           CNTR=4;
{          DO xmax_loop UNTIL CE;}
gsm11:     CALL ltp_computation;
           DM(I6,M5) = AX1;        {AX1 holds Nc for sub-window}
           CALL rpe_encoding;
xmax_loop:        NOP;
{?} IF NOT CE JUMP gsm11;
        JUMP finish;

{  This code implements the sub-window information for each of the 4
   speech sub-windows.}

speech_frame:I3=DM(speech_in);      {Only set input pointer once}
        I6=^lags;

           CNTR=4;
{          DO enc_subwindow UNTIL CE;}
gsm12:     CALL ltp_computation;   {Compute LTP coefficients}
           DM(I6,M5) = AX1;        {AX1 holds Nc for sub-window}
           CALL rpe_encoding;      {Encode and decode RPE sequence}
              I1=^Nc;              {Sub-window data can be sent}
              CNTR=17;             {17 coeffs per sub-window}
              CALL xmit_data;      {Copy to the output buffer}
enc_subwindow:        NOP;         {No CALL in last instr of DO}
{?} IF NOT CE JUMP gsm12;

   {All the coded variables have been sent to xmit_buffer}

finish:  CALL update_periodicity;  {VAD (06.32) routine}

           DIS AR_SAT;
           RTS;                       {Return to caller}
```

(listing continues on next page)

4 GSM Codec

```
xmit_data:  I0=DM(xmit_buffer);        {Copy coeffs to the output}
{          DO xmit UNTIL CE;}          {buffer}
gsm13:      AX0=DM(I1,M1);
xmit:              DM(I0,M1)=AX0;
{?} IF NOT CE JUMP gsm13;
            DM(xmit_buffer)=I0;
            RTS;                       {Return from Encoder}

{_____Subroutines for Encoder_____}

{  This section of code quantizes and codes the LAR value produced above
   as defined in section 4.2.7 of the recommendation}

lar_encoding:AX1 = I1;                 {Stores pointer to result}
          I5=^table_a;                 {This section of code computes}
          I4=^table_b;                 {the quantizing/coding of LARs}
          MX1=^table_mac;              {Pointers are set to various}
          MY1=^table_mic;              {data memory tables}
          AX0=256;                     {Used for rounding}
          CNTR=8;
{         DO quantize_lar UNTIL CE;}
gsm14:      MX0=PM(I5,M5);
            SI=I5;
            MY0=DM(I6,M5);
            MR=MX0*MY0 (SS), AY0=PM(I4,M5);    {temp=A[i]*LAR[i]}
            AF=MR1+AY0;                        {temp=A[i]*LAR[i]+B[i]}
            I5=MX1;
            AR=AX0+AF, AY0=PM(I5,M5);          {Round result}
            MX1=I5;
            SR=ASHIFT AR BY -9 (HI);           {LARc[i] = temp>>9}
            AR=SR1;
            I5=MY1;
            AF=AR-AY0, AY1=PM(I5,M5);          {Test min/max}
            MY1=I5;
            IF GT AR=PASS AY0;                 {Cap if above max}
            AF=AR-AY1;
            IF LT AR=PASS AY1;                 {of below min}
            AR=AR-AY1;                         {Subtract minimum value}
            I5=SI;
quantize_lar:    DM(I1,M1)=AR;                 {Save LARc[i]}
{?} IF NOT CE JUMP gsm14;
          RTS;
```

```
{  This subroutine computes the 8-pole short term lattice filter
   as defined in section 4.2.10 of the recommendation}

st_analysis:SR1=H#80;                    {Used for unbaised rounding}
{        DO st_compute UNTIL CE;}        {The counter is set by caller}
gsm15:     I5=^rp;                       {Point to decoded r-values}
           I2=^u;                        {Point to delay line}
           AR=DM(I0,M0);                 {Get filter input}
           AX0=AR;                       {Set sav=s[i], AX0 is sav}
           CNTR=8;                       {Compute all 8 poles}
{        DO st_loop UNTIL CE;}
gsm16:        MY0=DM(I5,M5);                      {Moved to dm}
              MR=SR1*MF (SS), MX1=DM(I2,M0);
              MR=MR+AR*MY0 (SS), AY0=MX1;
              AY1=AR, AR=MR1+AY0;                 {AR=temp}
              DM(I2,M1)=AX0, MR=SR1*MF (SS);      {u[i-1]=sav}
              MR=MR+MX1*MY0 (SS);
st_loop:      AX0=AR, AR=MR1+AY1;                 {AR=di, AX0=sav}
{?} IF NOT CE JUMP gsm16;
st_compute:   DM(I0,M1)=AR;                       {Write output over input}
{?} IF NOT CE JUMP gsm15;
        RTS;

{  This section of code computes the maximum cross-correlation value
   of the reconstructed short term signal dp() and the current
   sub-window as defined in section 4.2.11 of the recommendation}

ltp_computation:I0=I3;                   {Preserve I3 for now}
        SB=-6;                           {Maximum shift value}
        SI=DM(I0,M1);
        CNTR=sub_window_length;
{        DO find_dmax UNTIL CE;}
find_dmax:      SB=EXPADJ SI, SI=DM(I0,M1);{Find maximum of sub-window}
{?} IF NOT CE JUMP find_dmax;

        AY0=6;
        AX0=SB;
        AR=AX0+AY0;                      {Compute shift for scaling}
        DM(scal)=AR;                     {Save shift value}
        AR=-AR;
        SE=AR;
        I1=^wt;                          {Output to temporary array}
        I0=I3;                           {Preserve I3 for now}
        SI=DM(I0,M1);
        CNTR=sub_window_length;          {Scale entire sub-window}
{        DO init_wt UNTIL CE;}
gsm17:     SR=ASHIFT SI (HI), SI=DM(I0,M1);
init_wt:    DM(I1,M1)=SR1;
{?} IF NOT CE JUMP gsm17;
```

(listing continues on next page)

4 GSM Codec

```
          DIS AR_SAT;                                 {Can use saturation here}
          AX1=40;                                     {Mimimum value for Nc}
          I0=39;                                      {I0 holds Nc counter}
          AY0=0;                                      {Holds LSW of max value}
          AY1=0;                                      {Holds MSW of max value}
          I4=^dp+80;
          I2=^wt;
          I1=I2;
          CNTR=81;
{         DO cross_loop UNTIL CE;}
gsm18:    I5=I4;
          MR=0, MX0=DM(I1,M1), MY0=PM(I5,M5);
          CNTR=sub_window_length;
{         DO cross_corr UNTIL CE;}
cross_corr:          MR=MR+MX0*MY0 (SS), MX0=DM(I1,M1), MY0=PM(I5,M5);
{?} IF NOT CE JUMP cross_corr;
          AR=MR0-AY0, MY0=PM(I4,M6);
          AR=MR1-AY1+C-1;
          MODIFY(I0,M1);
          IF LT JUMP cross_loop;                      {Check for L_result < L_max}
          IF EQ AR=MR0-AY0;                           {If MSW=0, check LSW again}
          IF EQ JUMP cross_loop;                      {If LSW=0, the values are equal}
          AY0=MR0;                                    {Reset L_MAX to new value}
          AY1=MR1;                                    {in double precision}
          AX1=I0;                                     {AX1 holds current value for Nc}
cross_loop:        I1=I2;                             {Reset pointer into array}
{?} IF NOT CE JUMP gsm18;
          DM(Nc)=AX1;                                 {After loop, Nc is in AX1}

          SI=AY1;                                     {This section of code computes}
          AY1=6;                                      {the power of the reconstructed}
          AX0=DM(scal);                               {short term residual signal dp}
          AR=AX0-AY1;
          SE=AR;
          SR=ASHIFT SI (HI), AR=AY0;
          SR=SR OR LSHIFT AR (LO);
          SE=-3;
          AY0=^dp+120;                                {Use dp() array directly, do}
          AR=AY0-AX1, AY0=SR0;                        {not bother with temp array}
          AY1=SR1;
          I5=AR;
          MR=0, AR=PM(I5,M5);
          CNTR=sub_window_length;
{         DO power UNTIL CE;}
gsm19:    SR=ASHIFT AR (HI), AR=PM(I5,M5);            {Scale data}
          MY0=SR1;                                    {Copy to y-reg}
power:    MR=MR+SR1*MY0 (SS);                         {Compute L_power}
{?} IF NOT CE JUMP gsm19;
```

```
        AR=0;                           {This section of code computes}
        AF=PASS AY1;                    {and codes the LTP gain value}
        IF LT JUMP bc_found;            {L_max < 0, so bc=0}
        IF EQ AF=PASS AY0;
        IF EQ JUMP bc_found;            {L_max = 0, so bc=0}
        AR=3;
        AF=MR0-AY0;
        AF=MR1-AY1+C-1;
        IF LT JUMP bc_found;            {L_max > L_power, so bc=3}
        IF EQ AF=MR0-AY0;
        IF EQ JUMP bc_found;            {L_max = L_power, so bc=3}
        SE=EXP MR1 (HI);                {Normalize L_power}
        SE=EXP MR0 (LO)
        SR=NORM MR1 (HI), MR1=AY1;
        SR=SR OR NORM MR0 (LO), MR0=AY0;
        MY0=SR1, SR=NORM MR1 (HI);      {Normalize L_max, MY0 holds s}
        SR=SR OR NORM MR0 (LO);
        AY0=SR1, AF=PASS 0;             {AY0 holds R}
        I5=^table_dlb;                  {Check for each value of bc}
        AR=PASS 0, MX0=PM(I5,M5);
        MR=MX0*MY0 (SS), MX0=PM(I5,M5);
        AF=MR1-AY0;
        IF GE JUMP bc_found;
        AR=1;
        MR=MX0*MY0 (SS), MX0=PM(I5,M5);
        AF=MR1-AY0;
        IF GE JUMP bc_found;
        AR=2;
        MR=MX0*MY0 (SS);
        AF=MR1-AY0;
        IF GE JUMP bc_found;
        AR=3;

bc_found:   DM(bc)=AR;                  {AR holds the value of bc}
        ENA AR_SAT;                     {Re-enable ALU saturation}

{   This section of code computes the long term analysis filtering section
    as described in section 4.2.12 of the recommendation}

lt_analysis:AY0=^table_qlb;
        AR=AR+AY0;
        I5=AR;
        MY0=PM(I5,M4);
        AY0=^dp+120;
        AR=AY0-AX1;
        I4=AR;
        I5=^dpp;                        {Output array dpp()}
        I2=^wt+5;                       {The e-array goes into wt}
        MX1=H#80;
        MR=MX1*MF (SS), MX0=PM(I4,M5);
```

(listing continues on next page)

4 GSM Codec

```
          CNTR=sub_window_length;
{         DO calculate_e UNTIL CE;}
gsm20:    MR=MR+MX0*MY0 (SS), AY0=DM(I3,M1);   {Compute dpp[k]}
          AR=AY0-MR1, MX0=PM(I4,M5);           {Compute e[k]}
          DM(I5,M5)=MR1;                       {Save dpp()}
calculate_e:      DM(I2,M1)=AR, MR=MX1*MF (SS); {Save e() into wt()}
{?} IF NOT CE JUMP gsm20;

{ All the long term parameters (Nc, bc, mc) have been computed}

          RTS;

{ This subroutine computes, encodes and decodes the Residual Pulse
  Excitation sequence as defined in section 4.2.13 of the recommendation}

rpe_encoding:I0=^wt;                          {The beginning of wt must be}
          AX0=0;                              {cleared for use in the block}
          CNTR=5;                             {filter}
{         DO zero_start UNTIL CE;}
zero_start:       DM(I0,M1)=AX0;
{?} IF NOT CE JUMP zero_start;
          I0=^wt+45;                          {The end must also be cleared}
          CNTR=5;
{         DO zero_end UNTIL CE;}
zero_end:         DM(I0,M1)=AX0;
{?} IF NOT CE JUMP zero_end;

          DIS AR_SAT;
          I2=^wt;                             {wt will be reloaded with x()}
          CNTR=sub_window_length;
{         DO compute_x_array UNTIL CE;}
gsm21:    I0=I2;
          I4=^h;
          MR=0, MX0=DM(I0,M1), MY0=PM(I4,M5);
          MR0=8192;                           {Used for rounding}
          CNTR=11;                            {11-term filter}
{         DO compute_x UNTIL CE;}
compute_x:        MR=MR+MX0*MY0 (SS), MX0=DM(I0,M1), MY0=PM(I4,M5);
{?} IF NOT CE JUMP compute_x;
          AY0=MR0;                            {The output value must be}
          AR=MR0+AY0, AY0=MR1;                {Up-shifted with saturation}
          AY1=AR, AR=MR1+AY0+C;
          IF NOT AV JUMP done_2x;             {Check for overflow on 2x}
          AR=H#7FFF;                          {Overflow. manually saturate}
          AY1=H#8000;                         {output, and save value}
          IF AC AR=PASS AY1;
          JUMP compute_x_array;
done_2x:  AX1=AR, AR=PASS AY1;                {Compute 4x}
          AY0=AX1, AR=AR+AY1;
          ENA AR_SAT;                         {Automatic saturation can}
          AR=AX1+AY0+C;                       {be used on the last add}
          DIS AR_SAT;
```

242

```
compute_x_array:   DM(I2,M1)=AR;                    {Output writes over input}
{?} IF NOT CE JUMP gsm21;


{  This section of code computes the RPE Grid Selection as described
   in section 4.2.14 of the recommendation}

           AF=PASS 0;
           AY0=0;
           AY1=0;
           AX0=0;
           M0=3;                                     {Used for interleaving}
           I1=^wt;
           CNTR=4;
{          DO find_mc UNTIL CE;}
gsm22:     I2=I1;
           MR=0, SI=DM(I2,M0);                       {L_result=0, fetch first value}
           CNTR=13;
{          DO calculate_em UNTIL CE;}
gsm23:            SR=ASHIFT SI BY -2 (HI);{Downshift to avoid overflow}
                  MY0=SR1;                            {Copy to yop}
calculate_em:            MR=MR+SR1*MY0 (SS), SI=DM(I2,M0);{L_result is in MR}
{?} IF NOT CE JUMP gsm23;
           AR=MR0-AY0;
           AR=MR1-AY1+C-1;
           IF LT JUMP find_mc;                        {Check for L_result<EM}
           IF EQ AR=MR0-AY0;                          {If MSW=0, check LSW again}
           IF EQ JUMP find_mc;                        {L_result = EM}
           AY0=MR0;                                   {EM=L_result}
           AY1=MR1, AR=PASS AF;
           AX0=AR;                                    {Mc=m}
find_mc:   AF=AF+1, MX0=DM(I1,M1);
{?} IF NOT CE JUMP gsm22;
                                                      {Mc in AX0}

           ENA AR_SAT;
           DM(Mc)=AX0;
           AY0=^wt;
           I1=^wt;                                    {temp array will be reloaded}
           AR=AX0+AY0;                                {with xM()}
           I0=AR;
           AR=PASS 0;
           CNTR=13;
{          DO decimate UNTIL CE;}
gsm24:     AX0=DM(I0,M0);                             {Read every third value}
           AF=ABS AX0, DM(I1,M1)=AX0;                 {Check for maximum value}
           AF=AR-AF;
decimate:         IF LT AR=ABS AX0;                   {AR holds xmax}
{?} IF NOT CE JUMP gsm24;
           M0=0;                                      {Reset M0 to usual value}
```

(listing continues on next page)

4   GSM Codec

{***** The following code checks vad and stores xmax in oldxmax_buffer,
   if necessary. *****}

```
        AX0 = DM(vad);
        AF  = PASS AX0;
        IF NE JUMP xmax_speech;              {Yes - VAD, so do not store}
        SI  = AR;                            {Save xmax in SI}
```

{***** This section of code will write xmax into the oldxmax_buffer for use
in the next frame. Note that scaling also takes place. *****}

```
        AX0 = ^oldxmax_buffer;
        AY1 = ^oldxmax_buffer + 16;
        AR  = DM(oldxmax_pntr);
        AF  = AY1 - AR;
        IF LE AR = PASS AX0;                 {AR holds address}

        SR  = ASHIFT SI BY -4 (HI);          {SR1 holds scaled xmax}
        I5  = AR;
        DM(I5,M5) = SR1;                      {Write xmax to oldxmax_buffer}
        DM(oldxmax_pntr) = I5;
        AR  = SI;                            {Restore xmax}
```

{   This section of code computes the APCM quantization of the selected
RPE section as defined in section 4.2.15 of the recommendation}

```
xmax_speech:CALL quantize_xmax;       {input and output in AR}
        DM(xmaxc)=AR;                 {Save xmaxc}
        CALL get_xmaxc_pts;           {Compute exponent and mantissa}
                                      {Exponent in AY1}
        AY0=^table_nrfac;             {Mant in AR}
        MX0=AR, AR=AR+AY0;            {Now mant in MX0}
        I5=AR;
        MY0=PM(I5,M5);                {MYO holds temp2}
        AX0=6;
        AR=AX0-AY1;
        AY0=4;
        I0=^wt;                       {Temp array current holds xM()}
        I2=^xmc;
        SE=AR;                        {SE holds temp1}
        SI=DM(I0,M1);
        CNTR=13;
{       DO compute_xm UNTIL CE;}
gsm25:    SR=LSHIFT SI (HI), SI=DM(I0,M1);   {temp=xM[i]<<temp1}
          MR=SR1*MY0 (SS);                   {temp=temp*temp2}
          SR=ASHIFT MR1 BY -12 (HI);
          AR=SR1+AY0; {AR=xMc[i]}
compute_xm:      DM(I2,M1)=AR;               {Store xMc[i]}
{?} IF NOT CE JUMP gsm25;

        CALL rpe_decoding;                   {APCM inverse quantization}
```

```
{   This section of code updates the reconstructed short term residual
    signal dp() as defined in section 4.2.18 of the recommendation}

update_dp_code: I4=^dp;                          {I4 points to dp[-120]}
          I5=^dp+40;                             {I5 points to dp[-80]}
          CNTR=80;
{         DO update_dp UNTIL CE;        }
gsm26:        AX0=PM(I5,M5);
update_dp:        PM(I4,M5)=AX0;                 {dp[-120+k]=dp[-80+k]}
{?} IF NOT CE JUMP gsm26;
          I4=^dp+80;                             {I4 points to dp[-40]}
          I1=^ep;
          I5=^dpp;
          AX0=DM(I1,M1);
          AY0=DM(I5,M5);                         {Fetch first samples}
          CNTR=sub_window_length;
{         DO fill_dp UNTIL CE;}
gsm27:        AR=AX0+AY0, AX0=DM(I1,M1);
          AY0=DM(I5,M5);
fill_dp:      PM(I4,M5)=AR;                      {dp[-40+k]=ep[k]+dpp[k]}
{?} IF NOT CE JUMP gsm27;
          RTS;

{   This section of code computes the APCM quantization of the selected
    RPE section as defined in section 4.2.15 of the recommendation}

quantize_xmax:SI=AR, AF=PASS 0;               {This section of code quantizes}
          SR=ASHIFT AR BY -9 (HI);            {and codes xmax into xmaxc}
          CNTR=6;
{         DO get_exp UNTIL CE;}
gsm28:        AR=PASS SR1;                       {SR1 holds temp}
          IF GT AF=AF+1;                         {Increment exp until SR1=0}
get_exp:      SR=ASHIFT SR1 BY -1 (HI);
{?} IF NOT CE JUMP gsm28;

          AX1=5;
          AR=AX1+AF;                             {temp=exp+5}
          AR=-AR;                                {Use this for downshift}
          SE=AR, AR=PASS AF;                     {AR=exp}
          SR=LSHIFT AR BY 3 (HI);                {Place exponent}
          AY0=SR1, SR=ASHIFT SI (HI);            {Place mantissa}
          AR=SR1+AY0;                            {AR holds xmaxc}
          RTS;
```

(listing continues on next page)

4 GSM Codec

{_____Encoder and Voice Activity Detector Subroutines_____}

{ This section of code computes the reflection coefficients using the
   schur recursion as defined in section 4.2.5 of recommendation 6.10 and
   section 3.3.1 of recommendation 6.32}

```
schur_routine:I6=AY1;                        {This section of code prepares}
        AR=DM(I6,M5);                        {for the schur recursion}
        SE=EXP AR (HI), SI=DM(I6,M5);        {Normalize the autocorrelation}
        SE=EXP SI (LO);                      {sequence based on L_ACF[0]}
        SR=NORM AR (HI);
        SR=SR OR NORM SI (LO);
        AR=PASS SR1;                         {If L_ACF[0] = 0, set r to 0}
        IF EQ JUMP zero_reflec;
        I6=AY1;
        I5=MY1;
        AR=DM(I6,M5);
        CNTR=9;                              {Normalize all terms}
{       DO set_acf UNTIL CE;}
gsm29:  SR=NORM AR (HI), AR=DM(I6,M5);
        SR=SR OR NORM AR (LO), AR=DM(I6,M5);
set_acf:    DM(I5,M5)=SR1;
{?} IF NOT CE JUMP gsm29;

        I5=MY1;                              {This section of code creates}
        I4=^k+7;                             {the k-values and p-values}
        I0=^p;
        AR=DM(I5,M5);                        {Set P[0]=acf[0]}
        DM(I0,M1)=AR;
        CNTR=7;
{       DO create_k UNTIL CE;}              {Fill the k and p arrays}
gsm30:  AR=DM(I5,M5);
        DM(I0,M1)=AR;
create_k:       DM(I4,M6)=AR;
{?} IF NOT CE JUMP gsm30;
        AR=DM(I5,M5);
        DM(I0,M1)=AR;                        {Set P[8]=acf[8]}

        I5=M0;                               {Compute r-values}
        I6=7;                                {I6 used as downcounter}
        SR0=0;
        SR1=H#80;                            {Used in unbiased rounding}
        CNTR=7;                              {Loop through first 7 r-values}
{       DO compute_reflec UNTIL CE;}
gsm31:  I2=^p;                               {Reset pointers}
        I4=^k+7;
        AX0=DM(I2,M1);                       {Fetch P[0]}
        AX1=DM(I2,M2);                       {Fetch P[1]}
        MX0=AX1, AF=ABS AX1;                 {AF=abs(P[1])}
        AR=AF-AX0;
```

```
              IF LE JUMP do_division;       {If P[0]<abs(P[1]), r = 0}
              DM(I5,M5)=SR0;                {Final r =0}
              JUMP compute_reflec;
do_division:      CALL divide_routine;      {Compute r[n]=abs(P[1])/P[0]}
              AR=AY0, AF=ABS AX1;
              AY0=32767;
              AF=AF-AX0;                     {Check for abs(P[1])=P[0]}
              IF EQ AR=PASS AY0;             {Saturate if they are equal}
              IF POS AR=-AR;                 {Generate sign of r[n]}
              DM(I5,M5)=AR;                  {Store r[n]}
              MY0=AR, MR=SR1*MF (SS);
              MR=MR+MX0*MY0 (SS), AY0=AX0;   {Compute new P[0]} AR=MR1+AY0;
              DM(I2,M3)=AR;                  {Store new P[0]}
              CNTR=I6;                       {One less loop each iteration}
{             DO schur_recur UNTIL CE;}
gsm32:            MR=SR1*MF (SS), MX0=DM(I4,M4);
                 MR=MR+MX0*MY0 (SS), AY0=DM(I2,M2);
                 AR=MR1+AY0, MX1=AY0;    {AR=new P[m]}
                 MR=SR1*MF (SS);
                 MR=MR+MX1*MY0 (SS), AY0=MX0;
                 DM(I2,M3)=AR, AR=MR1+AY0;{Store P[m], AR=new K[9-m]}
schur_recur:         DM(I4,M6)=AR;          {Store new K[9-m]}
{?} IF NOT CE JUMP gsm32;
compute_reflec:      MODIFY(I6,M6);         {Decrement loop counter (I6)}
{?} IF NOT CE JUMP gsm31;
         I2=^p;                             {Compute r[8] outside of loop}
         AX0=DM(I2,M1);                     {Using same procedure as above}
         AX1=DM(I2,M2);
         AF=ABS AX1;
         CALL divide_routine;
         AR=AY0, AF=ABS AX1;
         AY0=32767;
         AF=AF-AX0;
         IF EQ AR=PASS AY0;
         AF=ABS AX1;
         AF=AF-AX0;                         {The test for valid r is here}
         IF GT AR=PASS 0;                   {r[8]=0 if P[0]<abs(P[1])}
         IF POS AR=-AR;
         DM(I5,M5)=AR;
         JUMP schur_done;

zero_reflec:AX0=0;                          {The r-values must be set to}
         I5=M0;                             {0 according to the recursion}
         CNTR=8;
{             DO zero_rs UNTIL CE;}
zero_rs:     DM(I5,M5)=AX0;
{?} IF NOT CE JUMP zero_rs;

schur_done: M0 = 0;
         RTS;
```

(listing continues on next page)

4 GSM Codec

```
{_____Divide Subroutine_____}
divide_routine:
        AY0=0;
        DIVS AF,AX0;
        CNTR=15;
{       DO div_loop UNTIL CE;}
div_loop:       DIVQ AX0;
{?} IF NOT CE JUMP div_loop;
        RTS;

{_____Decoder Subroutine_____}

{  This section of code implements the LPC-LTP-RPE decoder as defined in
   the GSM recommendation.}

dmr_decode: ENA AR_SAT;             {Enable ALU saturation mode}
        DM(recv_buffer)=I1;         {Save pointer to input coeff array}
        DM(speech_out)=I2;          {Save pointer to output speech array}
        MX1=H#4000;                 {This is used to set the MF register}
        MY1=H#100;                  {to H#80 so that it can be used in }
        MF=MX1*MY1 (SS);            {unbiased rounding in various places}

{*****  The code will now implement the comfort noise insertion as specified
    in GSM specification 6.31, section 3.1.  *****}

        AR  = PASS AX0;             {AX0 holds the SID signal}
        IF EQ JUMP start_dcd;
        CALL comfort_noise_generator;

{  Now, continue}

start_dcd: I1=^LARc;               {Copy the LARc array into proper place}
        CNTR=8;                     {there are 8 LARcs}
        CALL recv_data;            {This subroutine copies from input buff}


        CALL decode_LARc;          {Decode the LARcs to LARs}

        I3=DM(speech_out);         {Only set output pointer once!}

        CNTR=4;                     {Computations for 4 sub windows}
{       DO dcd_subwindow UNTIL CE;}
gsm33:      I1=^Nc;                 {Set pointer to start of sub-window}
            CNTR=17;                {data array 17 coefs per sub-window}
            CALL recv_data;         {Copy them from the input buffer}
            CALL get_xmaxc_pts;     {Decode xmaxc into exp and mantissa}
            CALL rpe_decoding;      {Decode xMc array into ep array}
            CALL lt_predictor;      {Compute drp for sub-window}
            CALL setup_wtr;         {Copy drp values in temp wtr}
dcd_subwindow:      NOP;            {No CALL in last instr of DO loop}
{?} IF NOT CE JUMP gsm33;
```

```
        I0=DM(speech_out);              {Set pointer to output array}
        I1=I0;                          {Set pointer to input/output}
        I6=^st_synthesis;              {Set pointer to st filter}
        I4=^old_LARrpp;                {Set pointer to old LARrpp}
        CALL st_filter;                 {Call short term filter manager}

{    4.3.5                                                              }

        I0=DM(speech_out);              {This section of code does the}
        MY0=28180;                      {pre-emp, up-scale and trunc}
        MX0=DM(msr);
        AY1=H#FFF8;                     {Same effect as down/up shift}
        MX1=H#80;                       {Used for unbaised rounding}
        MR=MX1*MF (SS);                 {Pre-load MR}
        CNTR=window_length;
{           DO post_process UNTIL CE;}
gsm34:      MR=MR+MX0*MY0 (SS), AY0=DM(I0,M0)   {De-emphasis filtering}
        AR=MR1+AY0;
        AF=PASS AR, MX0=AR;
        AR=AR+AF;                       {Upscale output}
        AR=AR AND AY1;                  {Spec does this with shifts}
post_process:    DM(I0,M1)=AR, MR=MX1*MF (SS);
{?} IF NOT CE JUMP gsm34;
        DM(msr)=MX0;

   {At this point, the buffer sr can be output to the speaker}

        DIS AR_SAT;
        RTS;                            {Return from Decoder}

{_____Subroutines for Decoder_____}

recv_data:  I0=DM(recv_buffer);        {This subroutine copies data}
{         DO recv UNTIL CE;}           {from the input coefficient}
gsm35:      AX0=DM(I0,M1);             {buffer to the appropraite }
recv:            DM(I1,M1)=AX0;         {location in memory while}
{?} IF NOT CE JUMP gsm35;
        DM(recv_buffer)=I0;             {maintaining pointer}
        RTS;

setup_wtr:  I5=^drp+120;               {This subroutine copies the}
        CNTR=40;                        {current sub-window data into}
{         DO copy_drp UNTIL CE;}       {a temporary array.  This temp}
gsm36:      AX0=DM(I5,M5);             {array will be used by the}
copy_drp:        DM(I3,M1)=AX0;         {short term synthesis filter}
{?} IF NOT CE JUMP gsm36;
        RTS;
```

(listing continues on next page)

4 GSM Codec

{ This section of code computes the short term synthesis filter as
   described in section 4.3.4 of the recommendation}

```
st_synthesis:MX1=H#80;                  {Used in un-biased rounding}
          M0=-3;                        {M0 is changed for this routine}
{         DO st_synth_compute UNTIL CE;}
gsm37:    I5=^rp+7;                      {Point to coefficient array}
          I2=^v+7;                       {Point to delay array}
          MY0=DM(I5,M6);                 {Moved from PM}
          MR=MX1*MF (SS), MX0=DM(I2,M2);
          AY0=DM(I1,M1);                 {AY0 holds sri, sri=wt[k]}
          CNTR=8;
{         DO st_synth_loop UNTIL CE;}
gsm38:        MR=MR+MX0*MY0 (SS);
              AY1=MX0, AR=AY0-MR1;                 {AR=sri}
              MR=MX1*MF (SS), AY0=AR;              {AY0=sri}
              MR=MR+AR*MY0 (SS), MX0=DM(I2,M3);
              AR=MR1+AY1, MY0=DM(I5,M6);           {AR=v[9-i]} st_synth_loop:
DM(I2,M0)=AR, MR=MX1*MF (SS);                      {Save v[9-i]}
{?} IF NOT CE JUMP gsm38;
              DM(I0,M1)=AY0;            {sr[k]=sri}
              MODIFY(I2,M3);            {Move pointer to delay line}
st_synth_compute: DM(I2,M0)=AY0;        {V[0]=sri}
{?} IF NOT CE JUMP gsm37;
          M0=0;                         {Reset M0 to usual value}
          RTS;
```

{ This section of code computes the long term synthesis filter as
   described in section 4.3.2 of the recommendation}

```
lt_predictor:AY1=DM(nrp);                      {Check the limits of Ncr}
          AR=DM(Nc);
          AY0=40;
          AF=AR-AY0;
          IF LT AR=PASS AY1;                   {Below min, so use last value}
          AY0=120;
          AF=AR-AY0;
          IF GT AR=PASS AY1;                   {Above max, so use last value}
          DM(nrp)=AR;
          AY0=^drp+120;
          AR=AY0-AR;
          I4=AR;
          I6=AY0;
          AY0=DM(bc);
          AX0=^table_qlb;
          AR=AX0+AY0;
          I5=AR;
          MX1=H#80;
          MR=MX1*MF (SS), MX0=DM(I4,M5);
          MY0=PM(I5,M4);   {brp}
```

```
          I2=^ep;
          CNTR=sub_window_length;
{         DO compute_drp UNTIL CE;}
gsm39:    MR=MR+MX0*MY0 (ss), AY0=DM(I2,M1);          {Compute drpp}
          AR=MR1+AY0, MX0=DM(I4,M5);                  {drp[k]=erp[k]+drpp}
compute_drp:     DM(I6,M5)=AR, MR=MX1*MF (SS);        {Store drp[k]}
{?} IF NOT CE JUMP gsm39;
          I4=^drp;                                    {I0 points to drp[-120]}
          I5=^drp+40;                                 {I1 points to drp[-80]}
          CNTR=120;
{         DO update_drp UNTIL CE;}
gsm40:    AX0=DM(I5,M5);
update_drp:      DM(I4,M5)=AX0;                       {drp[-120+k]=drp[-80+k]}
{?} IF NOT CE JUMP gsm40;
          RTS;


{_____Common Subroutines for Encoder and Decoder_____}

{  This section of code decodes the coded log area ratios as defined by
   section 4.2.8 of the recommendation}

decode_LARc:I2=^LARc;
          I1=^LARpp;
          I6=^table_mic;
          I4=^table_inva;
          I5=^table_b;
          SE=1;
          CNTR=8;
{         DO compute_larpp UNTIL CE;}
gsm41:    AX0=DM(I2,M1);
          AY0=PM(I6,M5);
          AR=AX0+AY0, SI=PM(I5,M5);
          SR=LSHIFT AR BY 10 (HI);
          AY1=SR1, SR=LSHIFT SI (HI);                 {AY1=temp1}
          AR=AY1-SR1, MY0=PM(I4,M5);                  {AR=temp1=temp1-temp2}
          MR0=H#8000;MR1=0;                           {Unbiased rounding}
          MR=MR+AR*MY0 (ss);                          {MR1=temp1=INVA[i]*temp1}
          AY0=MR1;
          AR=MR1+AY0;                                 {AR=LARpp[i]}
compute_larpp:   DM(I1,M1)=AR;                        {Store LARpp[i]}
{?} IF NOT CE JUMP gsm41;
          RTS;
```

(listing continues on next page)

4   GSM Codec

```
{  This section of code computes the mantissa and exponent parts of the
   xmaxc coefficient as described in section 4.2.15 of the recommendation}

get_xmaxc_pts:AR=DM(xmaxc);
        AY0=AR;
        AX0=15;
        SR=ASHIFT AR BY -3 (HI);
        AY1=1;
        AR=SR1-AY1;
        AF=AY0-AX0;
        IF LE AR=PASS 0;
        SR=LSHIFT AR BY 3 (HI);
        AY1=AR, AR=AY0-SR1;
        IF NE JUMP else_clause;            {Check if mant==0}
        AY1=-4;                            {Yes, so set mant and ex}
        AR=15;
        JUMP around_else;                  {Jump over else_clause}

else_clause:AY0=7;
        AF=AR-AY0;
        CNTR=3;
{       DO normalize_mant UNTIL CE;}
gsm42:      IF GT JUMP normalize_mant;
        SR=LSHIFT AR BY 1 (HI);
        AR=AY1-1;                          {Decrement exponent}
        AY1=AR, AF=PASS 1;                 {AY1=exp}
        AR=SR1+AF;                         {Increment mantissa}
normalize_mant:   AF=AR-AY0;
{?} IF NOT CE JUMP gsm42;

around_else:AY0=8;
        AR=AR-AY0;
        MX0=AR;                            {Mant must also be in MX0}
        RTS;

{  This section of code computes the reflection coefficients for the
   interpolated LARs as defined in section 4.2.9.2 of the recommendation}

make_rp: MX1=I6;                           {store I6}
        I5=^LARp;
        I6=^rp;
        CNTR=8;
{       DO compute_rp UNTIL CE; }
gsm43:      AX0=DM(I5,M5);
        AR=ABS AX0;
        AX1=AR;
        SR=LSHIFT AR BY 1 (HI);
        AX0=SR1;                           {AX0=temp<<1}
        SR=ASHIFT AR BY -2 (HI);
        AY0=26112;
```

```
            AR=SR1+AY0;                      {AR=temp>>2 + 26112}
            AY0=20070;
            AY1=11059;
            AF=AX1-AY0;
            IF LT AR=AX1+AY1;                {AR=temp+11059}
            AF=AX1-AY1;
            IF LT AR=PASS AX0;
            IF NEG AR=-AR;                    {Compute sign}
compute_rp:        DM(I6,M5)=AR;             {Store rp[i], Moved from PM}
{?} IF NOT CE JUMP gsm43;
            I6=MX1;                          {restore I6}
            RTS;
```

```
{  This section of code computes the interpolation of the LARpp() array
   and calls the subroutine to compute the reflection coefficients, and
   then the appropriate short term filter. This block is defined in section
   4.2.9.1 of the recommendation}
```

```
st_filter: SE=-2;                           {Compute the LARps for }
           I2=I4;                           {k_start = 0 to k_end = 12}
           I3=^LARpp;
           I5=^LARp;
           SI=DM(I3,M1);
           CNTR=8;
{          DO k_end_12 UNTIL CE;}
gsm44:     SR=ASHIFT SI (HI), SI=DM(I2,M1);
           AY0=SR1, SR=ASHIFT SI (HI);
           AF=SR1+AY0;
           SR=ASHIFT SI BY -1 (HI);
           AR=SR1+AF, SI=DM(I3,M1);
k_end_12:          DM(I5,M5)=AR;
{?} IF NOT CE JUMP gsm44;

           CALL make_rp;                    {Compute reflection coeffs}
           CNTR=13;                         {13 filter samples}
           CALL (I6);                       {Analysis or Synthesis}
           I5=^LARp;                        {Compute the LARps for}
           I2=I4;                           {k_start = 13 to k_end = 26}
           I3=^LARpp;
           SE=-1;
           SI=DM(I3,M1);
           CNTR=8;
{          DO k_end_26 UNTIL CE;}
gsm45:     SR=ASHIFT SI (HI), SI=DM(I2,M1);
           AY0=SR1, SR=ASHIFT SI (HI);
           AR=SR1+AY0, SI=DM(I3,M1);
k_end_26:          DM(I5,M5)=AR;
{?} IF NOT CE JUMP gsm45;
```

(listing continues on next page)

4 GSM Codec

```
        CALL make_rp;                   {Compute reflection coeffs}
        CNTR=14;                        {14 filter samples}
        CALL (I6);                      {Analysis or Synthesis}
        I5=^LARp;                       {Compute the LARps for}
        I2=I4;                          {k_start = 27 to k_end = 39}
        I3=^LARpp;
        SE=-2;
        SI=DM(I2,M1);
        CNTR=8;
{       DO k_end_39 UNTIL CE;}
gsm46:      SR=ASHIFT SI (HI), SI=DM(I3,M1);
            AY0=SR1, SR=ASHIFT SI (HI);
            AF=SR1+AY0;
            SR=ASHIFT SI BY -1 (HI);
            AR=SR1+AF, SI=DM(I2,M1);
k_end_39:          DM(I5,M5)=AR;
{?} IF NOT CE JUMP gsm46;


        CALL make_rp;                   {Compute reflection coeffs}
        CNTR=13;                        {13 filter samples}
        CALL (I6);
        I5=^LARp;                       {Compute the LARps for}
        I3=^LARpp;                      {k_start = 40 to k_end = 159}
        CNTR=8;
{       DO k_end_159 UNTIL CE;}
gsm47:      AX0=DM(I3,M1);
            DM(I5,M5)=AX0;
k_end_159:         DM(I4,M5)=AX0;       {LARpp(j-1)[i] = LARpp(j)[i]}
{?} IF NOT CE JUMP gsm47;


        CALL make_rp;                   {Compute reflection coeffs}
        CNTR=120;                       {120 filter samples}
        CALL (I6);
        RTS;

{   This section of code computes the inverse of the APCM quantization
    and the RPE grid positioning as defined in sections 4.2.16 and 4.2.17
    of the recommendation}

rpe_decoding:I0=^ep;
        AX0=0;                          {First set output ep() array}
        CNTR=sub_window_length;         {to 0s, so it can be filled}
{       DO zero_fill_ep UNTIL CE;}      {in the next section}
zero_fill_ep:    DM(I0,M1)=AX0;
{?} IF NOT CE JUMP zero_fill_ep;


        AX0=DM(mc);
        AY0=^ep;
        AR=AX0+AY0;
```

```
        I1=AR;                              {Point to start in ep() array}
        M0=3;
        AY0=^table_fac;
        AX0=MX0;
        AR=AX0+AY0;
        I5=AR;
        MY0=PM(I5,M4);                      {MY0 holds temp1}
        AX0=6;
        AR=AY1-AX0;
        AX1=AR, AF=AX0-AY1;
        AR=AF-1;
        SE=AR, AR=PASS 1;                   {SE holds temp2}
        SR=LSHIFT AR (HI), SE=AX1;
        AY1=SR1;                            {AY1 holds temp3}
        I0=^xmc;
        AY0=7;
        MX1=H#80;
        MR=MX1*MF (SS), SI=DM(I0,M1);
        CNTR=13;
{       DO inverse_apcm UNTIL CE;}
gsm48:      SR=LSHIFT SI BY 1 (HI);
            AR=SR1-AY0, SI=DM(I0,M1);       {AR=temp=xMc[i]<<1 - 7}
            SR=LSHIFT AR BY 12 (HI);        {SR1=temp=temp<<12}
            MR=MR+SR1*MY0 (SS);             {MR1=temp=temp1*temp}
            AR=MR1+AY1;                     {AR=temp=temp+temp3}
            SR=ASHIFT AR (HI);              {xMp[i]=temp>>temp2}
inverse_apcm:   DM(I1,M0)=SR1, MR=MX1*MF (SS);   {ep[Mc+(3*i)=xMp[i]}
{?} IF NOT CE JUMP gsm48;

   M0=0;                                    {Reset M0 to usual value}
   RTS;
{_____End of GSM0610 Code_____}
.ENDMOD;
```

Listing 4.2 Codec Routine (GSM0610.DSP)

4 GSM Codec

```
{_____

GSM0632.DSP
            Analog Devices INC. DSP Division
            One Technology Way, Norwood, MA 02062
            DSP Applications Hotline: (617) 461-3672
```

This subroutine implements the voice activity detection algorithm of
GSM specification 06.32 on the ADSP-210x family of DSPs. In line
comments reference various sections of this recommendation. It
is assumed that the reader is familiar with that document.

The code consists of two subroutines. VAD_ROUTINE is called by
the GSM encoder (06.10) after the autocorrelation is complete.
UPDATE_PERIODICITY is called by the GSM encoder after the subwindow
data is calculated.

This code is optimized to implement the Voice Activity Detection
in a minimal amount of Progam Memory space. Since the 21xx processors
can execute all of the GSM speech processing functions in much less
than 20ms, we have slightly increased execution time (less than .02ms)
in exchange for a decrease in code size.

Long words are stored as two successive 16 bit locations,
MSW first, LSW second.

This code has been successfully verified with the GSM 06.32 Digital
Test Sequences, dated March, 1990. The changes made to version 1.00
during validation are available in a separate document.

Release History:
```
___Date___  _Ver_ _____Comments_____
24-Oct-89   66    preliminary - waiting for test vectors
10-Jan-90   1.00  Second Release (waiting for VAD test vectors)
01-Nov-90   2.00  Third release. Validated with 06.32 test sequences
```

```
        Assembler Preprocessor Switches:
   -cp switch must always be used when assembling
   -Dalias switch aliases some variables to save RAM space

        Calling Parameters:
   M0=0;  M1=1;  M2=-1;  M3=2;  M4=0;  M5=1;  M6=-1;
   L0=0;  L1=0;  L2=0;   L3=0;  L4=0;  L5=0;  L6=0;

        Return Values: VAD
```

```
        Max Loop Nesting Depth: 2 levels
        Max PC Stack Nesting Depth: 3 levels

        Modes Assumed: AR_SAT enabled, M_MODE disabled
        ADSP-2101 Execution cycles:  2141 maximum

            vad_routine:  2055 cycles maximum
            update_periodicity: 86 cycles maximum
_____}

.MODULE   voice_activity_detection;

{_____Conditional Assembly_____}
{    Use (asm21 -cp -Dalias) to alias some variables to save RAM       }
#ifdef alias
    .INCLUDE        <var0632.ram>;
    .EXTERNAL        wt;                        {Working buffer for aliases}
    #define r_a_av1 wt+0
    #define vpar wt+0
    #define sacf wt+9
    #define sav0 wt+9
    #define L_coef wt+18
    #define L_av0 wt+36
    #define L_av1 wt+54
    #define L_work wt+54
#else
    .INCLUDE        <var0632.h>;
#endif {_____}
.ENTRY       vad_routine;
.ENTRY       update_periodicity;

.EXTERNAL    schur_routine;                     { found in GSM0610.DSP }
.EXTERNAL    divide_routine;                    { found in GSM0610.DSP }

.EXTERNAL    L_ACF;
.EXTERNAL    scaleauto;

.GLOBAL   vad, lags;
{  the following are GLOBAL for the reset routine only }
.GLOBAL   rvad, normrvad, L_sacf, L_sav0;
.GLOBAL   pt_sacf, pt_sav0, L_lastdm;
.GLOBAL   oldlagcount, veryoldlagcount, e_thvad, m_thvad, adaptcount;
.GLOBAL   burstcount, hangcount, oldlag;
```

(listing continues on next page)

4   GSM Codec

```
{_____3.1_____Adaptive Filtering and Energy Computation_____}

{               Test if L_ACF is equal to zero                    }

vad_routine:I6=^L_ACF;
        AR=DM(scaleauto);
        AR=PASS AR, AY0=DM(I6,M5);  {Get ms_ACF}
        IF LT AR=PASS 0;           {IF scaleauto<0 THEN: scalvad=0}

        SR=ASHIFT AR BY 1 (LO);
        AY1=SR0;                   {AY1=scalvad<<1}
        AR=PASS 0, AX0=DM(I6,M6);  {Get ls_ACF}
        DM(m_pvad)=AR;             {Init these anyways}
        DM(m_acf0)=AR;
        AR=-32768;
        DM(e_pvad)=AR;
        DM(e_acf0)=AR;
        AR=AX0 OR AY0, MR0=AY0;    {IF L_ACF[0]=0 THEN: goto 3.2}
        IF EQ JUMP acf_average;

{Outputs: scalvad<<1=AY1, ls_ACF[0]=AX0, I6=^L_ACF[0]}

{               Renormalization of the L_acf[0..8]                 }

        SE=EXP MR0 (HI), SI=AX0;   {Norm L_ACF[0]}
        SE=EXP SI (LO);
        AY0=SE;                    {Fix SE for >>19, take SR1}
        AX0=-3;
        AR=AX0-AY0;
        SE=AR;                     {SE=normacf-3}

        I5=^sacf;
        CNTR=9;
        DO norm_sacf UNTIL CE;
            SI=DM(I6,M5);
            SR=ASHIFT SI (HI), SI=DM(I6,M5);
            SR=SR OR LSHIFT SI (LO);
norm_sacf:        DM(I5,M5)=SR1;

{Outputs: scalvad<<1=AY1, -normacf=AY0}

{        Computation of e_acf and m_acf0                }

        I5=^sacf;
        AX0=32;
        AR=AX0+AY1;                {e_acf0=32+(scalvad<<1)+(-normacf)}
        AR=AR+AY0, SI=DM(I5,M5);   {get sacf[0]}
        DM(e_acf0)=AR;
        SR=ASHIFT SI BY 3 (LO);    {m_acf0=sacf[0]<<3}
        DM(m_acf0)=SR0;
```

258

```
{Outputs: scalvad<<1=AY1, e_acf0=AR, sacf[0]=SI, I5=^sacf[1]}

{          Computation of e_pvad and m_pvad                    }

        AY0=14;
        AF=AR+AY0, MX1=SI;
        AX0=DM(normrvad);                    {normrvad is stored as -normvad}
        I0=^rvad;                            {AF will be e_pvad}
        AF=AX0+AF, MY1=DM(I0,M1);            {get rvad[0] ahead of time}
                                             {get rvad[1], sacf[1]}
        MR=MX1*MY1 (SS), MX0=DM(I0,M1);      {sacf[0]*rvad[0]}
        MY0=DM(I5,M5);
        SR=ASHIFT MR1 BY -1 (HI);            { >> 1}
        SR=SR OR LSHIFT MR0 BY -1 (LO);
        MR0=SR0;
        MR1=SR1;
        CNTR=7;
        DO compute_pvad UNTIL CE;
            MR=MR+MX0*MY0 (SS), MX0=DM(I0,M1);
compute_pvad:     MY0=DM(I5,M5);
        MR=MR+MX0*MY0 (SS);
        AR=PASS MR1, AY0=MR0;
        IF LT JUMP msw_le;                   {IF ms_temp>=0}
        AR=AR OR AY0;
        IF NE JUMP gt_zero;                  {THEN IF L_temp==0}

msw_le:  MR1=0;                              {THEN: L_temp=1}
        MR0=1;

gt_zero: SE=EXP MR1 (HI);                    {SE= -NORM(L_temp)}
        SE=EXP MR0 (LO);

        SR=NORM MR1 (HI);                    {L_temp<<normprod, use SR0}
        SR=SR OR NORM MR0 (LO), AR=SE;

        AR=AR+AF;                            {e_pvad-normprod}
        DM(e_pvad)=AR;
        DM(m_pvad)=SR1;

{Outputs: scalvad<<1=AY1}
{_____3.2_____ACF Averaging_____}

acf_average:AX0=-10;
        AR=AX0+AY1;                  {Note that SE is neg for >>}
        SE=AR;                       {so SE is -(10-scalvad<<1)}

{Outputs: scalvad<<1=AY1}
```

(listing continues on next page)

4 GSM Codec

```
{           computation of L_av0[0..8] and L_av1[0..8]          }
            L6=72;                      {Circular buffers for L_sav0}
            L3=54;                      {and L_sacf, restore afterwards}
            M2=17;                      {Skip forward 9, 8.5 longs}
            M3=-35;                     {Skip back 17, -17.5 longs}
            I4=^L_ACF;                  {Restore Ms and Ls after use!}
            I0=^L_av0;
            I1=^L_sacf;
            I3=DM(pt_sacf);             {These pointers are updated using}
            I6=DM(pt_sav0);             {automatic circular buffers}
            I5=^L_av1;
            CNTR=9;
            DIS AR_SAT;

                DO acf_sum UNTIL CE;
                SI=DM(I4,M5);                           {L_temp=L_ACF[i]>>scal}
                SR=ASHIFT SI (HI), SI=DM(I4,M5);
                SR=SR OR LSHIFT SI (LO), AY1=DM(I1,M1);    {Get L_sacf[i]}

                AY0=DM(I1,M2);
                AF=SR0+AY0,    AY0=DM(I1,M1);  {Get L_sacf[i+9]}
                AR=SR1+AY1+C,  AX0=DM(I1,M2);
                AF=AX0+AF,     AY1=DM(I1,M1);  {Get L_sacf[i+18]}
                AR=AR+AY0+C,   AX0=DM(I1,M3);  {and skip back 17.5 longs}
                AF=AX0+AF,     DM(I3,M1)=SR1;  {L_sacf[pt_sacf]=L_temp}
                AR=AR+AY1+C,   DM(I3,M1)=SR0;
                AX1=AR, AR=PASS AF;
                DM(I0,M1)=AX1;              {L_av0[i]=sum}
                DM(I0,M1)=AR;

                AX0=DM(I6,M5);             {L_av1[i]=L_sav0[pt_sav0+i]}
                DM(I5,M5)=AX0;
                AX0=DM(I6,M6);
                DM(I5,M5)=AX0;

                DM(I6,M5)=AX1;             {L_sav0[pt_sav0+i]=sum}
acf_sum:        DM(I6,M5)=AR;

            ENA AR_SAT;
            DM(pt_sacf)=I3;                {Update pointers}
            DM(pt_sav0)=I6;

            L6=0;                          {Restore DAG regs}
            L3=0;
            M2=-1;
            M3=2;
```

GSM Codec   4

```
{_____3.3_____Predictor Values Computation_____}

{        3.3.1 Schur recursion                          }

        AY1=^L_av1; {in DM}                {Set calling parameters}
        MY1=^sacf;
        M0=^vpar;    {in DM}               {M0 is reset to 0 in subroutine}
        CALL schur_routine;                {Located in 06.10}

{Outputs: none}

{        3.3.2 Step up to obtain aav1[0..8]            }

        I6=^L_coef;
        I4=^vpar;

        AR=0x2000;                         {MSW 16384<<15}
        DM(I6,M5)=AR;                      {ms_coef[0]=16384<<15}
        AR=PASS 0, SI=DM(I4,M5);           {Get vpar[1]}
        DM(I6,M5)=AR;                      {ls_coef[0]=0}
        SR=ASHIFT SI BY 14 (LO);          {L_coef[1]=vpar<<14}
        DM(I6,M5)=SR1, AR=PASS 1;          {Setup AR as counter}
        DM(I6,M5)=SR0;
        AY0=AR;

{Outputs: AY0=1, AR=m counter=1, I6=^L_coef[2], I4=^vpar[2]}

{        Loop on the LPC analysis order                }

        M3=-2;                             {Restore Ms after use}
        M6=2;
        I5=^L_coef+2;
        CNTR=7;                            {7,6,5,4,3,2,1}
        DO m_loop UNTIL CE;
           I0=^L_coef+2;
           I1=I5;                          {Index for m-i}
           I2=^L_work;
           MODIFY(I5,M6);                  {Modify for next time thru}
           SR0=DM(I4,M5);                  {Get vpar[m]}
           CNTR=AR;                        {Loop m-1 times}
           DO v_mac UNTIL CE;
                MR1=DM(I0,M1);             {MR=L_coef[i]}
                MR0=DM(I0,M1);
                MY0=DM(I1,M3);             {Get L_coef[m-i]>>16}
                MR=MR+SR0*MY0 (SS);        {ms_coef[m-i]*vpar[m]}
                IF MV SAT MR;              {Saturate may not be needed}
                DM(I2,M1)=MR1;             {L_work=...}
v_mac:          DM(I2,M1)=MR0;
```

(listing continues on next page)

261

4 GSM Codec

```
        I2=^L_work;                    {L_work starts at [1] not [0]}
        I0=^L_coef+2;
        CNTR=AR;                       {Loop m-1 times}
        DO copy_row UNTIL CE;
              AX0=DM(I2,M1);
              DM(I0,M1)=AX0;
              AX0=DM(I2,M1);
copy_row:          DM(I0,M1)=AX0;

        SR=ASHIFT SR0 BY 14 (LO);      {L_coef[m]=vpar[m]<<14}
        DM(I6,M5)=SR1;
m_loop: DM(I6,M5)=SR0, AR=AR+AY0;      {Increment m counter}

    M3=2;                              {Restore DAG}
    M6=-1;

{Outputs: none}

{       Keep the aav1[0..8] for next section                        }

    I0=^L_coef;
    I2=^r_a_av1;                       {aav1, rav1 and aav1 are shared}
    SE=-19;
    CNTR=9;
    DO shift_aav1 UNTIL CE;
        SI=DM(I0,M3);
        SR=ASHIFT SI (HI);
shift_aav1:     DM(I2,M1)=SR0;         {aav1[i]=L_coef[i]>>19}


{Outputs: none}

{       Computation of the rav1[0..8]                      }
        I2=^r_a_av1;                   {rav1 here}
        I3=^L_work;
        CNTR=9;
        DO i_loop UNTIL CE;
            I0=^r_a_av1;
            I1=I2;
            MR=0, MX0=DM(I2,M1);       {Modify I2 with dummy read}
            SI=CNTR;
            CNTR=SI;                   {Loop 8-i times}
            DO k_loop UNTIL CE;
                  MX0=DM(I0,M1);
                  MY0=DM(I1,M1);
k_loop:           MR=MR+MX0*MY0 (SS);  {Sum(aav1[k]*aav1[k+i])}
```

```
              DM(I3,M1)=MR1;                    {Save L_work[i]}
i_loop:       DM(I3,M1)=MR0;

        I3=^L_work;
        I0=^r_a_av1;

        AR=DM(I3,M1);                           {SE=-NORM(L_work[0])}
        SE=EXP AR (HI), SI=DM(I3,M2);
        SE=EXP SI (LO), AY0=SI;

        AR=AR OR AY0, AX0=SE;
        IF NE AR=PASS AX0;                      {IF L_work==0 THEN: AR=SE}
                                                {ELSE: AR=0}
        DM(normrav1)=AR;                            {Save -normrav1 for 3.6}
        SE=AR;                                  {Keep -normrav1 for 3.4}
        CNTR=9;
        DO norm_rav1 UNTIL CE;
            SI=DM(I3,M1);
            SR=NORM SI (HI), SI=DM(I3,M1);
            SR=SR OR NORM SI (LO);
norm_rav1:      DM(I0,M1)=SR1;                  {rav1[i]=L_work<<normrav1}

{Outputs: -normrav1=SE}

{_____3.4_____Spectral Comparison_____}

{       Renormalize L_av0[0..8]                                           }
        I0=^L_av0;
        I1=^sav0;
        CNTR=9;

        SR0=DM(I0,M1);
        AY0=DM(I0,M2);
        AR=SR0 OR AY0, AY1=SE;                  {Save -normrav1 in AY1}
        IF NE JUMP else_norm;                   {IF sav0==0}

        AR=4095;                                {THEN: sav0[i]=4095}
        DO init_sav0 UNTIL CE;
init_sav0:      DM(I1,M1)=AR;
    JUMP endif_L_av0;
```

(listing continues on next page)

4 GSM Codec

```
else_norm:  SE=EXP SR0 (HI), SI=AY0;        {SE=-shift=NORM(L_av0[0]}
            SE=EXP SI (LO);

            AY0=-3;
            AR=SE;
            AR=AY0-AR;                       {AR=shift-3}
            SE=AR;
            DO norm_av0 UNTIL CE;            {sav0[i]=(L_av0[i]<<shift-3)>>16}
                SI=DM(I0,M1);
                SR=ASHIFT SI (HI), SI=DM(I0,M1);
                SR=SR OR LSHIFT SI (LO);
norm_av0:           DM(I1,M1)=SR1;

{Outputs: -normav1=AY1}

{       Compute partial sum of dm                                    }

endif_L_av0:I0=^sav0+1;
            I1=^r_a_av1+1;
            MR=0;                            {L_sump=0}
            CNTR=8;
            DO sump UNTIL CE;
                MX0=DM(I0,M1);
                MY0=DM(I1,M1);
sump:               MR=MR+MX0*MY0 (SS);

{Outputs: -normav1=AY1, L_sump=MR}

{       Compute division of partial sum by sav0[0]                   }

            AF=PASS 0;
            AR=ABS MR1, AY0=MR1;             {Set AS flag on L_sump for later}
            IF POS JUMP sump_ge;             {IF L_sump<0}

            DIS AR_SAT;
            AR=AF-MR0;                        {THEN: Negate L_sump}
            ENA AR_SAT;
            MR0=AR, AR=AF-MR1+C-1;
            MR1=AR;

sump_ge: AR=MR0 OR AY0;                       {IF L_temp==0}
         IF NE JUMP sump_ne;

            SE=0;                             {THEN: shift=0}
            MR=0;                             {      L_dm=0}
            JUMP endif_sump;

sump_ne: SI=DM(sav0);
         SR=ASHIFT SI BY 3 (LO);             {AY0=sav0[0]<<3}

            SE=EXP MR1 (HI), AY0=SR0;        {SE=-shift}
            SE=EXP MR0 (LO);
```

```
        SR=NORM MR1 (HI);                   {temp=(L_temp<<shift)>>16}
        SR=SR OR NORM MR0 (LO);

        AF=SR1-AY0, AX0=AY0;                {IF sav0[0]>=temp}
        IF GT JUMP divshift_1;

divshift_0: AF=PASS SR1;                    {THEN: will do temp/sav0[0]}
        AX1=0;                              {       lsw of L_dm=0}
        JUMP endif_sav0;

divshift_1: AX1=32768;                      {ELSE: lsw of L_dm=32768}
                                            {       do (temp-sav0[0])/sav0[0]}
endif_sav0: CALL divide_routine;            {Do divide AY0=AF/AX0}

        AF=PASS 0;
        AX0=0;                              {L_dm+temp, do the <<1 later}
        DIS AR_SAT;
        AR=AX1+AY0;
        SR0=AR, AR=AX0+C;

        IF POS JUMP sump_pos;               {IF L_sump<0, set by abs earlier}

        SR1=AR, AR=AF-SR0;                  {THEN: -L_dm}
        SR0=AR, AR=AF-SR1+C-1;

{Outputs: -normav1=AY1}

{         Renormalization and final computation of L_dm          }

sump_pos:   SR=LSHIFT SR0 BY 15 (LO);       {L_dm<<14+1, do the <<1 here}
            SR=SR OR ASHIFT AR BY 15 (HI);
            AR=SR1, SR=LSHIFT SR0 (LO);     {L_dm=L_dm>>shift}
            SR=SR OR ASHIFT AR (HI);
            MR0=SR0;
            MR1=SR1;

endif_sump: MX0=DM(r_a_av1);                {L_dm+(rav1[0]<<11) with sat}
            MY0=0x0400;                     {For <<11=2^(11-1) and DP add}
            MR=MR+MX0*MY0 (SS), SE=AY1;     {SE=-normav1}
            IF MV SAT MR;                   {Saturate L_dm just in case}

            SR=LSHIFT MR0 (LO);             {L_dm>>normrav1}
            SR=SR OR ASHIFT MR1 (HI);
```

(listing continues on next page)

4 GSM Codec

```
{Outputs: L_dm=SR}

{        Compute difference and save L_dm                      }
         I0=^L_lastdm+1;
         AY0=DM(I0,M2);
         AR=SR0-AY0, AY1=DM(I0,M0);      {L_temp=L_dm-L_lastdm}
         ENA AR_SAT;
         AX0=AR, AR=SR1-AY1+C-1;
         DIS AR_SAT;

         IF NOT AV JUMP exit_sat;        {IF overflow}

         AX0=0x0000;                     {THEN: saturate temp}
         IF LT JUMP exit_sat;            {IF >=0}
         AX0=0xFFFF;                     {THEN: saturate -full scale}

exit_sat:  DM(I0,M1)=SR1;               {L_lastdm=L_dm}
         DM(I0,M0)=SR0;

         IF GE JUMP temp_ge;            {IF L_temp<0}

         AX1=AR, AR=AF-AX0;            {THEN: -L_temp}
         AX0=AR, AR=AF-AX1+C-1;       {Can not overflow}

{Outputs: L_temp=AR AX0}

{        Evaluation of the stat flag                           }

temp_ge: AY0=3277;                      {L_temp-3277}
         AX1=AR, AR=AX0-AY0;
         ENA AR_SAT;
         AR=AX1-AF+C-1;

         IF GE AR=PASS 0;               {IF L_temp>=0,THEN: stat=0}
         IF LT AR=PASS 1;               {            ELSE: stat=1}
         DM(stat)=AR;

{Outputs: none}

{_____3.5_____Periodicity detection_____}

         AX0 = DM(oldlagcount);
         AY0 = DM(veryoldlagcount);
         AX1 = 4;
         AR  = 0;                        {AR = ptch = 0}
         AF  = AX0 + AY0;
         AF  = AF - AX1;                 {AF = temp - 4}
         IF GE AR = PASS 1;              {IF GE ptch = 1}
         DM(ptch) = AR;
```

```
{Outputs: none}

{_____3.6_____Threshold adaption_____}

{         Test to find if acf0 < pth                              }

         MR0 = 20;                        {MR0 = E_PLEV}
         MR1 = 25000;                     {MR1 = M_PLEV}
         AX0 = DM(e_acf0);
         AY0 = 19;                        {AY0 = E_PTH}
         AR  = AX0 - AY0;
         AR  = PASS AR;
         IF LT JUMP set_thvad;
         IF GT JUMP test_adapt;
         AX0 = DM(m_acf0);
         AY0 = 18750;                     {AY0 = M_PTH}
         AF  = AX0 - AY0;
         IF GE JUMP test_adapt;
set_thvad:  DM(e_thvad) = MR0;            {comp = 1}
         DM(m_thvad) = MR1;
         JUMP vvad_decision;              {jump to section 3.7}

{         Test to find if adaptation is needed                    }

test_adapt: AX0 = DM(ptch);              {comp = 0}
         AY0 = DM(stat);
         MR  = 0;
         AF  = PASS AX0;
         IF NE JUMP clr_adaptcount;
         AF  = PASS AY0;
         IF NE JUMP inc_adaptcount;
clr_adaptcount: DM(adaptcount) = MR0;    {comp = 1}
         JUMP vvad_decision;             {jump to section 3.7}

{         Increment adaptcount                                    }

inc_adaptcount: AY0 = DM(adaptcount);    {comp = 0}
         AY1 = 8;
         AR  = AY0 + 1;
         DM(adaptcount) = AR;
         AF  = AR - AY1;                 {AF = adaptcount - 8}
         IF LE JUMP vvad_decision;       {jump to section 3.7}
```

(listing continues on next page)

4   GSM Codec

```
{        Compute (thvad - thvad/dec)                                    }

         SE   = -5;
         AY1  = 16384;
         SI   = DM(m_thvad);
         SR   = ASHIFT SI (HI), AY0 = SI;
         AR   = AY0 - SR1;                     {AR=m_thvad - (m_thvad>>5) }
         AY0  = DM(e_thvad);
         AF   = AR - AY1, SR1 = AR;            {AF=m_thvad-16384, SR1=m_thvad}
         SE   = 1;
         IF LT SR = ASHIFT SR1 (HI);
         AR   = AY0;
         SI   = SR1;                           {SI = m_thvad}
         IF LT AR = AY0 - 1;                   {AR = e_thvad}

{outputs: m_thvad=SR1,SI;e_thvad=AR;}

{        Compute (pvad * fac)                               }

         SE   = -2;                            {shift >> 1 and format adjust}
         MX0  = 3;
         MY0  = DM(m_pvad);
         AY1  = DM(e_pvad);
         MR   = MX0 * MY0 (SS), AY0 = SI;   {AY0 = m_thvad}
         SR   = LSHIFT MR0 (LO), MR0 = AR;  {MR0 = e_thvad}
         AR   = AY1 + 1;                       {AR = e_temp}
         SR   = SR OR ASHIFT MR1 (HI), AY1 = AR;   {SR=L_temp, AY1=e_temp}
         AF   = PASS SR0;                      {L_temp can overflow 1 bit max}
         IF GE JUMP test_thvad;
         SR   = LSHIFT SR0 BY -1 (LO);         {SR0 = m_temp}
         AR   = AY1 + 1;                       {AR=e_temp}

{outputs:m_thvad=AY0,SI;e_thvad=MR0;m_pvad=MY0;m_temp=SR0;e_temp=AR}

{         Test to find if (thvad < pvad*fac)                        }

test_thvad: AY1 = MR0;                         {AY1 = e_thvad}
         MR1 = AR;                             {MR1=e_temp}
         AF   = AY1 - AR, AX0 = SR0;           {AF=e_thvad-e_temp}
         IF LT JUMP compute_min;
         IF GT JUMP pvad_margin;

         AF   = AY0 - SR0;                     {AF=m_thvad-m_temp}
         IF GE JUMP pvad_margin;

{outputs:m_temp=SR0,AX0;e_temp=AR,MR1;m_thvad=AY0,SI;e_thvad=MR0,AY1;m_pvad=MY0}

{        Compute minimum [comp=1]                                   }
```

```
compute_min:SR  = ASHIFT SI BY -4 (HI);        {SR1=m_thvad >> 4}
        DIS AR_SAT;
        AR  = SR1 + AY0;                       {AR = L_temp}
        ENA AR_SAT;
        AY0 = AR;
        IF NOT AV JUMP update_m_thvad;
        SR  = LSHIFT AR BY -1 (HI);            {SR1 = L_temp >> 1}
        AR  = AY1 + 1, AY0 = SR1;              {AR=ethvad+1,AY0=mthvad}
        AY1 = AR;                              {AY1 = e_thvad}
update_m_thvad: AF  = MR1 - AY1;               {AF = e_temp - e_thvad}
        IF GT JUMP pvad_margin;
        IF LT JUMP update_e_m;
        AF  = AX0 - AY0;                       {AF = m_temp - m_thvad}
        IF GE JUMP pvad_margin;
update_e_m: AY1 = MR1;                         {comp2=1, AY1 = e_thvad}
        AY0 = AX0;                             {AY0 = m_thvad}

{outputs:e_thvad=AY1; m_thvad=AY0; m_pvad=MY0}

{        Compute (pvad + margin) [comp=0,comp2=0]                    }

pvad_margin:DM(e_thvad) = AY1;
        DM(m_thvad) = AY0;
        AY0 = DM(e_pvad);
        MR1 = 19531;                           {MR1 = M_MARGIN}
        MR0 = 27;                              {MR0 = E_MARGIN}
        AR  = MR0 - AY0, AY1 = MY0;            {AR=E_MARGIN-e_pvad, AY1=m_pvad}
        IF EQ JUMP epvad_eq;
        IF LT JUMP epvad_greater;
swap_values: AR  = -AR, AX0 = AY1;             {MR1 = m_pvad}
        AY0 = MR0;                             {AY0 = E_MARGIN}
        AY1 = MR1;                             {AY1 = M_MARGIN}
        MR1 = AX0;
epvad_greater: SE  = AR;                       {AR  = -temp}
        SR  = ASHIFT MR1 (HI);                 {SR1 = temp}
        DIS AR_SAT;
        AR  = SR1 + AY1;                       {AR  = L_temp}
        ENA AR_SAT;
        SR1 = AR;                              {SR1 = m_temp}
        SE  = -1;
        IF AV SR = LSHIFT AR (HI);             {m_pvad > 0 always}
        AR  = AY0;
        IF AV AR = AY0 + 1;                    {AR  = e_temp}
        JUMP test_for_greater;
epvad_eq:   DIS AR_SAT;
        AR  = MR1 + AY1;                       {AR = m_pvad + M_MARGIN}
        ENA AR_SAT;
        SR  = LSHIFT AR BY -1 (HI);            {SR1 = m_temp}
        AR  = AY0 + 1;                         {AR  = e_temp}
```

(listing continues on next page)

4 GSM Codec

```
{outputs: m_temp=SR1; e_temp=AR}

{       Test to find if (thvad > (pvad+margin))                    }

test_for_greater:
        AY0 = DM(e_thvad);
        AY1 = DM(m_thvad);
        AF  = AY0 - AR;                 {AF = e_thvad-e_temp}
        IF GT JUMP update_thvad;
        IF LT JUMP update_rvad;
        AF  = AY1 - SR1;                {AF = m_thvad-m_temp}
        IF LE JUMP update_rvad;
update_thvad:DM(e_thvad) = AR;          {comp = 1}
        DM(m_thvad) = SR1;

{outputs: NONE}

{       Initialize new rvad                                        }

update_rvad:MX0 = DM(normrav1);         {comp = 0}
        DM(normrvad) = MX0;
        I0  = ^rvad;
        I1  = ^r_a_av1;                 {rav1, shared by rav1 and aav1}
        CNTR = 9;
        DO write_rvad UNTIL CE;
            MX0 = DM(I1,M1);
write_rvad:     DM(I0,M1) = MX0;

{outputs: NONE}

{       Set adaptcount                                             }

        MX0 = 9;
        DM(adaptcount) = MX0;

{_____3.7_____VAD decision_____}

vvad_decision:  AY0 = DM(e_pvad);
        AY1 = DM(m_pvad);
        AX0 = DM(e_thvad);
        AX1 = DM(m_thvad);
        AR  = AY0 - AX0;
        IF EQ AR = AY1 - AX1;
        AR  = PASS AR;
        AR  = 0;
        IF GT AR = PASS 1;

{outputs: vvad=AR}
```

```
{_____3.8_____VAD hangover decision_____}

        AY1 = DM(hangcount);
        AY0 = DM(burstcount);
        AX0 = AR,  AR  = PASS 0;            {AX0 = vvad}
        AF  = PASS AX0;
        IF NE AR = AY0 + 1;                 {AR = burstcount}
        MR1 = 5;
        AY0 = 3;
        AF  = AR - AY0;
        IF GE AR = PASS AY0;                {AR  = burstcount}
        DM(burstcount) = AR;
        AF  = PASS AF, AR = AY1;
        IF GE AR = PASS MR1;
        AF  = ABS AR, AY1 = AR;
        IF POS AR = AY1 - 1;
        MR1 = AR, AR = PASS AX0;            {MR1 = hangcount}
        IF POS AR = PASS 1;                 {AR = vad}
        DM(hangcount) = MR1;
        DM(vad) = AR;
        RTS;                                {Return to Main Speech transcoder}

{outputs: NONE}

{_____3.9_____Periodicity updating_____}

update_periodicity:
        AR  = 0;                           {lagcount = 0}
        AY0 = DM(oldlag);
        I1  = ^lags;
        CNTR = 4;
        DO update_lagcount UNTIL CE;
            AX1 = DM(I1,M1);               {AX1=lags[i],AF=oldlag-lags[i],}
            AF  = AY0 - AX1, AY1 = AR;      {AY1=lagcount}
            IF GT JUMP case_1;
case_2:     AR  = PASS AX1;
            JUMP find_smallag;
case_1:     AR = PASS AY0, AY0 = AX1;      {AY0 = minlag, AR = maxlag}
find_smallag:    CNTR = 3;                 {AR = smallag}
            DO compute_smallag UNTIL CE;
                AF  = AR - AY0;
```

(listing continues on next page)

4 GSM Codec

```
compute_smallag:         IF GE AR = PASS AF;      {AR = smallag}
          AF  = AY0 - AR;                         {AF = temp}
          AF  = AF - AR;                          {AF = temp - smallag}
          IF LT AR = AY0 - AR;
          AY0 = 2;
          AF  = AR - AY0, AR = AY1;
          IF LT AR = AY1 + 1;                     {AR=lagcount}
update_lagcount:AY0 = AX1;                        {AY0=oldlag}
        DM(oldlag) = AY0;
        AX0 = DM(oldlagcount);
        DM(oldlagcount) = AR;
        DM(veryoldlagcount) = AX0;
        RTS;                                      {Return to main speech transcoder}
.ENDMOD;
```

Listing 4.3 Voice Activity Detection Routine (GSM0632.DSP)

GSM Codec   4

```
{_____
GSM_SID.DSP
            Analog Devices Inc.  DSP Division
            One Technology Way, Norwood, MA, 02062
            DSP Applications: (617) 461-3672

   This code generates comfort noise as specified in GSM recommendation
   6.31, section 3.1. Interpolation of the generated values over
   several frames is not implemented.

   This subroutine is called from the dmr_decode routine when the
   frame to be decoded contains comfort noise parameters (silence
   descriptor frame). The frame of coefficients is over-written
   with the necessary LTP gain and lag values, and the pseudo-randomly
   generated grid position and RPE pulses, for each subwindow. The
   program then returns this properly formatted comfort noise frame
   for normal decoding.

   The pseudo-random number generator is adapted from the one found in
   Analog Devices DSP Applications Handbook 1, section 4.6.

   The pseudo-random number generator is also used by the substitution
   and muting sections of GSM_DTX.DSP.

   ADSP-2101 Execution cycles:  379 maximum

Release History:
___Date___  _Ver_ _____Comments_____
24-Aug-89   57    Incorporated random number generator
10-Jan-90   1.00  Second Release
01-Nov-90   2.00  Third release
_____}

.MODULE           Generate_Comfort_Noise;

.ENTRY            comfort_noise_generator, make_random;

.VAR/DM/RAM       seed_lsw, seed_msw;

.GLOBAL           seed_lsw, seed_msw;

{*****  This code generates comfort noise as specified in GSM recommendation
   6.31, section 3.1.  Interpolation of the generated values over several
   frames is not implemented. This code can be further optimized for
   the ADSP-2101. *****}
```

(listing continues on next page)

4 GSM Codec

```
comfort_noise_generator:

        M3  = 8;                         {I1 holds pointer to coeff}
        MODIFY(I1,M3);                   {Skip LAR values}
        M3  = 2;                         {Reset M3}

        MX0 = 40;
        MX1 = 120;                       {Constants to write to buffer}
        MY1 = 25;                        {Upper half of a}
        AX0 = 26125;                     {Lower half of a}
        SE  = -1;
        SR0 = DM(seed_lsw);
        SR1 = DM(seed_msw);

{For random numbers in the range:    0 to 3     AX1 = 0, MY0 = 2
                                     1 to 6     AX1 = 1, MY0 = 3}

        CNTR = 2;
        DO cn_update UNTIL CE;

          DM(I1,M1) = MX0;               {LTP lag (Ncr) }
          AR  = PASS 0;
          DM(I1,M1) = AR;                {LTP gain (bcr) }

          AX1 = 0;
          MY0 = 2;
          CNTR = 1;
          CALL make_random;             {RPE grid position (Mcr) }

          MODIFY(I1,M1);                 {skip block amplitude (Xmaxcr) }

          AX1 = 1;
          MY0 = 3;
          CNTR = 13;
          CALL make_random;             {RPE pulses 1 to 13 (Xmcr) }

          DM(I1,M1) = MX1;               {LTP lag (Ncr) }
          AR  = PASS 0;
          DM(I1,M1) = AR;                {LTP gain (bcr) }

          AX1 = 0;
          MY0 = 2;
          CNTR = 1;
          CALL make_random;             {RPE grid position (Mcr) }

          MODIFY(I1,M1);                 {skip block amplitude (Xmaxcr) }

          AX1 = 1;
          MY0 = 3;
          CNTR = 13;
          CALL make_random;             {RPE pulses 1 to 13 (Xmcr) }
```

```
cn_update:          DM(seed_lsw) = SR0;
         DM(seed_msw) = SR1;

         RTS;

make_random:DO gen_random UNTIL CE;
            MR  = SR1 * MY0 (UU);                    {Scale the seed}
            AY0 = MY0;
            AY1 = MR1;                               {Scaled seed in AY1}
            MR  = SR0 * MY1 (UU), MY0 = AX0;         {MR = x(lo) * a(hi)}
            MR  = MR + SR1 * MY0 (UU);               {MR = MR + x(hi)*a(lo)}
            AR  = PASS MR1, MR1 = MR0;
            MR2 = AR, AR  = AX1 + AY1;               {Offset the scaled seed}
            MR0 = H#FFFE;
            MR  = MR + SR0 * MY0 (UU), DM(I1,M1)=AR;  {MR=MR+x(lo)*a(lo)}
            SR  = ASHIFT MR2 BY 15 (HI);
            SR  = SR OR LSHIFT MR1 (HI);
gen_random:     SR  = SR OR LSHIFT MR0 (LO), MY0 = AY0;

         RTS;

.ENDMOD;
```

Listing 4.4 Comfort Noise Insertion Routine (GSM_SID.DSP)

4   GSM Codec

```
{_____

GSM_DTX.DSP
             Analog Devices Inc.  DSP Division
             One Technology Way, Norwood, MA  02062
             DSP Applications: (617) 461-3672

   This module contains routines for decoding a codeword that precedes the
   76 coefficients, classifying the frame, performing substitution and
   muting if necessary, and preparing the coefficients for decoding.

   The code is to be executed after the coefficient transfer is complete.
   It assumes that the coefficient buffer was overwritten only with
   GOOD SPEECH or VALID SID parameters. The code executes in the primary
   register set, before the dmr_decode routine is called.

   The 2-bit codeword classifies the frame as follows:
             00 — frame contains speech
             01 — unusable frame
             10 — frame contains valid comfort noise parameters (silence
                  descriptor (SID) frame)
             11 — invalid silence descriptor frame - substitute with previous
                  valid silence descriptor frame

   ADSP-2101 Computation Time:                      199 cycles maximum.

             state:                      max. cycles
             Good speech                     15
             Valid silence frame             39
             Invalid silence frame           42
             Unusable frame                  199

Release History:
__Date____  _Ver_ _____Comments_____
01-Nov-89   67    Initial implementation
10-Jan-90   1.00  Second Release
01-Nov-90   2.00  Third release

_____}


.MODULE          dtx_routine;
.VAR/PM/RAM/CIRC sil_fram_subwin[17];  { silence frame coeffs (06.11)}
.VAR/PM/RAM      sil_fram_lar[8];      { silence frame coeffs (06.11)}
.VAR/DM/RAM      valid_sid_buffer[9];  { valid coeffs from prior SID}
.VAR/DM/RAM      sub_n_mute;           { flag}
.VAR/DM/RAM      sid_inbuf;            { flag}
.VAR/DM/RAM      taf_count;            { counts frames between valid SID
                                         coeffs during Comfort Noise
                                         Insert}
```

GSM Codec   4

```
.EXTERNAL          make_random;
.EXTERNAL          seed_lsw, seed_msw;

.GLOBAL            sid_inbuf;
.GLOBAL            valid_sid_buffer;
.GLOBAL            sub_n_mute;
.GLOBAL            taf_count;

.ENTRY             decode_codeword;

{these are constants located in program memory ROM}
.INIT      sil_fram_subwin : H#2800, 0, H#100, 0, H#300, H#400, H#300,
                             H#400, H#400, H#300, H#300, H#300, H#300,
                             H#400, H#400, H#300, H#300;
                             {40, 0, 1, 0, 3, 4, 3, 4,
                             4, 3, 3, 3, 3, 4, 4, 3, 3;}
.INIT      sil_fram_lar :    H#2A00, H#2700, H#1500, H#A00, H#900,
                             H#400, H#300, H#200;
                             {42, 39, 21, 10, 9, 4, 3, 2;}

decode_codeword:I0  = ^valid_sid_buffer;
         AY0 = 2;
         MX1 = 1;
         MX0 = -24;
         MY0 = 0;

         AF  = PASS 1, AX0 = DM(I1,M1);   {AX0 = codeword}
         I4  = I1;                        {I4 is working pointer, save I1}
         AR  = AX0 AND AF;
         IF NE JUMP not_good_frame;

good_frame: DM(sub_n_mute) = MY0;
         DM(taf_count) = MX0;

         AR  = AX0 AND AY0;
         IF NE AR = PASS 1;

valid_sid:  DM(sid_inbuf) = AR;
         IF EQ RTS;                       {If good speech, return}
         CNTR = 8;
         M7  = 3;
         DO fill_valid_sid UNTIL CE;
             AR  = DM(I4,M5);
fill_valid_sid:   DM(I0,M1) = AR;         { save LAR values}

         MODIFY (I4,M7);
         M7  = 0;
         AR  = DM(I4,M4);
         DM(I0,M0) = AR;                  { save xmax value}
         RTS;
```

(listing continues on next page)

277

4 GSM Codec

```
not_good_frame: AR  = AX0 AND AY0;           { At this point, either UNUSABLE or}
        IF NE JUMP invalid_sid;              { INVALID SID frame}

unusable_frame: AX0 = DM(sub_n_mute);
        AX1 = DM(sid_inbuf);
        AF  = PASS AX0;
        IF NE JUMP check_xmax;        {JUMP if NOT first consecutive UNUSABLE}

        AF  = PASS AX1;
        IF EQ JUMP set_subnmut;       {JUMP if not generating comfort noise}

        AY0 = DM(taf_count);
        AF  = PASS AY0;
        IF LE JUMP inc_taf;           {JUMP if waiting for VALID SID frame}

set_subnmut:DM(sub_n_mute) = MX1;
        RTS;

inc_taf: AR  = AY0 + 1;
        DM(taf_count) = AR;
        RTS;

check_xmax: AF  = PASS 0;            { substitution and muting begins}
        M7  = 11;
        MODIFY (I4,M7);             { set pointer to xmax[1]}
        M7  = 17;
        AY0 = 4;
        CNTR = 4;
        DO dec_xmax UNTIL CE;
            AX0 = DM(I4,M4);
            AR  = AX0 - AY0;          { decrement xmax by 4}
            IF GE AF = PASS 1;
            IF LT AR = PASS 0;        { set minimum}
dec_xmax:         DM(I4,M7) = AR;    { write xmax}

        AR  = PASS AF;
        IF NE JUMP not_sil_frame;

writ_sil_frame:DM(sid_inbuf) = AR;  { if all four xmax < 4, insert silence}
        I0  = I1;
        I4  = ^sil_fram_lar;
        CNTR = 8;
        DO writ_sil_lar UNTIL CE;
            AR = PM(I4,M5);
writ_sil_lar:     DM(I0,M1) = AR;
        I4  = ^sil_fram_subwin;
        CNTR = 68;
        L4  = 17;
        DO writ_sil_subwin UNTIL CE;
            AR = PM(I4,M5);
```

```
writ_sil_subwin:  DM(I0,M1) = AR;
        L4  = 0;
        RTS;

not_sil_frame:  AR  = PASS AX1;              { AX1 = sid_inbuf}
        IF NE RTS;                           { if generating comfort noise, grid
                                               position determined elsewhere}

        I4  = I1;
        M1  = 10;                            { set-up}
        MODIFY (I1,M1);
        AX1 = 0;
        MY0 = 2;
        M1  = 17;

        SR0 = DM(seed_lsw);
        SR1 = DM(seed_msw);
        SE  = -1;
        MY1 = 25;
        AX0 = 26125;

        CNTR = 4;
        CALL make_random;
        M1  = 1;
        I1  = I4;
        RTS;

invalid_sid:DM(sub_n_mute) = MY0;            { frame contains INVALID SID parameters}
        DM(taf_count) = MX0;
        DM(sid_inbuf) = MX1;

        CNTR = 8;
        M7  = 3;
        DO writ_valid_sid UNTIL CE;          { replace 8 LARs with previous}
            AR  = DM(I0,M1);                 { valid values}
writ_valid_sid:  DM(I4,M5) = AR;
        MODIFY (I4,M7);
        M7  = 17;
        AR  = DM(I0,M0);
        DM(I4,M7) = AR;                      { replace xmax with previous}
        DM(I4,M7) = AR;                      { valid values}
        DM(I4,M7) = AR;
        DM(I4,M4) = AR;
        M7  = 2;
        RTS;

.ENDMOD;
```

Listing 4.5 Discontinuous Transmission Routine (GSM_DTX.DSP)

4 GSM Codec

{_____

DMR21xx.DSP

                    Analog Devices Inc. DSP Division
                    One Technology Way, Norwood, MA 02062
                    DSP Applications: (617) 461-3672

This module is a data acquisition shell for the digital mobile radio (GSM) speech processing functions, running on the ADSP-2101 or ADSP-2111 EZ_LAB. Sound from the microphone input is processed and echoed back to the speaker output.

The interrupt IRQ2 controls the state of the demonstration. There are five states, as follows:

State 0 — input is output directly in a talk-thru mode
         - no encoding, decoding, etc. take place
         - the voice activity flag is disabled

State 1 — speech is encoded and decoded in a talk-thru mode
         - This mode demonstrates the need for comfort noise
           insertion. The intelligibility of speech in a noisy
           background is reduced.
         - frames are encoded as speech or as comfort noise,
           dependent on the speech flag
         - frames are decoded as speech if the speech flag is
           active, otherwise output is muted
         - the voice activity flag is determined for each frame

State 2 — speech is encoded and decoded in a talk-thru mode
         - This mode is the normal operation of the GSM system.
         - frames are encoded and decoded as speech or as comfort
           noise, dependent on the speech flag
         - the voice activity flag is determined for each frame

State 3 — input is encoded and decoded in an example mode
         - each frame is encoded and decoded as a comfort noise
           (silence descriptor) frame
         - the voice activity flag is forced inactive

State 4 — continuously decodes the last valid silence descriptor frame
           (comfort noise insertion)
         - the voice activity flag is forced inactive

These five states are cycled through, entering the next state after an IRQ2 interrupt. State 0 is the initial state after reset.

In contrasting states 1 and 2, it is helpful to have a random noise source available to mix with the microphone input. This will show the adaptation of the voice activity detection threshold, and the loss of

280

intelligibility in state 1 compared to state 2 in a noisy environment. The muting in state 1 occurs immediately, unlike the gradual muting specified by GSM (which can take up to 320 ms). The code for immediate muting is added with the -Ddemo switch.

The FLAG_OUT signal of the ADSP-2101 or ADSP-2111 EZ_LAB board is configured to output the state of the Voice Activity Detector flag in states 1 and 2. A high output (LED on) signals that voice activity has been detected. This will not work when FLAG_OUT is used to control an AD28msp02.

This implementation allows serial port 0 to accept either 8-bit u-law or 16-bit linear data input, based on a C preprocessor switch. The u-law hardware companding is used with the codec provided on the EZ_LAB board. A 16-bit linear format is used with an AD28msp02 daughterboard plugged into the codec socket. The default format is 8-bit u-law.

This routine takes full advantage of the integration on the ADSP-2101 and ADSP-2111. It makes use of the IDLE function while waiting for the next frame of data. The transfer of the transmit/receive speech buffer takes place over serial port 0, using index register I7. If using the u-law codec, this is an autobuffered transfer. In order for the receive and transmit autobuffering to function synchronously, THIS IMPLEMENTATION REQUIRES RFS0 and TFS0 TO BE WIRED TOGETHER EXTERNALLY WHEN USING THE u-LAW CODEC. If an AD28msp02 is being used, autobuffering is NOT used. THIS IMPLEMENTATION REQUIRES RFS0 and TFS0 TO BE SEPARATE WHEN USING THE AD28msp02.

The Data Address Generator 2 registers I7, L7, M4, and M5 should NEVER, NEVER be altered in any routine. They are reserved for input and output data buffering, controlled by this shell program.

Release History:

Date	Ver	Comments
20-Jun-89	56	Initial release.
04-Jan-90	84	add routine for testing VAD - waiting for vectors
10-Jan-90	1.00	Second release
01-Nov-90	2.00	Third release - added 2111 and 28msp02 capability

Assembler Preprocessor Switches

-cp switch	must always be used when assembling
-Ddemo switch	enables functions necessary for the five-state demonstration
-Dtestvad	includes code to format coefficients for VAD and SP_FLAG testing
-Dadsp2111	must be used if running code on the ADSP-2111 microcomputer (default is ADSP-2101)

(listing continues on next page)

4 GSM Codec

```
    -Dmsp02                    changes incoming data format to 16 bit linear for
                               AD28msp02, disables autobuffering (default
                               is u-law codec, autobuffering enabled)
```

```
_____}
.MODULE/ABS=0      LPC_Codec_Shell;
.VAR/DM/RAM/CIRC   coeff_codeword, coeff_buffer[76];
                                        {Buffer for coeffs, codeword}
.VAR/DM/RAM/CIRC   speech_1[160];
.VAR/DM/RAM/CIRC   speech_2[160];           {Speech windows}

{_____Conditional Assembly_____}
{ use (asm21 -cp -Ddemo) for demonstration }

#ifdef       demo

.VAR/PM/RAM demo_codes[5];                    {codes for demonstration only}
.INIT       demo_codes: H#C00000, H#100000, H#200000,
                H#020100, H#030100;

#endif
{_____}
.ENTRY       start_dmr;

.EXTERNAL    dmr_encode, dmr_decode;
.EXTERNAL    reset_codec, decode_codeword;

.EXTERNAL    vad;
.EXTERNAL    sid_inbuf;

.EXTERNAL    sp_flag;
.EXTERNAL    taf_count;                       {temporary - for demonstration}

.GLOBAL      speech_1;
.GLOBAL      speech_2;
.GLOBAL      coeff_codeword;

reset_vector:      JUMP start_dmr; NOP; NOP; NOP;

{_____Conditional Assembly_____}
{ use (asm21 -cp -Ddemo) for demonstration }
#ifdef       demo
irq2:        JUMP next_demo; NOP; NOP; NOP;
#else
irq2:        RTI; NOP; NOP; NOP;
#endif
{_____}
```

```
{.....................Conditional Assembly..............................}
{ use (asm21 -cp -Dadsp2111) for use with ADSP-2111 }
#ifdef adsp2111
hipw:      NOP; NOP; NOP; NOP;
hipr:      NOP; NOP; NOP; NOP;
#endif {................................................................}

trans0:    RTI; NOP; NOP; NOP;

{.....................Conditional Assembly..............................}
{ use (asm21 -cp -Dmsp02) for use with AD28msp02 }
#ifdef msp02
recv0:     JUMP sample; NOP; NOP; NOP;
#else
recv0:     RTI; NOP; NOP; NOP;
#endif {................................................................}

trans1:    NOP; NOP; NOP; NOP;
revc1:     NOP; NOP; NOP; NOP;
timer_int: NOP; NOP; NOP; NOP;

start_dmr: ICNTL=B#10100;
           L0=0;    L1=0;    L2=0;    L3=0;
           L4=0;    L5=0;    L6=0;    L7=160;
           M0=0;    M1=1;    M2=-1;   M3=2;
           M4=0;    M5=1;    M6=-1;   M7=0;

           CALL reset_codec;

reg_setup: AX0 = 0;
           DM(0X3FFE) = AX0;                        { DM wait states }

{.....................Conditional Assembly..............................}
{ use (asm21 -cp -Dmsp02) for use with AD28msp02 }
#ifdef msp02

   { initialize 28msp02 - assumes 21xx rfs0, tfs0 separate }
         RESET FLAG_OUT;       { connected to data/~cntl }
         AX0 = 0x2A0F;         { ext sclk, ext rfs, int tfs}
         DM(0x3FF6) = AX0;     { control reg0 }
         AX0 = 0x1000;         { enable serial port0, keep flagout }
         DM(0x3FFF) = AX0;     { system control reg }

         IMASK = 0x10;

         AX0 = 0x20;           { ******* PWDD is inverted in early 28msp02 }
         TX0 = AX0;            { write control word to 28msp02 }
         IDLE;
         AX0 = 0x7C20;
         TX0 = AX0;            { write non-control word to 28msp02 }
         IDLE;
```

(listing continues on next page)

4   GSM Codec

```
        IMASK = 0;
        SET FLAG_OUT;           { connected to data/~cntl }
        AX0 = 0x0000;           { disable serial port0 }
        DM(0x3FFF) = AX0;       { system control reg }

#else

        AX0 = 2;
        DM(0X3FF5) = AX0;       { sclkdiv0 }
        AX0 = 255;
        DM(0X3FF4) = AX0;       { rfsdiv0 }
        AX0 = 0x6927;           { int sclk, int rfs, ext tfs }
        DM(0X3FF6) = AX0;       { control reg0 }
        AX0 = 0X0E77;
        DM(0x3FF3) = AX0;       { autobuffer reg0 }

#endif
{...................................................................................}

        I7=^speech_1;           { I7 is speech buffer pointer }

        AX0 = 0X1000;
        DM(0x3FFF) = AX0;       { system control reg }

{_____Conditional Assembly_____}
{ use (asm21 -cp -Ddemo) for demonstration - sets values for state 0}
#ifdef demo
        ENA SEC_REG;
        MR1 = 3; MR0 = 0; MY1 = 0; MY0 = 0; MX1 = 0; SI = 0;
        DIS SEC_REG;
#endif
{_____}

{......................Conditional Assembly................................}
{ use (asm21 -cp -Dadsp2111) for use with ADSP-2111 }
#ifdef adsp2111
        IMASK=0x88;
#else
        IMASK=0x28;
#endif
{...................................................................................}

{......................Conditional Assembly................................}
{ use (asm21 -cp -Dmsp02) for use with AD28msp02 }
#ifdef msp02
        ENA SEC_REG;
        MX0 = 0;                { reset sample counter }
        AX1 = 160;              { length of sample buffers speech_1,2 }
```

GSM Codec 4

```
code_1_loop:IDLE;                    { wait for next sample }
        AY1 = MX0;
        AR  = AX1 - AY1;             { check if buffer is full }
        IF NE JUMP code_1_loop;

        MX0 = 0;                     { buffer full, reset sample counter }
        DIS SEC_REG;
#else

code_1_loop:IDLE;                    { autobuffering counts samples }
#endif {.........................................................................}

        I7=^speech_2;                { swap speech output/input buffer }

{_____Conditional Assembly_____}
{ use (asm21 -cp -Ddemo) for demonstration }
#ifdef demo
        ENA SEC_REG;
        AF = PASS MR1;
        IF NE JUMP CODE_2_LOOP;
        M7 = MX1;
        DIS SEC_REG;
#endif {_____}

do_dmr_1:
{......................Conditional Assembly..............................}
{ use (asm21 -cp -Dmsp02) for use with AD28msp02 }
#ifndef msp02
        SE  = 2;                     { left-justify expanded u-law input }
        I0  = ^speech_1;
        CALL scale_routine;
#endif {.........................................................................}

        I0=^speech_1;
        I1=^coeff_buffer;
        CALL dmr_encode;

{_____Conditional Assembly_____}
{ use (asm21 -cp -Ddemo) for demonstration }
#ifdef demo
#ifndef msp02

        CALL vad_out;
#endif
#endif
```

(listing continues on next page)

4 GSM Codec

```
{_____}
            AR  = 2;                        {temporary}
            AX0 = DM(sp_flag);
            AF  = PASS AX0;                  {temporary}
            IF NE AR = PASS 0;               {temporary}
            DM(coeff_codeword) = AR;

{_____Conditional Assembly_____}
{ use (asm21 -cp -Dtestvad) to validate VAD and SP_FLAG }
#ifdef testvad
        CALL test_format;
#endif
{_____}

            {This is where the coefficient transfer will take place!!}

            I1=^coeff_codeword;
            I2=^speech_1;

{_____Conditional Assembly_____}
{ use (asm21 -cp -Ddemo) for demonstration }
#ifdef demo
        CALL set_codeword;          {routine sets coeff_codeword for demo}
#endif
{_____}

            CALL decode_codeword;
            AX0 = DM(sid_inbuf);

{_____Conditional Assembly_____}
{ use (asm21 -cp -Dtestvad) to validate VAD and SP_FLAG }
#ifdef testvad
        CALL test_unformat;
#endif
{_____}

            CALL dmr_decode;

{.........................Conditional Assembly...........................}
{ use (asm21 -cp -Dmsp02) for use with AD28msp02 }
#ifndef msp02
        SE  = -2;                   { right shift to 14 bits for u-law }
        I0  = ^speech_1;            { compression }
        CALL scale_routine;
#endif
{.........................................................................}
```

```
{.....................Conditional Assembly.............................}
{ use (asm21 -cp -Dmsp02) for use with AD28msp02 }
#ifdef msp02
        ENA SEC_REG;

code_2_loop:IDLE;                       { wait for next sample }
        AY1 = MX0;
        AR  = AX1 - AY1;                { check if buffer is full }
        IF NE JUMP code_2_loop;

        MX0 = 0;                        { buffer full, reset sample counter }
        DIS SEC_REG;
#else

code_2_loop:IDLE;                       { autobuffering counts samples }
#endif {.............................................................}

   I7=^speech_1;                        { swap speech output/input buffer }
{_____Conditional Assembly_____}
{ use (asm21 -cp -Ddemo) for demonstration }
#ifdef demo
        ENA SEC_REG;
        AF  = PASS MR1;
        IF NE JUMP CODE_1_LOOP;
        M7  = MX1;
        DIS SEC_REG;
#endif {_____}

do_dmr_2:

{.....................Conditional Assembly.............................}
{ use (asm21 -cp -Dmsp02) for use with AD28msp02 }
#ifndef msp02
        SE  = 2;                        { left-justify expanded u-law input }
        I0  = ^speech_2;
        CALL scale_routine;
#endif {.............................................................}

        I0=^speech_2;
        I1=^coeff_buffer;
        CALL dmr_encode;

{_____Conditional Assembly_____}
{ use (asm21 -cp -Ddemo) for demonstration }
#ifdef demo
#ifndef msp02
        CALL vad_out;
```

(listing continues on next page)

4 GSM Codec

```
#endif
#endif
{_____}

        AR  = 2;                        {temporary}
        AX0 = DM(sp_flag);
        AF  = PASS AX0;                 {temporary}

        IF NE AR = PASS 0;              {temporary}
        DM(coeff_codeword) = AR;

{_____Conditional Assembly_____}
{ use (asm21 -cp -Dtestvad) to validate VAD and SP_FLAG }
#ifdef testvad
        CALL test_format;
#endif

{_____}

        {This is where the coefficient transfer will take place!!}
        I1=^coeff_codeword;
        I2=^speech_2;

{_____Conditional Assembly_____}
{ use (asm21 -cp -Ddemo) for demonstration }
#ifdef demo
        CALL set_codeword;            {routine sets coeff_codeword for demo}
#endif
{_____}

        CALL decode_codeword;
        AX0 = DM(sid_inbuf);

{_____Conditional Assembly_____}
{ use (asm21 -cp -Dtestvad) to validate VAD and SP_FLAG }
#ifdef testvad
        CALL test_unformat;
#endif
{_____}

   CALL dmr_decode;

{.....................Conditional Assembly........................}
{ use (asm21 -cp -Dmsp02) for use with AD28msp02 }
#ifndef msp02
        SE = -2;                        { right shift to 14 bits for u-law }
        I0  = ^speech_2;                { compression }
        CALL scale_routine;
#endif
```

GSM Codec   4

```
{...........................................................................}
{.....................Conditional Assembly.................................}
{ use (asm21 -cp -Dmsp02) for use with AD28msp02 }
#ifdef msp02
        ENA SEC_REG;                                { sample counting done in sec regs }
#endif {.....................................................................}

        JUMP code_1_loop;

{_____Conditional Assembly_____}
{ use (asm21 -cp -Ddemo) for demonstration }
#ifdef demo
next_demo:  ENA SEC_REG;
            SE  = 2;
            AY0 = ^demo_codes;
            AR  = SI, AF = PASS 1;
            AY1 = 4;
            AR  = AR + AF;                           {increment current state}
            af  = ar - ay1;
            if gt ar = pass 0;
            SI  = AR, AR  = AR + AY0;                {offset pointer, save state}
            AX0 = I5;
            I5  = AR;
            SR0 = PM(I5,M4);                         {get demo state codeword}
            I5  = AX0;
            ay1 = 7;
            AR  = SR0 AND AY1;                       {extract force_vad_high, _low}
            MX1 = AR, SR = LSHIFT SR0 (LO);     {        talk_thru_flag}
            MR1 = SR1, SR = LSHIFT SR0 (LO);    {        mask_sp}
            MR0 = SR1, SR = LSHIFT SR0 (LO);    {        mask_taf}
            MY1 = SR1, SR = LSHIFT SR0 (LO);    {        force_codeword_high}
            MY0 = SR1;
            RTI;

set_codeword:   ENA SEC_REG;
                IMASK = 0;
                AY1 = 3;
                AF  = PASS 0;
                AY0 = DM(sp_flag);
                AR  = PASS AY0;
                IF EQ AF = PASS AY1;
                AR  = MR0 AND AF;                   {AR = masked sp_flag}
                AY1 = 2;
                AF  = PASS 1, AX0 = AR;
                AY0 = DM(taf_count);
                AR  = PASS AY0;
                IF GT AF = PASS AY1;
                AR  = MY1;
```

(listing continues on next page)

```
        AF  = AR AND AF;                      {AF = masked taf_count}
        AF  = AX0 OR AF, AR = MY0;
        AR  = AR OR AF;                       {AR = coeff_codeword}
        DM(coeff_codeword) = AR;
        AY0 = 1;
        AR  = AR - AY0;                       { check if unusable frame }
        IF NE JUMP set_cw_done;
        I4  = I1;                             { unusable frame - force }
        M7  = 12;                             { immediate muting for }
        MODIFY(I4,M7);                        { demonstration by setting }
        M7  = 17;                             { the four xmax values < 4 }
        CNTR = 4;                             { (in this case, = 0) }
        DO set_xmax_demo UNTIL CE;

set_xmax_demo:     DM(I4,M7) = AR;
                   M7  = 2;

{......................Conditional Assembly................................}
{ use (asm21 -cp -Dadsp2111) for use with ADSP-2111 }
#ifdef adsp2111

set_cw_done: IMASK=0x88;

#else

set_cw_done: IMASK=0x28;

#endif {................................................}

        DIS SEC_REG;
        RTS;
#endif
{_____}

{_____Conditional Assembly_____}
{ use (asm21 -cp -Dtestvad) to validate VAD and SP_FLAG }
#ifdef testvad

test_format:I1  = ^coeff_buffer;
            AX0 = DM(vad);
            AX1 = DM(sp_flag);
            CNTR = 2;
            DO add_bits UNTIL CE;
                AR  = H#8000;
                AF  = PASS AX0, AY0 = DM(I1,M0); IF EQ AR = PASS 0;
                AR  = AR OR AY0, AX0 = AX1;
                add_bits: DM(I1,M1) = AR;
            RTS;
```

```
test_unformat: AX1 = H#7FFF;
               AY0 = DM(I1,M0);
               AR  = AX1 AND AY0;
               DM(I1,M1) = AR;
               AY0 = DM(I1,M0);
               AR  = AX1 AND AY0;
               DM(I1,M2) = AR;
               RTS;
#endif
{_____}

{_____Conditional Assembly_____}
{ use (asm21 -cp -Ddemo) for demonstration }
#ifdef demo
#ifndef msp02

{this is temporary for outputting the voice activity flag for the demonstration}

vad_out: AX0 = DM(vad);
         AF  = PASS AX0;

         IF NE SET FLAG_OUT;

         IF EQ RESET FLAG_OUT;
         RTS;
#endif
#endif
{_____}

{......................Conditional Assembly..............................}
{ use (asm21 -cp -Dmsp02) for use with AD28msp02 }
#ifndef msp02
scale_routine: SI  = DM(I0,M1);
         CNTR = 160;
         DO shift_it UNTIL CE;
             SR  = ASHIFT SI (HI), SI = DM(I0,M2);
shift_it:    DM(I0,M3) = SR1;
         RTS;
```

(listing continues on next page)

4   GSM Codec

```
#endif
{........................................................}

{.....................Conditional Assembly............... .......}
{ use (asm21 -cp -Dmsp02) for use with AD28msp02 }
#ifdef msp02
sample:  ENA SEC_REG;
         AR  = DM(I7,M4);            { read buffer, do not move pointer }
         TX0 = AR;                   { write transmit data }
         AR  = RX0;                  { read received data }
         DM(I7,M5) = AR;             { write to buffer, increment pointer }
         AY0 = MX0;
         AR  = AY0 + 1;              { increment sample counter }
         MX0 = AR;
         RTI;
#endif {................................................................}

.ENDMOD;
```

Listing 4.6 Data Acquisition Shell Routine (DMR21xx.DSP)

Sub-Band ADPCM ■ 5

5.1   OVERVIEW

Pulse Code Modulation, or PCM (CCITT Recommendation G.711), is a method of digitizing analog wave forms to transmit speech signals. This quantization scheme provides 13 bits (µ-law) or 14 bits (A-law) of dynamic range in an 8-bit value. 13 or 14-bit dynamic range is the minimum requirement to accurately reproduce the full range of speech signals, therefore, µ-law and A-law encoding are widely used in telephony.

This A/D conversion can introduce quantization noise when the analog signals are quantized to digital values. The human ear is more sensitive to quantization noise when the noise component is relatively large compared to the size of the signal. Recommendation G.711 applies a non-uniform quantization function to adjust the data size in proportion to the input signal, thus reducing noise interference. As a result, smaller signals are approximated with greater accuracy.

Adaptive Differential Pulse Code Modulation, or ADPCM (CCITT Recommendation G.721), is more efficient to transmit than PCM. ADPCM uses an adaptive predictor to take advantage of the redundancies present in speech signals. It compares a signal sample with the previous sample and transmits the difference between the two. This reduces the number of bits needed to reproduce the speech. G.721 samples speech bandwidths of 200–3400 Hz at 8 kSa/s. The inputs and outputs of a G.721-based system are still PCM values. Although G.711 and G.721 are widely used, these methods are quality and bandwidth limited. Chapters 11 and 12 of *Digital Signal Processing Applications Using the ADSP-2100 Family, Volume 1*, briefly discuss PCM and ADPCM theory, and include program examples.

To improve the overall transmission quality and add a sub-carrier frequency, CCITT developed Sub-Band ADPCM (Recommendation G.722). Recommendation G.722 is a wideband audio recommendation (50 to 7000 Hz) that splits the frequency band into two sub-bands (0 to 4000 Hz and 4000 Hz to 8000 Hz), and applies ADPCM to the sub-bands independently. G.722 operates on linear samples of speech. The auxiliary data (non-encoded) channel is available for video transmission, and is used in applications such as teleconferencing.

5   Sub-Band ADPCM

This chapter describes a method to implement the G.722 algorithm with the ADSP-2100 Family of digital signal processors. To save memory space and to clarify the implementation, the program example (Listing 5.1) at the end of this chapter is written as a collection of subroutines. This format is efficient because the higher and lower sub-bands share most of the subroutines; in fact, many subroutines (such as `filtez` and `filtep`) are also shared by the encoder and the decoder of each sub-band.

5.2   SUB-BAND ADPCM ALGORITHM

CCITT Recommendation G.722 specifies the following six parts of the algorithm (see Figure 5.1):

- Transmit quadrature mirror filter (QMF)
- Lower sub-band encoder
- Higher sub-band encoder
- Lower sub-band decoder
- Higher sub-band decoder
- Receive quadrature mirror filter

The block diagram has two halves, transmit (encoder) and receive (decoder).The implementation of the multiplexer and demultiplexer are straightforward, and are not described in this chapter.

The subroutines included at the end of this chapter were verified against digital test sequences provided by CCITT for the standards, and are fully compatible with Recommendation G.722. When possible, the names of the subroutines and variables used in the algorithm match the names specified in the recommendation.

5.3   TRANSMIT PATH

This section describes the encoder and transmit path shown in Figure 5.1. The encoder operates at 64 kbits per second, with 16 kSa/s and 14 bits.

5.3.1   Transmit Quadrature Mirror Filter

The transmit quadrature mirror filter splits the frequency band into two sub-bands, higher and lower. It also decimates the input to the encoder from 16 kHz to 8 kHz. The filter is a 24 tap Finite Impulse Response filter, or FIR. The impulse response can be approximated as a simple delay function. The transmit quadrature mirror filter shares the same coefficients and 24 tap delay line with the receive QMF. Implementation of QMFs in

Sub-Band ADPCM   5

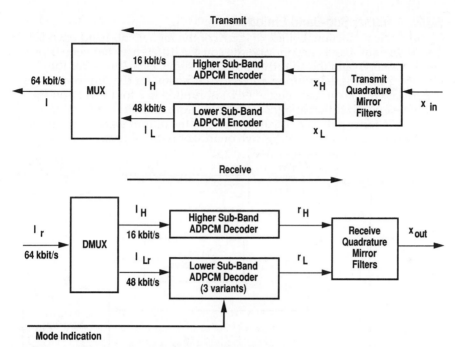

Figure 5.1 Sub-Band ADPCM Algorithm Block Diagram

ADSP-2100 family assembly language is computationally efficient because the data adress generators, or DAGs, use indirect addressing to fetch filter coefficients and data values in the same processor cycle. Also, you can use circular buffering to represent the tapped delay lines. The output variables of the filters, `xl(n)` and `xh(n)` (lower and higher sub-band signal components), are determined by the following equations:

```
xl(n) = xa + xb
xh(n) = xa - xb
```

where

```
xa = h2i * xin(j-2i)
xb = h2i+1 * xin(j - 2i - 1)
```

5　Sub-Band ADPCM

5.3.2　Higher Sub-Band Encoder

Figure 5.2 is a functional block diagram of the higher sub-band encoder. The higher sub-band encoder operates on the differences between input signal value xh and the adaptive predictor signal estimate. After the predicted value is determined and the subtraction for the difference signal is performed, the estimate signal (el) is applied to a four level non-linear adaptive quantizer that assigns six binary digits to yield the 48 kbits/s signal, Il. Since data is not truncated from the output signal, Ih, in the feedback loop, the inverse adaptive quantizer is also 4 levels.

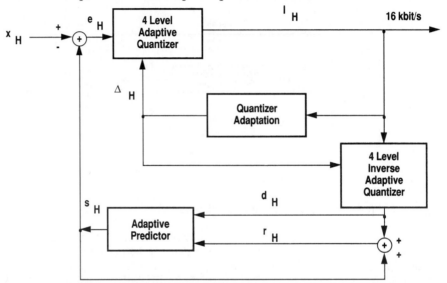

Figure 5.2 Higher Sub-Band Encoder Block Diagram

5.3.3　Lower Sub-Band Encoder

The lower sub-band encoder (shown in Figure 5.3) operates by estimating the difference in signal value between the predicted value and the actual input value. The structure of the adaptive predictor in the higher-band encoder is identical to the one in the lower sub-band encoder, but the names in memory of the adaptive predictor coefficients differ by an "l" or "h" to make the program more understandable. The number of bits required to represent the difference is smaller than the number of bits required to represent the total input signal. This difference is calculated by subtracting the predicted value from the input value:

```
el(n) = xl(n) - sl(n)
eh(n) = xh(n) - sh(n)
```

296

Sub-Band ADPCM 5

The predicted value (sl(n) or sh(n)) is produced by the adaptive predictor, which contains a second-order section to model poles, and a sixth-order section that models zeroes in the input signal. For every received sample, (xl(n) or xh(n)), upzero updates the six zero section coefficients of the predictor, uppol2 calculates the second pole section coefficient, and uppol1 calculates the first pole section coefficient.

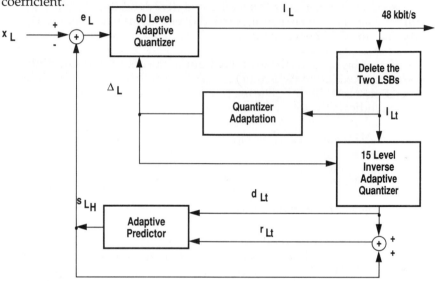

Figure 5.3 Lower Sub-Band Encoder Block Diagram

Operation is similar to the operation of the higher sub-band decoder except a 60-level adaptive quantizer is applied rather than a 4-level quantizer.

An important feature of the lower sub-band encoder is the feedback loop. The feedback loop is used for adaptation of the 60-level adaptive quantizer and to update the adaptive predictor. To do this, inside the feedback loop, the two least significant bits of Il are truncated to produce a 4-bit difference signal, Ilt . Since this value was already passed through the adaptive quantizer, an inverse adaptive quantizer produces dlt . This is a quantized difference signal that the adaptive predictor uses to produce sl (the estimate of the input signal) and update the adaptive predictor.

Four-bit operation (rather than 6-bit) leaves room for the auxiliary data channel in the lower sub-band encoder.

297

5   Sub-Band ADPCM

5.4   RECEIVE PATH

This section describes the decoder and receive path, shown in Figure 5.1. While the encoder operates at 64 kbits/s, the decoder accepts encoded signals at 64, 56 and 48 Kbits/s. The two lower bit rates correspond to the availability of an auxiliary data channel that uses either 8 or 16 Kbits/s. The auxiliary data channel is described as a data insertion device that is totally separate from the G.722 encoder and decoder. Bits from the auxiliary data channel are simply carried over the same transmission medium as the G.722 encoded data.

The different bit rates available at the input of the decoder (depending on whether the auxiliary data channel is used) are referred to as the "modes" of operation (see Table 5.1). During operation of the algorithm on-chip, you must indicate the desired mode.

MODE	7 kHz audio coding bit rate	Auxiliary data channel rate
1	64 Kbits/s	0 Kbits/s
2	56 Kbits/s	8 Kbits/s
3	48 Kbits/s	16 Kbits/s

Table 5.1 Decoder Modes Of Operation

5.4.1   Higher Sub-Band Decoder

The higher sub-band decoder (see Figure 5.4) is the simplest element of sub-band ADPCM. There are no choices to make for inverse adaptive quantizers or mode control to indicate word truncation. Instead, the input code word, Ih , is fed into the 4-level inverse adaptive quantizer (to obtain Dh) and into the quantizer adaptation segment in parallel. The adaptive predictor generates the signal estimate Sh and adds to this the output of the inverse adaptive quantizer to generate the decoder output signal, Rh .

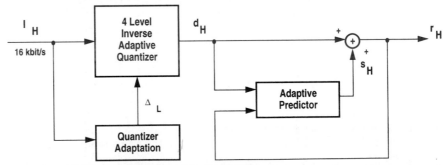

Figure 5.4 Higher Sub-Band Decoder Block Diagram

Sub-Band ADPCM   5

5.4.2   Lower Sub-Band Decoder

Figure 5.5 is a functional block diagram of the lower sub-band decoder. Generally, the higher and lower sub-band decoders and encoders share the same subroutine calls in almost the same order because they are similar in operation. In the lower sub-band decoder, however, the mode indication signal determines how many bits are truncated from the input codeword Ilr and which inverse adaptive quantizer is chosen in the feedback loop. Table 5.2 shows you the correlation between the Mode and the number of levels for the inverse adaptive quantizer.

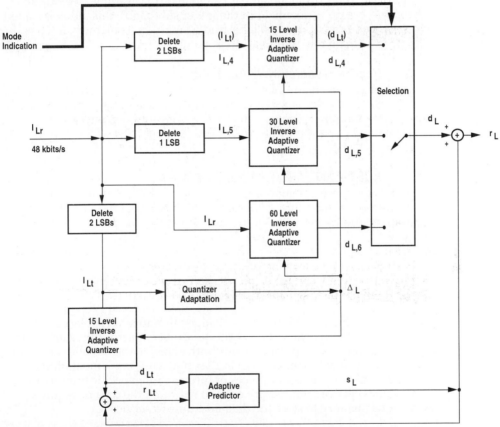

Figure 5.5 Lower Sub-Band Decoder Block Diagram

5   Sub-Band ADPCM

MODE	Inverse adaptive quantizer levels
1	60-level
2	30-level
3	15-level

Table 5.2 Inverse Adaptive Quantizer Modes Of Operation

In both the quantizers and inverse quantizers for the lower and higher sub-bands, indexed indirect memory access is used to read and write memory. An index of a data table is calculated (for example, see `quant1` subroutine, adaptive quantizer in the lower sub-band); this index is added to the base address of the data table and a single-cycle memory fetch is executed to obtain the desired address value.

5.4.3   Receive Quadrature Mirror Filter
The receive quadrature mirror filter interpolates the output of the decoder from 8 kHz to 16 kHz for input to the receive audio signal. The filter is a 24 tap finite impulse response filters whose impulse response can be approximated as a simple delay function.

5.5   ADSP-2100 FAMILY IMPLEMENTATION
The G.722 code implementation is a parameter set-up shell that calls subroutines corresponding to the subroutines listed in the CCITT recommendation. This is useful for the following reasons:

- Memory savings–both the encoder and decoder (and in some cases both sub-bands) use many of the same routines
- Easy transition from full-duplex to half-duplex implementation–copy the shell that includes the appropriate sub-routine calls

Circular buffering is used in the G.722 algorithm in several places: as delay lines in the receive and transmit quadrature mirror filters and in the adaptive predictor (a separate one for both encoder and decoder and both upper and lower sub-bands). The circular buffering implementation maintains only the necessary pieces of information (specific number of delay values). It does not require code to maintain the data or require extraneous memory to store the data and time to service it.

Sub-Band ADPCM 5

5.6 SUBROUTINE DESCRIPTIONS

This section contains brief descriptions of the subroutines used to implement CCITT Recommendation G.722.

5.6.1 reset_mem

This subroutine initializes the state variables required for correct operation of the algorithm. You must call the `reset_mem` routine before running the encoder or decoder. The `reset_mem` routine also does the following things:

- Initializes the linear and circular buffers required by the filters, encoder, and decoder
- Sets up pointers (data memory values) to the circular buffers so the index registers do not need to be dedicated to the circular buffers
- Set up modify and length registers that will remain constant for the remainder of the algorithm

5.6.2 filtez

This subroutine computes the output of the zero section of the adaptive predictor by multiplying the zero section coefficients by the quantized difference signal buffer values. Higher and lower sub-band encoders and decoders use this subroutine.

5.6.3 filtep

This subroutine computes the output of the pole section of the adaptive predictor by multiplying the pole section coefficients by the quantized reconstructed signal buffers. Higher and lower sub-band encoders and decoders use this subroutine.

5.6.4 quantl

This subroutine calculates the encoder output codeword based on the difference in signal value and the quantizer scale factor, `detl` (calculated in `scalel` below).

This subroutine fetches data words from the look-up tables included in the G.722 recommendation. It is necessary to first compute an index that locates the magnitude of the signal difference relative to the quantizer decision levels. This is accomplished in the `lll` loop. The decision levels (stored as a program memory data table) are multiplied by the quantizer scale factor and subtracted from the magnitude of the difference in signal value; if this value is less than zero, a flag is incremented to give the desired index. This index is then added to the address of the codeword data table and the correct six-bit codeword is chosen. Only the lower sub-band encoder uses this subroutine.

5 Sub-Band ADPCM

5.6.5   invqxl

This subroutine represents the `invqal` and `invqbl` sections of the algorithm. `invqal` computes the lower sub-band quantized difference in signal value for the adaptive predictor of the encoder and decoder. `invqbl` computes the quantized difference in signal value for the decoder output in the lower sub-band decoder. Since these two routines are identical except for the presence of the "mode" signal (decoder only), they are merged to save code space. Again, this subroutine is based on an indexed table look-up, and the choice of tables depends on the mode of operation. For the encoder, you supply a constant to choose the correct table (and number of bits to be truncated). For the decoder, the shell program mathematically determines which table is chosen according to the mode you define. Similar to `quantl`, an index is calculated and added to the indicated table as an offset.

5.6.6   logscl

This subroutine updates the logarithmic scale factor in the lower sub-band encoder and decoder. It is an indexed table look-up subroutine with limits imposed on the output value, `nbpl`.

5.6.7   scalel

This subroutine computes the quantizer scale factor in the lower sub band encoder and decoder. In addition to scaling output values, this subroutine performs an indexed table look-up.

5.6.8   upzero

This subroutine determines six zero section predictor coefficients. The output values (a buffer of size six) depend on the value and signs of the quantized difference in signal value, some leakage and gain constants, the delayed difference signal values, and old zero-section predictor coefficients.

5.6.9   uppol2

This subroutine generates the second pole predictor coefficient. It is determined from the sign and value of the partially reconstructed signal `pl`, some leakage and gain constants, and the old (delayed) pole predictor coefficients.

5.6.10   uppol1

This subroutine generates the first pole predictor coefficient. It depends on the delayed first pole predictor coefficient, some leakage and gain constants, `plt` and the old (delayed) pole predictor coefficients.

5.6.11   limit

This subroutine limits the output reconstructed signals for both lower and higher sub-band decoders.

5.6.12   quanth

This subroutine quantizes the difference in signal value in the higher sub-band encoder based on the magnitude of the signal, the higher sub-band quantizer scale factor, and a decision level indexed table look-up.

5.2.13   invqah

This subroutine computes the quantized difference in signal value in the higher sub-band encoder and decoder, based on the higher sub-band quantizer scale factor and the higher sub-band decoder output codeword, Ih, in indexed table look-up manner.

5.6.14   logsch

This subroutine determines the logarithmic quantizer scale factor in the higher sub-band encoder and decoder. This subroutine involves calculations for leakage and scale factors, and imposes some limits on the output signal.

Note: In addition to the routines mentioned above, several routines are performed in the shell itself. They are implemented in the shell because they are short, and because it saves two cycles (`call` and `rts`) for every execution. The following routines are implemented in the shell:

- SUBTRA
- RECONS
- PARREC

Also, the delay blocks DELAYZ, DELAYL, and DELAYA are implemented with the following two step process:

1. Variables (both single words and buffers) are given their initial value in `reset_mem` .

2. Variables are updated after processing through either the decoder or encoder with the newly computed value. They will contain the correct data for the next iteration through the system.

5 Sub-Band ADPCM

```
.module/ram/abs=0 g722;

/* ..................... variables for filters here ..................... */
.var/ram/dm/circ        tqmf_buf[23];
.var/ram/dm             accumab_ptr;
.var/ram/pm             coefs[24];
.init coefs:            <coeffs.dat>;
.var/ram/dm             xl;
.var/ram/dm/circ        accumc[11];
.var/ram/dm/circ        accumd[11];
.var/ram/dm             accumc_ptr;
.var/ram/dm             accumd_ptr;
.var/ram/dm             xh;
.var/ram/dm             xout1;
.var/ram/dm             xout2;
.var/ram/dm             xs;

/* ............... variables for encoder (hi and lo) here ............... */
.var/ram/dm             il;
.var/ram/dm             mode;
.var/ram/dm             szl;
.var/ram/dm             spl;
.var/ram/dm             sl;
.var/ram/dm             el;
.var/ram/dm             store_this;
.var/ram/pm             code4_table[0x20];
.init code4_table:<codword4.dat>;
.var/ram/pm             code5_table[0x40];
.init code5_table:<codword5.dat>;
.var/ram/pm             code6_table[0x80];
.init code6_table:<codword6.dat>;
.var/ram/pm             qq6_table[0x80];
.init qq6_table:        <quant6.dat>;
.var/ram/pm             qq5_table[0x10];
.init qq5_table:        <quant5.dat>;
.var/ram/pm             qq4_table[8];
.init qq4_table :       <quant4.dat>;
.var/ram/dm             delay_bpl[6];
.var/ram/dm             dltx_ptr;
.var/ram/dm             fbuf[6];
.var/ram/dm             tbuf[6];
.var/ram/dm/circ        delay_dltx[7];
.var/ram/dm             il4;
.var/ram/pm             wl_table[8];
.init wl_table:         <wl.dat>;
.var/ram/pm             ilb_table[32];
.init ilb_table:        <ilb.dat>;
.var/ram/dm             nbl;                    /* delay line */
.var/ram/dm             al1;
.var/ram/dm             al2;
```

```
.var/ram/dm              plt;
.var/ram/dm              plt1;
.var/ram/dm              plt2;
.var/ram/dm              rs;
.var/ram/dm              dlt;
.var/ram/dm              apl1;
.var/ram/dm              apl2;
.var/ram/dm              rlt;
.var/ram/dm              rlt1;
.var/ram/dm              rlt2;
.var/ram/pm              decis_levl[29];
.init decis_levl:        <q6shft3.dat>;
.var/ram/dm              detl;
.var/ram/pm              quant26bt_pos[30];
.init quant26bt_pos:     <quant6p.dat>;
.var/ram/pm              quant26bt_neg[30];
.init quant26bt_neg:     <quant6n.dat>;
.var/ram/dm              deth;
.var/ram/dm              sh;           /* this comes from adaptive predictor */
.var/ram/dm              eh;
.var/ram/pm              bit_out2[4];
.init bit_out2:          <bit_ih2.dat>;
.var/ram/dm              dh;
.var/ram/dm              ih;
.var/ram/dm              nbh;
.var/ram/dm              szh;
.var/ram/dm              sph;
.var/ram/dm              ph;
.var/ram/dm              yh;
.var/ram/dm              rh;
.var/ram/dm/circ         delay_dhx[7];
.var/ram/dm              delay_bph[6];
.var/ram/dm              dhx_ptr;
.var/ram/dm              ah1;
.var/ram/dm              ah2;
.var/ram/dm              aph1;
.var/ram/dm              aph2;
.var/ram/dm              ph1;
.var/ram/dm              ph2;
.var/ram/dm              rh1;
.var/ram/dm              rh2;

/* ..................... variables for decoder here ..................... */
.var/ram/dm              ilr;
.var/ram/dm              yl;
.var/ram/dm              rl;
.var/ram/dm              dec_deth;
.var/ram/dm              dec_detl;
.var/ram/dm              dec_dlt;
.var/ram/dm              dec_del_bpl[6];
.var/ram/dm              dec_dltx_ptr;
```

(listing continues on next page)

5 Sub-Band ADPCM

```
.var/ram/dm/circ        dec_del_dltx[7];
.var/ram/dm             dec_apl1;
.var/ram/dm             dec_apl2;
.var/ram/dm             dec_plt;
.var/ram/dm             dec_plt1;
.var/ram/dm             dec_plt2;
.var/ram/dm             dec_szl;
.var/ram/dm             dec_spl;
.var/ram/dm             dec_sl;
.var/ram/dm             dec_rlt1;
.var/ram/dm             dec_rlt2;
.var/ram/dm             dec_rlt;
.var/ram/dm             dec_al1;
.var/ram/dm             dec_al2;
.var/ram/dm             dl;
.var/ram/dm             dec_nbl;
.var/ram/dm             dec_yh;
.var/ram/dm             dec_dh;
.var/ram/dm             dec_nbh;

/* ..................... variables used in filtez ..................... */
.var/ram/dm             dec_del_bph[6];
.var/ram/dm             dec_dhx_ptr;
                                /* pointer for circ buffer index - hi sb dec */
.var/ram/dm/circ        dec_del_dhx[7];
.var/ram/dm             dec_szh;

/* ..................... variables used in filtep ..................... */
.var/ram/dm             dec_rh1;
.var/ram/dm             dec_rh2;
.var/ram/dm             dec_ah1;
.var/ram/dm             dec_aph1;
.var/ram/dm             dec_ah2;
.var/ram/dm             dec_aph2;
.var/ram/dm             dec_ph;
.var/ram/dm             dec_sph;
.var/ram/dm             dec_sh;
.var/ram/dm             dec_rh;
.var/ram/dm             dec_ph1;
.var/ram/dm             dec_ph2;
.var/ram/dm             x_num;

/* ................. starting with lower sub band encoder ................. */
/* ............... if in reset, initialize required memory ............... */

/* ..... encode: put input samples in my1 and mx0(calling parameters) ..... */
/* ................. my1 = first value, mx0 = second value ................. */
/* ............... returns il and ih stored together in ax0 ............... */
```

```
/* ............... decode: calling parameters: ilr and xh ................ */
/* ... return parameters: xout1 and xout2 (in ax0 and ax1 respectively) ... */
/* .......... note: supply mode signal to decoder also (in dm) .......... */

encode:   mstat = 0x0;
          i0 = dm(accumab_ptr);
          l0 = 23;
          m0 = 2;                   /* skipping through buffer with a stride of 2 */
          i5 = ^coefs;
          l5 = 0;
          m6 = 2;
          si = mx0;
          mr = 0, my0 = pm(i5,m6);
          cntr = 11;
          do e_loop until ce;

/* .... main multiply accumulate loop for even samples and coefficients .... */
          e_loop: mr = mr + mx0 * my0(ss), mx0 = dm(i0,m0), my0 = pm(i5,m6);
          dm(i0,m2) =my1, mr = mr + mx0 * my0(ss); /* final mult/accumulate */
                                               /* and write to delay line */

/* .. save mr here, want xa (contents of mr) to be at least 24 bits wide .. */
/* ...... so start moving mr outputs into alu regs for multiprecision ...... */
          sr0 = mr1;
          ay0 = mr0;                /* for multiprecis add in hight and lowt */
          cntr = 11;
          i5 = ^coefs+1;
          mr = 0,mx0 = dm(i0,m0), my0 = pm(i5,m6);
          do o_loop until ce;

/* ......... main loop for mult/accum odd inputs and coefficients ......... */
          o_loop: mr = mr + mx0 * my0(ss), mx0 = dm(i0,m0), my0 = pm(i5,m6);
          modify(i0,m0);
          modify(i0,m0);
          dm(i0,m2) = si, mr = mr + mx0 * my0(ss); /* final mult/accumulate */
                                               /* and write to delay line */

lowt:     ar = mr0 + ay0,  ay1 = sr0; /* add low precis word from loop first */
          ena ar_sat;
          ar = mr1 + ay1 + C;   /* need 16 bits of info, but to keep precise */
          dis ar_sat;           /* this is xl, needs to be limited */
          call chk_vals;
          dm(xl) = ar;

hight:    ar = ay0 - mr0, ay1 = sr0;
          ena ar_sat;
          ar = ay1 - mr1 + C -1;                    /* subtract with borrow */
          dis ar_sat;
          call chk_vals;
          dm(xh) = ar;
          dm(accumab_ptr) = i0;
```

(listing continues on next page)

5   Sub-Band ADPCM

```
/* ..... into regular encoder segment here - consider filters embedded ..... */
        mstat = 0x8;
        i1 = ^delay_bpl;
        i5 = dm(dltx_ptr);
        l5 = 7;
        l0 = 0;

/* ....... filtez - compute predictor output section - zero section ....... */
/* .... calling params: i1 points to delay_bpl, i2 points to delay_dltx .... */
/* .................... return parameters: mr1 (szl) .................... */
        call filtez;
        dm(szl) = ar;
        sr0 = dm(rlt1);
        my0 = dm(al1);
        ax0 = dm(rlt2);
        my1 = dm(al2);

/* ........ filtep - compute predictor output signal (pole section) ........ */
/* .................. calling params: sr0, my0, sr1, my1 .................. */
/* .................... return parameters : ar (spl) .................... */
        call filtep;

/* predic:compute the predictor output value in the lower sub_band encoder  */
/*  not a subroutine but a small piece of code to compute predictor output  */
/* .................. adding together szl + spl to form sl .................. */
        dm(spl) = ar;
        ay0 = dm(szl);
        ar = ar + ay0;
        dm(sl) = ar;
        ay0 = dm(xl);      /* this is subtra : xl - sl = el (diff. signal) */
        ar = ay0 - ar;
        dm(el) = ar;

/* ................ quantl: quantize the difference signal ................ */
/* ........ calling params: el(ar), detl (which has value at reset) ........ */
/* ............ return parameters: il (4 bit codeword) in ax0 ............ */
        call quantl;
        dm(il) = ax0;
        my0 = dm(detl);
        ay0 = 3;                      /* this is mode for block 4l */
        mr0 = ax0;
        ay1 = ^code4_table;           /* remember, this will change w/ invqbl */

/* invqxl: does both invqal and invqbl- computes quantized difference signal */
/* .............. for invqbl, truncate by 2 lsbs, so ay0 = 3 .............. */
/* .............. calling parameters: il(mr0), detl(my0) .............. */
/* ............ and ay1(address of correct table for codeword ............ */
/* ................... return paramters: dlt(mr1) ................... */
        call invqxl;
        modify(i5,m7);
        m6 = 0;
```

308

```
        dm(i5,m6) = mr1;
        mr0 = dm(nbl);
        ar = dm(il);

/* .... logscl: updates logarithmic quant. scale factor in low sub band .... */
/* ..... calling parameters: il (ilr in decoder) - in ar , nbl in mr0 ..... */
/* ..... return parameters: nbl used next time - note - same var name ..... */
        call logscl;
        dm(nbl) = ar;
        ay1 = 8;

/* ... scalel: compute the quantizer scale factor in the lower sub band ... */
/* calling params nbl(in ar) and 8(constant such that scalel can be scaleh */
/* ...................... return parameter: detl ........................ */
        call scalel;
        dm(detl) = sr0;
        ax0 = dm(i5,m5);
        ay0 = dm(szl);

/* parrec - simple addition to compute reconstructed signal for adaptive pred */
/* .................... no subroutine, just in place .................... */
/* ......... add predictor zero section + quantized diff signal .......... */
        ar = ax0 + ay0;
        dm(plt) = ar;

/* ... upzero: update zero section predictor coefficients (sixth order) ... */
/* ..... calling parameters: dlt(sr0); dlti(circ pointer for delaying ..... */
/* ................... dlt1, dlt2, ..., dlt6 from dlt .................... */
/* ........ bpli (linear_buffer in which all six values are delayed ........ */
/* .............. return params: updated bpli, delayed dltx .............. */
        i1 = ^delay_bpl;
        call upzero;
        ax0 = dm(al1);
        ay0 = ax0;
        mx0 = dm(al2);
        si = dm(plt);
        mr0 = dm(plt1);
        mr1 = dm(plt2);

/* . uppol2- update second predictor coefficient apl2 and delay it as al2 . */
/* ............. calling parameters: al1, al2, plt, plt1, plt2 ............. */
/* ................... return parameters: apl2 (in ar) ................... */
/* ................... note: apl2 is limited to +-.75 ................... */
        call uppol2;
        dm(apl2) = ar;
        dm(al2) = ar;
        mr0 = dm(plt1);
        mx0 = dm(al1);
        ay1 = dm(apl2);
        si = dm(plt);
```

(listing continues on next page)

5   Sub-Band ADPCM

```
/* .. uppol1 :update first predictor coefficient apl1 and delay it as al1 .. */
/* .............. calling parameters: al1, apl2, plt, plt1 ............... */
/* ................. note: wd3= .9375-.75 is always positive ............... */
        call uppol1;
        dm(apl1) =ar;
        dm(al1) = ar;

/* ...... recons : compute reconstructed signal for adaptive predictor ...... */
/* .................... parameters: sl(ax0), dlt(ay0) .................... */
/* .................... return parameters: rlt(ar) .................... */
        ax0 = dm(sl);
        ay0 = dm(store_this);
        ar = ax0 + ay0;
        dm(rlt) = ar;

/* . done with lower sub_band encoder; now implement delays for next time . */
        modify(i5,m5);
        ax0 = dm(rlt1);
        dm(rlt2) = ax0;
        dm(rlt1) = ar;
        ax0 = dm(plt1);
        dm(plt2) = ax0;
        ax0 = dm(plt);
        dm(plt1) = ax0;
        dm(dltx_ptr) = i5;    /* save i5 in dltx_ptr, restore next time */
hi_sb_enc:i1 = ^delay_bph;
        i5 = dm(dhx_ptr);

/* .................. filtez: calling params: ax0, ax1 .................. */
/* ...................... return params: ar(szh) ........................ */
        call filtez;
        dm(szh) = ar;
        sr0 = dm(rh1);
        my0 = dm(ah1);
        ax0 = dm(rh2);
        my1 = dm(ah2);

/* ............... filtep: calling parms: sr0, my0, sr1, my1 ............... */
/* ...................... return params: ar (sph) ........................ */
        call filtep;
        dm(sph) = ar;
        ay0 = dm(szh);
        ar = ar + ay0;
        dm(sh) = ar;                         /* predic: sh = sph + szh */
        ay0 = dm(xh);
        ar = ay0 - ar;
        dm(eh) = ar;                         /* subtra: eh = xh - sh */
        my0 = dm(deth);

/* ........ quanth: calling params: eh(ar), deth (has init. value) ........ */
```

```
/* ....................... return: ih in ax0 ......................... */
        call quanth;
        dm(ih) = ax0;
        ay0 = ax0;

/*  invqah: compute the quantized difference signal in th ehigher sub_band  */
/* ............. calling parameters: ih(in ax0); deth(in my0) ............. */
/* .................... return parameters: dh (in mr1) ................... */
        call invqah;
        modify(i5,m7);
        m6=0;
        dm(i5,m6) = mr1;
        ay0 = dm(ih);
        my0 = 0x7f00;
        mx0 = dm(nbh);

/* ... logsch: update logarithmic quantizer scale factor in hi sub band ... */
/* ...... calling paameters: ih(ay0), nbh(mx0), my0 has a constant ....... */
/* ............... return parameters: updated nbh (in ar) ............... */
        call logsch;
        dm(nbh) = ar;
        ay1 = 0xa;

/* ..... note : scalel and scaleh use same code, different parameters ..... */
        call scalel;
        dm(deth) = sr0;

/* .parrec - add pole predictor output to quantized diff. signal(in place . */
        ax0 = dm(i5,m5);
        ay0 = dm(szh);
        ar = ax0 + ay0;
        dm(ph) = ar;

/* ... upzero: update zero section predictor coefficients (sixth order) ... */
/* .......... calling parameters: dh(sr0); dhi(circ), bphi (circ) .......... */
/* ............... return params: updated bphi, delayed dhx .............. */
        i1 = ^delay_bph;
        call upzero;

/* ...... uppol2: update second predictor coef aph2 and delay as ah2 ...... */
/* ................ calling params: ah1, ah2, ph, ph1, ph2 ................ */
/* .................... return params:  aph2 (in ar) ..................... */
/* .................... note: aph2 is limited to +- .75 ................... */
        ax0 = dm(ah1);
        ay0 = ax0;
        mx0 = dm(ah2);
        si = dm(ph);
        mr0 = dm(ph1);
        mr1 = dm(ph2);
        call uppol2;
        dm(aph2) = ar;
        dm(ah2)  = ar;
```

(listing continues on next page)

5   Sub-Band ADPCM

```
/* .... uppol1:  update first predictor coef. aph2 and delay it as ah1 .... */
/* ............. note: wd3 = .9375 -.75 is always positive .............. */
        mr0 = dm(ph1);
        mx0 = dm(ah1);
        ay1 = dm(aph2);
        call uppol1;
        dm(apl1) = ar;
        dm(ah1) = ar;
        ax0 = dm(sh);
        ay0 = dm(store_this);
        ar = ax0 + ay0;
        dm(yh) = ar;

/* ...... limit determines the greatest and smallest magnitude of the ...... */
/* .................... reconstructed output signal ..................... */
/* ........ calling params: yl (in ar); return params: rh (in ar) ...... */
/* ...... done with higher sub-band encoder, now Delay for next time ...... */
        ax0 = dm(rh1);
        dm(rh2) = ax0;
        dm(rh1) = ar;
        ax0 = dm(ph1);
        dm(ph2) = ax0;
        ax0 = dm(ph);
        dm(ph1) = ax0;
        modify(i5,m5);
        dm(dhx_ptr) = i5;

/* ........... multiplexing ih and il to get signals together ........... */
        si = dm(ih);
        ar = dm(il);
        sr = lshift si by 6(lo);
        sr = sr or lshift ar by 0(lo);
        ax0 = sr0;

/* ............... multiplexed transmission word in ax0 ................ */
        rts;            /* done with encode */

/* ..................... LOWER SUB_BAND DECODER ...................... */
/* ....... expect to split transmitted word from ax0 into ilr and ih ....... */
decode: ay0 = 0x3f;
        ar = ax0 and ay0;
        dm(ilr) = ar;
        ay0 = 0xc0;
        ar = ax0 and ay0;
        sr = lshift ar by -6 (lo);
        dm(ih) = sr0;                       /* place ih in two lsb's of sr0 */

lo_sb_dec:
        mstat = 0x8;
        i1 = ^dec_del_bpl;
        i5 = dm(dec_dltx_ptr);
```

Sub-Band ADPCM   5

```
/* ........... filtez: compute predictor output for zero section ........... */
/* .. calling parmeters: addresses of zero section input and output bufs .. */
/* ................. return parameters: del_szl (in mr1) ................. */
        call filtez;
        dm(dec_szl) = ar;
        sr0 = dm(dec_rlt1);
        my0 = dm(dec_al1);
        ax0 = dm(dec_rlt2);
        my1 = dm(dec_al2);

/* ....... filtep: compute predictor output signal for pole section ....... */
/* ...... calling parameters: dec_rlt1, dec_rlt2, dec_al1 and dec_al2 ...... */
/* ................. return parameter: del_spl (in ar) .................. */
        call filtep;
        dm(dec_spl) = ar;
        ay0 = dm(dec_szl);
        ar = ar + ay0;
        dm(dec_sl) = ar;
        ay0 = 3;
        mr0 = dm(ilr);
        ay1 = ^code4_table;
        my0 = dm(dec_detl);

/* invqxl: compute quantized difference signal for adaptive predic in low sb */
/* ............... calling parameters: my0, mr0, ay1 , ay0 ............... */
/* ................. return parameters: mr1 (dec_dlt) ................. */
        call invqxl;
        modify(i5,m7);
        m6=0;
        dm(i5,m6) = mr1;
        ay0 = dm(mode);
        mr0 = ^code4_table;
        mr1 = ^code5_table;
        sr0 = ^code6_table;
        ax0 = 2;
        ax1 = 3;
        af = ay0 - 1;
        if eq ar = pass sr0;
        af = ay0 - ax0;
        if eq ar = pass mr1;
        af = ay0 - ax1;
        if eq ar = pass mr0;
        ay1 = ar;
        mr0 = dm(ilr);
        my0 = dm(dec_detl);
```

(listing continues on next page)

5   Sub-Band ADPCM

```
/* invqxl: compute quantized difference signal for decoder output in low sb */
/* ................ calling parameters: my0, mr0, ay1 , ay0 ................ */
/* ................... return parameters: mr1( dl) ..................... */
        call invqxl;
        dm(dl) = mr1;
        ay0 = dm(dec_sl);
        ar = mr1 + ay0;
        dm(yl) = ar;

/* ................. limit: calling parameters yl (ar) ................... */
/* .................... return parameters: rl (ar) ..................... */
        call limit;
        dm(rl) = ar;

/* .... logscl: quantizer scale factor adaptation in the lower sub-band .... */
/* ... calling parameters: dec_nbl (in mr1, dm(il4) calculated in invqxl ... */
        mr0 = dm(dec_nbl);
        ar = dm(ilr);
        call logscl;
        dm(dec_nbl) = ar;
        ay1 = 8;

/* ..... scalel: computes quantizer scale factor in the lower sub band ..... */
/* .. calling params:  updated dec_nbl, and ay1 for integer part scaling .. */
        call scalel;
        dm(dec_detl) = sr0;

/* . parrec - add pole predictor output to quantized diff. signal(in place . */
/* ................. for partially reconstructed signal ................. */
        ax0 = dm(i5,m5);
        ay0 = dm(dec_szl);
        ar = ax0 + ay0;
        dm(dec_plt) = ar;
        i1 = ^dec_del_bpl;

/* .......... upzero: update zero section predictor coefficients .......... */
/*calling params: dec_dlt(sr0),dec_dlti(circ buffer),dec_bpli(linear buffer)*/
/* ......... return parameters: updated dec_bpli, delayed dec_dlti ......... */
/* ..... note: am saving the index(i) register for circ buffers to mem ..... */
        call upzero;
        ax0 = dm(dec_al1);
        ay0 = ax0;
        mx0 = dm(dec_al2);
        si = dm(dec_plt);
        mr0 = dm(dec_plt1);
        mr1 = dm(dec_plt2);

/* . uppol2: update second predictor coefficient apl2 and delay it as al2 . */
/* . calling parameters: al1(ax0), al2(mx0), plt(si), plt1(mr0), plt2(mr1) . */
```

```
/* .................. return parameters: apl2 (in ar) ................... */
        call uppol2;
        dm(dec_apl2) = ar;
        dm(dec_al2) = ar;
        mr0 = dm(dec_plt1);
        mx0 = dm(dec_al1);
        ay1 = dm(dec_apl2);
        si = dm(dec_plt);

/* .......... uppol1: update first predictor coef. (pole setion) .......... */
/* calling params: dec_plt1 (mr0), dec_plt(si), dec_al1(mx0), dec_apl2(ay1) */
/* .................. return parameter: apl1 (in ar) ................... */
        call uppol1;
        dm(dec_apl1) = ar;
        dm(dec_al1)  = ar;
        ax0 = dm(dec_sl);
        ay0 = dm(store_this);

/* ...... recons : compute recontructed signal for adaptive predictor ...... */
/* .............. adding together  dec_sl(ax0), dec_dlt(ay0) .............. */
        ar = ax0 + ay0;
        dm(dec_rlt) = ar;

/* ... done with lower sub band decoder, implement delays for next time ... */
        modify(i5,m5);
        ax0 = dm(dec_rlt1);
        dm(dec_rlt2) = ax0;
        dm(dec_rlt1) = ar;
        ax0 = dm(dec_plt1);
        dm(dec_plt2) = ax0;
        ax0 = dm(dec_plt);
        dm(dec_plt1) = ax0;
        dm(dec_dltx_ptr) = i5;

/* ........................ HIGH SUB-BAND DECODER ........................ */
hi_sb_dec:
        i1 = ^dec_del_bph;
        i5 = dm(dec_dhx_ptr);

/* .......... filtez: compute predictor output for zero section .......... */
/* .. calling parameters: addresses of zero section input and output bufs .. */
/* .................. return parameters: dec_shl (in mr1) ................. */
        call filtez;
        dm(dec_szh) = ar;
        sr0 = dm(dec_rh1);
        my0 = dm(dec_ah1);
        ax0 = dm(dec_rh2);
        my1 = dm(dec_ah2);
```

(listing continues on next page)

5 Sub-Band ADPCM

```
/* ....... filtep: compute predictor output signal for pole section ....... */
/* ....... calling parameters: dec_rh1, dec_rh2, dec_ah1 and dec_ah2 ....... */
/* .................. return parameter: dec_sph (in ar) .................. */
        call filtep;
        dm(dec_sph) = ar;

/* predic:compute the predictor output value in the higher sub_band decoder */
/* .............. adding dec_szh and dec_sph to form dec_sh .............. */
        ay0 = dm(dec_szh);
        ar = ar + ay0;
        dm(dec_sh) = ar;
        ax0 = dm(ih);
        ay0 = ax0;
        my0 = dm(dec_deth);

/*  invqah: compute the quantized difference signal in th ehigher sub_band  */
/* ........... calling parameters: ih(in ax0); deth(in my0) ............. */
/* .................. return parameters: dec_dh (in mr1) .................. */
        call invqah;
        modify(i5,m7);
        m6=0;
        dm(i5,m6) = mr1;
        ay0 = dm(ih);
        my0 = 0x7f00;
        mx0 = dm(dec_nbh);

/* ... logsch: update logarithmic quantizer scale factor in hi sub band ... */
/* ..... calling parameters: ih(ay0), dec_nbh(mx0), my0 has a constant ..... */
/* .............. return parameters: updated dec_nbh (in ar) .............. */
        call logsch;
        dm(dec_nbh) = ar;
        ay1 = 0xa;

/* ... scalel: compute the quantizer scale factor in the higher sub band ... */
/* calling params: dec_nbl(in ar) and 10(constant so that scalel is re-used */
/* .................. return parameter: dec_deth(in sr0) .................. */
        call scalel;
        dm(dec_deth) = sr0;
        ax0 = dm(i5,m5);
        ay0 = dm(dec_szh);

/* ............. parrec: compute partially reconstructed signal ............. */
/* ............... add together ax0(dec_dh), ay0 (dec_szh) ............... */
        ar = ax0 + ay0;
        dm(dec_ph) =ar;
        i1 = ^dec_del_bph;

/* ......... upzero: update zero section predictor coefficients .......... */
/*  calling params: dec_dh (sr0), dec_dhi(circ buffer), dec_bph(linear buf  */
/* ......... return parameters: updated dec_bph, delayed dec_dhi .......... */
```

```
/* ..... note: am saving the index(i) register for circ buffers to mem ..... */
        call upzero;
        ax0 = dm(dec_ah1);
        ay0 = ax0;
        mx0 = dm(dec_ah2);
        si = dm(dec_ph);
        mr0 = dm(dec_ph1);
        mr1 = dm(dec_ph2);

/* . uppol2: update second predictor coefficient aph2 and delay it as ah2 . */
/* ........ calling parameters:dec_ah1(ax0),dec_ah2(mx0),dec_ph(si) ........ */
/* ...................... dec_ph1(mr0),dec_ph2(mr1 ...................... */
/* ................... return parameters: aph2 (in ar) ................... */
        call uppol2;
        dm(dec_aph2) = ar;
        dm(dec_ah2)  = ar;
        mr0 = dm(dec_ph1);
        mx0 = dm(dec_ah1);
        ay1 = dm(dec_aph2);

/* .......... uppol1: update first predictor coef. (pole setion) .......... */
/*calling parameters: dec_ph1 (mr0), dec_ph(si), dec_ah1(mx0), dec_aph2(ay1)*/
/* ................... return parameter: aph1 (in ar) ................... */
        call uppol1;
        dm(dec_aph1) = ar;
        dm(dec_ah1)  = ar;
        ax0 = dm(dec_sh);
        ay0 = dm(store_this);

/* ...... recons : compute reconstructed signal for adaptive predictor ...... */
/* ............... add  parameters: dec_sh(ax0), dec_dh(ay0) ............... */
/* ................... to get parameters: dec_yh(ar) ................... */
        ar = ax0 + ay0;
        dm(dec_yh) = ar;

/* ............... implementing delays for next time here ............... */
        ax0 = dm(dec_rh1);
        dm(dec_rh2) = ax0;
        dm(dec_rh1) = ar;                        /* ar has dec_yh */
        ax0 = dm(dec_ph1);
        dm(dec_ph2) = ax0;
        ax0 = dm(dec_ph);
        dm(dec_ph1) = ax0;
        modify(i5,m5);
        dm(dec_dhx_ptr) = i5;

/* ............ limit: limiting the output reconstructed signal ............ */
/* ................... calling params:dec_yh(in ar) ................... */
/* ................. return parameters: dec_rh(in ar) ................. */
        call limit;
        dm(rh) = ar;
```

(listing continues on next page)

5 Sub-Band ADPCM

```
/* ................... end of higher sub_band decoder ................... */
/* ............ start with receive quadrature mirror filters ............ */
recv_qmf: mstat = 0x0;
        i5 = ^coefs;
        l5 = 0;
        i0 = dm(accumc_ptr);
        m0 = 0;
        l0 = 11;
        i1 = dm(accumd_ptr);
        l1 = 11;
        m6 = 2;
        ena ar_sat;
        ax0 = dm(rl);
        ay0 = dm(rh);
        ar = ax0 + ay0;                  /* xs in af */
        dm(xs) = ar;
        ar = ax0 - ay0;                  /* xd in ar */
        dis ar_sat;
        mx0 = ar;
        si = ar;
        cntr = 11;
        mr = 0, my0 = pm(i5,m6);
        do accumc_loop until ce;
                accumc_loop: mr = mr + mx0 * my0(ss), mx0 = dm(i0,m3),
                my0 = pm(i5,m6);
        mr = mr + mx0 * my0 (ss);
        modify(i0,m1);
        sr = ashift mr1 by -15(hi);
        sr = sr or lshift mr0 by -15(lo);
        dm(i0,m2) = si;
        ar = pass sr0;
        call chk_vals;
        dm(xout1) = ar;                  /* could leave this in a register */
        i5 = ^coefs +1;
        mr = 0, my0 = pm(i5,m6);
        cntr = 11;
        mx0 = dm(xs);
        si = mx0;
        do accumd_loop until ce;
                accumd_loop: mr = mr + mx0 * my0(ss), mx0 = dm(i1,m3),
                my0 = pm(i5,m6);
        mr = mr + mx0 * my0(ss);
        modify(i1,m1);
        sr = ashift mr1 by -15(hi);
        sr = sr or lshift mr0 by -15(lo);
        dm(i1,m2) = si;
        ar =pass sr0;
        call chk_vals;
        dm(xout2) = ar;
        dm(accumc_ptr) = i0;
        dm(accumd_ptr) = i1;
        rts;
```

```
reset_mem: ax0 = 1;
           dm(rs) = ax0;
           ax0 = 0x8;
           dm(deth) = ax0;
           dm(dec_deth) = ax0;
           ax0 = 0x20;
           dm(detl) = ax0;
           dm(dec_detl) = ax0;
           ax0 = 0;
           dm(nbl)  = ax0;
           dm(al1)  = ax0;
           dm(al2)  = ax0;
           dm(plt1) = ax0;
           dm(plt2) = ax0;
           dm(rlt1) = ax0;
           dm(rlt2) = ax0;
           dm(nbh)  = ax0;
           dm(ah1)  = ax0;
           dm(ah2)  = ax0;
           dm(ph1)  = ax0;
           dm(ph2)  = ax0;
           dm(rh1)  = ax0;
           dm(rh2)  = ax0;
           dm(dec_rlt1) = ax0;
           dm(dec_rlt2) = ax0;
           dm(dec_al1)  = ax0;
           dm(dec_al2)  = ax0;
           dm(dec_nbl)  = ax0;
           dm(dec_plt1) = ax0;
           dm(dec_plt2) = ax0;
           dm(dec_rh1)  = ax0;
           dm(dec_rh2)  = ax0;
           dm(dec_ah1)  = ax0;
           dm(dec_ah2)  = ax0;
           dm(dec_nbh)  = ax0;
           dm(dec_ph1)  = ax0;
           dm(dec_ph2)  = ax0;
```

(listing continues on next page)

5 Sub-Band ADPCM

```
/* ................. reserved regs for c run-time model ................. */
        m1 = 1;
        m5 = 1;
        m3 = -1;
        m7 = -1;
        m2 = 0;
        i5 = ^delay_dltx;
        l5 = 7;
        cntr = 7;
        do init_circ0 until ce;
                init_circ0: dm(i5,m5) = 0;
        i5 = ^delay_dhx;
        cntr = 7;
        do init_circ1 until ce;
                init_circ1: dm(i5,m5) = 0;
        i5 = ^dec_del_dltx;
        cntr = 7;
        do init_circ2 until ce;
                init_circ2: dm(i5,m5) = 0;
        i5 = ^dec_del_dhx;
        cntr = 7;
        do init_circ3 until ce;
                init_circ3: dm(i5,m5) = 0;
        i0 = ^delay_bpl;
        l0 =0;
        i1 = ^delay_bph;
        l1 = 0;
        i5 = ^dec_del_bpl;
        l5 = 0;
        i6 = ^dec_del_bph;
        l6 = 0;
        cntr = 0x6;
        do init_lin until ce;
                dm(i0,m1) = 0;
                dm(i1,m1) = 0;
                dm(i5,m5) = 0;
        init_lin: dm(i6,m5) = 0;
        i0 = ^tbuf;
        i1 = ^fbuf;
        cntr = 6;
        do init_temp_bufs until ce;
                dm(i0,m1) = 0;
        init_temp_bufs: dm(i1,m1) = 0;   /* initialize temporary buffers */
```

```
/* .... save circ buffer index ptrs in mem, may need them in the future .... */
/* .... set up permanent length and index registers for encoder/decoder .... */
/* set up pointers for circular buffers, restore at the end of encode/decode */
        ax0 = ^delay_dltx;
        dm(dltx_ptr) = ax0;
        ax0 = ^delay_dhx;
        dm(dhx_ptr) = ax0;
        ax0 = ^dec_del_dltx;
        dm(dec_dltx_ptr) = ax0;
        ax0 = ^dec_del_dhx;
        dm(dec_dhx_ptr) = ax0;

/* ............. set up pointers for circ. buffers in filters ............. */
/* ................... initialize circ buffers in mem ................... */
        i0 = ^tqmf_buf;
        l0=0;
        cntr = 23;
        do init_tqmf until ce;
                init_tqmf: dm(i0,m1) = 0;
        i1 = ^accumc;
        l1=0;
        i5 = ^accumd;
        l5=0;
        cntr = 11;
        do init_fil until ce;
                dm(i5,m5) = 0;
                init_fil: dm(i1,m1) = 0;

        ax0 = ^tqmf_buf;      /* these  are input values */
        dm(accumab_ptr) = ax0;
        ax0 = ^accumc;
        dm(accumc_ptr) = ax0;
        ax0 = ^accumd;
        dm(accumd_ptr) = ax0;
        l5 = 7;                       /* final set up for length register */
        m1 = 1;
        m5 = 1;
        m2 = 0;
        m3 = -1;
        m7 = -1;
        l1 = 0;
        rts;
```

(listing continues on next page)

5 Sub-Band ADPCM

```
filtez: mr = 0, ay0 = dm(i5,m5);                    /* i1 points to bp1 */
                           /* i2 points to delay buffer(dltx) */
        i0 = ^fbuf;      /* io is temp buffer for adding delay line values */
        ar = pass ay0, my0 = dm(i1,m1);
        ar = ar + ay0;
        cntr = 6;
        do m_loop until ce;
                mr = ar * my0(ss), ay0 = dm(i5,m5);
                dm(i0,m1) = mr1;
                ar = pass ay0, my0 = dm(i1,m1);
        m_loop: ar = ar + ay0;

        i0 = ^fbuf;
        ar = pass 0;
        cntr = 6;
        do filt1 until ce;
                ay0 = dm(i0,m1);
        filt1: ar = ar + ay0;
        rts;

filtep: ay1 = sr0;
        ar = sr0 + ay1;                  /* add rlt1 + rlt1 */
        mr = ar * my0(ss), ay1 = ax0;    /* multiply by al1 */
        ay0 = mr1;                       /* save wd1 in ay0 */
        ar = ax0 + ay1, ay0 = mr1;       /* ar = rlt2 + rlt2,wd1 in ay0 */
        mr = ar  * my1(ss);              /* wd2 * al2 in mr1 */
        ar = mr1 + ay0;                  /* wd1 + wd2 */
        rts;

quantl: sr = ashift ar by -15(lo);
        af = pass sr0;
        if eq jump cont;
        ay0 = 0x7fff;
        af = ay0 - ar, ax0 = ay0; ar = ax0 and af;

cont:   i6 = ^decis_levl;
        ay1 = 0;
        ay0 = ar;
        my0 = dm(detl);
        af = pass 0, mx0 = pm(i6,m5);
        cntr = 0x1d;
        do lll until ce;
                mr = mx0 * my0(ss), mx0 = pm(i6,m5);
                ar = ay0 - mr1;
                if lt af = af + 1;
        lll: ar = pass af;
        /* i0 has mil starting from 1 */
        ay1 = 0x1e;                           /* process i0 now from ay1 */
        ar = ay1 - ar;
        af = pass af;                         /* if el is greater than table values */
```

```
        if eq ar = pass ay1;              /* mil gets 30 */
        i6 = ar;
        ar = ^quant26bt_neg;
        ay0 = ^quant26bt_pos;
        af = pass sr0;
        if eq ar = pass ay0;
        af = pass ar;
        ar = af - 1;            /* offset by 1 to start addressing from 0 */
        m6 = ar;
        modify(i6,m6);
        ax0 = pm(i6,m6);
        rts;

        /* invqxl is either invqbl or invqal depending on params passed */
invqxl: ar = ay0 - 1;                     /* ay0 is passed in to indicate */
        ar = -ar;                         /* how many bits to shift by */
        se = ar;
        sr = lshift mr0(lo);
        sr = ashift sr0 by 1(lo);
        ar = sr0 + ay1;
        i6 = ar;
        mr1 = ^qq6_table;
        mr0 = ^qq5_table;
        sr0 = ^qq4_table;
        mr2 = 2;
        sr1 = 3;
        af = ay0 - 1,ax0 = pm(i6,m5);    /* save value from table here */
        if eq ar = pass mr1;
        af = ay0 - mr2,ax1 = pm(i6,m5);  /* save sign from table here */
        if eq ar = pass mr0;
        af = ay0 - sr1;
        if eq jump offset_0;
        af = pass ax0;
        ar = ar + af;
        af = pass ar;
                                          /* work around here; qq4 starts */
                                          /* at offset 0, qq5 & qq6 at 1 */
        ar = af -1;                       /* need this for qq5 & qq6, not 4 */
        jump get_sign;

offset_0: ar = pass sr0;
        af = pass ax0;
        ar = ar + af;                /* no offset for qq4, values start at 0 */

get_sign: i6 = ar;
        ar = pm(i6,m5);
        sr = ashift ar by 3(lo);          /* now add sign */
        af = pass ax1, ar= sr0;           /* if its neg, negate value */
        if lt ar = -ar;
        mr = ar * my0(ss);                /* round off here, check it out */
        rts;
```

(listing continues on next page)

5 Sub-Band ADPCM

```
logscl:  my0 = 0x7f00;       /* compensating for scale factor 32512 */
         mr = mr0 * my0(SS);            /* wd in mr1 */
         sr = lshift ar by -2(lo);
         sr = ashift sr0 by 1(lo);
         ay1 = ^code4_table;
         ar = sr0 + ay1;
         i6 = ar;
         m6 = 0;
         ax0 = pm(i6,m6);
         ay0 = ^wl_table;  /* use value from code4_table as index */
         ar = ax0 + ay0;    /* into wl_table */
         i6 = ar;
         ay1 = pm(i6,m6);  /* address for wl(il4) here */
         ar = mr1 + ay1;    /* nbpl here  */
         ay0 = 0x4800;
         af = pass ar;
         if lt af = pass 0;        /* limiting ar - if >18432 */
                                   /* nbpl gets 18432 */
         ar = ar - ay0;            /* if < 0 gets 0 */
         if gt af = pass ay0;
         ar = pass af;
         rts;                      /* this is new delay value */

scalel:  si = ar;
         sr = ashift ar by -6(hi);
         ay0 = 0x1f;
         ar = sr1 and ay0;             /* and with 31 - ar has WD1 */
         mr0 = ar;                     /* this is wd1 in mr0 */
         sr = ashift si by -11(lo);    /* this gives wd2 in sr0 */
         ar =  ay1 - sr0;              /* ay1 has 8 for scalel */
                                       /* and 10 scaleh */
         ar = -ar;
         se = ar;                /* se gets 8 - wd2 */
         ay1= ^ilb_table;
         ar = mr0 + ay1;
         i6 = ar;                /* use wd1 as index in ilb_table */
         ar = pm(i6,m5);       /* wd3 = ilb(wd1) >> (8-wd2) */
         sr = ashift ar (lo);
         sr = ashift sr0 by 2(lo);
         rts;

upzero:  ay0 = ax0;
         se = -15;
         mx0 = 0x7f80;
         ar = pass ay0;
         if eq jump wdi_over;
         ay0 = 0x80;
```

```
wdi_over: sr = ashift ar(lo),si = dm(i5,m5);
          ay1 = sr0;
          cntr = 6;
          do upzero_1 until ce;
                  sr = ashift si(lo), si = dm(i5,m5);
                  ax1 = sr0;
                  af = pass ay0;
                  ar = ax1 - ay1, my0 = dm(i1,m2);
                  if ne af = -af;
                  mr = mx0 *my0(ss);
                  ar = mr1 + af;
          upzero_1:dm(i1,m1) = ar;
          dm(store_this) = si;
          rts;

uppol2:   ar = ax0 + ay0;    /* mx0 has al2, ay0,ax0 have al1 */
                             /* si has plt,mr0 has plt1,mr1 has plt2 */
          af = pass ar;
          ar = ar + af;
          se = -15;
          sr = ashift si(lo), ay0 = ar;        /* wd1 in ay0 */
          ay1 = sr0;                           /* sg0 in ay1 */
          sr = ashift mr0(lo);                 /* sg1 in sr0 */
          ar = sr0 xor ay1;
          ar = ay0;
          if eq ar = -ay0;                     /* wd2 in ar */
          sr = ashift ar by -7(lo);
          ax0 = sr0;                           /* wd2 in ax0 */
          ax1 = 0x80;
          sr = ashift mr1(lo);                 /* sg2 in sr0 */
          ar = sr0 xor ay1;
          ar = ax1;
          if ne ar = - ax1;
          af = pass ar;
          ar = ax0 + af;                       /* wd2 + wd3 = wd4 */
          my0 = 0x7f00;
          mr = mx0 * my0 (ss);
          ay1 = mr1;
          ar = ar + ay1;                       /* apl2 = wd4 + wd5 */
          ay0 = 0x3000;
          ar = abs ar;
          af = ar - ay0;
          if gt ar = pass ay0;    /* note: apl2 limited to ± .75 */
          if neg ar = -ar;
          rts;
```

(listing continues on next page)

5 Sub-Band ADPCM

```
uppol1:   ay0 = 0xc0;
          sr = ashift si by -15(lo);
          af = pass sr0;                     /* sg0 in af */
          sr = ashift mr0 by -15(lo);        /* sg1 in sr0 */
          ar = sr0 xor af;
          ar = ay0;
          if ne ar = - ay0;
          my0 = 0x7f80;
          mr = mx0 *  my0(ss), ay0 = ar;
          ar = mr1 + ay0;          /* apl1 before limits = wd1 + w2 */
          mr0 = ar;
          ax0 = 0x3c00;
          ar = ax0 - ay1;
          ay1 = ar;                /* wd3 in ar, ay1 has apl2 */
          ar = abs mr0;            /* note: wd3 is always positive, */
                                   /* so abs value works */
          af = ar - ay1;
          if gt ar = pass ay1;
          if neg ar = -ar;
          rts;

limit:    ax1 = ar;
          af = pass ax1;
          ay1 = 0x3fff;
          ar = ax1 - ay1;
          ar = pass ar;
          if gt af = pass ay1;
          ay1 = 0xc000;
          ar = ax1- ay1;
          ar = pass ar;
          if lt af = pass ay1;
          ar = pass af;
          rts;

quanth:   sr = ashift ar by -15(lo);        /* sr0 has sih */
          af = pass sr0;
          if eq jump continue;
          ay0 = 0x7fff;
          ax0 = 0x7fff;
          af = ay0 - ar;
          ar = ax0 and af;

continue: af = pass 0,ay0 = ar;
          ay1 = 2;
          mx0 = 0x11a0;           /* q2(564) <<3 */
          mr = mx0 * my0 (ss);    /* this means mr1 will be 2.14 */
          ar = mr1 - ay0;         /* (q2(1)<<3*deth)-wd;if gt 0, */
                                  /* mih is 1) */
          if le af = pass 1;      /* af has mih */
          ar = pass af;
          af = pass 0, ax1 = ar;  /* clear af for processing */
```

```
            ar = pass sr0;
            if eq af = pass ay1;
            ar = ax1 + af;
            ay1 = ^bit_out2;
            ar = ar + ay1;
            i6 = ar;
            ax0 = pm(i6,m5);
            rts;

/* ...... INVQAH: inverse adaptive quantizer for the higher sub-band ..... */
        invqah: mr0 = 0x650;
            mr1 = 0x1cf0;
            ar = pass ay0;
            if eq af = - mr1; /* save sih as flag for passing later */
            ar = ay0 -1;
            if eq af = - mr0;
            ay0 = ar;
            ar = ay0 -1;
            if eq af = pass mr1;
            ay0 = ar;
            ar = ay0 -1;
            if eq af = pass mr0;
            ar = pass af;
            mr = ar * my0 (ss);
            rts;

        logsch: mr0 = 0xff2a;      /* these correspond to wh for 1 & 2 */
            mr1 = 0x31e;
            ar = pass ay0;
            if eq af = pass mr1;    /* save sih here as flag */
                                    /* for passing later */
            ar = ay0 -1;
            if eq af = pass mr0;
            ay0 = ar;
            ar = ay0 -1;
            if eq af = pass mr1;
            ay0 = ar;
            ar = ay0 -1;
            if eq af = pass mr0;            /* af has wh(ih2) */
            mr = mx0 * my0(ss);
            ar = mr1 + af;
            ay0 = 0x5800;
            ar = abs ar;
            af = ar - ay0;
            if gt ar = pass ay0;
            if neg ar = pass 0;
            rts;
```

(listing continues on next page)

5   Sub-Band ADPCM

```
chk_vals:ax1 = 0xc000;
        ax0 = 0x3fff;
        if av jump chk_ov;
        af = pass ar;
        ar = abs ar;
        if pos jump chk_pos;
        ar = ax0 + af;               /* if abs val is neg, execute this code */
        if lt af = pass ax1;
        ar = pass af;
        rts;

chk_pos: ar = af - ax0;
        if gt af = pass ax0;
        ar = pass af;
        rts;

chk_ov:  ar = pass ax0;
        if lt ar = pass ax1;
        rts;

.endmod;
```

Listing 5.1 Implementation Of The G.722 Algorithm

5.7   BENCHMARKS

Table 5.3 contains typical benchmarks for implementing Sub-band ADPCM (CCITT Recommendation G.722).

Memory Usage:	PM RAM		DM RAM	
	1312 Locations		208 Locations	

DSP	Processor Speed	Number of Cycles		Execution Time	Processor Loading
ADSP-2101	20 MHz	Encoder	821	41.05 μs	34.8%
		Decoder	742	37.10 μs	27.2%
		Total	1563	78.15 μs	62%
ADSP-2171	33 MHz	Encoder	821	24.63 μs	19.7%
		Decoder	742	22.26 μs	17.8%
		Total	1563	46.89 μs	37.5%

Table 5.3 Typical Benchmark Performance

Speech Recognition ■ 6

6.1 OVERVIEW

This chapter describes the basic framework for speech recognition using ADSP-2100 Family Digital Signal Processors. Although there are many techniques available for speech recognition, this chapter focuses on a single LPC-based technique. This technique takes advantage of the flexible architecture, computational power, and integration of these processors. It also takes full advantage of the family's development tools, which support a modular design that can be easily tested and quickly modified. The modular code design lets you customize the code efficiently for each individual application. For this reason, the programming examples in this chapter have not been optimized since the memory space and speed requirements are specific for each system.

Because of advances in speech recognition and processing, you can design systems that users control through speech. Today, systems and applications that have limited vocabularies are available with a recognition accuracy nearing 100%. The increasing speed and integration of digital signal processors make portable speech processing and recognition units possible. State of the art DSPs have serial ports, substantial memory, and analog interfaces on a single chip, letting you design single-chip solutions for many speech processing applications.

Speech recognition research and development has several goals. Simplifying the interface between user and machine is one major goal. Just as many users consider the mouse an improvement to the user interface on a personal computer, machine speech recognition and understanding has the potential to greatly simplify the way people work with machines. Examples of this emerging technology include dialing telephones and controlling consumer electronics through voice-activation. As voice input and output become further integrated into the everyday machines, many advances will be possible.

6   Speech Recognition

Analog Devices is at the forefront of this emerging technology. With the powerful ADSP-2100 family of DSPs, ADI has asserted its leadership and commitment to this field. For example, the ADSP-21msp50 is a complete system for speech processing. It contains an analog interface, two serial ports, one parallel communication port, expansive on-chip memory, and the superior signal processing architecture of Analog Devices Digital Signal Processors.

6.2   SPEECH RECOGNITION SYSTEMS

Speech recognition systems fall into two categories:

- *Speaker dependent systems* that are used (and often trained) by one person

- *Speaker independent systems* that can be used by anyone

Regardless of the type of system, the theory behind speech recognition is relatively simple. First, the DSP acquires an input word and compares it to a library of stored words. Then, the DSP selects the library word that most closely matches the unknown input word. The selected word is the recognition result. Systems that follow this model have two distinct phases: *training phase* and *recognition phase*.

To help you understand the processes used to develop the speech recognition system implemented in this chapter, this section also briefly describes the theory of voice production and modeling.

6.2.1   Voice Production & Modeling

You can separate human speech production into two distinct sections: sound production and sound shaping. *Sound production* is caused by air passing across the vocal chords (as in "a", "e", and "o") or from a constriction in the vocal tract (as in "sss", "p", or "sh"). Sound production using the vocal chords is called *voiced speech*; *unvoiced speech* is produced by the tongue, lips, teeth, and mouth. In signal processing terminology, sound production is called *excitation*.

Sound shaping is a combination of the vocal tract, the placement of the tongue, lips, teeth, and the nasal passages. For each fundamental sound, or phoneme, of English, the shape of the vocal tract is somewhat different, leading to a different sound. In signal processing terminology, sound shaping is called *filtering*.

Speech Recognition   6

An efficient method of modeling human speech is to separate the speech into its components: an excitation (the sound production) and a filter (the sound shaping). When you need to compress speech for transmission, each part can be efficiently coded and transmitted independently. The coded parameters can then be decoded and synthesized to reconstruct the original speech.

In most speech processing applications, the two parts of speech have an equal importance. For speech recognition, however, they do not. The excitation changes drastically from person to person, and it changes according to the speaker's gender, physical and emotional state. The sound shaping, or filtering, is less sensitive to these factors. For this reason, in a basic speech recognition system, you only need to consider the filter.

A robust and efficient method exists for estimating the sound shaping filter. Called linear predictive coding, or LPC, it estimates the filter characteristics, or the spectral envelope of the sound shaping. By using only the LPC generated coefficients, redundant and unnecessary information is removed from the speech signal, leaving just the essential information for speech recognition. For a more detailed explanation of LPC, refer to *Chapter 10 of Digital Signal Processing Applications Using the ADSP-2100 Family, Volume 1*.

Since many different sounds are strung together to form a single word, many sets of LPC filter coefficients are necessary to represent the word. A series of coefficient sets is stored to represent the sound-shaping filter at each particular place the word is sampled. This is possible because speech is a slowly-varying signal. If the speech is processed in short enough time slices, or frames, the sound-shaping filter is approximately constant for the duration of that frame. A series of LPC coefficient sets generated from a series of frames then represents the word, with each frame representing a time-slice of the speech.

A word is stored as a series of frames, with each time-slice of speech represented as a feature vector, using a set of LPC coefficients. In an isolated word system, the beginning and ending points of the word can be detected automatically, therefore only the word itself is captured and stored.

You can build a recognition library from these captured words. Each word to be recognized is stored in a library. For speaker-independent systems, multiple copies of each word may be stored to represent different ways of saying the same word. Once the library is built, the system training is complete, and the task of recognition can begin.

6   Speech Recognition

6.2.2   Training Phase

When you train a system to recognize words, you first create a library of stored words. The training phase changes depending on the type of speech recognition system. The system compares the input words against this library to find the closest match.

In a speaker dependent system, ideally the user and trainer are the same person. In this situation, these systems offer the best performance because the input words will be fairly consistent. Also, the recognition library can be relatively small because of the limited number of speech samples required to recognize the input words. Because of accents, dialects, and other variations in speech, the performance of speaker-dependent systems degrades when one person trains the system and another person uses it.

Speaker independent systems are usually trained with speech from many people. This process can be more involved than training speaker-dependent systems because you need more speech samples, (perhaps several hundred, or a thousand samples for each word) to train the system. Speaker independent systems typically require larger memories to hold the larger library.

Although the number of required samples may vary depending on the type of speech recognition system, fundamentally the training process remains the same. Figure 6.1 shows a functional block diagram of the training phase of the speech recognition system implemented in this chapter.

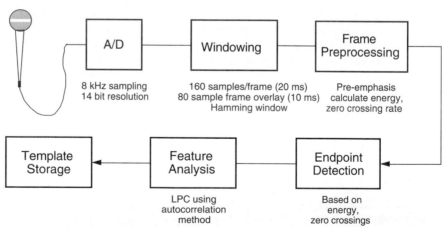

Figure 6.1 Speech Training System Block Diagram

Speech Recognition   6

6.2.3   Recognition Phase

Figure 6.2 shows a block diagram of the recognition phase. Notice that the two phases of the speech recognition system share the same word acquisition functions.

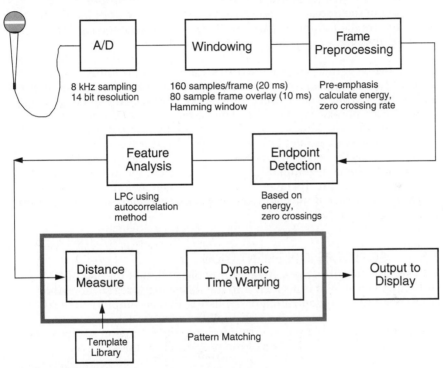

Figure 6.2 Speech Recognition System Block Diagram

During speech recognition, the DSP compares an unknown input word to a library of stored words. Then, for the recognition result, it selects the library word that is most similar to the unknown word. The method implemented in this chapter is a template-based, dynamic time warping (DTW) system. Each library word, stored as a series of feature vectors containing LPC-derived coefficients, is referred to as a template. Since the time you take to completely utter a word changes, dynamic time warping aligns the time axes of the unknown word to a library template. By lengthening or shortening sections of the unknown word, the system attains a "best fit" between words. With DTW, different durations for words have little effect on the recognition accuracy. Dynamic time warping is described in Section 6.3.2.3.

333

6 Speech Recognition

The unknown word and the library word are represented as a series of feature vectors. To compare the words, a measure of similarity between the words is necessary. At the most basic level, the system must measure the similarity between two feature vectors. This is referred to as a distance, or distortion, measure. Many distortion measures are proposed and evaluated in the published literature. Two of the most popular distortion measures are the *Itakura log-likelihood ratio* and the *bandpass cepstral distortion measure.*

During the recognition stage, the distortion measure is integrated into the dynamic time warping routine. The "best fit" between the unknown word and a library word is calculated; the system compares the unknown input word to each library word in turn. The system maintains a recognition word score for each library word. For a single template-per-word system (usually speaker-dependent systems), typically, the system chooses the lowest score as the recognition result. For speaker-independent systems where more than one template-per-word is stored, the lowest scores for each library word are averaged. This results in an average recognition word score for each library word. The system still selects the lowest score as the recognition result.

6.3   SOFTWARE IMPLEMENTATION

This section describes the software implementation of the speech recognition system; it is divided into the following three sections that correspond to the organization of the listing that accompany the text.

The software implementation is divided into the following sections that correspond to the organization of the program examples.

- Word Acquisition and Analysis

- Word Recognition

- Main Shell Routines

Speech Recognition 6

6.3.1 Word Acquisition & Analysis

This section describes the functions necessary for word acquisition. These functions are divided into a receive shell, frame analysis, word endpoint detection and coefficient conversion. Each of these functions is contained in a separate subroutine.

The input and output of data samples is interrupt-driven using serial port 0. The interrupt routine is in the main shell. The data is sampled at 8 kHz, and is sectioned into 20 ms frames (160 samples). Each frame overlaps the previous frame by 10 ms (80 samples). This leads to 100 frames per second. As currently configured, a word can not exceed one second.

6.3.1.1 Receive Shell

The subroutine *get_word* in the receive shell is called when a word is acquired. This routine returns a word with I0 pointing to the starting vector of the word, and AX0 holding a count of the number of vectors in the word.

After the software initializes the necessary variables, pointers, and buffers, the receive shell enables data acquisition. A circular buffer of 240 locations is used as the sample input buffer. Each frame of samples is 160 samples long, with consecutive frames overlapping by 80 samples. The code at `code_1_loop` counts the samples. The first time through this loop, the loop exits after a full frame of 160 samples is acquired. After the first iteration, it exits after 80 samples. The result is an 80 sample overlap for consecutive frames.

The 160 samples are copied into a separate frame buffer for processing, since the frame analysis destroys the input data. A frame pointer points to the beginning of each frame in the 240 location circular input buffer. Next, the software analyzes the frame and converts the coefficients. (Sections 6.3.1.2, *Frame Analysis* and 6.3.1.4, *Coefficient Conversion* describe these processes.) The resulting coefficients are written into the feature vector buffer at the current location.

Before word acquisition is complete, the system must detect the word's endpoints (see Section 6.3.1.3, *Endpoint Detection*). Several options exist based on the results from this subroutine call. If the system detects the word's start, or possible start, the vector pointer is advanced and the vector count is incremented. The system compares the length of the word to a maximum length, and, if this maximum is exceeded, forces the word to end. If the word does not exceed the maximum length, the system acquires additional frames and the repeats the process.

6 Speech Recognition

If the system detects the word's end, it stops sampling data. It pads the beginning of the word with the five previous feature vectors to insert noise before the word. Then, the routine returns.

If the system fails to detect the beginning or end of a word, the vector pointer and count are reset. The feature vector is written into a circular start buffer, and can be used to pad the beginning of the next word. The code then jumps to the start, and acquires more frames.

At several places in the shell, code is included for conditional assembly. If you use this code, the first four features of the feature vector are the energy, change in energy, zero crossing rate (ZCR), and change in zero crossing rate. For most applications, this information is not necessary. If your application requires this information, the included code adds processing to the receive shell. When the end of the word is detected, the energy values are scaled based on the maximum value determined. Then, the change in energy and the change in ZCR values are determined for each feature vector.

6.3.1.2 Frame Analysis

The subroutine analyzes the frames with an LPC analysis that uses auto correlation and the Schur recursion. It requires pointers to the data frame and the output buffer as inputs. The routine returns eight reflection coefficients.

The subroutine calculates the energy of the frame after scaling and offset filtering the coefficients. A sum of the magnitude of the energy measure is used. Before summing, the magnitudes are scaled to prevent overflows. The subroutine also calculates the zero crossing rate. One zero crossing occurs each time the input data changes sign and passes beyond the noise threshold. Each crossing increases the zero crossing rate by 205. Using this value, a 4000 Hz input results in a zero crossing rate of 32595, taking advantage of the full data width.

Pre-emphasis takes place after these calculations, and uses a coefficient of -28180. Next, a Hamming window is multiplied by the frame of data. Finally, the auto correlation and the Schur recursion complete the frame analysis.

6.3.1.3 Endpoint Detection

The endpoint detector is a variation of an endpoint detector proposed by Rabiner (see the references for the source). It is based on the energy (sum of magnitudes) and the zero crossing rate (ZCR) of each frame of input data. The subroutine determines the word's endpoints by comparing these values to several thresholds. These thresholds adapt to steady background noise levels. Several flags are returned from this routine to indicate a definite word start, a possible word start, or the end of a word.

There are two types of thresholds for the energy and ZCR, possible thresholds and word start (WS) thresholds. *Possible thresholds* are set just above the background noise levels, and for this reason, they may be exceeded occasionally by spurious background noise. The *word start thresholds* are set relatively high so they are exceeded only when the system is sure a word is being spoken. Setting WS thresholds to high, however, causes the detector to miss some softly spoken words. It may be necessary to experiment with threshold levels to achieve the best results.

There are two additional thresholds. The *minimum word length threshold* is set to the minimum number of frames per word. This should be long enough to avoid isolating background noise spikes, but not too long. The *threshold time* is the length of silence that must be detected before a word end is determined. This is necessary to allow silence in the middle of words (especially preceding stops, like "t" or "p").

When searching for the start of a word, the algorithm first compares the frame energy and zero crossing rate to the WS thresholds. If the frame energy or ZCR exceeds the threshold, the word start flag is asserted, and the system starts storing frames. If the threshold is not exceeded, the possible thresholds are compared. If the frame energy or ZCR exceeds the possible thresholds, the possible start flag is set and the system starts storing frames. For this to be considered the actual start of a word, however, the WS thresholds must be exceeded before the frame energy and ZCR fall below the possible thresholds.

Once a word is determined, the algorithm searches for the end of the word. The subroutine finds the end of the word when the energy and ZCR fall below the possible thresholds for longer than the threshold time. When this happens, the word end flag is set.

6 Speech Recognition

6.3.1.4 Coefficient Conversion

The LPC analysis performed on the incoming frames of data produces eight reflection coefficients for each 160 samples of input speech. While this data compression is outstanding, for recognition purposes, the reflection coefficients are not the best features to represent the speech. There are two widely used representations; the predictor coefficients of the LPC analysis and the cepstral coefficients. The *predictor coefficients* are the parameters of the all-pole filter that is being modeled. These predictor coefficients are often referred to as α_k. The *cepstral coefficients* are parameters of the impulse response of the log power spectrum of the input speech. In this case, the cepstral coefficients are solved for recursively from the predictor coefficients, and are referred to as c_k.

Coefficient conversion immediately follows the frame analysis, but happens before feature vector storage. The conversion module is separated into several subroutine calls, each with a specific function. The implementation is in floating-point, with a 16-bit mantissa and 16-bit exponent. This method lets you ignore scaling issues, speeding the code development. The floating-point routines are adapted from the routines in Chapter 3, *Floating-Point Arithmetic*, in *Digital Signal Processing Using the ADSP-2100 Family*, Volume 1, and are called throughout the module.

The first subroutine called from the conversion shell, `k_to_alpha`, converts the fixed-point reflection coefficients (k's) to the floating-point predictor coefficients (α_k). The conversion is accomplished using a the following recursion

$$a_i^{(i)} = k_i$$
$$a_j^{(i)} = a_j^{(i-1)} + k_i a_{i-j}^{(i-1)} \quad 1 \le j \le i-1$$

which is solved recursively for i = 1, 2, ..., p. The final results are found from

$$a_j = a_j^{(p)} \quad 1 \le j \le p$$

For the current system, p = 8.

Two buffers are used to store the temporary results of the recursion, one for even values of i and one for odd values. These buffers alternate as the input buffer and the result buffer at each stage of the recursion, until the final result is contained in the even buffer (since p = 8).

At the completion of this subroutine, all of the predictor coefficients have been calculated. Many of the popular distortion measures use these parameters in the recognition calculations, such as the Itakura log-likelihood ratio. If the predictor coefficients are the desired features, conversion from floating-point back to a fixed-point representation finishes the routine.

The present system uses a cepstral representation. The following recursion is also used to convert from the predictor coefficients to the cepstral coefficients.

$$c_1 = -a_1$$

$$c_k = -a_k - \sum_{i=1}^{k-1} a_i c_{k-i} \left(\frac{k-i}{k} \right) \quad 1 \le k \le p$$

$$c_k = -\sum_{i=1}^{p} a_i c_{k-i} \left(\frac{k-i}{k} \right) \quad p < k$$

The implementation of this algorithm in subroutine `alpha_to_cep` is straightforward, and is commented in the code. From the eight predictor coefficients, twelve cepstral coefficients are calculated. These twelve coefficients are also used in several well-known distortion measures, and can be used directly following conversion to a fixed-point representation.

You can obtain better performance by using a window in the cepstral domain to weight each coefficient. Several different weights are described and evaluated in the literature, including weighting by the inverse variance of the coefficients or weighting by a raised sine function. The weighting chosen for this implementation is shown below:

$$w(k) = 1 + 6 \sin\left(\frac{\pi k}{12} \right) \quad 1 \le k \le 12$$

The subroutine `weight_cep`, used to weight the cepstral coefficients, is also straightforward. The weighting values are initialized in a separate buffer, making them easier to modify.

6   Speech Recognition

The next subroutine *normalize_cep*, normalizes the twelve cepstral coefficients to the length of the entire vector. Normalization is necessary for the cepstral projection distortion measure. The length of the vector is the square-root of the sum of each coefficient squared. To square each coefficient, multiply it by itself. These values are then accumulated in a temporary variable. The square-root subroutine calculates an approximate square-root of the mantissa. This subroutine is adapted from a subroutine in Chapter 4, *Function Approximation*, of *Digital Signal Processing Applications Using the ADSP-2100 Family*, Volume 1, and calculates an approximate square-root of the mantissa. If the exponent is even, it is divided by two, giving the correct floating-point result. If the exponent is odd, it is incremented and divided by two, and the mantissa is scaled by

$$\frac{1}{\sqrt{2}}$$

This results in the appropriate value. Each cepstral coefficient is then scaled by this calculated length, using a floating-point divide routine.

The final step is to convert the floating-point cepstral coefficients back to fixed-point using `cep_to_fixed` . The results are written over the original input buffer.

6.3.2   Isolated Word Recognition

Following the word acquisition, one of two things happens. If the system is in the training mode, the word is stored in the library, and a record is kept of its location and length. In the recognition mode, this unknown word is compared to each template in the library, and the recognition result value is returned.

6.3.2.1   Library Routines

The library routines store and catalog the acquired words. The words are stored in a template library that occupies the external program memory of the ADSP-2100 Family processor (as currently implemented). The most important function of these routines is to store a new word in the template library. The code uses several variables to organize the library. These include a variable for catalog size (tracks of the number of words in the library); a library catalog is built as words are added. Two values are stored for each library template. The first value represents a pointer to the starting location of the template. The second value represents the length of the template, and it is stored as the number of vectors in the word. A final variable records the location of the next available catalog entry.

Speech Recognition 6

While template storage and library catalog maintenance are the most important functions of the library routines, the code contains other optional routines including playback. If the features representing the library are reflection coefficients, the words are played back through the speaker. Several variations of playback are available. A single template may be played, all the templates in the library may be played in order, or the library templates may be played in any order. This final routine is useful to play the library templates after recognition, beginning with the most probable word and ending with the least.

6.3.2.2 Comparison

A comparison routine compares an unknown word to the full library. Several comparison routines exist, differing only in the distance measure used for the recognition. In the implemented system, four different people are used for training, with each person's speech stored in a different bank of program memory.

The unknown word is compared to each of the four template banks separately. Each bank has its own library catalog, storing the location and length of each entry in the bank. Using a bank's catalog, the comparison subroutine initializes values needed to compare the unknown word to each of the specified bank's templates, in order. The comparison includes dynamic time warping and the distortion measure (see Section 6.3.2.3, *Dynamic Time Warping*, for more information). The comparison subroutine must be called once for each bank used.

The result of the comparison between the unknown word and a template is the word distance score. A buffer must be specified to hold these double-precision results, msw followed by lsw. These word distance scores are stored in the same order as the words stored in the library. A different buffer is used for each bank. After an unknown word is compared to each template in all four banks, the results are stored in four separate distance buffers.

Since all banks contain the same vocabulary in the same order, four word distance scores exist for each template. A *K-Nearest Neighbor* routine averages the results for each word. The implemented algorithm finds the two lowest scores of the four scores for each vocabulary word. These two are then summed, resulting in the final word distance score. This final word distance score is found for each word of the vocabulary. Using the K-Nearest Neighbor decision algorithm, the speech recognition becomes speaker-independent.

6 Speech Recognition

6.3.2.3 *Dynamic Time Warping*

The speech recognition code contains a complete system to perform dynamic time warping, or DTW, between two words. DTW dynamically matches two patterns of different lengths. In this case, each pattern represents a word, and each pattern is represented by a time sequence of feature vectors taken from a moving window of speech. The DTW algorithm aligns the time axis of the library word with the time axis of the unknown word, leading to the lowest possible word distance score.

The constraints used in this implementation were suggested by Itakura. This example tries to match unknown word(x), of length N, to a library word(y), of length, M. The indices x and y refer to a particular time frame of speech data, represented by a feature vector. A distance matrix can be calculated to represent the distance between an x (unknown word) feature vector and all y (library word) feature vectors, evaluated for $0 <= x <= N$. Each point of the distance matrix has a value that is the distance between a single x feature vector and a single y feature vector. The specific distance measure used between feature vectors is arbitrary. The distance matrix is the only thing DTW needs.

To warp the time axis of the library word to the time axis of the unknown word, several constraints must be set. The starting point of the warping is (0,0), and the ending point must always be (N,M). The minimum slope of the warp is 1/2, and the maximum slope is 2. Finally, two consecutive slopes of 0 are not allowed. Figure 6.3 shows a diagram of a distance matrix with these constraints.

As this diagram shows, most of the distance matrix is invalid when the slope and warping constraints are imposed. Significant execution time is saved if only valid warping paths are considered, and only vector distances within the warp boundaries are calculated.

To determine the boundaries of the warping, the points A and B (or x_A and x_B), shown in the diagram, must be calculated. The following equations represent these two points:

$$x_A = \frac{1}{3}(2M - N)$$

$$x_B = \frac{2}{3}(2N - M)$$

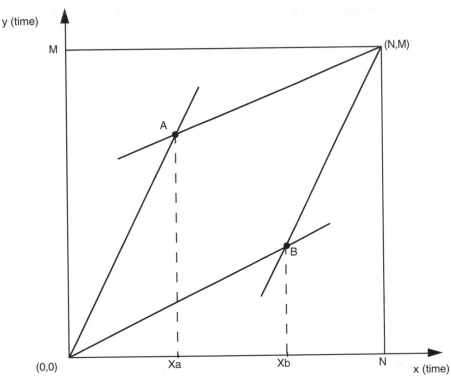

Figure 6.3 Distance Matrix With Slope Constraints

Since the actual processing is performed only at points where x and y are integers, the values of x_A and x_B are rounded down to the nearest integer in all cases, without loss of accuracy.

The values of x_A and x_B must be in the range of $0 < x_A < N$ and $0 < x_B < N$. This imposes a constraint on the lengths of the unknown word and the library word. The equations for this requirement are:

$$2M - N \geq 3$$

$$2N - M \geq 2$$

If this relation is not met, the two words cannot be warped together in this implementation.

6   Speech Recognition

Finally, the minimum and maximum y values must be determined for each x value. The equations for this are:

y minimum

$$= \frac{1}{2}x \quad 0 \le x \le x_B$$

$$= 2x + (M - 2N) \quad x_B < x \le N$$

y maximum

$$= 2x \quad 0 \le x \le x_A$$

$$= \frac{1}{2}x + \left(M - \frac{1}{2}N\right) \quad x_A < x \le N$$

The warping can be broken into two or three sections, based on the relationship of x_A and x_B. x_A can be less than, greater than, or equal to x_B. Each of these cases has different boundaries for each section, as summarized below in Table 6.1.

Section	$x_A < x_B$	$x_B < x_A$	$x_A = x_B$
1	0 <= x <= x_A	0 <= x <= x_B	0 <= x <= x_A, x_B
2	x_A < x <= x_B	x_B < x <= x_A	x_A, x_B < x <= N
3	x_B < x <= N	x_A < x <= N	none

Table 6.1 Time Warping Boundaries

For each case, the boundaries of y are different, but the warping is the same. The DTW finds the path of minimum word distance through the distance matrix, while considering the given constraints. This is done sequentially, beginning at x = 0 and ending at x = N. The following recursion shows the path through the matrix that is subject to warping constraints.

$$D(x,y) = d(x,y) + \min\left[D(x-1,y), D(x-1,y-1), D(x-1,y-2)\right] \quad 0 \le x \le N$$

$D(x,y)$ represents the (intermediate) word distance score at (x,y), and $d(x,y)$ is the value (vector distance) at point (x,y).

Since the recursion only involves the values in columns (x-1) and x, the complete distance matrix does not need to be calculated before the recursion begins. Instead, two buffers are set up. The intermediate sum

Speech Recognition   6

buffer contains the values of D(x-1,y) for all y, organized as msw, lsw, warp value. The vector distance buffer contains the values of d(x,y) for all allowable y, organized as msw, lsw, empty location. The warp value in the intermediate sum buffer is the previous warp value (from column (x-2) to (x-1)), and is required to determine the allowable warp from column x-1 to x.

When the time warping commences from column x-1 to column x, the values in the intermediate sum buffer are examined to determine the minimum intermediate sum present in the allowed warping path. This minimum is then added to the value of the vector distance and placed in the vector distance buffer, along with the slope of the warp used. Figure 6. 4 shows the allowable paths. After the warping is complete for all values of y (y minimum–y maximum) the vector distance buffer contains the current intermediate sums. Before the next column is processed, these values must be copied into the intermediate sum buffer.

The recursion continues until x=N, when the vector distance buffer contains the final word distance score.

A single exception exists to the constraints on the warping path. It specifies that a warp of 0 is not allowed for two consecutive warps. However, since only integer indices are considered for (x,y), a case exists

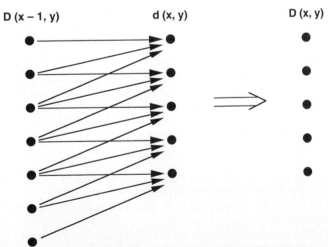

Figure 6.4 Time Warping Paths Between Intermediate Sums & Vector Distances

6   Speech Recognition

where the best option is to allow two consecutive 0 warps. You can saturate the intermediate distance calculated at the exception point, but this option yields an unknown effect on the recognition.

The warping path is constrained by saturating the intermediate sum values that, if selected, result in illegal warps. This insures that these paths are selected.

6.3.2.4   Ranking

After the unknown input word is compared to the complete template library, the routine returns a single buffer containing a word distance score for each vocabulary word. The ranking routine then compares the scores for each template. The routine finds the smallest distance contained in the buffer and places a pointer to the corresponding library catalog entry in the candidate order buffer. The buffer is filled, storing the most probable recognition candidate first, the second next, and so on. The least word distance score is considered the most likely recognition result. The location of each word's entry in the catalog library is the stored value. A separate buffer contains the candidate's number in the library (first, second, tenth, etc.), stored in the same order. The returned buffer for candidate order contains pointers to library catalog entries of template words, in order from the least word distance score to the greatest word distance score. The second buffer contains each candidate's number in the library, stored in the same order.

You will probably need to make minor modifications to the code so it will be compatible with your particular application.

6.3.3   Main Shell Routines

This speech recognition system has two different main shell routines. The *executive shell* is used for the initial training of the system, and can be used for testing the resulting templates. The *demonstration shell* is used after the system is fully trained, and it is designed to demonstrate hands-free dialing for a telephone application, although the recognition system can be used for any application.

Only one of these shells is used at a time. Since the code is written in a modular fashion, and includes many subroutine calls, this scheme is possible. Both shells contain an interrupt table, initialization functions, and an interrupt routine used to process samples. The interrupt sample routine has an output flag to select whether data is being input or output, since both are not done at the same time. The additional features of each shell are described in more detail in the following sections.

Speech Recognition 6

6.3.3.1 *Executive Shell*

The executive shell (EXECSHEL.DSP), shown in Listing 6.1, calls the functions necessary for speech recognition: getting an input word, adding a word to the library, and recognizing a word. Figure 6.5 shows the link file menu tree used by EXECSHEL.DSP. The interface is the minimum necessary to accomplish the tasks. If the LPC reflection coefficients are used as features, this shell can call routines to playback a single word or the entire library. If another representation is used, the recognized word can be output to a display.

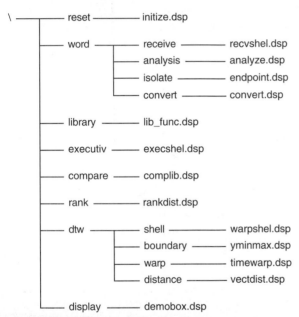

Figure 6.5 EXECSHEL.DSP Link File Menu Tree

The code is organized into one main loop. On reset, the system gets an input word. This word is either added to the recognition library or compared for recognition. An interrupt must be asserted before the word is spoken if the word will become a library template. Note that the interrupt is only enabled during the word acquisition routine.

6 Speech Recognition

6.3.3.2 Demonstration Shell

The demonstration shell (DEMOSHEL.DSP), shown in Listing 6.2, calls some of the functions necessary for speech recognition: getting an input word and recognizing a word. It also calls many display routines. Figure 6.6 shows the link file menu tree used by DEMOSHEL.DSP. The interface is designed for a demonstration of hands-free dialing. The specific display routines can be changed to communicate with any desired display.

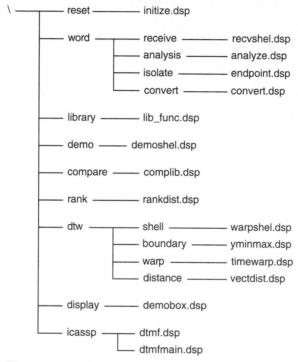

Figure 6.6 DEMOSHEL.DSP Link File Menu Tree

The code is organized to reflect the different stages of dialing a telephone. Using a fifteen word vocabulary (the letter "o", zero, one, two, three, four, five, six, seven, eight, nine, dial, delete, phonecall, scratch), the demonstration accepts a spoken phone number as isolated words, then dials. After reset, the demonstration continuously gets input words for recognition. The command, "phonecall," alerts the processor that a phone number is about to be spoken. Once "phonecall" is recognized, the demonstration moves to the next stage.

348

Speech Recognition 6

The demonstration then accepts the first digit of the number, and moves into the final state of the system. In the last state, the processor adds each digit to the phone number, as it is spoken. When the command "dial" is recognized, the software boots the next boot page, which consists of a dialing routine.

If a mistake is made during recognition, the command "delete" removes the preceding digit from the phone number. Repeatedly speaking "delete" continues to erase digits. To reset the demonstration use the command "scratch" to return the demonstration to its initial state, where it waits for the command "phonecall."

This demonstration shell performs a basic calling routine. It could serve as a framework for an actual implementation. Functions that might be added include: local or long distance dialing, memory dialing, and so on.

6.4   HARDWARE IMPLEMENTATION
The speech recognition uses a hardware platform designed specifically for this application. This expansion board is connected to the ADSP-2101 EZ-LAB® Demonstration Board through the EZ-LAB connector. Figure 6.7 is the schematic diagram for this circuit board.

6.5   LISTINGS
This section contains the listings for this chapter.

6 Speech Recognition

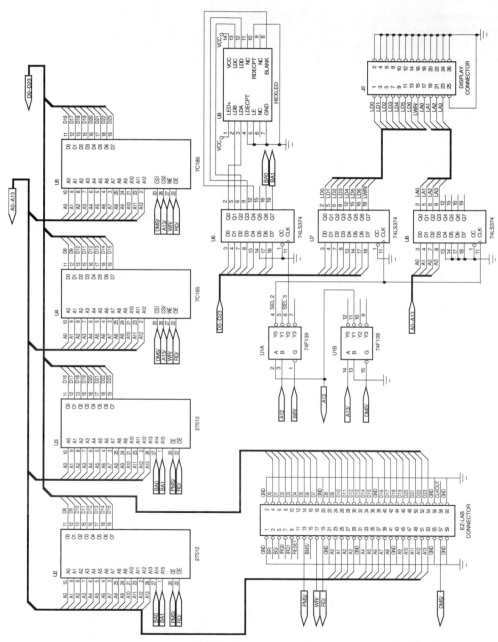

Figure 6.7 Speech Recognition System Circuit Board Schematic Diagram

Speech Recognition   6

```
.MODULE/ABS=0/RAM/BOOT=0        executive_shell;

.VAR/DM/RAM        flag;
.VAR/DM/RAM        output_flag;
.VAR/DM/RAM        unknown_feature_dimension;
.VAR/DM/RAM        library_feature_dimension;

.GLOBAL            output_flag;
.GLOBAL            unknown_feature_dimension;
.GLOBAL            library_feature_dimension;
.GLOBAL            flag;

.EXTERNAL          get_word;
.EXTERNAL          put_in_library;
{.EXTERNAL         play_library;}
{.EXTERNAL         play_single;}
{.EXTERNAL         coarse_compare;}
{.EXTERNAL         fine_compare;}
{.EXTERNAL         full_compare;}
{.EXTERNAL         shuffle_play;}
.EXTERNAL          rank_candidates;
.EXTERNAL          reset_recog;
.EXTERNAL          display_digit;
{.EXTERNAL         add_a_digit;}
{.EXTERNAL         display_number;}
.EXTERNAL          set_bank_select;
.EXTERNAL          show_bank;
.EXTERNAL          cepstral_compare;
{%%%%%%%%%%%%%%%%%%%%%%%%%%%%%%%%%%%%%%%%%%%%%%%%%%%%%%%%%%%%%%%%%%%%%%%%%%%%%}
{%%%%%%%%%%%%%%%%%%%%%%%%%%%%%%%%%%%%%%%%%%%%%%%%%%%%%%%%%%%%%%%%%%%%%%%%%%%%%}
{_____main shell for speech recognition_____}
{%%%%%%%%%%%%%%%%%%%%%%%%%%%%%%%%%%%%%%%%%%%%%%%%%%%%%%%%%%%%%%%%%%%%%%%%%%%%%}
{%%%%%%%%%%%%%%%%%%%%%%%%%%%%%%%%%%%%%%%%%%%%%%%%%%%%%%%%%%%%%%%%%%%%%%%%%%%%%}

reset_vector:      JUMP start; NOP; NOP; NOP;
irq2:              JUMP toggle_flag; NOP; NOP; NOP;
trans0:            NOP; NOP; NOP; NOP;
recv0:             JUMP sample; NOP; NOP; NOP;
trans1:            NOP; NOP; NOP; NOP;
recv1:             NOP; NOP; NOP; NOP;
timer_int:         NOP; NOP; NOP; NOP;

start:             IMASK=0;
                   ICNTL=B#00100;
                   L0=0;   L1=0;   L2=0;   L3=0;
                   L4=0;   L5=0;   L6=0;   L7=0;
                   M0=0;   M1=1;   M2=-1;  M3=2;
                   M4=0;   M5=1;   M6=-1;  M7=2;
```

(listing continues on next page)

351

6 Speech Recognition

```
reg_setup:   AX0 = 0;
             DM(0X3FFE) = AX0;        { DM wait states }
             AX0 = 2;
             DM(0X3FF5) = AX0;        { sclkdiv0 with 12.288 MHz input }
             AX0 = 255;
             DM(0X3FF4) = AX0;        { rfsdiv0 }
             AX0 = 0X6927;
             DM(0X3FF6) = AX0;        { control reg0 }
             AX0 = 0X1004;
             DM(0X3FFF) = AX0;        { system control reg }

             CALL reset_recog;
             AR = 4;
             CALL set_bank_select;
             CALL show_bank;

recognition:
             IMASK = 0x20;
             CALL get_word;
             AY0 = DM(flag);
             AF  = PASS AY0;
             IF NE JUMP build_library;

             CALL cepstral_compare;
             CALL rank_candidates;    { buffer pointer returned in AY0 }

             CALL display_digit;

             AX0 = 1;                 { play the top three candidates}
{            CALL shuffle_play;}
             JUMP recognition;

build_library:
             CALL put_in_library;
{            CALL play_library;}
             AX0 = 0;
             DM(flag) = AX0;
             JUMP recognition;

{_____toggle record/recognize flag_____}
toggle_flag:
             ENA SEC_REG;
             MR0 = DM(flag);
             AR  = NOT MR0;
             DM(flag) = AR;
             RTI;
```

Speech Recognition   6

```
{_____process sample_____}
sample:  ENA SEC_REG;

         AR  = DM(output_flag);
         AR  = PASS AR;
         IF EQ JUMP get_input;

send_output:
         SR1 = DM(I7,M5);
         SR  = ASHIFT SR1 BY -2 (HI);
         TX0  = SR1;
         JUMP inc_count;

get_input:
         AR=RX0;
         TX0 = AR;
         DM(I7,M5)=AR;            {Save sample}

inc_count:
         AY0=MX0;
         AR=AY0+1;
         MX0=AR;
         RTI;

.ENDMOD;
```

Listing 6.1 Executive Shell Subroutine (EXECSHEL.DSP)

6 Speech Recognition

```
.MODULE/ABS=0/RAM/BOOT=0          demonstration_shell;

.VAR/DM/RAM flag;
.VAR/DM/RAM output_flag;
.VAR/DM/RAM unknown_feature_dimension;
.VAR/DM/RAM library_feature_dimension;

.GLOBAL     flag;
.GLOBAL     output_flag;
.GLOBAL     unknown_feature_dimension;
.GLOBAL     library_feature_dimension;

.EXTERNAL   get_word;
.EXTERNAL   put_in_library;
{.EXTERNAL  play_library;}
{.EXTERNAL  play_single;}
{.EXTERNAL  coarse_compare;}
{.EXTERNAL  fine_compare;}
{.EXTERNAL  full_compare;}
{.EXTERNAL  shuffle_play;}
.EXTERNAL   rank_candidates;
.EXTERNAL   reset_recog;
.EXTERNAL   init_catalog;
.EXTERNAL   catalog_size;
.EXTERNAL   inc_bank_select;
.EXTERNAL   show_bank;
.EXTERNAL   set_local_call;
.EXTERNAL   set_long_distance;
.EXTERNAL   digit_count;
.EXTERNAL   display_number;
.EXTERNAL   display_digit;
.EXTERNAL   add_a_digit;
.EXTERNAL   display_numpls;
.EXTERNAL   display_dial;
.EXTERNAL   reset_display;
.EXTERNAL   timed_display;
.EXTERNAL   reset_timed;
.EXTERNAL   cepstral_compare;

{%%%%%%%%%%%%%%%%%%%%%%%%%%%%%%%%%%%%%%%%%%%%%%%%%%%%%%%%%%%%%%%%%%%%%%%%%%%%%%%%%%%%}
{%%%%%%%%%%%%%%%%%%%%%%%%%%%%%%%%%%%%%%%%%%%%%%%%%%%%%%%%%%%%%%%%%%%%%%%%%%%%%%%%%%%%}
{_____main shell for speech recognition demonstration_____}
{%%%%%%%%%%%%%%%%%%%%%%%%%%%%%%%%%%%%%%%%%%%%%%%%%%%%%%%%%%%%%%%%%%%%%%%%%%%%%%%%%%%%}
{%%%%%%%%%%%%%%%%%%%%%%%%%%%%%%%%%%%%%%%%%%%%%%%%%%%%%%%%%%%%%%%%%%%%%%%%%%%%%%%%%%%%}
```

```
reset_vector:
            JUMP start; NOP; NOP; NOP;
irq2:       JUMP next_bank; NOP; NOP; NOP;
trans0:     RTI; NOP; NOP; NOP;
recv0:      JUMP sample; NOP; NOP; NOP;
trans1:     NOP; NOP; NOP; NOP;
recv1:      NOP; NOP; NOP; NOP;
timer_int:  JUMP timed_display; NOP; NOP; NOP;

start:   IMASK=0;
         ICNTL=B#00100;
         L0=0;    L1=0;    L2=0;    L3=0;
         L4=0;    L5=0;    L6=0;    L7=0;
         M0=0;    M1=1;    M2=-1;   M3=2;
         M4=0;    M5=1;    M6=-1;   M7=2;

reg_setup:
         AX0 = 0;
         DM(0X3FFE) = AX0;                 { DM wait states }
         AX0 = 2;
         DM(0X3FF5) = AX0;                 { sclkdiv0 with 12.288 MHz input }
         AX0 = 255;
         DM(0X3FF4) = AX0;                 { rfsdiv0 }
         AX0 = 0X6927;
         DM(0X3FF6) = AX0;                 { control reg0 }
         AX0 = 0X1003;
         DM(0X3FFF) = AX0;                 { system control reg }

         CALL reset_recog;
         CALL reset_display;
         {CALL play_library;}

{_____wait for (phonecall) while displaying intro_____}

phone_call:
         IMASK = 0x21;
         ENA TIMER;
         CALL get_word;
         DIS TIMER;
         CALL cepstral_compare;
         CALL rank_candidates;             { buffer pointer returned in AY0 }

         { failsafe feature }
         IF NOT FLAG_IN JUMP its_a_call;

         { is it (phonecall)? }
         AX0 = 14;
         AF  = AX0 - AY1;
         IF NE JUMP phone_call;
```

(listing continues on next page)

6 Speech Recognition

```
        { decrement catalog_size to remove (phonecall) }
its_a_call:
        AY0 = DM(catalog_size);
        AR  = AY0 - 1;
        DM(catalog_size) = AR;

{_____wait for digit while displaying (number please?)_____}

first_digit:
        CALL display_numpls;        { display }
        AX0 = 0;
        DM(digit_count) = AX0;
        CALL set_local_call;

        CALL get_word;
        CALL cepstral_compare;
        CALL rank_candidates;       { buffer index returned in AY1 }

        { is it (question_mark)?}
        AX0 = 14;
        AF  = AX0 - AY1;
        IF GT JUMP chk_scratch1;
        AY1 = 15;
        CALL display_digit;
        JUMP first_digit;

        { is it (scratch)?}
chk_scratch1:
        AX0 = 12;
        AF  = AX0 - AY1;
        IF EQ JUMP catsiz_reset;

        { is it (dial) or (delete)? }
        AX0 = 11;
        AF  = AX0 - AY1;
        IF EQ JUMP first_digit;
        AX0 = 13;
        AF  = AX0 - AY1;
        IF EQ JUMP first_digit;

        { is it (one)? }
        AX0 = 1;
        AF  = AX0 - AY1;
        IF EQ CALL set_long_distance;
        CALL add_a_digit;           { increment digit_count }
        CALL display_digit;         { display digit }
```

Speech Recognition   6

```
{_____collect and display remaining digits, wait for (dial)_____}
more_digits:
        CALL display_number;        { display number }
        CALL get_word;
        CALL cepstral_compare;
        CALL rank_candidates;       { buffer pointer returned in AY0 }

        { failsafe feature }
        IF NOT FLAG_IN JUMP dial_number;

        { is it (question_mark)?}
        AX0 = 14;
        AF  = AX0 - AY1;
        IF GT JUMP chk_scratch2;
        AY1 = 15;
        CALL display_digit;
        JUMP more_digits;

        { is it (scratch)?}
chk_scratch2:
        AX0 = 12;
        AF  = AX0 - AY1;
        IF EQ JUMP catsiz_reset;

        { is it (dial)? }
        AX0 = 11;
        AF  = AX0 - AY1;
        IF EQ JUMP dial_number;

        { is it (delete)? }
        AX0 = 13;
        AF  = AX0 - AY1;
        IF NE JUMP its_a_digit;
        AY0 = DM(digit_count);
        AR  = AY0 - 1;
        IF EQ JUMP first_digit;
        DM(digit_count) = AR;
        JUMP more_digits;

its_a_digit:
        CALL add_a_digit;           { increment digit_count }
        CALL display_digit;         { display digit }

        JUMP more_digits;
```

(listing continues on next page)

6   Speech Recognition

```
{_____reset catalog_size_____}

catsiz_reset:
        AY0 = DM(catalog_size);
        AR  = AY0 + 1;
        DM(catalog_size) = AR;
        CALL reset_timed;
        JUMP phone_call;

{_____boot code to dial the number_____}

dial_number:
        CALL display_dial;
        AR  = 0X025B;
        DM(0X3FFF) = AR;      { boot page 1 }

{_____enable new template library_____}

next_bank:
        ENA SEC_REG;
        CALL inc_bank_select;
        CALL show_bank;
        CALL init_catalog;
        RTI;

{_____process sample_____}

sample: ENA SEC_REG;
        AR  = DM(output_flag);
        AR  = PASS AR;
        IF EQ JUMP get_input;

send_output:
        SR1 = DM(I7,M5);

        SR  = ASHIFT SR1 BY -2 (HI);
        TX0  = SR1;
        JUMP inc_count;

get_input:
        AR=RX0;
        DM(I7,M5)=AR;          {Save sample}
inc_count:
        AY0=MX0;
        AR=AY0+1;
        MX0=AR;
        RTI;

.ENDMOD;
```

Listing 6.2 Demonstration Shell Subroutine (DEMOSHEL.DSP)

Speech Recognition   6

{_____

     Analog Devices Inc., DSP Division
     One Technology Way, Norwood, MA 02062
     DSP Applications Assistance: (617) 461-3672

INITIZE.DSP

This routine performs necessary initialization of data memory variables for
the speech recognition system. There are several assembly switches switches
available. The -Dplayback switch is used when reflection coefficients are
stored as features and templates are to be output to a speaker. The -Dinit_lib
switch is used to initialize a rom copy of the library catalogs. One of the
two remaining switches MUST be used. The -Drecord switch is used with
external program ram to record new templates. The -Ddemo switch is used in a
rom-based system for the demonstration and recognition accuracy testing.

The conditional assembly options and a description of each follows. At a
minimum, assembly must include:

    asm21 INITIZE -cp   -Drecord   used when recording new templates

OR

    asm21 INITIZE -cp   -Ddemo   used when demonstrating system,
                                          templates and catalog already stored in
                                          rom or initialized with switch

The other options are: -Dplayback   allows playback of library templates
                                          if reflection coefficients are the
                                          stored features
                       -Dinit_lib   used to initialize the template library
                                          catalog with data contained in the file
                                          catalog.dat

_____ }

```
.MODULE/RAM/BOOT=0        initialize;

.VAR/PM/RoM/SEG=EXT_PM   catalog_init[32];

.EXTERNAL    threshold_time;
.EXTERNAL    min_word_length;
.EXTERNAL    ws_energy_thresh;
.EXTERNAL    ws_zcr_thresh;
.EXTERNAL    ps_energy_thresh;
.EXTERNAL    ps_zcr_thresh;
```

(listing continues on next page)

6 Speech Recognition

```
.EXTERNAL    flag;
.EXTERNAL    unknown_feature_dimension;
.EXTERNAL    library_feature_dimension;

.EXTERNAL    catalog_size;
.EXTERNAL    library_catalog;
.EXTERNAL    next_catalog_entry;
.EXTERNAL    template_library;
.EXTERNAL    catalog_init;
{.......................................................................}
{ conditional assembly use -Dplayback }
#ifdef playback
.EXTERNAL    voiced_energy_thresh;
.EXTERNAL    spseed_lsw;
.EXTERNAL    spseed_msw;
.EXTERNAL    synth_train;
#endif
{.......................................................................}

.ENTRY       reset_recog;
{.......................................................................}
{ conditional assembly use -Ddemo }
#ifdef demo
.ENTRY       init_catalog;     { necessary when multiple template banks used  }
#endif
{.......................................................................}

{%%%%%%%%%%%%%%%%%%%%%%%%%%%%%%%%%%%%%%%%%%%%%%%%%%%%%%%%%%%%%%%%%%%%%%%%%}
{%%%%%%%%%%%%%%%%%%%%%%%%%%%%%%%%%%%%%%%%%%%%%%%%%%%%%%%%%%%%%%%%%%%%%%%%%}

{_____initialize data memory variables_____}

{%%%%%%%%%%%%%%%%%%%%%%%%%%%%%%%%%%%%%%%%%%%%%%%%%%%%%%%%%%%%%%%%%%%%%%%%%}
{%%%%%%%%%%%%%%%%%%%%%%%%%%%%%%%%%%%%%%%%%%%%%%%%%%%%%%%%%%%%%%%%%%%%%%%%%}

reset_recog:
            AX0 = 15;                       { variables from endpoint detection }
            DM(threshold_time) = AX0;
            AX0 = 30;
            DM(min_word_length) = AX0;
            AX0 = 1000;
            DM(ws_energy_thresh) = AX0;
            AX0 = 11000;
            DM(ws_zcr_thresh) = AX0;
            AX0 = 250;
            DM(ps_energy_thresh) = AX0;
            AX0 = 5500;
            DM(ps_zcr_thresh) = AX0;

            AX0 = 0;                        { shell variables }
            DM(flag) = AX0;
            AX0 = 12;
            DM(unknown_feature_dimension) = AX0;
            AX0 = 12;
            DM(library_feature_dimension) = AX0;
```

```
{.........................................................................}
{ conditional assembly use -Dplayback }
#ifdef playback

        AX0 = 3072;                           { synthesis variables }
        DM(voiced_energy_thresh) = AX0;
        AX0 = 15381;
        DM(spseed_lsw) = AX0;
        AX0 = 7349;
        DM(spseed_msw) = AX0;
#endif {.................................................................}

{_____EITHER this section_____}

{.........................................................................}
{ conditional assembly use -Drecord }
#ifdef record
        AX0 = 0;                              { for recording new templates }
        DM(catalog_size) = AX0;
        AX0 = ^template_library;
        DM(library_catalog) = AX0;
        AX0 = ^library_catalog;
        DM(next_catalog_entry) = AX0;
#endif
{.........................................................................}

{_____OR this section_____}

{.........................................................................}
{ conditional assembly use -Dinit_lib }
#ifdef init_lib
.INIT   catalog_init : <catalog.dat>;    { for initializing external pm }
#endif {.................................................................}
{.........................................................................}
{ conditional assembly use -Ddemo }
#ifdef demo
init_catalog:      I0  = ^catalog_size;    { necessary when multiple template }
                   I4  = ^catalog_init;    { banks are used }
                   CNTR = 32;
                   DO copy_catalog UNTIL CE;
                        AX0 = PM(I4,M5);
copy_catalog:           DM(I0,M1) = AX0;
#endif
{.........................................................................}

{_____}

        RTS;

.ENDMOD;
```

Listing 6.3 Data Variable Initialization Routine (INITIZE.DSP)

6   Speech Recognition

```
.MODULE/RAM/BOOT=0        receive_shell;

.CONST            vector_buffer_length = 1200;   {enough for one second}

.VAR/DM/RAM        LPC_coeff_buffer[12];
.VAR/DM/RAM/CIRC   input_buffer[240];
.VAR/DM/RAM        frame_buffer[160];
.VAR/DM/RAM        frame_pntr;
.VAR/DM/RAM        vector_pntr;
.VAR/DM/RAM        feature_vector_buffer[vector_buffer_length];
.VAR/DM/RAM        vector_count;
.VAR/DM/RAM/CIRC   start_buffer[60];
.VAR/DM/RAM        start_buffer_pntr;

.ENTRY     get_word;

.EXTERNAL     analyze_frame;
.EXTERNAL     frame_energy, frame_zcr;
.EXTERNAL     word_start_flag, poss_start_flag;
.EXTERNAL     word_end_flag, find_endpoints;
.EXTERNAL     unknown_feature_dimension;
.EXTERNAL     output_flag;
.EXTERNAL     convert_coeffs;

{%%%%%%%%%%%%%%%%%%%%%%%%%%%%%%%%%%%%%%%%%%%%%%%%%%%%%%%%%%%%%%%%%%%%%%%%%%%%%%}
{%%%%%%%%%%%%%%%%%%%%%%%%%%%%%%%%%%%%%%%%%%%%%%%%%%%%%%%%%%%%%%%%%%%%%%%%%%%%%%}

{_____receive input word_____}

{%%%%%%%%%%%%%%%%%%%%%%%%%%%%%%%%%%%%%%%%%%%%%%%%%%%%%%%%%%%%%%%%%%%%%%%%%%%%%%}
{%%%%%%%%%%%%%%%%%%%%%%%%%%%%%%%%%%%%%%%%%%%%%%%%%%%%%%%%%%%%%%%%%%%%%%%%%%%%%%}

{_____initialize buffers and analysis variables_____}

get_word:AX0 = 0;

        DM(vector_count) = AX0;
        DM(output_flag) = AX0;
        DM(word_end_flag) = AX0;
        DM(word_start_flag) = AX0;
        DM(poss_start_flag) = AX0;

        I0  = ^frame_buffer;
        CNTR = 160;
{         DO dmloop1 UNTIL CE;}
dmloop1:    DM(I0,M1) = AX0;
          IF NOT CE JUMP dmloop1;

        I0  = ^input_buffer;
        CNTR = 240;
{         DO dmloop2 UNTIL CE;}
```

```
dmloop2:    DM(I0,M1) = AX0;
            IF NOT CE JUMP dmloop2;

            I0  = ^start_buffer;
            CNTR = 60;
{           DO dmloop3 UNTIL CE;}
dmloop3:    DM(I0,M1) = AX0;
            IF NOT CE JUMP dmloop3;

            AX0 = ^input_buffer;
            DM(frame_pntr) = AX0;
            AX0 = ^feature_vector_buffer + 60;
            DM(vector_pntr) = AX0;
            AX0 = ^start_buffer;
            DM(start_buffer_pntr) = AX0;

            RESET FLAG_OUT;

            I7  = ^input_buffer;         {I7 is speech buffer pointer}
            L7  = 240;

            ENA SEC_REG;
            MX0=0;                       {sample recvd counter}
            AX1=160;

            AX0 = IMASK;
            AY0 = 0X08;
            AR  = AX0 OR AY0;
            IMASK = AR;                  {0X28;}
{_____acquire data frame_____}
code_1_loop:
            AY1=MX0;
            AR=AX1-AY1;
            IF NE JUMP code_1_loop;
            AX1 = 80;
            MX0 = 0;
            DIS SEC_REG;

{_____copy data to frame buffer_____}

            I0 = DM(frame_pntr);
            I1 = ^frame_buffer;
            CNTR = 160;
            L0 = 240;
{           DO copy_frame UNTIL CE;}
RSHELL1:    AX0 = DM(I0,M1);
copy_frame: DM(I1,M1) = AX0;
            IF NOT CE JUMP RSHELL1;
            {L0 = 0;}
```

(listing continues on next page)

6 Speech Recognition

```
{_____update frame pointer_____}

        I0 = DM(frame_pntr);
        M3 = 80;
        {L0 = 240;}
        MODIFY (I0,M3);
        M3 = 2;
        L0 = 0;
        DM(frame_pntr) = I0;

{_____frame analysis_____}

do_dmr_1:
        I0=^frame_buffer;
        I1=^LPC_coeff_buffer;
        CALL analyze_frame;

{_____feature conversion_____}

        I4  = ^LPC_coeff_buffer;
        CALL convert_coeffs;

{_____store feature vector_____}

        I0  = DM(vector_pntr);
        I1  = ^LPC_coeff_buffer;

{.............................................................................}
   { conditional assembly }
   #ifdef eight_features
        AX0 = DM(frame_energy);
        AX1 = DM(frame_zcr);
        DM(I0,M3) = AX0;                    {M3 = 2}
        DM(I0,M3) = AX1;

        AY0 = DM(unknown_feature_dimension);
        AR  = 4;                            {E, delta_E, zcr, delta_zcr}
        AR  = AY0 - AR;
        CNTR = AR;
   #else
        CNTR = DM(unknown_feature_dimension);
   #endif
{.............................................................................}

{       DO write_feature UNTIL CE;}
RSHELL2:          AY0 = DM(I1,M1);
write_feature:    DM(I0,M1) = AY0;
                  IF NOT CE JUMP RSHELL2;
```

Speech Recognition 6

```
{_____speech test_____}

        CALL find_endpoints;

{_____update vector pointer, count_____}

test_sp_flags:
        AX0 = DM(word_start_flag);
        AY0 = DM(poss_start_flag);
        AR  = AX0 OR AY0;
        IF EQ JUMP test_word_end;
        SET FLAG_OUT;

update_vp:
        I0  = DM(vector_pntr);
        M3  = DM(unknown_feature_dimension);
        MODIFY(I0,M3);
        DM(vector_pntr) = I0;
        M3  = 2;

update_vc:
        AY0 = DM(vector_count);
        AY1 = 100;
        AR  = AY0 + 1;
        AF  = AY1 - AR;
        IF LE JUMP word_end;
        DM(vector_count) = AR;

        JUMP more_frames;

{_____word test_____}

test_word_end:
        RESET FLAG_OUT;
        AX0 = DM(word_end_flag);
        AF  = PASS AX0;
        IF EQ JUMP reset_vpvc;

word_end:
        IMASK = 0;

{_____copy start_buffer to feature_vector_buffer_____}

        I0  = ^feature_vector_buffer;
        I1  = DM(start_buffer_pntr);
        L1  = 60;
        CNTR = 60;
        DO copy_start UNTIL CE;
                AX0 = DM(I1,M1);
copy_start:     DM(I0,M1) = AX0;

        L1  = 0;
```

(listing continues on next page)

365

6 Speech Recognition

```
{.....................................................................}
   { conditional assembly }
   #ifdef eight_features

{_____scale energy in word template_____}

        I0   = ^feature_vector_buffer;
        CNTR = DM(vector_count);
        M3   = DM(unknown_feature_dimension);
        AF   = PASS 0, AX0 = DM(I0,M3);
        DO find_max_energy UNTIL CE;
           AR  = AF - AX0, AX0 = DM(I0,M3);
find_max_energy:
        xIF LT AF = PASS AX0;
        AR  = AF + 1;

        I0   = ^feature_vector_buffer;
        CNTR = DM(vector_count);
        AX0 = AR;
        DO scale_energy UNTIL CE;
           AY1 = DM(I0,M0);
           AY0 = 0;
           DIVS AY1,AX0;
           CNTR = 15;
           DO scale_divloop UNTIL CE;
scale_divloop:
        DIVQ AX0;
scale_energy:
        DM(I0,M3) = AY0;
        M3   = 2;

{_____calculate delta energy,zcr in word template_____}

        I0   = ^feature_vector_buffer;
        M2   = DM(unknown_feature_dimension);
        AY1  = DM(vector_count);
        AR   = AY1 - 1;

        AY0 = 0;
        I1   = I0;
        AX0 = DM(I1,M1);                    { read energy }
        DM(I1,M1) = AY0;                    { store delta energy }
        AX1 = DM(I1,M1);                    { read zero-crossings }
        DM(I1,M1) = AY0;                    { store delta zero-crossings }
```

```
        CNTR = AR;
        DO compute_deltas UNTIL CE;
            MODIFY(I0,M2);
            I1  = I0;
            AY0 = DM(I1,M1);                { read energy }
            AR  = AY0 - AX0, AX0 = AY0;
            DM(I1,M1) = AR;                 { store delta energy }
            AY1 = DM(I1,M1);               { read zero-crossings }
            AR  = AY1 - AX1, AX1 = AY1;
compute_deltas:
            DM(I1,M0) = AR;                { store delta zero-crossings }
        M2  = -1;

    #endif
{.....................................................................}

{_____word template complete - return_____}

        L7  = 0;
        I0  = ^feature_vector_buffer;
        AX0 = DM(vector_count);

        RTS;

{_____reset vector pointer and count_____}

reset_vpvc:
        AX0 = ^feature_vector_buffer + 60;
        DM(vector_pntr) = AX0;
        AX0 = 0;
        DM(vector_count) = AX0;

{_____store vector in start buffer_____}

        I0  = DM(start_buffer_pntr);
        L0  = 60;
        I1  = ^LPC_coeff_buffer;

{.....................................................................}

    { conditional assembly }
    #ifdef eight_features
        AX0 = DM(frame_energy);
        AX1 = DM(frame_zcr);
        DM(I0,M3) = AX0;                    {M3 = 2}
        DM(I0,M3) = AX1;

        AY0 = DM(unknown_feature_dimension);
        AR  = 4;                            {E, delta_E, zcr, delta_zcr}
        AR  = AY0 - AR;
        CNTR = AR;
```

(listing continues on next page)

6　Speech Recognition

```
    #else
         CNTR = DM(unknown_feature_dimension);
    #endif
{......................................................................}

{        DO write_start UNTIL CE;}
RSHELL9: AY0 = DM(I1,M1);
write_start:
         DM(I0,M1) = AY0;
         IF NOT CE JUMP RSHELL9;

         DM(start_buffer_pntr) = I0;
         L0  = 0;

{_____jump to get more frames_____}

more_frames:
         ENA SEC_REG;
         JUMP code_1_loop;

{======================================================================}

.ENDMOD;
```

Listing 6.4 Receive Word Routine (RECVSHEL.DSP)

Speech Recognition 6

```
.MODULE/RAM/BOOT=0        LPC_analysis;

.ENTRY    analyze_frame;

.CONST    analysis_window_length = 160;
.CONST    input_scaler = 1;                    { 1 for u-law, 2 for A-law}
                                               { assumes right justified input}

.CONST    zcr_noise_threshold = 15;

.VAR/PM/RoM/SEG=EXT_PM
                  hamming_dat[analysis_window_length];
.VAR/DM/RAM       frame_energy;
.VAR/DM/RAM       frame_zcr;
.VAR/DM/RAM       spL_ACF[18];                 {9 long (32-bit) words}
.VAR/DM/RAM       r[8];
.VAR/DM/RAM       k[8];
.VAR/DM/RAM       acf[9];
.VAR/DM/RAM       p[9];
{.VAR/DM/RAM      z1, L_z2_h, L_z2_l, mp;}
.VAR/DM/RAM       spscaleauto;                 {Used in pre-emphasis save}
.VAR/DM/RAM       speech_in, xmit_buffer;

.INIT             hamming_dat : <hammdat.dat>;

.GLOBAL           spL_ACF, spscaleauto;
{.GLOBAL          mp, L_z2_l, L_z2_h, z1;}
.GLOBAL           frame_energy, frame_zcr;

{%%%%%%%%%%%%%%%%%%%%%%%%%%%%%%%%%%%%%%%%%%%%%%%%%%%%%%%%%%%%%%%%%%%%%%%%%%%}
{%%%%%%%%%%%%%%%%%%%%%%%%%%%%%%%%%%%%%%%%%%%%%%%%%%%%%%%%%%%%%%%%%%%%%%%%%%%}

{_____Analysis Subroutine_____}

{%%%%%%%%%%%%%%%%%%%%%%%%%%%%%%%%%%%%%%%%%%%%%%%%%%%%%%%%%%%%%%%%%%%%%%%%%%%}
{%%%%%%%%%%%%%%%%%%%%%%%%%%%%%%%%%%%%%%%%%%%%%%%%%%%%%%%%%%%%%%%%%%%%%%%%%%%}

analyze_frame:
        ENA AR_SAT;                    {Enable ALU saturation}
        DM(speech_in)=I0;              {Save pointer to input window}
        DM(xmit_buffer)=I1;            {Save pointer to coeff window}
        MX1=H#4000;                    {This multiply will place the}
        MY1=H#100;                     {vale of H#80 in MF that will}
        MF=MX1*MY1 (SS);               {be used for unbiased rounding}
```

(listing continues on next page)

6 Speech Recognition

```
{_____downscaling and offset compensation_____}

          I0=DM(speech_in);              {Get pointer to input data}
          I1=I0;                         {Set pointer for output data}
          SE=-15;                        {Commonly used shift value}
          MX1=H#80;                      {Used for unbaised rounding}
          AX1=16384;                     {Used to round result}
          MY0=32735;                     {Coefficient value}
          AY1=H#7FFF;                    {Used to mask lower L_z2}

          MY1 = 0;
          MR  = 0;                       {Since frames now overlapped}
{         MY1=DM(z1);}
{         MR0=DM(L_z2_l);}
{         MR1=DM(L_z2_h);}
          DIS AR_SAT;                    {Cannot do saturation}
          AR=MR0 AND AY1, SI=DM(I1,M1);  {Fill the pipeline}
          CNTR=analysis_window_length;

{         DO offset_comp UNTIL CE;}
AN1:      SR=LSHIFT SI BY input_scaler (HI);{assumes right justified}
          AX0=SR1, SR=ASHIFT MR1 (HI);   {Get upper part of L_z2 (msp)}
          SR=SR OR LSHIFT MR0 (LO);      {Get LSB of L_z2 (lsp)}
          MR=MX1*MF (SS), MX0=SR0;       {Prepare MR, MX0=msp}
          MR=MR+AR*MY0 (SS), AY0=MY1;    {Compute temp}
          AR=AX0-AY0, AY0=MR1;           {Compute new s1}
          SR=ASHIFT AR BY 15 (LO);       {Compute new L_s2}
          AR=SR0+AY0, MY1=AX0;           {MY1 holds z1, L_s2+temp is in}
          AF=SR1+C, AY0=AR;              {SR in double precision}
          MR=MX0*MY0 (SS);               {Compute msp*32735}
          SR=ASHIFT MR1 BY -1 (HI);      {Downshift by one bit }
          SR=SR OR LSHIFT MR0 BY -1 (LO);{before adding to L_s2}
          AR=SR0+AY0, AY0=AX1;           {Compute new L_z2 in }
          MR0=AR, AR=SR1+AF+C;           {double precision MR0=L_z2}
          MR1=AR, AR=MR0+AY0;            {MR1=L_z2, round result }
          SR=LSHIFT AR (LO);             {and downshift for output}
          AR=MR1+C, SI=DM(I1,M1);        {Get next input sample}
          SR=SR OR ASHIFT AR (HI);
offset_comp:
          DM(I0,M1)=SR0, AR=MR0 AND AY1;{Store result, get next lsp}
          IF NOT CE JUMP AN1;
{         DM(L_z2_l)=MR0;}               {Save values for next call}
{         DM(L_z2_h)=MR1;}
{         DM(z1)=MY1;}
          ENA AR_SAT;                    {Re-enable ALU saturation}
```

Speech Recognition   6

```
{_____energy calculation_____}

        I0   = DM(speech_in);
        AF   = PASS 0, AX0 = DM(I0,M1);
        CNTR = analysis_window_length;
{       DO calc_energy UNTIL CE;}
AN2:    AR   = ABS AX0;
        SR   = ASHIFT AR BY -7 (HI);
calc_energy:AF = SR1 + AF, AX0 = DM(I0,M1);
        IF NOT CE JUMP AN2;
        AR   = PASS AF;
        DM(frame_energy) = AR;


{_____zero crossing calculation_____}

        I0   = DM(speech_in);
        AF   = PASS 0, AX0 = DM(I0,M1);
        AR   = ABS AX0;               {set either POS or NEG}
        AX1  = 205;                   {temporary - improves scaling}
        AY0  = zcr_noise_threshold;
        CNTR = analysis_window_length;
{       DO calc_zcr UNTIL CE;}
AN3:    IF POS JUMP last_was_pos;
last_was_neg:
        AR  = AX0 - AY0, AX0 = DM(I0,M1);
        IF GE AF = AX1 + AF;
        JUMP calc_zcr;
last_was_pos:
        AR  = AX0 + AY0, AX0 = DM(I0,M1);
        IF LT AF = AX1 + AF;
calc_zcr:
        AR  = ABS AR;
        IF NOT CE JUMP AN3;
        AR = PASS AF;
        DM(frame_zcr) = AR;


{_____pre-emphasis filter_____}

        MX0 = 0;
{       MX0=DM(mp);}                  {Get saved value for mp}
        MY0=-28180;                   {MY0 holds coefficient value}
        MX1=H#80;                     {These are used for biased}
        MR=MX1*MF (SS);               {rounding}
        SB=-4;                        {Maximum scale value}
        I0=DM(speech_in);             {In-place computation}
        CNTR=analysis_window_length;
{       DO pre_emp UNTIL CE;}
```

(listing continues on next page)

6   Speech Recognition

```
AN4:            MR=MR+MX0*MY0 (SS), AY0=DM(I0,M0);
                AR=MR1+AY0, MX0=AY0;
                SB=EXPADJ AR;                   {Check for maximum value}
pre_emp:        DM(I0,M1)=AR, MR=MX1*MF (SS); {Save filtered data}
                IF NOT CE JUMP AN4;
{               DM(mp)=MX0;}
                AY0=SB;                         {Get exponent of max value}
                AX0=4;                          {Add 4 to get scale value}
                AR=AX0+AY0;
                DM(spscaleauto)=AR;             {Save scale for later}

{_____hamming windowing_____}

                I0 = DM(speech_in);
                I1 = I0; {output}
                I5 = ^hamming_dat;
                MX0 = DM(I0,M1);
                MY0 = PM(I5,M5);
                MX1 = H#80;
                MR  = MX1 * MF (SS);
                CNTR = analysis_window_length;
{               DO window_frame UNTIL CE;}
AN5:            MR  = MR + MX0 * MY0 (SS), MX0 = DM(I0,M1);
                MY0 = PM(I5,M5);
window_frame:
                DM(I1,M1) = MR1, MR = MX1 * MF (SS);
                IF NOT CE JUMP AN5;

{_____dynamic scaling_____}

                IF LE JUMP auto_corr;           {If 0 scale, only copy data}
                AF=PASS 1;
                AR=AF-AR;
                SI=16384;
                SE=AR;
                I0=DM(speech_in);
                I1=I0;                          {Output writes over the input}
                SR=ASHIFT SI (HI);
                AF=PASS AR, AR=SR1;             {SR1 holds temp for multiply}
                MX1=H#80;                       {Used for unbiased rounding}
                MR=MX1*MF (SS), MY0=DM(I0,M1);  {Fetch first value}
                CNTR=analysis_window_length;
{               DO scale UNTIL CE;}
AN6:            MR=MR+SR1*MY0 (SS), MY0=DM(I0,M1);  {Compute scaled data}
scale:          DM(I1,M1)=MR1, MR=MX1*MF (SS);      {Save scaled data}
                IF NOT CE JUMP AN6;
```

Speech Recognition   6

```
{_____autocorrelation_____}

auto_corr:
        I1=DM(speech_in);              {This section of code computes}
        I5=I1;                         {the autocorr section for LPC}
        I2=analysis_window_length;     {I2 used as down counter}
        I6=^spL_ACF;                   {Set pointer to output array}
        CNTR=9;                        {Compute nine terms}
{       DO corr_loop UNTIL CE;}
AN7:    I0=I1;                         {Reset pointers for mac loop}
        I4=I5;
        MR=0, MX0=DM(I0,M1);           {Get first sample}
        CNTR=I2;                       {I2 decrements once each loop}
{       DO data_loop UNTIL CE;}
AN8:    MY0=DM(I4,M5);
data_loop:      MR=MR+MX0*MY0 (SS), MX0=DM(I0,M1);
        IF NOT CE JUMP AN8;
        MODIFY(I2,M2);                 {Decrement I2, Increment I5}
        MY0=DM(I5,M5);
        DM(I6,M5)=MR1;                 {Save double precision result}
corr_loop:  DM(I6,M5)=MR0;            {MSW first}
        IF NOT CE JUMP AN7;

        I0=DM(speech_in);              {This section of code rescales}
        SE=DM(spscaleauto);            {the input data}
        I1=I0;                         {Output writes over input}
        SI=DM(I0,M1);
        CNTR=analysis_window_length;
{       DO rescale UNTIL CE;}
AN9:    SR=ASHIFT SI (HI), SI=DM(I0,M1);
rescale:    DM(I1,M1)=SR1;
        IF NOT CE JUMP AN9;

{_____schur recursion_____}

set_up_schur:
        AY1 = ^spL_ACF; {in DM}
        MY1 = ^acf;
        M0  = ^r;
        CALL schur_routine;

{_____output reflection coefficients_____}

transmit_lar:
        I1 = ^r;                       {The quantized LAR values}
        CNTR=8;                        {can now be sent}
        CALL xmit_data;                {Copy to the output buffer}
```

(listing continues on next page)

6   Speech Recognition

```
      {All the coded variables have been sent to xmit_buffer}

finish:   DIS AR_SAT;
          RTS;                           {Return to caller}

xmit_data:
          I0=DM(xmit_buffer);            {Copy coeffs to the output}
{         DO xmit UNTIL CE;}             {buffer}
AN10:        AX0=DM(I1,M1);
xmit:        DM(I0,M1)=AX0;
             IF NOT CE JUMP AN10;
          DM(xmit_buffer)=I0;
          RTS;                           {Return from Encoder}

{_____Encoder and Voice Activity Detector Subroutines_____}

{  This section of code computes the reflection coefficients using the
   schur recursion }

schur_routine:
          I6=AY1;                        {This section of code prepares}
          AR=DM(I6,M5);                  {for the schur recursion}
          SE=EXP AR (HI), SI=DM(I6,M5);  {Normalize the autocorrelation}
          SE=EXP SI (LO);                {sequence based on spL_ACF[0]}
          SR=NORM AR (HI);
          SR=SR OR NORM SI (LO);
          AR=PASS SR1;                   {If spL_ACF[0] = 0, set r to 0}
          IF EQ JUMP zero_reflec;
          I6=AY1;
          I5=MY1;
          AR=DM(I6,M5);
          CNTR=9;                        {Normalize all terms}
{         DO set_acf UNTIL CE;}
AN11:        SR=NORM AR (HI), AR=DM(I6,M5);
             SR=SR OR NORM AR (LO), AR=DM(I6,M5);
set_acf:     DM(I5,M5)=SR1;
             IF NOT CE JUMP AN11;

          I5=MY1;                        {This section of code creates}
          I4=^k+7;                       {the k-values and p-values}
          I0=^p;
          AR=DM(I5,M5);                  {Set P[0]=acf[0]}
          DM(I0,M1)=AR;
          CNTR=7;
{         DO create_k UNTIL CE;}         {Fill the k and p arrays}
```

```
AN12:        AR=DM(I5,M5);
             DM(I0,M1)=AR;
create_k:    DM(I4,M6)=AR;
             IF NOT CE JUMP AN12;
           AR=DM(I5,M5);
           DM(I0,M1)=AR;                  {Set P[8]=acf[8]}

           I5=M0;                         {Compute r-values}
           I6=7;                          {I6 used as downcounter}
           SR0=0;
           SR1=H#80;                      {Used in unbiased rounding}
           CNTR=7;                        {Loop through first 7 r-values}
{          DO compute_reflec UNTIL CE;}
AN13:        I2=^p;                       {Reset pointers}
             I4=^k+7;
             AX0=DM(I2,M1);               {Fetch P[0]}
             AX1=DM(I2,M2);               {Fetch P[1]}
             MX0=AX1, AF=ABS AX1;         {AF=abs(P[1])}
             AR=AF-AX0;
             IF LE JUMP do_division;      {If P[0]<abs(P[1]), r = 0}
             DM(I5,M5)=SR0;               {Final r =0}
             JUMP compute_reflec;
do_division:
             CALL divide_routine;         {Compute r[n]=abs(P[1])/P[0]}
             AR=AY0, AF=ABS AX1;
             AY0=32767;
             AF=AF-AX0;                    {Check for abs(P[1])=P[0]}
             IF EQ AR=PASS AY0;            {Saturate if they are equal}
             IF POS AR=-AR;                {Generate sign of r[n]}
             DM(I5,M5)=AR;                 {Store r[n]}
             MY0=AR, MR=SR1*MF (SS);
             MR=MR+MX0*MY0 (SS), AY0=AX0;  {Compute new P[0]}
             AR=MR1+AY0;
             DM(I2,M3)=AR;                 {Store new P[0]}
             CNTR=I6;                      {One less loop each iteration}
{            DO schur_recur UNTIL CE;}
AN14:          MR=SR1*MF (SS), MX0=DM(I4,M4);
               MR=MR+MX0*MY0 (SS), AY0=DM(I2,M2);
               AR=MR1+AY0, MX1=AY0;    {AR=new P[m]}
               MR=SR1*MF (SS);
               MR=MR+MX1*MY0 (SS), AY0=MX0;
               DM(I2,M3)=AR, AR=MR1+AY0; {Store P[m], AR=new K[9-m]}
schur_recur:   DM(I4,M6)=AR;              {Store new K[9-m]}
               IF NOT CE JUMP AN14;
```

(listing continues on next page)

6 Speech Recognition

```
compute_reflec:
        MODIFY(I6,M6);                  {Decrement loop counter (I6)}
        IF NOT CE JUMP AN13;
    I2=^p;                              {Compute r[8] outside of loop}
    AX0=DM(I2,M1);                      {Using same procedure as above}
    AX1=DM(I2,M2);
    AF=ABS AX1;
    CALL divide_routine;
    AR=AY0, AF=ABS AX1;
    AY0=32767;
    AF=AF-AX0;
    IF EQ AR=PASS AY0;
    AF=ABS AX1;
    AF=AF-AX0;                          {The test for valid r is here}
    IF GT AR=PASS 0;                    {r[8]=0 if P[0]<abs(P[1])}
    IF POS AR=-AR;
    DM(I5,M5)=AR;
    JUMP schur_done;

zero_reflec:
    AX0=0;                              {The r-values must be set to}
    I5=M0;                             {0 according to the recursion}
    CNTR=8;
{       DO zero_rs UNTIL CE;}
zero_rs:    DM(I5,M5)=AX0;
        IF NOT CE JUMP ZERO_RS;

schur_done:
    M0 = 0;
    RTS;

{_____Divide Subroutine_____}

divide_routine:
    AY0=0;
    DIVS AF,AX0;
    CNTR=15;
{       DO div_loop UNTIL CE;}
div_loop:   DIVQ AX0;
        IF NOT CE JUMP DIV_LOOP;
    RTS;

.ENDMOD;
```

Listing 6.5 Frame Analysis Routine (ANALYZE.DSP)

Speech Recognition 6

```
.MODULE/RAM/BOOT=0        detect_endpoints;

.VAR/RAM/DM        word_start_flag;
.VAR/RAM/DM        ws_energy_thresh;
.VAR/RAM/DM        ws_zcr_thresh;
.VAR/RAM/DM        silence_time;

.VAR/RAM/DM        poss_start_flag;
.VAR/RAM/DM        ps_energy_thresh;
.VAR/RAM/DM        ps_zcr_thresh;

.VAR/RAM/DM        min_word_length;
.VAR/RAM/DM        threshold_time;
.VAR/RAM/DM        word_end_flag;
.VAR/RAM/DM        speech_count;

.GLOBAL        word_end_flag, word_start_flag, poss_start_flag;
.GLOBAL        threshold_time;
.GLOBAL        min_word_length;
.GLOBAL        ws_energy_thresh;
.GLOBAL        ws_zcr_thresh;
.GLOBAL        ps_energy_thresh;
.GLOBAL        ps_zcr_thresh;

.EXTERNAL      frame_energy, frame_zcr;

.ENTRY         find_endpoints;

{%%%%%%%%%%%%%%%%%%%%%%%%%%%%%%%%%%%%%%%%%%%%%%%%%%%%%%%%%%%%%%%%%%%%%%%%%%%%%}
{%%%%%%%%%%%%%%%%%%%%%%%%%%%%%%%%%%%%%%%%%%%%%%%%%%%%%%%%%%%%%%%%%%%%%%%%%%%%%}

{_____Endpoint Detection Subroutine_____}

{%%%%%%%%%%%%%%%%%%%%%%%%%%%%%%%%%%%%%%%%%%%%%%%%%%%%%%%%%%%%%%%%%%%%%%%%%%%%%}
{%%%%%%%%%%%%%%%%%%%%%%%%%%%%%%%%%%%%%%%%%%%%%%%%%%%%%%%%%%%%%%%%%%%%%%%%%%%%%}

find_endpoints:
        MX0 = 0;
        MX1 = 1;
        AX0 = DM(word_start_flag);
        AR  = PASS AX0;
        IF EQ JUMP find_word_start;

{===================== find end of word =====================================}

find_word_end:
        AX0 = DM(ps_energy_thresh);
        AX1 = DM(ps_zcr_thresh);
        CALL comp_energy_and_zcr;
        IF NE JUMP set_word_start;    {still speech, return to shell}
```

(listing continues on next page)

377

6   Speech Recognition

```
{_____no longer speech_____}

        AY0 = DM(silence_time);
        AR  = AY0 + 1;                   { increment silence time }
        DM(silence_time) = AR;
        AY0 = DM(threshold_time);        { if threshold time exceeded,}
        AR  = AR - AY0;                  { assume end of word }

{_____silence inside word_____}

        IF LT JUMP inc_sp_count;         { returns to shell }

{_____end of word reached_____}
end_of_word:
        AX0 = DM(speech_count);
        AY0 = DM(min_word_length);
        AR  = AY0 - AX0;

word_too_short:
        IF GT JUMP reset_vars;           { returns to shell }

word_length_ok:
        DM(word_start_flag) = MX0;
        DM(word_end_flag) = MX1;
        RTS;

{===================== find start of word ===================================}

find_word_start:
        AX0 = DM(ws_energy_thresh);
        AX1 = DM(ws_zcr_thresh);
        CALL comp_energy_and_zcr;
        IF NE JUMP set_word_start;       { returns to shell }

{_____check for possible starting point_____}
not_word_start:
        AX0 = DM(ps_energy_thresh);
        AX1 = DM(ps_zcr_thresh);
        CALL comp_energy_and_zcr;
        IF EQ JUMP reset_vars;           { returns to shell }

{_____possible starting point found_____}
poss_word_start:
        DM(poss_start_flag) = MX1;
        JUMP inc_sp_count;               { returns to shell }
```

378

Speech Recognition 6

```
{=============================================================================}

{============== set variables for word start and increment speech count =====}

{=============================================================================}

set_word_start:
            DM(word_start_flag) = MX1;
            DM(poss_start_flag) = MX0;
            DM(silence_time) = MX0;

inc_sp_count:
            AY0 = DM(speech_count);
            AR = AY0 + 1;
            DM(speech_count) = AR;
            RTS;

{=============================================================================}

{============== reset variables to find new starting endpoint ===============}

{=============================================================================}

reset_vars:
            DM(poss_start_flag) = MX0;
D           M(word_start_flag) = MX0;
            DM(word_end_flag) = MX0;
            DM(silence_time) = MX0;
            DM(speech_count) = MX0;
            RTS;

{=============================================================================}

{============== compare frame energy and zcr with thresholds ================}

{=============================================================================}

comp_energy_and_zcr:                            { inputs: AX0 = energy threshold }
            AY0 = DM(frame_energy);             {          AX1 = zcr threshold    }
            AY1 = DM(frame_zcr);
            AF  = PASS 0;
            AR  = AY0 - AX0;
            IF GT AF = PASS 1;                  { test frame_energy }
            AR  = AY1 - AX1;
            IF GT AF = PASS 1;                  { test frame_zcr }
            AR  = PASS AF;                      { AR will be NE for (poss_) speech }
            RTS;

{=============================================================================}

.ENDMOD;
```

Listing 6.6 Endpoint Detection Routine (ENDPOINT.DSP)

6 Speech Recognition

{_____

       Analog Devices Inc., DSP Division
       One Technology Way, Norwood, MA 02062
       DSP Applications Assistance: (617) 461-3672

CONVERT.DSP

The purpose of the routines in this module is to derive different feature
representations from the LPC reflection coefficients. Separate routines exist
to convert from reflection (k) to predictor (alpha) coefficients, from
predictor to cepstral (c) coefficients, and to weight and normalize the
cepstral vector.

For several of these conversion recursions, scaling to prevent overflows would
reduce the significance of results. A pseudo-floating-point number format is
used to alleviate scaling concerns during processing. The inputs are in 1.15
fixed-point format, floating-point is used in processing, and the results are
returned in 1.15 fixed-point format. The pseudo-floating-point format has a
one word (16 bit) mantissa and a one word (16 bit) exponent, stored as
mantissa followed by exponent.

The floating-point routines are adapted from Applications Handbook, volume 1.
All of the routines have been optimized for this particular application.

A more detailed description of the algorithms implemented can be found in the
Application Note.
_____}

```
.MODULE/RAM/BOOT=0       coefficient_conversion;
.VAR/DM/RAM i_odd_a[16];        { temporary buffers for storage of intermediate}
.VAR/DM/RAM i_even_a[16];       {    predictor coefficient values }
.VAR/DM/RAM cepstral_coeff[24];     { temporary storage of cepstral coeffs }
.VAR/DM/RAM temp_mant;              { temporary mantissa storage }
.VAR/DM/RAM temp_exp;              { temporary exponent storage }
.VAR/PM/RAM sqrt_coeff[5];          { used in square root approximation }

.INIT    sqrt_coeff : H#5D1D00, H#A9ED00, H#46D600, H#DDAA00, H#072D00;

.VAR/PM/RAM weighting_coeff[12];     { used to weight (window) cepstral }
                                     {    coeff for improved recognition }

{ weighting of ( 1 + 6*sin(pi*k/12) ) }
.INIT        weighting_coeff : 0X2EAE00, 0X492400, 0X5FDD00, 0X714D00,
                               0X7C4200, 0X7FFF00, 0X7C4200, 0X714D00,
                               0X5FDD00, 0X492400, 0X2EAE00, 0X124900;
```

Speech Recognition 6

```
{ weighting of ( 1 + 6.5*sin(pi*k/16) ) }
{.INIT   weighting_coeff :    0X000000, 0X3B8400, 0X4EB200, 0X5F8200,
                              0X6D4D00, 0X778E00, 0X7DDE00, 0X7FFF00,
                              0X7DDE00, 0X778E00, 0X6D4D00, 0X5F8200;
}

.CONST   cepstral_order = 12;

.ENTRY   convert_coeffs;

{%%%%%%%%%%%%%%%%%%%%%%%%%%%%%%%%%%%%%%%%%%%%%%%%%%%%%%%%%%%%%%%%%%%%%%%%%%}
{%%%%%%%%%%%%%%%%%%%%%%%%%%%%%%%%%%%%%%%%%%%%%%%%%%%%%%%%%%%%%%%%%%%%%%%%%%}

{_____Conversion Shell_____}

{%%%%%%%%%%%%%%%%%%%%%%%%%%%%%%%%%%%%%%%%%%%%%%%%%%%%%%%%%%%%%%%%%%%%%%%%%%}
{%%%%%%%%%%%%%%%%%%%%%%%%%%%%%%%%%%%%%%%%%%%%%%%%%%%%%%%%%%%%%%%%%%%%%%%%%%}

{       required inputs:      I4  -> ^reflection_coefficients             }

convert_coeffs:
        I6  = I4;            { I6 points to output buffer }
        CALL k_to_alpha;     { convert from reflection to predictor }
        CALL alpha_to_cep;   {              predictor  to cepstral }
        CALL weight_cep;     { weight cepstral coefficients }
        CALL normalize_cep;  { normal coeffs to length of vector }
        CALL cep_to_fixed;   { convert back to fixed-point }
        RTS;

{%%%%%%%%%%%%%%%%%%%%%%%%%%%%%%%%%%%%%%%%%%%%%%%%%%%%%%%%%%%%%%%%%%%%%%%%%%}
{%%%%%%%%%%%%%%%%%%%%%%%%%%%%%%%%%%%%%%%%%%%%%%%%%%%%%%%%%%%%%%%%%%%%%%%%%%}

{_____Convert from reflection to predictor coefficients_____}

{%%%%%%%%%%%%%%%%%%%%%%%%%%%%%%%%%%%%%%%%%%%%%%%%%%%%%%%%%%%%%%%%%%%%%%%%%%}
{%%%%%%%%%%%%%%%%%%%%%%%%%%%%%%%%%%%%%%%%%%%%%%%%%%%%%%%%%%%%%%%%%%%%%%%%%%}

{       required inputs:      I4  -> ^reflection_coefficients             }

{_____}
{        floating point implementation                                  }

{ for each stage of this recursion:
                        I1 -> pointer to result buffer
                        I0 -> points to last data in previous
                              results buffer
                        I2 -> points to start of previous
                              results buffer                 }
```

(listing continues on next page)

6 Speech Recognition

```
k_to_alpha:
        M3  = -3;

        I1  = ^i_odd_a;                  { i = 1 }
        MY0 = DM(I4,M5);                 { read k1 }
        DM(I1,M1) = MY0;                 { a1(1) mantissa = k1 (M1 = 1) }
        DM(I1,M2) = 0;                   { a1(1) exponent = 0 (M2 = -1) }

        I0  = I1;
        I1  = ^i_even_a;                 { i = 2 }
        MY0 = DM(I4,M5);                 { read k2 }
        AX0 = 0;                         { k2 exponent = 0 }
        AX1 = MY0;                       { k2 mantissa }
        AY1 = DM(I0,M1);                 { a1(1) mantissa }
        AY0 = DM(I0,M1);                 { a1(1) exponent }
        CALL fpm;                        { floating-point multiply }
        AX0 = AR, AR = pass SR1;
        AX1 = AR;
        CALL fpa;                        { floating-point add }
        DM(I1,M1) = SR1;                 { a1(2) = a1(1) + k2*a1(1) mantissa}
        DM(I1,M1) = AR;                  {                           exponent }
        DM(I1,M1) = MY0;                 { a2(2) mantissa = k2 }
        DM(I1,M2) = 0;                   { a2(2) exponent = 0 }
          I0  = I1;

        I3  = 2;                         { used as stage counter }
        CNTR = 3;       { two recursions done. two per loop. 2+2*3=8 total }
{       DO conver_recur UNTIL CE;}
loopa:      I1  = ^i_odd_a;             { output pointer }
            I2  = ^i_even_a;            { input pointer }
            CNTR = I3;
            CALL recursion;
            MODIFY(I3,M1);              { increment stage counter }

            I1  = ^i_even_a;            { output pointer }
            I2  = ^i_odd_a;             { input pointer }
            CNTR = I3;
            CALL recursion;
conver_recur:
            MODIFY(I3,M1);             { increment stage counter }
            IF NOT CE JUMP loopa;

        M3 = 2;
        RTS;
{==============================================================================}
```

```
{_____Conversion Recursion_____}

{============================================================================}

{   floating point implementation    }

recursion:
        MY0 = DM(I4,M5);                        { read k }
{       DO recur_routine UNTIL CE;}
loopb:      AX0 = 0;                            { k exponent }
            AX1 = MY0;                          { k mantissa }
            AY1 = DM(I0,M1);                    { a mantissa }
            AY0 = DM(I0,M3);                    { a exponent }
            CALL fpm;                           { k * a }
            AX0 = AR, AR = pass SR1;
            AX1 = AR;
            AY1 = DM(I2,M1);
            AY0 = DM(I2,M1);
            CALL fpa;                           { a + k*a }
            DM(I1,M1) = SR1;                    { write new result mantissa }
recur_routine:
            DM(I1,M1) = AR;                     {                    exponent }
            IF NOT CE JUMP loopb;

        DM(I1,M1) = MY0;                        { ai(i) = ki mantissa }
        DM(I1,M2) = 0;                          {             exponent }
        I0 = I1;
        RTS;

{%%%%%%%%%%%%%%%%%%%%%%%%%%%%%%%%%%%%%%%%%%%%%%%%%%%%%%%%%%%%%%%%%%%%%%%%%%%%%%%}
{%%%%%%%%%%%%%%%%%%%%%%%%%%%%%%%%%%%%%%%%%%%%%%%%%%%%%%%%%%%%%%%%%%%%%%%%%%%%%%%}
{%%%%%%%%%%%%%%%%%%%%%%%%%%%%%%%%%%%%%%%%%%%%%%%%%%%%%%%%%%%%%%%%%%%%%%%%%%%%%%%}

{_____convert from predictor to cepstral coefficients_____}

{%%%%%%%%%%%%%%%%%%%%%%%%%%%%%%%%%%%%%%%%%%%%%%%%%%%%%%%%%%%%%%%%%%%%%%%%%%%%%%%}
{%%%%%%%%%%%%%%%%%%%%%%%%%%%%%%%%%%%%%%%%%%%%%%%%%%%%%%%%%%%%%%%%%%%%%%%%%%%%%%%}

alpha_to_cep:M7  = -3;

   { c1 = -a1 }
        I5  = ^i_even_a;                        { holds predictor coefficients }
        I1  = ^cepstral_coeff;
        AX1 = DM(I5,M5);
        AR  = -AX1;                             { negate mantissa only }
        DM(I1,M1) = AR;                         { store c1 mantissa }
        AY0 = DM(I5,M5);
        DM(I1,M1) = AY0;                        { store c1 exponent }
```

(listing continues on next page)

6 Speech Recognition

```
{ c2 = - (a2 + 1/2*c1*a1) }
      I5  = ^i_even_a + 2;
      AX1 = DM(I5,M5);              { read a2 }
      AX0 = DM(I5,M7);
      DM(temp_mant) = AX1;         { preload with a2 mantissa }
      DM(temp_exp) = AX0;          { preload with a2 exponent }

      I1  = ^cepstral_coeff;
      AX1 = 0x4000;                { 1/2 in 1.15 format }
      CALL scale_n_sum;     {accumulates sum of products in temp_mant,exp}

      AY0 = AR, AR = -SR1;         { negate mantissa }
      DM(I1,M1) = AR;              { store c2 mantissa }
      DM(I1,M1) = AY0;             { store c2 exponent }

{ c3 = - (a3 + 2/3*c2*a1 + 1/3*c1*a2) }
      I5  = ^i_even_a + 4;
      I1  = ^cepstral_coeff;
      AX1 = DM(I5,M5);             { read a3 }
      AX0 = DM(I5,M7);
      DM(temp_mant) = AX1;         { preload with a3 mantissa }
      DM(temp_exp) = AX0;          { preload with a3 exponent }
      AX1 = 0x2AAA;                { 1/3 in 1.15 format }
      CALL scale_n_sum;

      AX1 = 0x5555;                { 2/3 in 1.15 format }
      CALL scale_n_sum;

      AY0 = AR, AR = -SR1;         { negate mantissa }
      DM(I1,M1) = AR;              { store c3 mantissa }
      DM(I1,M1) = AY0;             { store c3 exponent }

{ c4 = - (a4 + 3/4*c3*a1 + 1/2*c2*a2 + 1/4*c1*a3) }
      I5  = ^i_even_a + 6;
      I1  = ^cepstral_coeff;
      AX1 = DM(I5,M5);             { read a4 }
      AX0 = DM(I5,M7);
      DM(temp_mant) = AX1;
      DM(temp_exp) = AX0;
      AX1 = 0x2000;                { 1/4 in 1.15 format }
      CALL scale_n_sum;

      AX1 = 0x4000;                { 1/2 in 1.15 format }
      CALL scale_n_sum;

      AX1 = 0x6000;                { 3/4 in 1.15 format }
      CALL scale_n_sum;

      AY0 = AR, AR = -SR1;
      DM(I1,M1) = AR;              { store c4 mantissa }
      DM(I1,M1) = AY0;             { store c4 exponent }
```

```
{ c5 = - (a5 + 4/5*c4*a1 + 3/5*c3*a2 + 2/5*c2*a3 + 1/5*c1*a4) }
      I5  = ^i_even_a + 8;
      I1  = ^cepstral_coeff;
      AX1 = DM(I5,M5);                    { read a5 }
      AX0 = DM(I5,M7);
      DM(temp_mant) = AX1;
      DM(temp_exp) = AX0;
      AX1 = 0x1999;
      CALL scale_n_sum;

      AX1 = 0x3333;
      CALL scale_n_sum;

      AX1 = 0x4CCC;
      CALL scale_n_sum;

      AX1 = 0x6666;
      CALL scale_n_sum;

      AY0 = AR, AR = -SR1;
      DM(I1,M1) = AR;
      DM(I1,M1) = AY0;

{ c6 = - (a6 + 5/6*c5*a1+2/3*c4*a2+1/2*c3*a3+1/3*c2*a4+1/6*c1*a5) }
      I5  = ^i_even_a + 10;
      I1  = ^cepstral_coeff;
      AX1 = DM(I5,M5);                    { read a6 }
      AX0 = DM(I5,M7);
      DM(temp_mant) = AX1;
      DM(temp_exp) = AX0;
      AX1 = 0x1555;
      CALL scale_n_sum;

      AX1 = 0x2AAA;
      CALL scale_n_sum;

      AX1 = 0x4000;
      CALL scale_n_sum;

      AX1 = 0x5555;
      CALL scale_n_sum;

      AX1 = 0x6AAA;
      CALL scale_n_sum;

      AY0 = AR, AR = -SR1;
      DM(I1,M1) = AR;
      DM(I1,M1) = AY0;
```

(listing continues on next page)

6　Speech Recognition

```
{ c7 = -(a7 + 6/7*c6*a1+5/7*c5*a2+4/7*c4*a3+3/7*c3*a4+2/7*c2*a5+1/7*c1*a6) }
        I5  = ^i_even_a + 12;
        I1  = ^cepstral_coeff;
        AX1 = DM(I5,M5);                    { read a7 }
        AX0 = DM(I5,M7);
        DM(temp_mant) = AX1;
        DM(temp_exp) = AX0;

        AX1 = 0x1249;
        CALL scale_n_sum;

        AX1 = 0x2492;
        CALL scale_n_sum;

        AX1 = 0x36DB;
        CALL scale_n_sum;

        AX1 = 0x4924;
        CALL scale_n_sum;

        AX1 = 0x5B6D;
        CALL scale_n_sum;

        AX1 = 0x6DB6;
        CALL scale_n_sum;

        AY0 = AR, AR = -SR1;
        DM(I1,M1) = AR;
        DM(I1,M1) = AY0;

{c8=-(a8+7/8*c7*a1+3/4*c6*a2+5/8*c5*a3+1/2*c4*a4+3/8*c3*a5+1/4*c2*a6+1/8*c1*a7)}
        I5  = ^i_even_a + 14;
        I1  = ^cepstral_coeff;
        AX1 = DM(I5,M5);                    { read a8 }
        AX0 = DM(I5,M7);
        DM(temp_mant) = AX1;
        DM(temp_exp) = AX0;

        AX1 = 0x1000;
        CALL scale_n_sum;

        AX1 = 0x2000;
        CALL scale_n_sum;

        AX1 = 0x3000;
        CALL scale_n_sum;

        AX1 = 0x4000;
        CALL scale_n_sum;
```

```
        AX1 = 0x5000;
        CALL scale_n_sum;

        AX1 = 0x6000;
        CALL scale_n_sum;

        AX1 = 0x7000;
        CALL scale_n_sum;

        AY0 = AR, AR = -SR1;
        DM(I1,M1) = AR;
        DM(I1,M1) = AY0;
{c9=-(8/9*c8*a1+7/9*c7*a2+2/3*c6*a3
            +5/9*c5*a4+4/9*c4*a5+1/3*c3*a6+2/9*c2*a7+1/9*c1*a8)}
        I5  = ^i_even_a + 14;
        I1  = ^cepstral_coeff;
        AX0 = 0;
        DM(temp_mant) = AX0;
        DM(temp_exp) = AX0;

        AX1 = 0x0E38;
        CALL scale_n_sum;

        AX1 = 0x1C71;
        CALL scale_n_sum;

        AX1 = 0x2AAA;
        CALL scale_n_sum;

        AX1 = 0x38E3;
        CALL scale_n_sum;

        AX1 = 0x471C;
        CALL scale_n_sum;

        AX1 = 0x5555;
        CALL scale_n_sum;

        AX1 = 0x638E;
        CALL scale_n_sum;

        AX1 = 0x71C7;
        CALL scale_n_sum;

        AY0 = AR, AR = -SR1;
        DM(I1,M1) = AR;
        DM(I1,M1) = AY0;
```

(listing continues on next page)

6 Speech Recognition

```
{c10=-(9/10*c9*a1+4/5*c8*a2+7/10*c7*a3
              +3/5*c6*a4+1/2*c5*a5+2/5*c4*a6+3/10*c3*a7+2/10*c2*a8))}
          I5   = ^i_even_a + 14;
          I1   = ^cepstral_coeff + 2;
          AX0 = 0;
          DM(temp_mant) = AX0;
          DM(temp_exp) = AX0;

          AX1 = 0x1999;
          CALL scale_n_sum;

          AX1 = 0x2666;
          CALL scale_n_sum;

          AX1 = 0x3333;
          CALL scale_n_sum;

          AX1 = 0x4000;
          CALL scale_n_sum;

          AX1 = 0x4CCC;
          CALL scale_n_sum;

          AX1 = 0x5999;
          CALL scale_n_sum;

          AX1 = 0x6666;
          CALL scale_n_sum;

          AX1 = 0x7333;
          CALL scale_n_sum;

          AY0 = AR, AR = -SR1;
          DM(I1,M1) = AR;
          DM(I1,M1) = AY0;

{c11=-(10/11*c10*a1+9/11*c9*a2+8/11*c8*a3
              +7/11*c7*a4+6/11*c6*a5+5/11*c5*a6+4/11*c4*a7+3/11*c3*a8))}
          I5   = ^i_even_a + 14;
          I1   = ^cepstral_coeff + 4;
          AX0 = 0;
          DM(temp_mant) = AX0;
          DM(temp_exp) = AX0;

          AX1 = 0x22EB;
          CALL scale_n_sum;

          AX1 = 0x2E8B;
          CALL scale_n_sum;
```

```
        AX1 = 0x3A2E;
        CALL scale_n_sum;

        AX1 = 0x45D1;
        CALL scale_n_sum;

        AX1 = 0x5174;
        CALL scale_n_sum;

        AX1 = 0x5D17;
        CALL scale_n_sum;

        AX1 = 0x68BA;
        CALL scale_n_sum;

        AX1 = 0x745D;
        CALL scale_n_sum;

        AY0 = AR, AR = -SR1;
        DM(I1,M1) = AR;
        DM(I1,M1) = AY0;

{c12=-(11/12*c11*a1+5/6*c10*a2+3/4*c9*a3
            +2/3*c8*a4+7/12*c7*a5+1/2*c6*a6+5/12*c5*a7+1/3*c4*a8)}
        I5  = ^i_even_a + 14;
        I1  = ^cepstral_coeff + 6;
        AX0 = 0;
        DM(temp_mant) = AX0;
        DM(temp_exp) = AX0;

        AX1 = 0x2AAA;
        CALL scale_n_sum;

        AX1 = 0x3555;
        CALL scale_n_sum;

        AX1 = 0x4000;
        CALL scale_n_sum;

        AX1 = 0x4AAA;
        CALL scale_n_sum;

        AX1 = 0x5555;
        CALL scale_n_sum;

        AX1 = 0x6000;
        CALL scale_n_sum;

        AX1 = 0x6AAA;
        CALL scale_n_sum;
```

(listing continues on next page)

```
        AX1 = 0x7555;
        CALL scale_n_sum;

        AY0 = AR, AR = -SR1;
        DM(I1,M1) = AR;
        DM(I1,M1) = AY0;

        M7  = 2;
        RTS;

{===============================================================================}

{_____scale the product and add to running sum_____}

{===============================================================================}

{  required inputs:        AX1 -> scale value                               }

scale_n_sum:
        AX0 = 0;                      { scale factor has exponent = 0 }
        AY1 = DM(I1,M1);              { read cepstral coefficient mant, exp }
        AY0 = DM(I1,M1);              { now points to next coefficient }
        CALL fpm;                     { scale factor * cepstral coeff }
        AX0 = AR;
        AX1 = SR1;

        AY1 = DM(I5,M5);              { read predictor coefficient mant, exp }
        AY0 = DM(I5,M7);              { now points to previous coefficient }
        CALL fpm;                     { predictor coeff * (scal * cepstral) }
        AX0 = AR;
        AX1 = SR1;
        AY1 = DM(temp_mant);
        AY0 = DM(temp_exp);
        CALL fpa;                     { accumulate product with prev results }
        DM(temp_mant) = SR1;          { store new results }
        DM(temp_exp) = AR;
        RTS;

{%%%%%%%%%%%%%%%%%%%%%%%%%%%%%%%%%%%%%%%%%%%%%%%%%%%%%%%%%%%%%%%%%%%%%%%%%%%%%%%%%}
{%%%%%%%%%%%%%%%%%%%%%%%%%%%%%%%%%%%%%%%%%%%%%%%%%%%%%%%%%%%%%%%%%%%%%%%%%%%%%%%%%}
{%%%%%%%%%%%%%%%%%%%%%%%%%%%%%%%%%%%%%%%%%%%%%%%%%%%%%%%%%%%%%%%%%%%%%%%%%%%%%%%%%}

{_____weight cepstral coefficients_____}

{%%%%%%%%%%%%%%%%%%%%%%%%%%%%%%%%%%%%%%%%%%%%%%%%%%%%%%%%%%%%%%%%%%%%%%%%%%%%%%%%%}
{%%%%%%%%%%%%%%%%%%%%%%%%%%%%%%%%%%%%%%%%%%%%%%%%%%%%%%%%%%%%%%%%%%%%%%%%%%%%%%%%%}
```

```
weight_cep: I0  = ^cepstral_coeff;
            I5  = ^weighting_coeff;

            CNTR = cepstral_order;
{           DO weighting UNTIL CE;}
top_weight: AX1 = PM(I5,M5);
            AX0 = 0;                    { weighting coeff exponent = 0 }
            AY1 = DM(I0,M1);
            AY0 = DM(I0,M2);
            CALL fpm;
            DM(I0,M1) = SR1;            { store results }
weighting:  DM(I0,M1) = AR;
            IF NOT CE JUMP top_weight;

            RTS;

{%%%%%%%%%%%%%%%%%%%%%%%%%%%%%%%%%%%%%%%%%%%%%%%%%%%%%%%%%%%%%%%%%%%%%%%%%}
{%%%%%%%%%%%%%%%%%%%%%%%%%%%%%%%%%%%%%%%%%%%%%%%%%%%%%%%%%%%%%%%%%%%%%%%%%}
{%%%%%%%%%%%%%%%%%%%%%%%%%%%%%%%%%%%%%%%%%%%%%%%%%%%%%%%%%%%%%%%%%%%%%%%%%}

{_____convert cepstral coefficients to 1.15 fixed point_____}

{%%%%%%%%%%%%%%%%%%%%%%%%%%%%%%%%%%%%%%%%%%%%%%%%%%%%%%%%%%%%%%%%%%%%%%%%%}
{%%%%%%%%%%%%%%%%%%%%%%%%%%%%%%%%%%%%%%%%%%%%%%%%%%%%%%%%%%%%%%%%%%%%%%%%%}

cep_to_fixed:I0  = ^cepstral_coeff;
            AY0 = 0;                    { exponent bias value }
            CNTR = cepstral_order;
  {         DO scale_cep UNTIL CE;}
scale_loop: SI  = DM(I0,M1);
            AX0 = DM(I0,M1);
            CALL fixone;               { converts to 1.15 format }
scale_cep:  DM(I6,M5) = SR1;           { I6 points to result buffer }
            IF NOT CE JUMP scale_loop;
            RTS;

{%%%%%%%%%%%%%%%%%%%%%%%%%%%%%%%%%%%%%%%%%%%%%%%%%%%%%%%%%%%%%%%%%%%%%%%%%}
{%%%%%%%%%%%%%%%%%%%%%%%%%%%%%%%%%%%%%%%%%%%%%%%%%%%%%%%%%%%%%%%%%%%%%%%%%}
{%%%%%%%%%%%%%%%%%%%%%%%%%%%%%%%%%%%%%%%%%%%%%%%%%%%%%%%%%%%%%%%%%%%%%%%%%}

{_____normalize cepstral coefficients to length of vector_____}

{%%%%%%%%%%%%%%%%%%%%%%%%%%%%%%%%%%%%%%%%%%%%%%%%%%%%%%%%%%%%%%%%%%%%%%%%%}
{%%%%%%%%%%%%%%%%%%%%%%%%%%%%%%%%%%%%%%%%%%%%%%%%%%%%%%%%%%%%%%%%%%%%%%%%%}
```

(listing continues on next page)

6 Speech Recognition

```
normalize_cep:
            I0   = ^cepstral_coeff;
            AX0 = 0;
            DM(temp_mant) = AX0;
            DM(temp_exp) = AX0;
            CNTR = cepstral_order;
{           DO mag_sqd UNTIL CE;}
top_mag_sqd:
            AX1 = DM(I0,M1);
            AX0 = DM(I0,M1);
            AY1 = AX1;
            AY0 = AX0;
            CALL fpm;                   { square cepstral coefficient }
            AX0 = AR;
            AX1 = SR1;
            AY1 = DM(temp_mant);
            AY0 = DM(temp_exp);
            CALL fpa;                   { accumulate squared values }
            DM(temp_mant) = SR1;
mag_sqd:    DM(temp_exp) = AR;
            IF NOT CE JUMP top_mag_sqd;

            CALL sqrt;                  {find square root of mag sqrd mantissa }
            AY0 = DM(temp_exp);         {find square root of mag sqrd exponent }
            AX0 = 0X1;
            AF  = AX0 AND AY0;
            IF EQ JUMP exp_sqrt;        { is exponent even or odd ? }

            AR  = AY0 + 1;              {exponent odd, must add one and scale }
            AY0 = AR;                   { mantissa by 1/(SQRT(2)) }
            MY0 = 0X5A82;
            MR  = SR1 * MY0 (SS);
            SR1 = MR1;                  { scaled mantissa }

exp_sqrt:   DM(temp_mant) = SR1;
            SI  = AY0;
            SR  = ASHIFT SI BY -1 (HI);   { divide exponent by two }
            DM(temp_exp) = SR1;

            I0   = ^cepstral_coeff;  { normalize all cepstral coefficients }
            CNTR = cepstral_order;   {    to length of cepstral vector }
{           DO normalization UNTIL CE;}
top_norm:   AX1 = DM(I0,M1);
            AX0 = DM(I0,M2);
            AY1 = DM(temp_mant);
            AY0 = DM(temp_exp);
            CALL fpd;                   { floating-point divide (coeff/length) }
            DM(I0,M1) = SR1;
```

```
normalization:    DM(I0,M1) = AR;
                  IF NOT CE JUMP top_norm;
                  RTS;
```

```
{%%%%%%%%%%%%%%%%%%%%%%%%%%%%%%%%%%%%%%%%%%%%%%%%%%%%%%%%%%%%%%%%%%%%%%%%%%%%%}
{%%%%%%%%%%%%%%%%%%%%%%%%%%%%%%%%%%%%%%%%%%%%%%%%%%%%%%%%%%%%%%%%%%%%%%%%%%%%%}

{_____floating point multiply_____}

{%%%%%%%%%%%%%%%%%%%%%%%%%%%%%%%%%%%%%%%%%%%%%%%%%%%%%%%%%%%%%%%%%%%%%%%%%%%%%}
{%%%%%%%%%%%%%%%%%%%%%%%%%%%%%%%%%%%%%%%%%%%%%%%%%%%%%%%%%%%%%%%%%%%%%%%%%%%%%}
{
   Floating-Point Multiply
        Z = X * Y

   Calling Parameters
        AX0 = Exponent of X
        AX1 = Fraction of X
        AY0 = Exponent of Y
        AY1 = Fraction of Y
        MX0 = Excess Code

   Return Values
        AR = Exponent of Z
        SR1 = Fraction of Z
}

fpm:    MX0 = 0;                        { set exponent bias = 0 }
        AF=AX0+AY0, MX1=AX1;            {Add exponents}
        MY1=AY1;
        AX0=MX0, MR=MX1*MY1 (SS);       {Multiply mantissas}
        IF MV SAT MR;                   {Check for overflow}
        SE=EXP MR1 (HI);
        AF=AF-AX0, AX0=SE;              {Subtract bias}
        AR=AX0+AF;                      {Compute exponent}
        SR=NORM MR1 (HI);              {Normalize}
        SR=SR OR NORM MR0 (LO);
        RTS;
```

```
{%%%%%%%%%%%%%%%%%%%%%%%%%%%%%%%%%%%%%%%%%%%%%%%%%%%%%%%%%%%%%%%%%%%%%%%%%%%%%}
{%%%%%%%%%%%%%%%%%%%%%%%%%%%%%%%%%%%%%%%%%%%%%%%%%%%%%%%%%%%%%%%%%%%%%%%%%%%%%}

{_____floating point addition_____}

{%%%%%%%%%%%%%%%%%%%%%%%%%%%%%%%%%%%%%%%%%%%%%%%%%%%%%%%%%%%%%%%%%%%%%%%%%%%%%}
{%%%%%%%%%%%%%%%%%%%%%%%%%%%%%%%%%%%%%%%%%%%%%%%%%%%%%%%%%%%%%%%%%%%%%%%%%%%%%}
```

(listing continues on next page)

6 Speech Recognition

```
{
    Floating-Point Addition
          z = x + y

    Calling Parameters
          AX0 = Exponent of x
          AX1 = Fraction of x
          AY0 = Exponent of y
          AY1 = Fraction of y

    Return Values
          AR = Exponent of z
          SR1 = Fraction of z
}

fpa:      AF=AX0-AY0;                   {Is Ex > Ey?}
          IF GT JUMP shifty;            {Yes, shift y}
          SI=AX1, AR=PASS AF;           {No, shift x}
          SE=AR;
          SR=ASHIFT SI (HI);
          JUMP add;
shifty:   SI=AY1, AR=-AF;
          SE=AR;
          SR=ASHIFT SI (HI), AY1=AX1;
          AY0=AX0;
add:      AR=SR1+AY1;                   {Add fractional parts}
          IF AV JUMP work_around;
          SE=EXP AR (HI);
          AX0=SE, SR=NORM AR (HI);      {Normalize}
          AR= AX0+AY0;                  {Compute exponent}
          RTS;
work_around:
          AX1 = 0X08;                   { work around for HIX }
          AY1 = ASTAT;
          AX0 = AR, AR = AX1 AND AY1;
          SR  = LSHIFT AR BY 12 (HI);
          SE  = 1;
          AR  = AX0;
          AX0 = SE, SR = SR OR NORM AR (HI);
          AR  = AX0 + AY0;
          RTS;

{%%%%%%%%%%%%%%%%%%%%%%%%%%%%%%%%%%%%%%%%%%%%%%%%%%%%%%%%%%%%%%%%%%%%%%%%%%%%%%%%%%}
{%%%%%%%%%%%%%%%%%%%%%%%%%%%%%%%%%%%%%%%%%%%%%%%%%%%%%%%%%%%%%%%%%%%%%%%%%%%%%%%%%%}

{_____floating point conversion_____}

{%%%%%%%%%%%%%%%%%%%%%%%%%%%%%%%%%%%%%%%%%%%%%%%%%%%%%%%%%%%%%%%%%%%%%%%%%%%%%%%%%%}
{%%%%%%%%%%%%%%%%%%%%%%%%%%%%%%%%%%%%%%%%%%%%%%%%%%%%%%%%%%%%%%%%%%%%%%%%%%%%%%%%%%}
```

```
{
    Convert two-word floating-point to 1.15 fixed-point

    Calling Parameters
            AX0 = exponent              [16.0 signed twos complement]
            AY0 = exponent bias         [16.0 signed twos complement]
            SI = mantissa               [1.15 signed twos complement]

    Return Values
            SR1 = fixed-point number    [1.15 signed twos complement] }

.ENTRY    fixone;
fixone:   AR=AX0-AY0;                   {Compute unbiased exponent}
          IF GT JUMP overshift;         { positive exponent would
                                          overflow so saturate }
          SE=AR;
          SR=ASHIFT SI (HI);            {Shift fractional part}
          RTS;
overshift:  AR  = SI;
            AF  = PASS AR;
            ENA AR_SAT;
            AR  = AR + AF;              { saturate positive or negative }
            DIS AR_SAT;
            SR1 = AR;
            RTS;

{%%%%%%%%%%%%%%%%%%%%%%%%%%%%%%%%%%%%%%%%%%%%%%%%%%%%%%%%%%%%%%%%%%%%%%%%%%}
{%%%%%%%%%%%%%%%%%%%%%%%%%%%%%%%%%%%%%%%%%%%%%%%%%%%%%%%%%%%%%%%%%%%%%%%%%%}

{_____square root routine_____}

{%%%%%%%%%%%%%%%%%%%%%%%%%%%%%%%%%%%%%%%%%%%%%%%%%%%%%%%%%%%%%%%%%%%%%%%%%%}
{%%%%%%%%%%%%%%%%%%%%%%%%%%%%%%%%%%%%%%%%%%%%%%%%%%%%%%%%%%%%%%%%%%%%%%%%%%}

{
    Square Root
          y = sqrt(x)

    Calling Parameters
          SR1 = x in 1.15 format
          M5 = 1
          L5 = 0

    Return Values
          SR1 = y in 1.15 format
}

{ most of the error checking has been removed from this routine }
```

(listing continues on next page)

6 Speech Recognition

```
.CONST    base=H#0D49;

sqrt:     I5=^sqrt_coeff;                           {Pointer to coeff. buffer}
          MY0=SR1, AR=PASS SR1;
          IF EQ RTS;                                { if x=0 then y=0 }
          MR=0;
          MR1=base;                                 {Load constant value}
          MF=AR*MY0 (RND), MX0=PM(I5,M5);           {MF = x**2}
          MR=MR+MX0*MY0 (SS), MX0=PM(I5,M5);        {MR = base + C1*x}
          CNTR=4;
          DO approx UNTIL CE;
             MR=MR+MX0*MF (SS), MX0=PM(I5,M5);
approx:      MF=AR*MF (RND);
          SR=ASHIFT MR1 BY 1 (HI);
          RTS;

{%%%%%%%%%%%%%%%%%%%%%%%%%%%%%%%%%%%%%%%%%%%%%%%%%%%%%%%%%%%%%%%%%%%%%%%%%%%%%%%%%}
{%%%%%%%%%%%%%%%%%%%%%%%%%%%%%%%%%%%%%%%%%%%%%%%%%%%%%%%%%%%%%%%%%%%%%%%%%%%%%%%%%}

{_____floating point divide_____}

{%%%%%%%%%%%%%%%%%%%%%%%%%%%%%%%%%%%%%%%%%%%%%%%%%%%%%%%%%%%%%%%%%%%%%%%%%%%%%%%%%}
{%%%%%%%%%%%%%%%%%%%%%%%%%%%%%%%%%%%%%%%%%%%%%%%%%%%%%%%%%%%%%%%%%%%%%%%%%%%%%%%%%}

{
   Floating-Point Divide
        z = x / y

   Calling Parameters
        AX0 = Exponent of x
        AX1 = Fraction of x
        AY0 = Exponent of y
        AY1 = Fraction of y

   Return Values
        AR = Exponent of z
        SR1 = Fraction of z
}

fpd:      MX0 = 0;
          SR0=AY1, AR=ABS AX1;
          SR1=AR, AF=ABS SR0;
          SI=AX1, AR=SR1-AF;                        {Is Fx > Fy?}
          IF LT JUMP divide;                        {Yes, go divide}
          SR=ASHIFT SI BY -1 (LO);                  {No, shift Fx right}
          AF=PASS AX0;
          AR=AF+1, AX1=SR0;                         {Increase exponent}
          AX0=AR;
```

```
divide:   AF=AX0-AY0, AX0=MX0;
          MR=0;
          AR=AX0+AF, AY0=MR1;
          AF=PASS AX1, AX1=AY1;                  {Add bias}
          DIVS AF, AX1;                          {Divide fractions}
          DIVQ AX1; DIVQ AX1; DIVQ AX1; DIVQ AX1; DIVQ AX1;
          DIVQ AX1; DIVQ AX1; DIVQ AX1; DIVQ AX1; DIVQ AX1;
          DIVQ AX1; DIVQ AX1; DIVQ AX1; DIVQ AX1; DIVQ AX1;
          MR0=AY0, AF=PASS AR;
          SI=AY0, SE=EXP MR0 (HI);
          AX0=SE, SR=NORM SI (HI);               {Normalize}
          AR=AX0+AF;                             {Compute exponent}
          RTS;

{%%%%%%%%%%%%%%%%%%%%%%%%%%%%%%%%%%%%%%%%%%%%%%%%%%%%%%%%%%%%%%%%%%%%%%%%%}

.ENDMOD;
```

Listing 6.7 Coefficient Conversion Routine (CONVERT.DSP)

6 Speech Recognition

{_____

        Analog Devices Inc., DSP Division
        One Technology Way, Norwood, MA 02062
        DSP Applications Assistance: (617) 461-3672
_____

LIB_FUNC.DSP

The routines in this module perform two different operations on templates. A routine is present that adds a currently stored unknown word into the template library, updating the library catalog and catalog size in the process.

The other operation allows a library template to be played back through an external speaker. Since this is (currently) only possible when the stored features are the reflection coefficients, conditional assembly is used for these routines - they can easily be removed.

The conditional assembly options and a description of each follows. At a minimum, assembly must include:

```
        asm21 LIB_FUNC -cp
```

The other options are: -Dplayback allows playback of library templates
                                      if reflection coefficients are the
                                        stored features
                 -Dinit_lib used to initialize the template library
                                        with data contained in the file
                                        library.dat

_____}

```
.MODULE/RAM/BOOT=0      library_functions;

.ENTRY    put_in_library;

.VAR/DM/RAM catalog_size, library_catalog[30], next_catalog_entry; {32 total}
.VAR/PM/RoM template_library[14144];  { 14K-(32 from above)-(160 hamm coeff)}

.GLOBAL      catalog_size, library_catalog, template_library;
.GLOBAL      next_catalog_entry;

.EXTERNAL    library_feature_dimension;
.EXTERNAL    unknown_feature_dimension;
```

Speech Recognition   6

```
{........................................................................}
{ conditional assembly use -Dplayback }
#ifdef playback
.ENTRY        shuffle_play, play_library, play_single;

.VAR/DM/RAM current_selection;
.VAR/DM/RAM shuffle_pntr;

.EXTERNAL    output_template;
#endif {................................................................}
{........................................................................}
{ conditional assembly use -Dinit_lib }
#ifdef init_lib
.INIT         template_library : <library.dat>;
#endif {................................................................}

{%%%%%%%%%%%%%%%%%%%%%%%%%%%%%%%%%%%%%%%%%%%%%%%%%%%%%%%%%%%%%%%%%%%%%%%%%%}
{%%%%%%%%%%%%%%%%%%%%%%%%%%%%%%%%%%%%%%%%%%%%%%%%%%%%%%%%%%%%%%%%%%%%%%%%%%}

{_____Store Template in Library_____}

{%%%%%%%%%%%%%%%%%%%%%%%%%%%%%%%%%%%%%%%%%%%%%%%%%%%%%%%%%%%%%%%%%%%%%%%%%%}
{%%%%%%%%%%%%%%%%%%%%%%%%%%%%%%%%%%%%%%%%%%%%%%%%%%%%%%%%%%%%%%%%%%%%%%%%%%}

{        required inputs:    I0 - location of word to be stored
                            AX0 - length of word to be stored }

put_in_library:
            I1  = DM(next_catalog_entry);
            AY0 = DM(I1,M1);
            I5  = AY0;          { start of next library location }
            DM(I1,M1) = AX0;   { store template length (# of vectors) }
            DM(next_catalog_entry) = I1;

            M3  = DM(unknown_feature_dimension);
            CNTR = AX0;                     { AX0 holds length }
            DO store_vectors UNTIL CE;
                I2  = I0;                   { I0 points to word }
                CNTR = DM(library_feature_dimension);
                DO store_features UNTIL CE;
                    MX0 = DM(I2,M1);
store_features:         PM(I5,M5) = MX0;
store_vectors:    MODIFY(I0,M3);
            M3  = 2;
            AX0 = I5;
            DM(I1,M0) = AX0;            { set start of next template }

            AY0 = DM(catalog_size);
            AR  = AY0 + 1;             { increment catalog size }
            DM(catalog_size) = AR;

            RTS;
```

(listing continues on next page)

6   Speech Recognition

```
{..........................................................................}
{ conditional assembly use -Dplayback }
#ifdef playback

{%%%%%%%%%%%%%%%%%%%%%%%%%%%%%%%%%%%%%%%%%%%%%%%%%%%%%%%%%%%%%%%%%%%%%%%%%%%%}
{%%%%%%%%%%%%%%%%%%%%%%%%%%%%%%%%%%%%%%%%%%%%%%%%%%%%%%%%%%%%%%%%%%%%%%%%%%%%}

{_____Play All Library Templates_____}

{%%%%%%%%%%%%%%%%%%%%%%%%%%%%%%%%%%%%%%%%%%%%%%%%%%%%%%%%%%%%%%%%%%%%%%%%%%%%}
{%%%%%%%%%%%%%%%%%%%%%%%%%%%%%%%%%%%%%%%%%%%%%%%%%%%%%%%%%%%%%%%%%%%%%%%%%%%%}

play_library:
          AX0 = DM(catalog_size);
          AR  = PASS AX0;
          IF LE RTS;

          AR  = ^library_catalog;
          DM(current_selection) = AR; CNTR = AX0;
          DO playit UNTIL CE;
              I3  = DM(current_selection);
              CALL play_single;

playit:        NOP;
          RTS;

{%%%%%%%%%%%%%%%%%%%%%%%%%%%%%%%%%%%%%%%%%%%%%%%%%%%%%%%%%%%%%%%%%%%%%%%%%%%%}
{%%%%%%%%%%%%%%%%%%%%%%%%%%%%%%%%%%%%%%%%%%%%%%%%%%%%%%%%%%%%%%%%%%%%%%%%%%%%}

{_____Play Single Library Template_____}

{%%%%%%%%%%%%%%%%%%%%%%%%%%%%%%%%%%%%%%%%%%%%%%%%%%%%%%%%%%%%%%%%%%%%%%%%%%%%}
{%%%%%%%%%%%%%%%%%%%%%%%%%%%%%%%%%%%%%%%%%%%%%%%%%%%%%%%%%%%%%%%%%%%%%%%%%%%%}

{       required inputs:     I3  - location of library_catalog entry }

play_single:MX0 = DM(I3,M1);
          I5  = MX0;                   { start of library word }
          AX0 = DM(I3,M1);             { length }
          AX1 = DM(library_feature_dimension);
          DM(current_selection) = I3;

          CALL output_template;
          IMASK = 0;
          RTS;
```

Speech Recognition   6

```
{%%%%%%%%%%%%%%%%%%%%%%%%%%%%%%%%%%%%%%%%%%%%%%%%%%%%%%%%%%%%%%%%%%%%%%%%%}
{%%%%%%%%%%%%%%%%%%%%%%%%%%%%%%%%%%%%%%%%%%%%%%%%%%%%%%%%%%%%%%%%%%%%%%%%%}

{_____Play Ranked Library Templates in order_____}

{%%%%%%%%%%%%%%%%%%%%%%%%%%%%%%%%%%%%%%%%%%%%%%%%%%%%%%%%%%%%%%%%%%%%%%%%%}
{%%%%%%%%%%%%%%%%%%%%%%%%%%%%%%%%%%%%%%%%%%%%%%%%%%%%%%%%%%%%%%%%%%%%%%%%%}

{        required inputs:     AX0 - # of templates to play, in order
                             AY0 - shuffled order of templates }

shuffle_play:
            MR0 = DM(catalog_size);
            AR  = PASS MR0;
            IF LE RTS;

            DM(shuffle_pntr) = AY0;
            CNTR = AX0;
            DO play_shuffled UNTIL CE;
                  I1  = DM(shuffle_pntr);
                  AX0 = DM(I1,M1);
                  I3  = AX0;
                  DM(shuffle_pntr) = I1;

                  CALL play_single;

play_shuffled:    NOP;

            RTS;

#endif
{...........................................................................}

.ENDMOD;
```

Listing 6.8 Library Functions Routine (LIB_FUNC.DSP)

6 Speech Recognition

```
.MODULE/RAM/BOOT=0      coarse_distance;

.VAR/DM/RAM          unknown_addr, unknown_length;
.VAR/DM/RAM          candidate_distance1[30];
.VAR/DM/RAM          candidate_distance2[30];
.VAR/DM/RAM          candidate_distance3[30];
.VAR/DM/RAM          candidate_distance4[30];
.VAR/DM/RAM          current_compare, result_buffer;
.VAR/DM/RAM          distance_routine;

.EXTERNAL            catalog_size, library_catalog;
{.EXTERNAL           coarse_euclidean, fine_euclidean;}
.EXTERNAL            warp_words;
{.EXTERNAL           full_euclidean;}
.EXTERNAL            cepstral_projection;
.EXTERNAL            set_bank_select;
.EXTERNAL            inc_bank_select;
.EXTERNAL            show_bank;
.EXTERNAL            blank_hex_led;
.EXTERNAL            init_catalog;

.GLOBAL             candidate_distance1;
.GLOBAL             candidate_distance2;
.GLOBAL             candidate_distance3;
.GLOBAL             candidate_distance4;

{.ENTRY             coarse_compare, fine_compare, full_compare;}
.ENTRY              cepstral_compare;

{%%%%%%%%%%%%%%%%%%%%%%%%%%%%%%%%%%%%%%%%%%%%%%%%%%%%%%%%%%%%%%%%%%%%%%%%%%%%%%%}
{%%%%%%%%%%%%%%%%%%%%%%%%%%%%%%%%%%%%%%%%%%%%%%%%%%%%%%%%%%%%%%%%%%%%%%%%%%%%%%%}
{                                                                           }
{%%%%%%%%%%%%%%%%%%%%%%%%%%%%%%%%%%%%%%%%%%%%%%%%%%%%%%%%%%%%%%%%%%%%%%%%%%%%%%%}
{%%%%%%%%%%%%%%%%%%%%%%%%%%%%%%%%%%%%%%%%%%%%%%%%%%%%%%%%%%%%%%%%%%%%%%%%%%%%%%%}

{coarse_compare: AX1 = ^coarse_euclidean;
        JUMP compare;
}

{%%%%%%%%%%%%%%%%%%%%%%%%%%%%%%%%%%%%%%%%%%%%%%%%%%%%%%%%%%%%%%%%%%%%%%%%%%%%%%%}
{%%%%%%%%%%%%%%%%%%%%%%%%%%%%%%%%%%%%%%%%%%%%%%%%%%%%%%%%%%%%%%%%%%%%%%%%%%%%%%%}
{                                                                           }
{%%%%%%%%%%%%%%%%%%%%%%%%%%%%%%%%%%%%%%%%%%%%%%%%%%%%%%%%%%%%%%%%%%%%%%%%%%%%%%%}
{%%%%%%%%%%%%%%%%%%%%%%%%%%%%%%%%%%%%%%%%%%%%%%%%%%%%%%%%%%%%%%%%%%%%%%%%%%%%%%%}
{fine_compare: AX1 = ^fine_euclidean;
        JUMP compare;
}
```

Speech Recognition   6

```
{%%%%%%%%%%%%%%%%%%%%%%%%%%%%%%%%%%%%%%%%%%%%%%%%%%%%%%%%%%%%%%%%%%%%%%%%%%%}
{%%%%%%%%%%%%%%%%%%%%%%%%%%%%%%%%%%%%%%%%%%%%%%%%%%%%%%%%%%%%%%%%%%%%%%%%%%%}
{                                                                         }
{%%%%%%%%%%%%%%%%%%%%%%%%%%%%%%%%%%%%%%%%%%%%%%%%%%%%%%%%%%%%%%%%%%%%%%%%%%%}
{%%%%%%%%%%%%%%%%%%%%%%%%%%%%%%%%%%%%%%%%%%%%%%%%%%%%%%%%%%%%%%%%%%%%%%%%%%%}
{full_compare: AX1 = ^full_euclidean;
        JUMP compare;
}


{%%%%%%%%%%%%%%%%%%%%%%%%%%%%%%%%%%%%%%%%%%%%%%%%%%%%%%%%%%%%%%%%%%%%%%%%%%%}
{%%%%%%%%%%%%%%%%%%%%%%%%%%%%%%%%%%%%%%%%%%%%%%%%%%%%%%%%%%%%%%%%%%%%%%%%%%%}
{                                                                         }
{%%%%%%%%%%%%%%%%%%%%%%%%%%%%%%%%%%%%%%%%%%%%%%%%%%%%%%%%%%%%%%%%%%%%%%%%%%%}
{%%%%%%%%%%%%%%%%%%%%%%%%%%%%%%%%%%%%%%%%%%%%%%%%%%%%%%%%%%%%%%%%%%%%%%%%%%%}
{       required inputs:    I0  -> location of unknown word
                           AX0 -> length of unknown word
}
cepstral_compare:
        DM(unknown_addr) = I0;
        DM(unknown_length) = AX0;

        AR  = 0;
        CALL set_bank_select;       { set memory bank }
        CALL show_bank;             { display memory bank }
        CALL init_catalog;          { initialize library catalog }

        { initialize results pointer, distance measure }
        I1  = ^candidate_distance1;
        AX1 = ^cepstral_projection;
        CALL compare;

        CALL inc_bank_select;       { set memory bank }
        CALL show_bank;             { display memory bank }
        CALL init_catalog;          { initialize library catalog }

        { initialize results pointer, distance measure }
        I1  = ^candidate_distance2;
        AX1 = ^cepstral_projection;
        CALL compare;

        CALL inc_bank_select;       { set memory bank }
        CALL show_bank;             { display memory bank }
        CALL init_catalog;          { initialize library catalog }

        { initialize results pointer, distance measure }
        I1  = ^candidate_distance3;
        AX1 = ^cepstral_projection;
        CALL compare;
```

(listing continues on next page)

6 Speech Recognition

```
        CALL inc_bank_select;              { set memory bank }
        CALL show_bank;                    { display memory bank }
        CALL init_catalog;                 { initialize library catalog }

        { initialize results pointer, distance measure }
        I1  = ^candidate_distance4;
        AX1 = ^cepstral_projection;

        CALL compare;

        CALL knn_routine;

        CALL blank_hex_led;
        RTS;

{%%%%%%%%%%%%%%%%%%%%%%%%%%%%%%%%%%%%%%%%%%%%%%%%%%%%%%%%%%%%%%%%%%%%%%%%%%%%%%%%%}
{%%%%%%%%%%%%%%%%%%%%%%%%%%%%%%%%%%%%%%%%%%%%%%%%%%%%%%%%%%%%%%%%%%%%%%%%%%%%%%%%%}
{                                                                               }
{%%%%%%%%%%%%%%%%%%%%%%%%%%%%%%%%%%%%%%%%%%%%%%%%%%%%%%%%%%%%%%%%%%%%%%%%%%%%%%%%%}
{%%%%%%%%%%%%%%%%%%%%%%%%%%%%%%%%%%%%%%%%%%%%%%%%%%%%%%%%%%%%%%%%%%%%%%%%%%%%%%%%%}
{       required inputs:     I1  -> location of results buffer
                            AX1 -> pointer to distance measure
}

compare: AY0 = DM(catalog_size);
         AR  = PASS AY0;
         IF LE RTS;

         CNTR = AY0;
         DM(distance_routine) = AX1;
         AR  = ^library_catalog;
         DM(current_compare) = AR;
         DM(result_buffer) = I1;            { stored as msw, lsw for each }

         DO calc_dist UNTIL CE;
             I0  = DM(unknown_addr);
             AX0 = DM(unknown_length);
             I3  = DM(current_compare);
             I6  = DM(distance_routine);
             I5  = DM(result_buffer);

             CALL warp_words;

             I1  = DM(result_buffer);
             MODIFY(I1,M3);
             DM(result_buffer) = I1;

             I3  = DM(current_compare);
             MODIFY(I3,M3);
calc_dist: DM(current_compare) = I3;

         RTS;
```

404

Speech Recognition 6

```
{%%%%%%%%%%%%%%%%%%%%%%%%%%%%%%%%%%%%%%%%%%%%%%%%%%%%%%%%%%%%%%%%%%%%%%%%%%%}
{%%%%%%%%%%%%%%%%%%%%%%%%%%%%%%%%%%%%%%%%%%%%%%%%%%%%%%%%%%%%%%%%%%%%%%%%%%%}
{%%%%%%%%%%%%%%%%%%%%%%%%%%%%%%%%%%%%%%%%%%%%%%%%%%%%%%%%%%%%%%%%%%%%%%%%%%%}
{                                                                         }
{%%%%%%%%%%%%%%%%%%%%%%%%%%%%%%%%%%%%%%%%%%%%%%%%%%%%%%%%%%%%%%%%%%%%%%%%%%%}
{%%%%%%%%%%%%%%%%%%%%%%%%%%%%%%%%%%%%%%%%%%%%%%%%%%%%%%%%%%%%%%%%%%%%%%%%%%%}
knn_routine:
        AY0 = DM(catalog_size);
        AR  = PASS AY0;
        IF LE RTS;

        CNTR = AY0;
        I0  = ^candidate_distance1;
        I1  = ^candidate_distance2;
        I2  = ^candidate_distance3;
        I3  = ^candidate_distance4;
        I4  = ^candidate_distance1;
        DO compute_knn UNTIL CE;
          I5  = I0;
comp12:   AX1 = DM(I0,M1);
          AX0 = DM(I0,M2);
          AY1 = DM(I1,M1);
          AY0 = DM(I1,M2);
          AR  = AY0 - AX0;
          AR  = AY1 - AX1 + C -1;
          IF GE JUMP comp23;
          I5  = I1;
          AX1 = AY1;
          AX0 = AY0;

comp23:   AY1 = DM(I2,M1);
          AY0 = DM(I2,M2);
          AR  = AY0 - AX0;
          AR  = AY1 - AX1 + C -1;
          IF GE JUMP comp34;
          I5  = I2;
          AX1 = AY1;
          AX0 = AY0;

comp34:   AY1 = DM(I3,M1);
          AY0 = DM(I3,M2);
          AR  = AY0 - AX0;
          AR  = AY1 - AX1 + C -1;
          IF GE JUMP store_best;
          I5  = I3;
          AX1 = AY1;
          AX0 = AY0;
```

(listing continues on next page)

6 Speech Recognition

```
store_best:  SR1 = AX1;
             SR0 = AX0;
             AX1 = H#7FFF;
             DM(I5,M5) = AX1;
             AX0 = H#FFFF;
             DM(I5,M5) = AX0;

comp12_2:    AX1 = DM(I0,M1);
             AX0 = DM(I0,M1);
             AY1 = DM(I1,M1);
             AY0 = DM(I1,M1);
             AR  = AY0 - AX0;
             AR  = AY1 - AX1 + C -1;
             IF GE JUMP comp23_2;
             AX1 = AY1;
             AX0 = AY0;

comp23_2:    AY1 = DM(I2,M1);
             AY0 = DM(I2,M1);
             AR  = AY0 - AX0;
             AR  = AY1 - AX1 + C -1;
             IF GE JUMP comp34_2;
             AX1 = AY1;
             AX0 = AY0;

comp34_2:    AY1 = DM(I3,M1);
             AY0 = DM(I3,M1);
             AR  = AY0 - AX0;
             AR  = AY1 - AX1 + C -1;
             IF GE JUMP sum_top_2;
             AX1 = AY1;
             AX0 = AY0;

sum_top_2:   AY1 = SR1;
             AY0 = SR0;
             AR  = AX0 + AY0;
             ENA AR_SAT;
             AX0 = AR, AR  = AX1 + AY1 + C;
             DIS AR_SAT;
             DM(I4,M5) = AR;
compute_knn:DM(I4,M5) = AX0;

        RTS;

{%%%%%%%%%%%%%%%%%%%%%%%%%%%%%%%%%%%%%%%%%%%%%%%%%%%%%%%%%%%%%%%%%%%%%%%%%%%%%}

.ENDMOD;
```

Listing 6.9 Word Comparison Routine (COMPLIB.DSP)

Speech Recognition   6

{_____

      Analog Devices Inc., DSP Division
      One Technology Way, Norwood, MA 02062
      DSP Applications Assistance: (617) 461-3672
_____

RANKDIST.DSP

The routine in this module will rank the recognition results based upon
distances contained in the candidate distance buffer. All the words of the
catalog are ranked (up to 15). The results are stored in two forms.
Candidate_order contains a pointer to the catalog entry for each word, and
order_number contains the number (zero to fourteen) each word has in the
library. Both these buffers have the best recognition candidate first, then
the second best, and so on.

_____ }


```
.MODULE/RAM/BOOT=0        rank_candidates_distances;

.VAR/DM/RAM        candidate_order[15], order_number[15];

.EXTERNAL          candidate_distance1;
.EXTERNAL          catalog_size;
.EXTERNAL          library_catalog;

.ENTRY             rank_candidates;

{%%%%%%%%%%%%%%%%%%%%%%%%%%%%%%%%%%%%%%%%%%%%%%%%%%%%%%%%%%%%%%%%%%%%%%%%%%}
{%%%%%%%%%%%%%%%%%%%%%%%%%%%%%%%%%%%%%%%%%%%%%%%%%%%%%%%%%%%%%%%%%%%%%%%%%%}

{_____Rank Candidates by Distance_____}

{%%%%%%%%%%%%%%%%%%%%%%%%%%%%%%%%%%%%%%%%%%%%%%%%%%%%%%%%%%%%%%%%%%%%%%%%%%}
{%%%%%%%%%%%%%%%%%%%%%%%%%%%%%%%%%%%%%%%%%%%%%%%%%%%%%%%%%%%%%%%%%%%%%%%%%%}

rank_candidates:
        MR0 = DM(catalog_size);
        AR  = PASS MR0;
        IF LE RTS;

        I1  = ^order_number;
        I2  = ^candidate_order;
        AX1 = H#7FFF;
        AX0 = H#FFFF;
        CNTR = MR0;                      { MR0 holds catalog_size }
        DO sort_candidates UNTIL CE;
            I0  = ^candidate_distance1;
            I3  = ^library_catalog;
            AF  = PASS 0;                { initialize library entry # }
            CNTR = MR0;
            DO least_dist UNTIL CE;
```

(listing continues on next page)

407

6 Speech Recognition

```
                AY1 = DM(I0,M1);
                AY0 = DM(I0,M1);
                AR  = AY0 - AX0;
                AR  = AY1 - AX1 + C - 1;
                IF GE JUMP inc_word_count;

                MX0 = I3;
                DM(I2,M0) = MX0;    { store library_catalog pointer}

                AR  = PASS AF;
                DM(I1,M0) = AR;     { store library entry/order # }
                AX1 = AY1;          { update threshold }
                AX0 = AY0;
                MR1 = I0;           { save till loop complete }

inc_word_count:   AF  = AF + 1;
least_dist:       MODIFY(I3,M3);

                MODIFY(I2,M1);          { pointer to candidate_order }
                MODIFY(I1,M1);          { pointer to order_number }
                I0  = MR1;
                MODIFY(I0,M2);
                MODIFY(I0,M2);
                AX1 = H#7FFF;
                AX0 = H#FFFF;           { effectively removes the least}
                DM(I0,M1) = AX1;        { distance found from the }
sort_candidates:
                DM(I0,M1) = AX0;        { candidate_distance buffer }

          AY0 = ^candidate_order;
          AY1 = DM(order_number);
          RTS;

.ENDMOD;
```

Listing 6.10 Word Ranking Routine (RANKDIST.DSP)

Speech Recognition   6

```
{_____

        Analog Devices Inc., DSP Division
        One Technology Way, Norwood, MA  02062
        DSP Applications Assistance:  (617) 461-3672
```

WARPSHELL.DSP

The routine contained in this module will calculate the distance between a
library template and an unknown word. It accomplishes this using dynamic time
warping and a selected distance measure. The resulting distance is a double
precision value, stored as msw, lsw.

The routine first calculates several necessary values. As it begins the
warping, it branches into one of three code regions, depending on the
relationship between Xa and Xb.

Each of these regions is divided into two or three different warping sections.
Each section has a different set of warping constraints, so y_min and y_max
must be calculated differently. Pointers to the correct min/max routines are
initialized in each section.

Following this, the warp_section routine performs the actual Dynamic Time
Warping for the current section. Y values are calculated each time through the
loop. The x_coordinate is incremented each time through the loop.

Update_sums copies previous results, stored in the vector_distance_buffer,
into the intermediate sums buffer. A new column of distance is then calculated
and stored in the vector distance buffer. Finally, the time warping occurs,
and the x value is incremented for the next loop.

```
        _____}
```

```
.MODULE/RAM/BOOT=0        dtw_shell;

.VAR/DM/RAM        Xa, Xb;       { points where warping constraint changes }
.VAR/DM/RAM        M, N;        { # of vectors in library template,unknown word}
.VAR/DM/RAM        x_coordinate;      { current x value }
.VAR/DM/RAM        y_min_routine, y_max_routine;
.VAR/DM/RAM        library_word_start, x_vector_pntr;
.VAR/DM/RAM        old_y_min;
.VAR/DM/RAM        vector_distance_buffer[180];
                               {filled with distance matrix col.}
.VAR/DM/RAM        intermediate_sum_buffer[192]; {minimum sum of possible
                                          warping paths, stored as msw,
                                          lsw, previous warping slope }
.VAR/DM/RAM        result_pntr, distance_measure;
```

(listing continues on next page)

409

6 Speech Recognition

```
.EXTERNAL    pre_Xa_y_max, post_Xa_y_max;
.EXTERNAL    pre_Xb_y_min, post_Xb_y_min;
.EXTERNAL    y_min, y_max, y_range;
.EXTERNAL    calc_y_range;
.EXTERNAL    build_vd_buff;
.EXTERNAL    compute_warp, update_sums;
.EXTERNAL    unknown_feature_dimension;

.GLOBAL      M, N;
.GLOBAL      x_vector_pntr;
.GLOBAL      old_y_min;
.GLOBAL      vector_distance_buffer, intermediate_sum_buffer;

.ENTRY       warp_words;

{%%%%%%%%%%%%%%%%%%%%%%%%%%%%%%%%%%%%%%%%%%%%%%%%%%%%%%%%%%%%%%%%%%%%%%%%%%%%%%%%}
{%%%%%%%%%%%%%%%%%%%%%%%%%%%%%%%%%%%%%%%%%%%%%%%%%%%%%%%%%%%%%%%%%%%%%%%%%%%%%%%%}

{_____use dynamic time warping to compare two words_____}

{%%%%%%%%%%%%%%%%%%%%%%%%%%%%%%%%%%%%%%%%%%%%%%%%%%%%%%%%%%%%%%%%%%%%%%%%%%%%%%%%}
{%%%%%%%%%%%%%%%%%%%%%%%%%%%%%%%%%%%%%%%%%%%%%%%%%%%%%%%%%%%%%%%%%%%%%%%%%%%%%%%%}

{        required inputs:
                    I3  -> location of word in template catalog
                    I0  -> location of unknown word
                    AX0 -> length (# of vectors) of unknown word
                    I6  -> pointer to distance routine
                    I5  -> location of results (msw, lsw)              }

warp_words:
        DM(x_vector_pntr) = I0;
        AY0 = AX0;
        AR  = AY0 - 1;              { axis starts at 0, not 1 }
        DM(N) = AR;

        MX0 = DM(I3,M1);            { read lib template location }
        DM(library_word_start) = MX0;
        AY0 = DM(I3,M2);           { read lib template length }
        AX0 = AR, AR  = AY0 - 1;   { y axis starts at 0, not 1 }
        DM(M) = AR;

        DM(distance_measure) = I6;
        DM(result_pntr) = I5;
```

Speech Recognition 6

```
{_____check if ( 2*M-N < 3 )_____}

        AY0 = AR;
        AR  = AR + AY0, AY1 = AX0;
        AR  = AR - AY1;
        MX0 = AR;                       { MX0 = (2 * M) - N }
        AY1 = 3;
        AR  = AR - AY1;
        IF LT JUMP cannot_warp;

{_____check if ( 2*N-M < 2 )_____}

        AF  = PASS AX0;
        AR  = AX0 + AF;
        AR  = AR - AY0;                 { AR = (2 * N) - M }
        AY1 = 2;
        AF  = AR - AY1;
        IF LT JUMP cannot_warp;
        AY1 = AR;
        AR  = AR + AY1;
        MX1 = AR;                       { MX1 = 2*( (2*N) - M ) }

{_____compute Xa, Xb_____}

        MY0 = H#2AAB;                   { MY0 = (1/3) that always rounds down }
        MR  = MX0 * MY0 (UU);
        DM(Xa) = MR1;                   { Xa = (1/3) * ( (2*M) - N ) }
        AY1 = MR1, MR  = MX1 * MY0 (UU);
        DM(Xb) = MR1;                   { Xb = (2/3) * ( (2*N) - M ) }

{_____setup for warping_____}

        AX0 = 0;
        DM(y_min) = AX0;
        AX0 = 1;
        DM(y_range) = AX0;
        DM(x_coordinate) = AX0;

{_____calculate distance between first vectors (x,y = 0)_____}

        I0  = DM(x_vector_pntr);
        I4  = DM(library_word_start);
        I6  = DM(distance_measure);
        CALL (I6);
        I3  = ^vector_distance_buffer;
        DM(I3,M1) = SR1;
        DM(I3,M3) = SR0;
```

(listing continues on next page)

6 Speech Recognition

```
        I0  = DM(x_vector_pntr);
        M3  = DM(unknown_feature_dimension);
        MODIFY(I0,M3);                          { I0 points to next x vector }
        M3  = 2;
        DM(x_vector_pntr) = I0;
```

```
{_____find relationship of Xa, Xb_____}
```

```
        AX0 = DM(Xa);
        AY0 = DM(Xb);
        AR  = AX0 - AY0;                   { AR = Xa - Xb }
        IF GT JUMP Xa_gt_Xb;
        IF LT JUMP Xb_gt_Xa;
```

```
{===================== match words - Xa equal to Xb =====================}
```

```
Xa_eq_Xb:
        CNTR = DM(Xa);         {# of x vectors in first section is Xa }

        AX0 = ^pre_Xa_y_max;
        DM(y_max_routine) = AX0;
        AX0 = ^pre_Xb_y_min;
        DM(y_min_routine) = AX0;
        CALL warp_section;
        AX0 = DM(Xb);
        AY0 = DM(N);
        AR  = AY0 - AX0;
        CNTR = AR;             { # of x vectors in final section is (N-Xb) }
        AX0 = ^post_Xa_y_max;
        DM(y_max_routine) = AX0;
        AX0 = ^post_Xb_y_min;
        DM(y_min_routine) = AX0;
        CALL warp_section;

        JUMP write_result;
```

```
{===================== match words - Xb greater than Xa =====================}
```

```
Xb_gt_Xa:
        CNTR = DM(Xa);         {# of x vectors in first section is Xa }

        AX0 = ^pre_Xa_y_max;
        DM(y_max_routine) = AX0;
        AX0 = ^pre_Xb_y_min;
        DM(y_min_routine) = AX0;
        CALL warp_section;

        AX0 = DM(Xa);
        AY0 = DM(Xb);
        AR  = AY0 - AX0;
        CNTR = AR;             {# of x vectors in middle section is (Xb-Xa) }
```

```
        AX0 = ^post_Xa_y_max;
        DM(y_max_routine) = AX0;
        AX0 = ^pre_Xb_y_min;
        DM(y_min_routine) = AX0;
        CALL warp_section;

        AX0 = DM(Xb);
        AY0 = DM(N);
        AR  = AY0 - AX0;
        CNTR = AR;              {# of x vectors in final section is (N-Xb) }
        AX0 = ^post_Xa_y_max;
        DM(y_max_routine) = AX0;
        AX0 = ^post_Xb_y_min;
        DM(y_min_routine) = AX0;
        CALL warp_section;

        JUMP write_result;

{=================== match words - Xa greater than Xb ======================}

Xa_gt_Xb:CNTR = DM(Xb);        {# of x vectors in first section is Xb }

        AX0 = ^pre_Xa_y_max;
        DM(y_max_routine) = AX0;
        AX0 = ^pre_Xb_y_min;
        DM(y_min_routine) = AX0;
        CALL warp_section;

        AX0 = DM(Xb);
        AY0 = DM(Xa);
        AR  = AY0 - AX0;
        CNTR = AR;              {# of x vectors in middle section is (Xa-Xb) }
        AX0 = ^pre_Xa_y_max;
        DM(y_max_routine) = AX0;
        AX0 = ^post_Xb_y_min;
        DM(y_min_routine) = AX0;
        CALL warp_section;

        AX0 = DM(Xa);
        AY0 = DM(N);
        AR  = AY0 - AX0;
        CNTR = AR;              {# of x vectors in final section is (N-Xa) }
        AX0 = ^post_Xa_y_max;
        DM(y_max_routine) = AX0;
        AX0 = ^post_Xb_y_min;
        DM(y_min_routine) = AX0;
        CALL warp_section;

        JUMP write_result;
```

(listing continues on next page)

6   Speech Recognition

```
{===================== cannot warp due to M, N ==============================}

cannot_warp:
        I3  = ^vector_distance_buffer;
        AX0 = H#7FFF;                    { sets word distance score to }
        DM(I3,M1) = AX0;                 {   maximum double-precision }
        AX0 = H#FFFF;                    {   value. }
        DM(I3,M1) = AX0;

{=================== write word distance result and return =================}

write_result:
        I3  = ^vector_distance_buffer;
        I0  = DM(result_pntr);
        AX0 = DM(I3,M1);
        DM(I0,M1) = AX0;
        AX0 = DM(I3,M1);
        DM(I0,M1) = AX0;

        RTS;
{=========================================================================}
{_____warp an entire section_____}
{=========================================================================}

warp_section:
          DO section UNTIL CE;
          SR0 = DM(y_min);
          DM(old_y_min) = SR0;

          SR0 = DM(x_coordinate);        { current x value }
          I6  = DM(y_max_routine);
          CALL (I6);                     { calculate maximum y value }
          I6  = DM(y_min_routine);
          CALL (I6);                     { calculate minimum y value }

          CALL update_sums;             { copy previous results to sums buffer }

          CALL calc_y_range;            { calculate range of y for warp}

          I6  = DM(distance_measure);
          I5  = DM(library_word_start);

          CALL build_vd_buff;           {compute distance matrix column}

          CALL compute_warp;            {warp sums buffer into current column}
```

```
            I0  = DM(x_vector_pntr);
            M3  = DM(unknown_feature_dimension);
            MODIFY(I0,M3);                    {next unknown(x) feature vector}
            M3  = 2;
            DM(x_vector_pntr) = I0;
            AY0 = DM(x_coordinate);
            AR  = AY0 + 1;                    {increment x value counter}
section:    DM(x_coordinate) = AR;
        RTS;

{%%%%%%%%%%%%%%%%%%%%%%%%%%%%%%%%%%%%%%%%%%%%%%%%%%%%%%%%%%%%%%%%%%%%%%%%%%%%%%%}

.ENDMOD;
```

Listing 6.11 Library Template/Word Distance Routine (WARPSHEL.DSP)

6 Speech Recognition

```
{_____
            Analog Devices Inc., DSP Division
            One Technology Way, Norwood, MA  02062
            DSP Applications Assistance:  (617) 461-3672
   _____
```

```
YMINMAX.DSP
The routines in this module calculate the minimum, maximum, and range of the
ycoordinate for dynamic time warping, using Itakura warping constraints. The
points Xa and Xb are the x-coordinate locations where the constraining slope
changes for the upper and lower boundaries, respectively.
   _____}


.MODULE/RAM/BOOT=0       find_y_bounds;

.VAR/DM/RAM y_min, y_max, y_range;

.GLOBAL     y_min, y_max, y_range;

.EXTERNAL   N, M;

.ENTRY      pre_Xb_y_min, post_Xb_y_min;
.ENTRY      pre_Xa_y_max, post_Xa_y_max;
.ENTRY      calc_y_range;

{%%%%%%%%%%%%%%%%%%%%%%%%%%%%%%%%%%%%%%%%%%%%%%%%%%%%%%%%%%%%%%%%%%%%%%%%%%%%}
{%%%%%%%%%%%%%%%%%%%%%%%%%%%%%%%%%%%%%%%%%%%%%%%%%%%%%%%%%%%%%%%%%%%%%%%%%%%%}
{_____routines to find y_min, y_max_____}
{%%%%%%%%%%%%%%%%%%%%%%%%%%%%%%%%%%%%%%%%%%%%%%%%%%%%%%%%%%%%%%%%%%%%%%%%%%%%}
{%%%%%%%%%%%%%%%%%%%%%%%%%%%%%%%%%%%%%%%%%%%%%%%%%%%%%%%%%%%%%%%%%%%%%%%%%%%%}

{       required inputs:     SR0 -> current x_coordinate
}

{_____ y_min = .5 * x _____}

pre_Xb_y_min:
            MR  = 0, MX0 = SR0;
            MY0 = H#4000;                { .5 in fixed-point format }
            MR0 = H#8000;
            MR  = MR + MX0 * MY0 (UU);
            DM(y_min) = MR1;
            RTS;

{_____ y_min = 2(x-N) + M _____}

post_Xb_y_min:
            AY0 = DM(N);
            AR  = SR0 - AY0;        { AR = x_coordinate - N }
            AY0 = AR;
            AR  = AR + AY0;         { AR = 2 * (x_coordinate - N) }
```

416

```
            AY1 = DM(M);
            AR  = AR + AY1;            { AR = 2 * (x_coordinate - N) + M }
            DM(y_min) = AR;
            RTS;

{_____ y_max = 2 * x _____}

pre_Xa_y_max:
            AY0 = SR0;
            AR  = SR0 + AY0;
            DM(y_max) = AR;

            RTS;
{_____ y_max = .5*(x-N) + M _____}

post_Xa_y_max:
            AY0 = DM(N);
            AR  = SR0 - AY0;
            MR  = 0;
            MR1 = DM(M);
            MY0 = H#4000;             { .5 in fixed-point format }
            MR  = MR + AR * MY0 (SS);
            DM(y_max) = MR1;
            RTS;

{%%%%%%%%%%%%%%%%%%%%%%%%%%%%%%%%%%%%%%%%%%%%%%%%%%%%%%%%%%%%%%%%%%%%%%%%%%%%%%%%%}
{%%%%%%%%%%%%%%%%%%%%%%%%%%%%%%%%%%%%%%%%%%%%%%%%%%%%%%%%%%%%%%%%%%%%%%%%%%%%%%%%%}

{_____routine to find y_range_____}

{%%%%%%%%%%%%%%%%%%%%%%%%%%%%%%%%%%%%%%%%%%%%%%%%%%%%%%%%%%%%%%%%%%%%%%%%%%%%%%%%%}
{%%%%%%%%%%%%%%%%%%%%%%%%%%%%%%%%%%%%%%%%%%%%%%%%%%%%%%%%%%%%%%%%%%%%%%%%%%%%%%%%%}

{_____ y_range = y_max - y_min + 1 _____}

calc_y_range:
            AX0 = DM(y_max);
            AY0 = DM(y_min);
            AF  = AX0 - AY0;
            AR  = AF + 1;
            DM(y_range) = AR;
            RTS;

{%%%%%%%%%%%%%%%%%%%%%%%%%%%%%%%%%%%%%%%%%%%%%%%%%%%%%%%%%%%%%%%%%%%%%%%%%%%%%%%%%}

.ENDMOD;
```

Listing 6.12 Y Coordinate Range Routine (YMINMAX.DSP)

6 Speech Recognition

{_____

      Analog Devices Inc., DSP Division
      One Technology Way, Norwood, MA 02062
      DSP Applications Assistance: (617) 461-3672
_____

TIMEWARP.DSP

Two distinct routines are contained in this module, both dealing with dynamic time warping. The first update_sums copies the current vector_distance_buffer contents into the intermediate_sum_buffer. The future illegal warping paths are removed by saturating the boundary distances before and after copying, using the y_offset value. y_offset measures the difference between the minimum y value of two adjacent columns.

The second routine, compute_warp, will perform the dynamic time warping between two columns. One column is the intermediate_sum_buffer, which contains the results of all previous warps. The other column is the vector_distance_buffer, which contains a column of the distance matrix. A column consists of the distances between a single unknown and many library template feature vectors.

The sum buffer is warped into the distance buffer. The previous warping path is examined in each case to prevent an illegal warp, and the accumulated sums are stored into the distance buffer.

The y_offset is the difference (in the y direction) between the y_min of the (x)th column and y_min (stored as old_y_min) of the (x-1)th column.
_____}

```
.MODULE/RAM/BOOT=0       dynamic_time_warping;

.EXTERNAL         intermediate_sum_buffer, vector_distance_buffer;
.EXTERNAL         y_min, old_y_min, y_range;

.ENTRY            compute_warp;
.ENTRY            update_sums;

{=========================================================================}
{=========================================================================}

{_____move vector_distance to intermediate_sum_____}

{=========================================================================}
{=========================================================================}
```

```
update_sums:
        I1  = ^intermediate_sum_buffer;
        AX1 = H#7FFF;
        AX0 = H#FFFF;
        DM(I1,M1) = AX1;            { initialize D(x-1,y-2) }
        DM(I1,M3) = AX0;
        DM(I1,M1) = AX1;            { initialize D(x-1,y-1) }
        DM(I1,M3) = AX0;            { leaves I1 pointing to D(x-1,y) }

        AR  = DM(y_min);
        AY0 = DM(old_y_min);
        AR  = AR - AY0;             { AR is y_offset }

        AY1 = -3;
        AY0 = -6;
        AF  = PASS 1, MR0 = AR;     { MR0 is y_offset }
        AR  = PASS 0;
        AF  = AF - MR0;
        IF LT AR = PASS AY0;
        IF EQ AR = PASS AY1;        { now AR holds real offset value }

        M3  = AR;
        MODIFY(I1,M3);              { now I1 points to }
        M3  = 2;                    { ( D(x-1,y) + real offset value ) }

        I5  = ^vector_distance_buffer;
        CNTR = DM(y_range);
        DO copy_sum UNTIL CE;
           CNTR = 3;
           DO copy_parts UNTIL CE;
                   AY0 = DM(I5,M5);
copy_parts:        DM(I1,M1) = AY0;
copy_sum:  NOP;

        DM(I1,M1) = AX1;            { initialize D(x-1,y_max+1) }
        DM(I1,M3) = AX0;
        DM(I1,M1) = AX1;            { initialize D(x-1,y_max+2) }
        DM(I1,M3) = AX0;

        RTS;
```

(listing continues on next page)

6 Speech Recognition

```
{=======================================================================}
{=======================================================================}

{_____warp one column in x dimension_____}

{=======================================================================}
{=======================================================================}

{        required inputs:           y_range = (y_max - y_min + 1)        }

{_____setup_____}

compute_warp:
        M0   = -5;
        M1   = -4;
        M2   =  1;
        M3   =  8;
        M7   =  2;
        I5   = ^vector_distance_buffer;
        I1   = ^intermediate_sum_buffer + 6;     { points to D(x-1,y) }

        CNTR = DM(y_range);
        DO warp_buffer UNTIL CE;

{_____compare D(x-1,y-1), D(x-1,y)_____}

compare_warp_1:
        AF   = PASS 0, AX1 = DM(I1,M2);       {init warp_value in AF }

        AX0  = DM(I1,M2);                     { read D(x-1,y) }

        AR   = DM(I1,M0);                     { read old_warp_value }

        AY1  = DM(I1,M2);
        AR   = PASS AR, AY0 = DM(I1,M1);      { read D(x-1,y-1) }
        IF NE JUMP do_comparison;            {jump if old_warp_value NE 0}

        AR   = DM(y_range);                   { if old_warp_value = 0, only  }

        AF   = PASS AR;                       { allow consecutive warps of 0 }

        AR   = CNTR;                          { when it's the first warp of  }

        AR   = AR - AF;                       { the column                   }

        IF NE JUMP set_warp_1;
        AF   = PASS 0;                        { reset state }
```

```
do_comparison:
          AR  = AX0 - AY0;
          AR  = AX1 - AY1 + C - 1;
          IF LT JUMP compare_warp_2;

set_warp_1:
          AF  = PASS 1, AX1 = AY1;              { change warp_value }
          AX0 = AY0;                            { move D(x-1,y-1) to AXn }

{_____compare D(x-1,y-2), minimum[D(x-1,y),D(x-1,y-1)]_____}

compare_warp_2:
          AY1 = DM(I1,M2);
          AY0 = DM(I1,M3);                      { read D(x-1,y-2) }
          AR  = AX0 - AY0;
          AR  = AX1 - AY1 + C - 1;
          IF LT JUMP compute_sum;

set_warp_2: AX1 = AY1;
          AY1 = 2;                              { change warp_value }
          AF  = PASS AY1, AX0 = AY0;            { move D(x-1,y-2) to AXn }

{_____compute sum of d(x,y), minimum[D(x-1,y),D(x-1,y-1),D(x-1,y-2)]_____}

compute_sum:AY1 = DM(I5,M5);
          AY0 = DM(I5,M4);                      { read d(x,y) }
          AR  = AX0 + AY0;

          ena ar_sat;

          DM(I5,M6) = AR, AR  = AX1 + AY1 + C; { write D(x,y) lsw }

          dis ar_sat;

{          if av trap;}
          DM(I5,M7) = AR, AR  = PASS AF;        { write D(x,y) msw }
warp_buffer:DM(I5,M5) = AR;                     { write D(x,y) warp_value }

{_____restore state_____}

       M0  = 0;
       M1  = 1;
       M2  = -1;
       M3  = 2;
       M7  = 2;
       RTS;

{============================================================================}

.ENDMOD;
```

Listing 6.13 Dynamic Time Warping Routine (TIMEWARP.DSP)

6   Speech Recognition

{_____
            Analog Devices Inc., DSP Division
            One Technology Way, Norwood, MA 02062
            DSP Applications Assistance: (617) 461-3672
_____

VECTDIST.DSP

This routine will calculate the distances necessary to fill one column of the
distance matrix. It uses a single vector of the unknown word (x dimension) and
the correct range of the library template vectors (y dimension). The resulting
distances are stored in the vector_distance_buffer.

There are two implemented distortion measures, the full euclidean and the
cepstral projection. A pointer to the selected measure is passed to this
module in I6. Additional distortion measures can easily be added, following
the structure of the current measures.
_____ }

```
.MODULE/RAM/BOOT=0        build_vector_distance;

.EXTERNAL         y_min, y_range;
.EXTERNAL         x_vector_pntr, vector_distance_buffer;
.EXTERNAL         library_feature_dimension;

.ENTRY            build_vd_buff;
.ENTRY            full_euclidean;
.ENTRY            cepstral_projection;

{%%%%%%%%%%%%%%%%%%%%%%%%%%%%%%%%%%%%%%%%%%%%%%%%%%%%%%%%%%%%%%%%%%%%%%%%%%%%%%}
{%%%%%%%%%%%%%%%%%%%%%%%%%%%%%%%%%%%%%%%%%%%%%%%%%%%%%%%%%%%%%%%%%%%%%%%%%%%%%%}

{_____calculate distance matrix column_____}

{%%%%%%%%%%%%%%%%%%%%%%%%%%%%%%%%%%%%%%%%%%%%%%%%%%%%%%%%%%%%%%%%%%%%%%%%%%%%%%}
{%%%%%%%%%%%%%%%%%%%%%%%%%%%%%%%%%%%%%%%%%%%%%%%%%%%%%%%%%%%%%%%%%%%%%%%%%%%%%%}

{      required inputs:    I5 -> start of library template
                          I6 -> start of distance measure routine}

{_____calculate offset from start of library template_____}

build_vd_buff:
        MX0 = DM(y_min);
        SI  = DM(library_feature_dimension);
        SR  = ASHIFT SI BY -1 (HI);       { (# of features)/2 }
        MY0 = SR1;
        MR  = MX0 * MY0 (UU);             { 2 * y_min * (# features)/2 }

        M7  = MR0;
        MODIFY(I5,M7);                    { I5 now points to y_min feature vector}
```

Speech Recognition   6

```
{_____setup_____}

        M7  = DM(library_feature_dimension);
        I3  = ^vector_distance_buffer;
        CNTR = DM(y_range);

{_____calculate vector distances and store_____}

        DO build_buffer UNTIL CE;
            I0  = DM(x_vector_pntr);          { location of unknown vector }
            I4  = I5;                         { location of library template vector }

            CALL (I6);                        { calls distance measure routine }

            DM(I3,M1) = SR1;                  { store distance msw }
            DM(I3,M3) = SR0;                  { store distance lsw, skip warp_value }
build_buffer:
            MODIFY(I5,M7);          { I5 points to next library template vector }

{_____reset state and return_____}

        M7  = 2;
        RTS;

{%%%%%%%%%%%%%%%%%%%%%%%%%%%%%%%%%%%%%%%%%%%%%%%%%%%%%%%%%%%%%%%%%%%%%%%%%%%%%%%}
{%%%%%%%%%%%%%%%%%%%%%%%%%%%%%%%%%%%%%%%%%%%%%%%%%%%%%%%%%%%%%%%%%%%%%%%%%%%%%%%}

{_____distance measure routines_____}

{%%%%%%%%%%%%%%%%%%%%%%%%%%%%%%%%%%%%%%%%%%%%%%%%%%%%%%%%%%%%%%%%%%%%%%%%%%%%%%%}
{%%%%%%%%%%%%%%%%%%%%%%%%%%%%%%%%%%%%%%%%%%%%%%%%%%%%%%%%%%%%%%%%%%%%%%%%%%%%%%%}

{       required inputs:    I0  -> start of DM vector
                            I4  -> start of PM vector}

{===================== full euclidean distance ===========================}

full_euclidean:
        CNTR = 12;
        MR  = 0, AX0 = DM(I0,M1), AY0 = PM(I4,M5);
        DO full_sumdiffsq UNTIL CE;
            AR  = AX0 - AY0, AX0 = DM(I0,M1);          { calculate difference }

            MY0 = AR;
```

(listing continues on next page)

6   Speech Recognition

```
full_sumdiffsq:
        MR   = MR + AR * MY0 (SS), AY0 = PM(I4,M5); {accumulat difsq}

        SR  = ASHIFT MR2 BY 5 (HI);      { leaves seven bits for warping}
        SR  = SR OR LSHIFT MR1 BY -11 (HI);
        SR  = SR OR LSHIFT MR0 BY -11 (LO);
        RTS;

{====================== cepstral projection distance ======================}

cepstral_projection:
        CNTR = 12;
        MR  = 0, MX0 = DM(I0,M1);
        MY0 = PM(I4,M5);
        DO dot_product UNTIL CE;            {calculates negative of dot product}

        MR  = MR - MX0 * MY0 (SS), MX0 = DM(I0,M1);
dot_product:
        MY0 = PM(I4,M5);

        SR  = ASHIFT MR2 BY 5 (HI);      { leaves seven bits for warping}
        SR  = SR OR LSHIFT MR1 BY -11 (HI);
        SR  = SR OR LSHIFT MR0 BY -11 (LO);
        RTS;

{%%%%%%%%%%%%%%%%%%%%%%%%%%%%%%%%%%%%%%%%%%%%%%%%%%%%%%%%%%%%%%%%%%%%%%%%%%%%%}

.ENDMOD;
```

Listing 6.14 Vector Distance Routine (VECTDIST.DSP)

Speech Recognition 6

```
{_____}
{ DISBOOT.DSP                                     3-10-90        }
{                                                                }
{_____}
.MODULE/RAM/BOOT=0      display;
.PORT                   led_and_bank;
.PORT                   display_base;
.INCLUDE                <vocab.h>;
.VAR/DM                 bank_select;           {stored PM EPROM bank, 1 of 4}
.VAR/DM                 hex_led;               {stores value of hex led}
.VAR/DM/STATIC          digit_count;           {passed to boot page 2}
.VAR/DM/STATIC          phone_number[16];      {passed to boot page 2}
.GLOBAL                 digit_count, phone_number;
.VAR/DM/STATIC          long_distance_flag;    {passed to boot page 2}
.GLOBAL                 long_distance_flag;
.VAR/DM                 timed_pntr;

.ENTRY     display_digit;
.ENTRY     add_a_digit;
.ENTRY     display_text;
.ENTRY     clear_display;
.ENTRY     display_number;
.ENTRY     set_local_call;
.ENTRY     set_long_distance;
.ENTRY     show_bank;
.ENTRY     inc_bank_select;
.ENTRY     set_bank_select;
.ENTRY     display_numpls;
.ENTRY     display_dial;
.ENTRY     reset_display;
.ENTRY     timed_display;
.ENTRY     reset_timed;
.ENTRY     blank_hex_led;
{_____}
{          Wait_Some                      @ 12.28 MHz           }
{ Count = (desired time in sec)/(5*cycletime).                  }
{ AY0 = lsw of count                                            }
{ AY1 = msw of count                                            }
{ alters: AX0,AY0,AX0,AR                                        }
```

(listing continues on next page)

6 Speech Recognition

```
wait_four:      CALL wait_two;              {wait four seconds}
wait_two:       CALL wait_one;              {wait two seconds}
wait_one:       CALL wait_half;             {wait one second}
wait_half:      CALL wait_quarter;          {half second}
wait_quarter:   CALL wait_eighth;           {quarter second}
wait_eighth:    CALL wait_sixteenth;        {eighth second}
wait_sixteenth: CALL wait_thirtysec;        {sixteenth second}
wait_thirtysec: AY0=0X2c00;                 {lsw of count for 1/32 sec}
                AY1=0X0001;                 {msw of count for 1/32 sec}
wait_some:      AX0=0;                      {for borrow}
time_loop:          AR=AY0-1;
                    AY0=AR, AR=AY1-AX0+C-1;
                    AY1=AR;
                    AR=AR OR AY0;
                    IF NE JUMP time_loop;
            RTS;
{_____}
{                       Display_Text                                 }
{ I4 = ^ascii text buffer in PM                                      }
{ Format of text buffer:<# characters, ascii data;>                  }
{ alters: I4,L4,I2,L2,AR,AY0,AY1                                     }

display_text:
            CALL clear_display;
            L4=0;
            AY0=PM(I4,M5);                  {get # characters}
            CALL display_spaces;           {display leading spaces}
            CNTR=AY0;                       {#characters to display}
char_loop:      AR=PM(I4,M5);              {get character}
                CALL disp_char;            {display one character}
                IF NOT CE JUMP char_loop;
            RTS;

{_____}
{                   Display One Character                            }
{ AR = ascii character                                               }
{ I2 = display pointer, decremented by one                           }
{ alters: AY0,AR,I2                                                  }

disp_char:  AY1=0x0080;
            AR=AR OR AY1;                              {WR high}
            DM(I2,M0)=AR, AR=AR XOR AY1;
            DM(I2,M0)=AR;                              {WR low}
            AR=AR OR AY1;
            NOP;
            DM(I2,M2)=AR;                              {WR high}
            RTS;
```

426

Speech Recognition   6

```
{_____}
{           Clear the ASCII display with N spaces               }
{ CNTR = number of spaces                                       }
{ I2 = returned with the current characters location            }
{ alters: I2,L2,AR,AY0                                          }

clear_display:    AY0 = -16;                        {Entry to clear entire display}

display_spaces:   I2=^display_base + 15;            {Entry to clear leading spaces}
                  L2=0;
                  AR=16;
                  AR=AR-AY0;                         {center the word}
                  IF LE JUMP spaces_done;
                  SR=LSHIFT AR BY -1 (LO);           {SR0=(16-#characters)/2}
                  AR  = PASS SR0;
                  IF LE JUMP spaces_done;
                  CNTR=SR0;
                  AR=0x0020;                         {space}
clear_loop:           CALL disp_char;
                      IF NOT CE JUMP clear_loop;
spaces_done:      RTS;

{_____}
{                    Display Number                             }
{ Displays digit_count characters from digit buffer in DM.      }
{ Format of text buffer:<# characters, ascii data;>            }
{ alters: I4,L4,I2,L2,AR,AY0,AY1                                }
{ Modified to inset dash after 3 digits.                        }

display_number:   CALL clear_display;
                  I4=^phone_number;
                  L4=0;
                  AY0=DM(digit_count);               {get # digits}
                  CALL display_spaces;               {display leading spaces}
                  CNTR=AY0;                          {#characters to display}
                  AF=PASS 0;                         {counts digits}
                  AY0=0x30;                          {ascii 0 offset}
digd_loop:            AX0=3;
                      AR=AX0-AF;
                      IF EQ CALL display_dash;
                      AF=AF+1;
                      AR=DM(I4,M5);                  {get digit}
                      AR=AR+AY0;                     {offset for ascii}
                      CALL disp_char;                {display one character}
                      IF NOT CE JUMP digd_loop;
              RTS;

display_dash:
              AR=0x2d;                               {ascii dash}
              CALL disp_char;
              RTS;
```

(listing continues on next page)

6   Speech Recognition

```
{_____}
{                          Display a Digit                            }
{ AY1 = index of digit to display                                     }

display_digit:
          AR=^word_catalog;
          AR=AR+AY1;
          I4=AR;
          AR  = PM(I4,M4);
          I4  = AR;
          CALL display_text;
          CALL wait_quarter;
          CALL wait_sixteenth;
          RTS;

{_____}
{                  Add a Digit to the Phone Number                    }
{ Adds a digit to phone_number and increments digit_count             }
{ AY1 = digit to add                                                  }
{ alters: AY0,AR,I4                                                   }

add_a_digit:
          AY0=DM(digit_count);
          AR=AY0+1;
          AY0=12;
          AF=AR-AY0;
          IF GE RTS;
          DM(digit_count)=AR;
          AY0=^phone_number;
          AR=AR+AY0;
          I4=AR;
          MODIFY(I4,M6);              {^ + # digits - 1 to get address}
          AR  = 10;
          AR  = AR - AY1;
          IF NE AR = PASS AY1;    { (oh) is tenth in the list }
          DM(I4,M4) = AR;
          RTS;

{_____}
{                          Set Local Call                             }

set_local_call:
          AR=0;
          DM(long_distance_flag)=AR;
          RTS;

{_____}
{                      Set Long Distance Call                         }

set_long_distance:
          AR=1;
          DM(long_distance_flag)=AR;
          RTS;
```

428

Speech Recognition 6

```
{_____}
{                      Blank the HEX LED                      }
{ alters: SR0                                                 }

blank_hex_led:
          SR0=0x0010;
          CALL set_hex_led;
          RTS;

{_____}
{                    Display to HEX LED                       }
{ SR0 = Hex value to display                                  }
{ alters: AR, SR                                              }

show_bank:    SR0=DM(bank_select);          {entry to display bank}
set_hex_led:dm(hex_led)=SR0;                {normal entry}
          AR=DM(bank_select);
          SR=SR OR LSHIFT AR BY 8 (LO);
          DM(led_and_bank)=SR0;
          RTS;

{_____}
{               Set the PM EPROM Bank Select                  }
{ AR = bank value                                             }
{ alters: SR,AY0                                              }

inc_bank_select:
          AY0=DM(bank_select);              {entry to inc bank}
          AR=AY0+1;
          AY0 = 3;
          AR  = AR AND AY0;

set_bank_select:
          DM(bank_select)=AR;               {normal entry}
          SR0=DM(hex_led);
          SR=SR OR LSHIFT AR BY 8 (LO);
          DM(led_and_bank)=SR0;
          RTS;

{_____}
{               display particular text                       }

display_numpls:
          I4   = ^num_;
          JUMP display_text;

display_dial:
          I4   = ^dial;
          JUMP display_text;
```

(listing continues on next page)

6   Speech Recognition

```
{_____}
{                   initialize display variables                 }
reset_display:
        AR  = 0;
        DM(bank_select) = AR;
        CALL show_bank;
reset_timed:
        CALL clear_display;
        AR  = H#FF;
        DM(0X3FFB) = AR;                        { TSCALE }
        I4  = ^timed_catalog;
        AR  = PM(I4,M5);
        DM(0X3FFC) = AR;                        { TCOUNT }
        AX0 = PM(I4,M5);
        AR  = PM(I4,M5);
        DM(0X3FFD) = AR;                        { TPERIOD }
        DM(timed_pntr) = I4;
        I4  = AX0;
        CALL display_text;
        RTS;

{_____}
{             display opening on timer interrupts                }
timed_display:
        ENA SEC_REG;
        MR1 = I4;                               { save state }
        MR0 = L4;                               { save state }

        I4  = DM(timed_pntr);
        L4  = 48;
        MY0 = PM(I4,M5);
        MY1 = PM(I4,M5);
        DM(timed_pntr) = I4;
        L4  = 0;

        DM(0X3FFD) = MY1;                       { TPERIOD }
        I4  = MY0;
        MY0 = I2;                               { save state }
        MY1 = L2;                               { save state }
        CALL display_text;

        I4  = MR1;                              { restore state }
        L4  = MR0;                              { restore state }
        I2  = MY0;                              { restore state }
        L2  = MY1;                              { restore state }
        RTI;

{_____}

.ENDMOD;
```

Listing 6.15 Display Driver Routine (DEMOBOX.DSP)

Speech Recognition   6

```
{   DTMF Signal Generator

    ADSP-2101 EZ-LAB demonstration

    Analog Devices, Inc.
    DSP Division
    P.O.Box 9106
    Norwood, MA 02062
}

.module/boot=1     DTMF_Dialer;
.ENTRY     eight_khz;
.ENTRY     new_digit;
.ENTRY     dm_inits;
.ENTRY     make_tones;
.ENTRY     make_silence;

{ sine routine variables}
.VAR/PM   sin_coeff[5];
.INIT     sin_coeff: H#324000, H#005300, H#AACC00, H#08B700, H#1CCE00;

{ dynamic scratchpad variables }
.var      hertz1, hertz2,       { row and col frequency in Hertz }
          sum1, sum2,           { row and col phase accumulators }
          sin1, sin2;           { returned values from calling sin }

.VAR/DM   maketones_or_silence;

{ fixed variables to be loaded from booted PM }
.var      scale,                { attenuation of each sine before summing }
          hz_list[32];          { lookup table for digit row,col freqs }

{ NOTE *** put all fixed DM inits into PM and copy over into DM !! *** NOTE }

.const    PM_copy_length=33;

.var/pm   PM_scale, PM_hz_list[32];

{altered so that A == dial tone}
.init     PM_scale:        h#FFFF;
.init     PM_hz_list[00]:h#03AD,h#0538,h#02B9,h#04B9,h#02B9,h#0538,h#02B9,h#05C5;
.init     PM_hz_list[08]:h#0302,h#04B9,h#0302,h#0538,h#0302,h#05C5,h#0354,h#04B9;
.init     PM_hz_list[16]:h#0354,h#0538,h#0354,h#05C5,h#01B8,h#015E,h#0302,h#0661;
.init     PM_hz_list[24]:h#0354,h#0661,h#03AD,h#0661,h#03AD,h#04B9,h#03AD,h#05C5;
```

(listing continues on next page)

6 Speech Recognition

```
{──────────────────────────────────────────────────────────}
{                    Eight KHz Interrupt Routine            }
eight_khz:
          ENA SEC_REG;
          AR=DM(maketones_or_silence);     {0 = quite, 1 = maketones}
          AR=PASS AR;
          IF EQ JUMP quiet;

          se=dm(scale);
tone1:    ay0=dm(sum1);
          si=dm(hertz1);                    { freq stored as Hz in DM }
          sr=ashift si by 3 (hi);
          my0=h#4189;                       { mult Hz by .512 * 2 }
          mr=sr1*my0(rnd);                  { i.e. mult by 1.024 }
          sr=ashift mr1 by 1 (hi);
          ar=sr1+ay0;
          dm(sum1)=ar;
          ax0=ar;
          call boot_sin;
          sr=ashift ar (hi);                { scale value in SE }
          dm(sin1)=sr1;

tone2:    ay0=dm(sum2);
          si=dm(hertz2);                    { freq stored as Hz in DM }
          sr=ashift si by 3 (hi);
          my0=h#4189;                       { mult Hz by .512 * 2 }
          mr=sr1*my0(rnd);                  { i.e. mult by 1.024 }
          sr=ashift mr1 by 1 (hi);
          ar=sr1+ay0;
          dm(sum2)=ar;
          ax0=ar;
          call boot_sin;
          sr=ashift ar (hi);                { scale value in SE }
          dm(sin2)=sr1;

add_em:   ax0=dm(sin1);
          ay0=dm(sin2);
          ar=ax0+ay0;

sound:    sr=ashift ar by -2 (hi);         { compand 14 LSBs only! }
          tx0=sr1;                          { send "signal" sample to SPORT }
             DIS SEC_REG;
          rti;

quiet:    ar=0;
          tx0=ar;                           { send "silence" sample to SPORT }
             DIS SEC_REG;
          rti;
```

Speech Recognition   6

```
{————————————————————————————————————————————————————————}
{————— S U B R O U T I N E S —————————————————————————————}
{————————————————————————————————————————————————————————}

{————————————————————————————————————————————————————————}
{                    Change the Digit                     }
{ AX0 = digit                                             }
new_digit:
        ay0=h#000F;
        ar=ax0 and ay0;
        sr=lshift ar by 1 (hi);
        ay0=^hz_list;
        ar=sr1+ay0;
        i1=ar;
        ax0=dm(i1,m1);              { look up row freq for digit }
        dm(hertz1)=ax0;
        ax0=dm(i1,m1);              { look up col freq for digit }
        dm(hertz2)=ax0;
            SI=0;
            DM(sum1)=SI;
            DM(sum2)=SI;
        rts;

{————————————————————————————————————————————————————————}
{                 Maketones or Makesilence                }
make_tones:
        AR=1;
        DM(maketones_or_silence)=AR;
        RTS;

make_silence:
        AR=0;
        DM(maketones_or_silence)=AR;
        RTS;

{————————————————————————————————————————————————————————}
{                     Initialize PM                       }
dm_inits:
        i0=^scale;
        m1=1;
        l0=0;
        i4=^PM_scale;
        m5=1;
        l4=0;
        cntr=PM_copy_length;
        do boot_copy until ce;
            si=pm(i4,m5);
            sr=lshift si by 8 (hi);
            si=px;
            sr=sr or lshift si by 0 (hi);
boot_copy:  dm(i0,m1)=sr1;
        rts;
```

(listing continues on next page)

6 Speech Recognition

```
{_____}
{
    Sine Approximation
         Y = boot_sin(x)

    Calling Parameters
         AX0 = x in scaled 1.15 format
         M7 = 1
         L7 = 0

    Return Values
         AR = y in 1.15 format

    Altered Registers
         AY0,AF,AR,MY1,MX1,MF,MR,SR,I7

    Computation Time
         25 cycles
}

boot_sin:
         M5=1;
         L7=0;
         I7=^sin_coeff;                      {Pointer to coeff. buffer}
         AY0=H#4000;
         AR=AX0, AF=AX0 AND AY0;             {Check 2nd or 4th quad.}
         IF NE AR=-AX0;                      {If yes, negate input}
         AY0=H#7FFF;
         AR=AR AND AY0;                      {Remove sign bit}
         MY1=AR;
         MF=AR*MY1 (RND), MX1=PM(I7,M5);     {MF = x2}
         MR=MX1*MY1 (SS), MX1=PM(I7,M5);     {MR = C1x}
         CNTR=3;
         DO approx UNTIL CE;
             MR=MR+MX1*MF (SS);
approx:      MF=AR*MF (RND), MX1=PM(I7,M5);
         MR=MR+MX1*MF (SS);
         SR=ASHIFT MR1 BY 3 (HI);
         SR=SR OR LSHIFT MR0 BY 3 (LO);      {Convert to 1.15 format}
         AR=PASS SR1;
         IF LT AR=PASS AY0;                  {Saturate if needed}
         AF=PASS AX0;
         IF LT AR=-AR;                       {Negate output if needed}
         RTS;
.ENDMOD;
```

Listing 6.16 DTMF Signal Generator Routine (DTMF.DSP)

Speech Recognition 6

```
{_____}
{               Dial Phone Number and Display                }
{_____}

.MODULE/RAM/BOOT=1/ABS=0        dial_n_display;

.PORT          led_and_bank;
.PORT          display_base;
.EXTERNAL      eight_khz, new_digit, dm_inits, make_tones, make_silence;
.EXTERNAL      digit_count, phone_number, long_distance_flag;
.INCLUDE       <vocab.h>;

reset_vec:     CALL dm_inits;
               CALL init_control_regs;
               JUMP start;
               NOP;
irq2:          RTI;NOP;NOP;NOP;
s0_tx:         RTI;NOP;NOP;NOP;
s0_rx:         JUMP eight_khz;NOP;NOP;NOP;
s1_tx_irq1:    RTI;NOP;NOP;NOP;
s1_rx_irq0:    RTI;NOP;NOP;NOP;
timer_exp:     JUMP timeout;NOP;NOP;NOP;

start:         L0=0;L1=0;L2=0;L3=0;
               L4=0;L5=0;L6=0;L7=0;
               M0=0;M1=1;M2=-1;M3=2;            {standard setup}
               M4=0;M5=1;M6=-1;M7=0;

               CALL make_silence;
               ICNTL=0x00111;
               IMASK=b#001001;                  {enable timer & rx0 interrupt}

               AX0=0xA;                          {dial tone}
               CALL new_digit;
               CALL make_tones;
               CALL wait_one;
               CALL wait_half;
               CALL wait_quarter;
               CNTR=DM(digit_count);
               AR=0;
               DM(digit_count)=AR;
               I4=^phone_number;
```

(listing continues on next page)

435

6   Speech Recognition

```
each_tone:          AX0=DM(I4,M5);
                    CALL new_digit;
                    CALL make_tones;

                    AY0=DM(digit_count);      {display sucessive digits}
                    AR=AY0+1;
                    DM(digit_count)=AR;
                    CALL display_number;

                    CALL wait_sixteenth;
                    CALL make_silence;
                    CALL wait_eighth;
                    IF NOT CE JUMP each_tone;

            CALL wait_two;
            I4=^gsm_;
            CALL display_text;
            AR=0x029b;                        {boot page 2}
            DM(0x3FFF)=AR;

{_____Now Go To Boot Page One_____}

{_____}
{                             Subroutines                                     }
{_____}

{_____}
{                    Wait using timer interrupt                               }
wait_four:          CALL wait_two;
wait_two:           CALL wait_one;
wait_one:           CALL wait_half;
wait_half:          CALL wait_quarter;
wait_quarter:       CALL wait_eighth;
wait_eighth:        CALL wait_sixteenth;
wait_sixteenth:     AY0=0xFF;
                    AY1=0x0BC4;
wait_timer:         DM(0x3FFB)=AY0;           {TSCALE}
                    DM(0x3FFC)=AY1;           {TCOUNT}
                    DM(0x3FFD)=AY1;           {TPERIOD}
                    AY0=0;
                    ENA TIMER;
wait_here:          AR=PASS AY0;
                    IF EQ JUMP wait_here;
                    DIS TIMER;
                    RTS;
```

Speech Recognition   6

```
{_____}
{                    Timer Interrupt Handler                 }

timeout: AY0=0xFFFF;                    {set the timer expired flag}
        RTI;

{_____}
{                      Display_Text                          }
{ I4 = ^ascii text buffer in PM                              }
{ Format of text buffer:<# characters, ascii data;>          }
{ alters: I4,L4,I2,L2,AR,AY0,AY1                             }

display_text:
        CALL clear_display;
        L4=0;
        AY1=PM(I4,M5);                  {get # characters}
        AR=16;
        AR=AR-AY1;                      {center the word}
        SR=LSHIFT AR BY -1 (LO);        {SR0=(16-#characters)/2}
        CNTR=SR0;
        CALL display_spaces;            {display leading spaces}
        CNTR=AY1;                       {#characters to display}
char_loop:      AR=PM(I4,M5);           {get character}
                CALL disp_char;         {display one character}
                IF NOT CE JUMP char_loop;
        RTS;

{_____}
{                   Display One Character                    }
{ AR = ascii character                                       }
{ I2 = display pointer, decremented by one                   }
{ alters: AY0,AR,I2                                          }

disp_char:
        AY0=0x0080;
        AR=AR OR AY0;                   {WR high}
        DM(I2,M0)=AR, AR=AR XOR AY0;
        DM(I2,M0)=AR;                   {WR low}
        AR=AR OR AY0;
        NOP;
        DM(I2,M2)=AR;                   {WR high}
        RTS;
```

(listing continues on next page)

437

6 Speech Recognition

```
{_____}
{            Clear the ASCII display with N spaces                }
{ CNTR = number of spaces                                         }
{ I2 = returned with the current characters location              }
{ alters: I2,L2,AR,AY0                                            }

clear_display:
            CNTR=16;                    {Entry to clear entire display}

display_spaces:
            I2=^display_base + 15;   {Entry to clear leading spaces}
            AR=CNTR;
            AR=PASS AR;
            IF EQ JUMP exit_clear;   {Return if no leading zeros}
            L2=0;
            AR=0x0020;                  {space}
clear_loop:     CALL disp_char;
                IF NOT CE JUMP clear_loop;
            RTS;
exit_clear: POP CNTR;
            RTS;

{_____}
{                      Display Number                             }
{ Displays digit_count characters from digit buffer in DM.        }
{ Format of text buffer:<# characters, ascii data;>              }
{ alters: I4,L4,I2,L2,AR,AY0,AY1                                 }

display_number:
            CALL clear_display;
            I4=^phone_number;
            L4=0;
            AY1=DM(digit_count);     {get # digits}
            CNTR=3;
            CALL display_spaces;     {display leading spaces}
            CNTR=AY1;                {#characters to display}
            AF=PASS 0;               {counts digits}
            AY1=0x30;                {ascii 0 offset}
digd_loop:      AX0=3;
                AR=AX0-AF;
                IF EQ CALL display_dash;
                AF=AF+1;
                AR=DM(I4,M5);     {get digit}
                AR=AR+AY1;        {offset for ascii}
                CALL disp_char;   {display one character}
                IF NOT CE JUMP digd_loop;
            RTS;

display_dash:
            AR=0x2d;                    {ascii dash}
            CALL disp_char;
            RTS;
```

```
{_____}
{                       Init Control Registers                }
{ Set Up SPORTS and TIMER on EZ-LAB board after RESET         }
{ used for ADSP-2101 EZ-LAB demonstrations                    }
{ Altered Registers: I0,M1,L0                                 }

init_control_regs:
            L0=0;
            M1=1;
            I0=h#3FEF;          {point to last DM-mapped control registers }

{ h#3FEF }  DM(I0,M1)=H#0000;          {SPORT1 AUTOBUFF DISABLED}
{ h#3FF0 }  DM(I0,M1)=H#0000;          {SPORT1 RFSDIV NOT USED}
{ h#3FF1 }  DM(I0,M1)=H#0000;          {SPORT1 SCLKDIV NOT USED}
{ h#3FF2 }  DM(I0,M1)=H#0000;          {SPORT1 CNTL DISABLED}
{ h#3FF3 }  DM(I0,M1)=H#0000;          {SPORT0 AUTOBUFF DISABLED}
{ h#3FF4 }  DM(I0,M1)=   255;          {RFSDIV for 8 kHz interrupt rate}
{ h#3FF5 }  DM(I0,M1)=     2;          {SCLKDIV=2 makes 2.048 MHz
                                        with 12.288 MHz xtal}

 { h#3FF6 } DM(I0,M1)=H#6927;          {Int SCLK,
                                        RFS req, TFS req,
                                        Int RFS, Int TFS,
                                        u_law, 8-bit PCM }
{ h#3FF7 }  DM(I0,M1)=H#0000;          {TRANSMIT MULTICHANNELS}
{ h#3FF8 }  DM(I0,M1)=H#0000;
{ h#3FF9 }  DM(I0,M1)=H#0000;          {RECEIVE MULTICHANNELS}
{ h#3FFA }  DM(I0,M1)=H#0000;
{ h#3FFB }  DM(I0,M1)=H#0000;          {TIMER NOT USED, CLEARED}
{ h#3FFC }  DM(I0,M1)=H#0000;
{ h#3FFD }  DM(I0,M1)=H#0000;
{ h#3FFE }  DM(I0,M1)=H#0000;          {DM WAIT STATES}
{ h#3FFF }  DM(I0,M1)=H#101B;          {SPORT0 ENABLED}
                                       {BOOT PAGE 0, 3 PM WAITS}
                                       {3 BOOT WAITS}

            rts;

{_____}

.ENDMOD;
```

Listing 6.17 Automatic Dialing Routine (DTMFMAIN.DSP)

6   Speech Recognition

6.6   REFERENCES

Atal, B.S. June 1974. "Effectiveness of Linear Prediction Characteristics of the Speech Wave for Automatic Speaker Identification and Verification," *Journal of the Acoustical Society of America*, vol. 55, No. 6, pp. 1304-1312.

Gray, A.H. and J. D. Markel. October 1976. "Distance Measures for Speech Processing," *IEEE Transactions on Acoustics, Speech, and Signal Processing*, vol. ASSP-24, No. 5, pp. 380-391.

Gray, R.M., A. Buzo, A. H. Gray, and Y. Matsuyama. August 1980. "Distortion Measures for Speech Processing," *IEEE Transactions on Acoustics, Speech, and Signal Processing*, vol. ASSP-28, No. 4, pp. 376-376.

Itakura, F. February 1975. "Minimum Prediction Residual Principle Applied to Speech Recognition," *IEEE Transactions on Acoustics, Speech, and Signal Processing*, vol. ASSP-23, No. 1, pp. 67-72.

Juang, B.H., L. R. Rabiner, and J. G. Wilpon. July 1987. "On the Use of Bandpass Liftering in Speech Recognition," *IEEE Transactions on Acoustics, Speech, and Signal Processing*, vol. ASSP-35, No. 7, pp. 947-954.

Makhoul, J. April 1975. "Linear Prediction: A Tutorial Review," *Proceedings of the IEEE*, vol. 63, No. 4, pp. 561-580.

Mansour, D. and B. H. Juang. November 1989. "A Family of Distortion Measures Based Upon Projection Operation for Robust Speech Recognition," *IEEE Transactions on Acoustics, Speech, and Signal Processing*, vol. 37, No. 11, pp. 1659-1671.

Nocerino, N., F. K. Soong, L. R. Rabiner, and D. H. Klatt. December 1985. "Comparative Study of Several Distortion Measures for Speech Recognition," *Speech Communication*, vol. 4, pp. 317-331.

Paliwal, K.K. 1982. "On the Performance of the Quefrency-Weighted Cepstral Coefficients in Vowel Recognition," *Speech Communication*, vol. 1, pp. 151-154.

Rabiner, L.R., S. E. Levinson, A. E. Rosenberg, and J. G. Wilpon. August 1979. "Speaker Independent Recognition of Isolated Words Using Clustering Techniques," *IEEE Transactions on Acoustics, Speech, and Signal Processing*, vol. ASSP-27, No. 4, pp. 336-349.

Speech Recognition   6

Rabiner. L.R. and M. R. Sambur. February 1975. "An Algorithm for Determining the Endpoints of Isolated Utterances," *The Bell System Technical Journal*, vol. 54, No. 2, pp. 297-315.

Rabiner, L.R. and R. W. Schafer. 1978. *Digital Processing of Speech Signals*, Prentice-Hall, Inc., Englewood Cliffs, New Jersey.

Rabiner, L.R. and J. G. Wilpon. 1987. "Some Performance Benchmarks for Isolated Word Speech Recognition Systems," *Computer Speech and Language*, vol. 2, pp. 343-357.

Schroeder, M.R. April 1981. "Direct (Nonrecursive) Relations Between Cepstrum and Predictor Coefficients," *IEEE Transactions on Acoustics, Speech, and Signal Processing*, vol. ASSP-29, No. 2, pp. 297-301.

Tohkura, Y. April 1986. "A Weighted Cepstral Distance Measure for Speech Recognition," *Proceedings of ICASSP 1986*, pp. 761-764.

Discrete Cosine ■ 7
Transform

7.1   OVERVIEW

The Discrete Cosine Transform, or DCT, transforms data into a format that can be easily compressed. The characteristics of the DCT make it ideally suited for image compression algorithms. These algorithms let you minimize the amount of data needed to recreate a digitized image.

Reducing digitized images into the least amount of data possible has the following advantages:

- Less memory required to store images
- Channel bandwidth efficiency increased when you transmit images
- Less time may be needed to analyze images

Performing the DCT on a digitized image creates a data array that can be compressed by data compaction algorithms. Then, data can be stored or transmitted in its compacted form. The image quality depends on the amount of quantization used in the compaction algorithm. To reproduce the original image, the data is retrieved from memory, uncompacted, and an inverse DCT is performed.

Some of today's most popular image data compression applications include:

- Teleconferencing using motion-compensated video codecs
- ISDN multimedia communications including voice, video, text, and images
- Video channel transmission using commercial geosynchronous telecommunications satellites
- Digital facsimile transmission using dedicated equipment and personal computers

This chapter describes a basic implementation of the DCT.

7 Discrete Cosine Transform

7.2   BACKGROUND

Several image data compression algorithms use the DCT to remove spatial data redundancies in two-dimensional (2D) data. Images are subdivided into smaller, two-dimensional blocks. These blocks are then processed independently of the neighboring blocks.

Figure 7.1 illustrates how a two-dimensional discrete cosine transform is performed on a block of data. In general, the two dimensional, discrete cosine transform (2D DCT) transforms an (n x n) data array into an (n x n) result array. First the DCT transforms the columns, then it transforms the rows. The resulting data elements are called the *transform coefficients*, or *DCT coefficients*. For example, if you use 8 x 8 blocks of 8-bit input data, an 8-point DCT is performed on each row in the block. This creates a new 8 x 8 block of data. Next, an 8-point DCT is performed on each column of the new block. This generates an 8 x 8 block of 12-bit output values. These 64 output values are the DCT coefficients.

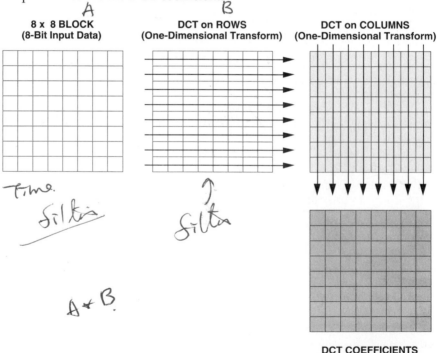

Figure 7.1 A Two-Dimensional Discrete Cosine Transform

444

Discrete Cosine Transform 7

After the DCT is calculated, the data can be reduced to concentrate the important information into a few of the coefficients, leaving the remaining coefficients equal to zero, or otherwise "insignificant." Typically, the (n x n) result array is quite sparse; this is the desired energy compaction effect.

When you transmit only the coefficients with large values, the total volume of data is reduced. You can use several methods to choose which coefficients to transmit. Once the coefficients are chosen and quantized, you can use additional algorithms, such as Huffman coding or run-length coding algorithms, to achieve a higher data compression ratio.

Either the DCT or the discrete Fourier transform (DFT) could be used in image compression algorithms, however, the characteristics of the DCT make it better suited for execution on ADSP-2100 family processors. Table 7.1 highlights the important differences between these two transforms.

Cosine Transform	Fourier Transform
Real arithmetic only, real data and coefficients	Requires complex arithmetic, complex data, complex coefficients
Excellent image energy compaction	Very good image energy compaction
Phase information not available	Phase, magnitude available
Blocking artifact not as apparent	Blocking artifact noticeable
Assumes data outside window is mirror-image of data inside window	Assumes data outside window is a duplicate of data window, shifted

Table 7.1 Cosine Transform vs. Fourier Transform Characteristics

Often, a receiver can reconstruct a complex image with relatively few retained transform coefficients. Depending on the number of Fourier coefficients retained in DFT compression, the reconstructed image may exhibit visible block boundaries because of the Gibbs phenomenon. This is called the blocking artifact. Figure 7.2 demonstrates how images reconstructed from DCT coefficients exhibit less blocking artifact than those reconstructed from Fourier coefficients.

7 Discrete Cosine Transform

Discrete Fourier Transform

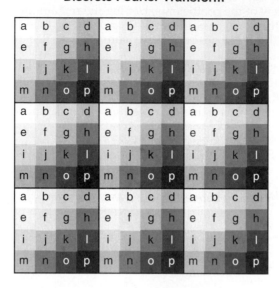

Discrete Cosine Transform

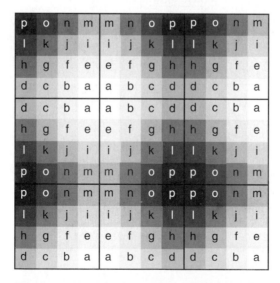

Figure 7.2 The DCT Reduces The Blocking Artifact

446

Discrete Cosine Transform   7

The letters in Figure 7.2 correspond to pixel intensity. The top half of Figure 7.2 shows four 4 x 4 blocks after the DFT is performed. Because of the way the DFT transforms a block, it expects the neighboring blocks to be exact copies. Notice that the differing intensity of adjacent pixels in neighboring blocks gives the appearance of borders between the blocks, thus increasing blocking artifacts.

The bottom half of Figure 7.2 shows four blocks after the DCT is performed. The DCT expects neighboring blocks to be mirror images. In this example, adjacent pixels across the borders of the 4 x 4 blocks appear to have the same intensity, thereby reducing blocking artifact.

Computationally, the DCT is more efficient than the DFT because it is a completely real transform and does not require complex variables or arithmetic. You can increase the computational speed of the DCT by using a fast cosine transform (FCT or FDCT) algorithm. This chapter describes implementation of the DCT using a fast algorithm first presented by H. S. Hou (see references).

Table 7.2 lists benchmark times for executing the DCT using ADSP-2100 family microcomputers.

DSP	Processor Speed	8x8	16x16
		2492 cycles	10046 cycles
ADSP-2101/2111	20 MHz	0.1246 ms	0.5023 ms
ADSP-2171	33 MHz	0.0755 ms	0.3044 ms

Table 7.2 Benchmark Times For Executing The DCT

This chapter includes programming examples for the one-dimensional DCT and the two-dimensional DCT. One-dimensional DCTs are used more often for speech compression applications, while two-dimensional transforms are commonly used for image data compression. The two-dimensional transform is implemented by performing one-dimensional transforms on each row of an image, then performing one-dimensional transforms on each column (refer to Figure 7.1). The routines printed in this chapter are in-place routines. In other words, the results of the DCT computation are written over the input data buffer values.

7   Discrete Cosine Transform

The formal, mathematical definition for the one-dimensional cosine transform is:

$$F(u) = \frac{2c(u)}{N} \sum_{m=0}^{N-1} f(m) \cos\left(\frac{(2m+1)u\pi}{2N}\right), \text{ where } u = 0, 1, 2,..., N\text{-}1$$

where

$$c(u) = \frac{\sqrt{2}}{2}, \text{ for } u = 0$$

$$c(u) = 1, \text{ for } u = 1, 2, 3, ..., N\text{-}1$$

Often, textbooks define c(u) at u=0 as one divided by the square-root of two. The square-root of two divided by two yields the same result and is used to simplify calculations. The output coefficient F(0) is often referred to as the DC component, or the DC term.

The two-dimensional DCT is mathematically defined as:

$$F(u, v) = \frac{4c(u, v)}{N^2} \sum_{m=0}^{N-1} \sum_{n=0}^{N-1} f(m, n) \cos\left(\frac{(2m+1)u\pi}{2N}\right) \cos\left(\frac{(2n+1)v\pi}{2N}\right)$$

for u,v = 0,1,2,...,N-1

where

$$c(u, v) = \frac{1}{2}, \text{ for } u = v = 0$$

$$c(u, v) = 1, \text{ for } u, v = 1, 2, ..., N\text{-}1$$

Two-dimensional transforms are equivalent to 2N-pt one-dimensional transforms.

Discrete Cosine Transform   7

7.3   COMPUTATIONAL METHODS

There are many fast algorithms for accelerating the computation of the DCT. Most algorithms can be classified into one of three categories:

- indirect computation
- direct matrix factorization
- recursive computation

Indirect computation involves doubling the length of an N-pt sequence to a 2N-pt sequence with its mirror image. The result is geometrically even waveform about its centerpoint. A 2N-pt FFT is then performed on that sequence and its results are multiplied by a complex exponential phase-shift vector. When correctly executed, the result is the DCT. However, this process requires lengthy FFT calculations that involve complex numbers.

Implementing the one-dimensional DCT described in the above equation, reordering the input and output sequences, and ignoring the DC scale coefficient c(u) yields a simple matrix-multiply operation. Using this as a starting point, the *matrix factorization* techniques yielded several fast algorithms that required significantly less arithmetic operations than the full-matrix multiply. Most importantly, these techniques do not require complex operations or many data moves, unlike the FFT method.

The drawback to matrix factorization methods is that a new value of N requires factoring the matrix again. This yields a different solution and compromises flexibility.

Probably the best (numerically and arithmetically efficient) method is a *recursive method* proposed by H. S. Hou in 1987.

7.4   HOU'S FAST DISCRETE COSINE ALGORITHM

Hou's Fast Discrete Cosine Transform Algorithm (FDCT) is numerically stable, fast and recursive. Similar to the Cooley-Tukey FFT algorithm, It generates the next larger DCT from two identical, smaller DCTs. This deviates from direct factorization algorithms that factor the desired N-pt DCT matrix. In that, the higher order matrices are generated from lower order DCT matrices instead. Refer to Hou's paper (see the references) for a tutorial on the DCT in general and his algorithmic implementation.

Hou's algorithm can be efficiently implemented on ADSP-2100 family processors because of the DSP's internal architecture. This architecture provides an ALU, MAC, and barrel shifter connected in parallel through

7   Discrete Cosine Transform

the internal result bus. The results of any ALU, MAC or shifter operation is available to any of these computational units on the next processor cycle. Also, during any ALU, MAC, or shifter operation, two new operands (one from data memory, one from program memory) can be fetched. This means that you can perform a multiply accumulate instruction and fetch a new cosine value (from PM) and new data value (from DM) during the same instruction cycle.

Zero-overhead looping and the recursive nature of Hou's algorithm, gives two-dimensional DCTs an extraordinarily fast execution time. Because you can mix index and modify registers in both data address generators, you do not need to sequence data in memory. You choose the correct modify register after a data access so the index register points to the next desired data value. The resulting assembly code is easy to read, self-documenting, and can be separated into small, manageable modules. (For more information, see Section 7.5, "Zig-Zag Scanning of DCT Coefficients", and Section 7.6, "Zig-Zag Scanning and ADSP-21xx Processors")

The programming examples presented in this chapter perform the cosine transform in-place. This means the input data values are read from a buffer (arranged in normal, sequential order) and the results are written back to the same buffer in the same order. Hou's method was chosen because it simplifies two-dimensional transforms and it uses relatively little data memory space. If you do not need an in-place DCT, change the pointer to point to a different output buffer before the bit-reversing routine is called.

Figure 7.3 illustrates that an N=16 DCT can be recursively separated into smaller DCTs (N=8, N=4, and N=2.) This diagram was drawn with the same notation that Hou used to demonstrate the recursive nature of this method, and to extend the diagram to N=16.

Discrete Cosine Transform 7

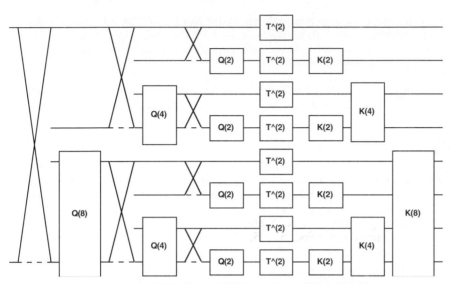

Figure 7.3. Implementation Of An N=16 DCT Using Hou's Matrix Notation

Figure 7.4 shows the equivalent signal flow graph of a fast 16-pt DCT. The code in Listings 7.1–7.10 implements this structure. The first routine, called DIF16, reads the input data and performs eight, decimation in frequency, radix-2, real butterflies with real cosine coefficients. The results are stored in a 16-word buffer called TMP. Next, the DIF8 routine computes two sets of four DIF real butterflies on the TMP buffer, in-place. Then DIF4 computes four sets of two DIF butterflies on TMP, in place. This is followed by DIF2, which performs eight sets of a single DIF butterfly on TMP, in-place. Referring to the diagram shown in Figure 7.3, the section just described implements the $Q(8)$, $Q(4)$, $Q(2)$, and $T^{\wedge}(2)$ blocks. The next subroutines that are called in Listing 7.1, implement the $K(2)$, $K(4)$, $K(8)$ blocks, and the last subroutine, called DCBREV (in Listings 7.1 and 7.10), scales the DC term by the reciprocal of the square root of two and performs the bit-reversing required to write the final TMP buffer values to the original input buffer locations in normal order. The TMP buffer is used to hold the in-place results of each subroutine between successive subroutine calls.

Very little memory is required to compute the DCT. In addition to the input data buffer, only 16 data memory locations are used for a 16-pt DCT (8 locations for an 8-pt DCT). These locations comprise the TMP buffer. Listings 7.10 and 7.18 use two additional data memory locations for the

7   Discrete Cosine Transform

two-dimensional implementations of the 16-pt and 8-pt transforms, respectively. These two locations, called XADR and XADR2, store pointers to the correct row and columns of a two-dimensional array during the two-dimensional transform.

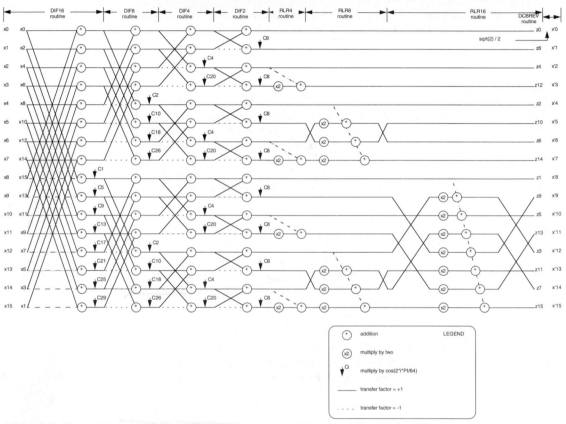

Figure 7.4. Signal Flow Graph For A Fast DCT

Discrete Cosine Transform   7

These listings also use program memory efficiently. Only 15 real cosine coefficients are stored in program memory for the 16-pt transforms, and seven real cosine coefficients for the 8-pt transforms. The instructions require few program locations because they are stored in short, recursively called subroutines. Notice that the same subroutines are used when the two-dimensional transforms are performed on a row or a column, even though the data elements are spaced differently in data memory during a row DCT and a column DCT.

Listings 7.1–7.10 use the 16-pt transforms and Listings 7.11–7.18 use the 8-pt transforms. Most subroutines are modified versions of the 16-pt subroutines. For example, the 8-pt DCT does not use an equivalent of DIF16 (Listing 7.2), so input data must be read from memory in the DIF8_8 routine (Listing 7.12). Bit-reversing the TMP buffer is performed over an 8-location buffer in DCBREV_8 (Listing 7.17) instead of the 16-location buffer used in DCBREV (Listing 7.9). Also, the routines that perform in-place operations on TMP (Listing 7.12–7.17) are shortened since the TMP buffer is only eight locations long in the 8-pt case.

7.5   ZIG-ZAG SCANNING OF DCT COEFFICIENTS

Processing 8 x 8 data blocks and getting 8 x 8 result blocks does not reduce data. The reduction occurs after the DCT is computed. All 64 DCT coefficients are passed through a quantization stage. The compression ratio dictates the amount of quantization. After quantization, you still have 64 DCT coefficients, but more values are equal to zero or to the same value as "neighboring" coefficients. The next step is to perform run-length coding on the quantized DCT coefficients. This process reduces the data. To efficiently run-length code the quantized DCT coefficients, you must transform the 8 x 8 block with a two-dimensional transform so similar values appear together frequently. Transforming the 8 x 8 block by rows and columns, example, is not spatially efficient for compacted information. Various scanning standards specify and use two-dimensional Zig-Zag scanning.

7 Discrete Cosine Transform

Figure 7.5 is an example of zig-zag scanning of quantized coefficient addresses.

00	01	02	03	04	05	06	07
08	09	10	11	12	13	14	15
16	17	18	19	20	21	22	23
24	25	26	27	28	29	30	31
32	33	34	35	36	37	38	39
40	41	42	43	44	45	46	47
48	49	50	51	52	53	54	55
56	57	58	59	60	61	62	63

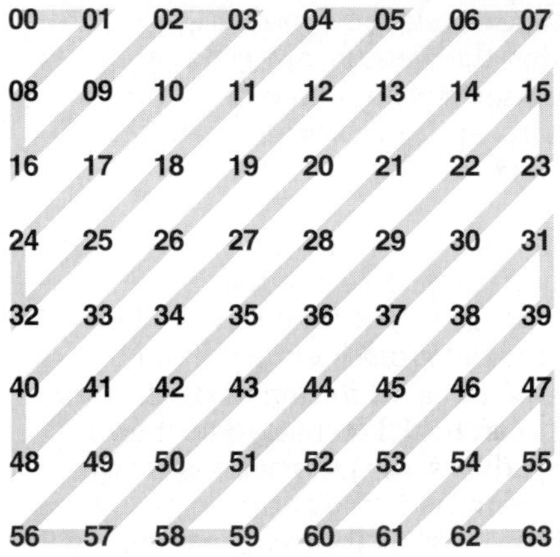

Figure 7.5. Zig-Zag Scanning of Quantized Coefficient Addresses

In this example, scanning starts at address 00 and continues through address 63.

7.6   ZIG-ZAG SCANNING & ADSP-21XX PROCESSORS

The data address generators (DAGs) in the ADSP-21xx Family core architecture are used for indirect data addressing. Each DAG uses four sets of registers consisting of three types of dedicated address registers for address computation. The three types of address registers are *index* (I) registers, *modify* (M) registers, and *length* (L) registers. Address computations are independent of arithmetic computations by the processor's ALU, MAC, and SHIFTER. During the same instruction cycles, the processor can perform arithmetic computations and data addressing in parallel.

Index registers hold absolute addresses that point to memory locations. Modify register contents are automatically added to index register contents during an indirect address operation. As a result, the index register points to the next desired data element for the next instruction cycle. Length registers are used with dedicated modulus arithmetic to make sure the index pointers stay in the circular buffers when circular addressing is used. Each of the four index registers can hold the complete,

Discrete Cosine Transform   7

absolute address for any addressable location in memory. When an instruction specifies an index register, it also specifies which modify register to use. The length register is not specified by the instruction since length and index registers are associated.

In an instruction, you can specify one of four possible modify registers with an index register, and you can change the address modify value without wasting cycles to change the modify register contents. You can do this by selecting a different M register to be used for a given I register. For additional information about using DAG resisters, refer to the *ADSP-2100 Family User's Manual*.

While following the zig-zag scanning trace in Figure 7.5, you may have noticed that between data elements there are four different address offset amounts. You can effectively zig-zag scan an 8 x 8 block of data stored in two dimensional format without adding addressing overhead and you eliminate the need to copy data to other memory locations for straightforward addressing. Also, you can combine indirect addressing operations with arithmetic computations. This means you can perform zig-zag scanning of quantized DCT coefficients in parallel with other required operations. You can add zig-zag scanning to your program without adding execution time and code space.

7.7   LISTINGS
This section contains the listing for this chapter.

7 Discrete Cosine Transform

```
{ ONE DIMENSIONAL, FAST, DISCRETE COSINE TRANSFORM, 16 POINTS
      Implementation:
            as described by Hsieh S. Hou in IEEE Transactions on Acoustics,
            Speech, and Signal Processing, Vol. ASSP-35, No. 10, October 1987

      Target Processor:
            ADSP-2100 family of DSP processors from Analog Devices, Inc.

      Execution Benchmark:
            318 instruction cycles — ADSP-2101 — 15.90 us at CLKIN=20.00Mhz

      Memory Storage Requirement:
            272 PM = 257 program memory code, 15 program memory data (coefficients)
            32 DM = 16 data memory scratch pad, 16 data memory (16-pt vector)
            Note: resulting transform coefficients written over original input data
            Assumes: unsigned 8-bit input data, signed 16-bit output coefficients

      Release History: 27-March-1989, extensively revised: 17-July-1989
            Revised: 23-July-1989 Revised for ADSP-2101: 28-July-1993

      Analog Devices, Inc., DSP Division, P.O.Box 9106, Norwood, MA 02062, USA }

.module/ram/abs=0        fast_16pt_dct;
.var/pm/ram              cosvals[15];                 { cosine coefficients }
.var/circ/abs=0x3800     tmp[16];                     { temporary scratch memory }
.var                     x[16];                       { 16pt vector to transform }
.global    tmp;
.external  DIF16, DIF8, DIF4, DIF2, RLR4, RLR8, RLR16, DC_AND_BREV;
.init      x: <fhex.dat>;
.init cosvals[00]:  h#7F6200, h#70E200, h#513300, h#252800,
                    h#F37500, h#C3AA00, h#9D0E00, h#858300;
.init  cosvals[08]: h#7D8A00, h#471C00, h#E70800, h#959300;
.init  cosvals[12]: h#764100, h#CF0500;
.init  cosvals[14]: h#5A8200;


{ interupt vector table }
rest: jump setup; rti; rti; rti;                      { jump here on reset }
            rti; rti; rti; rti;                       { irq2 interrupt vector }
            rti; rti; rti; rti;                       { sport0 tx interrupt vector }
            rti; rti; rti; rti;                       { sport0 rx interrupt vector }
            rti; rti; rti; rti;                       { sport1 tx interrupt vector }
            rti; rti; rti; rti;                       { sport1 rx interrupt vector }
            rti; rti; rti; rti;                       { timer interrupt vector }
```

Discrete Cosine Transform   7

```
setup:    l0=0; l1=0; l2=0; l3=0; l5=0; l6=0;
dct16:    i2=^x; i3=^x+15; i6=^cosvals;
          m2=2; m3=-2; m5=1; m6=1; m7=-3; se=1;
          call DIF16;
          call DIF8;
          call DIF4;
          call DIF2;
          call RLR4;
          call RLR8;
          call RLR16;
          i5=^x;
          call DC_AND_BREV;
idwait: idle;
          jump idwait;
.endmod;
```

Listing 7.1 One-Dimensional Fast Discrete Cosine Transform (16 Points) Routine

7 Discrete Cosine Transform

```
.module/ram do_DIF16;                      { 1 16-way DIF }
.external tmp;
.entry DIF16;

DIF16:
   i0=^tmp;
   i1=^tmp+8;
   m1=1;

   ax1=dm(i3,m3);

   af=pass ax1, ax0=dm(i2,m2);
   ar=ax0+af, ax1=dm(i3,m3), my0=pm(i6,m6);
   ar=ax0-af, dm(i0,m1)=ar;
   mr=ar*my0(ss);

   cntr=6;
   do x16 until ce;
       af=pass ax1, ax0=dm(i2,m2);
       ar=ax0+af, ax1=dm(i3,m3), my0=pm(i6,m6);
       ar=ax0-af, dm(i0,m1)=ar;
x16:       mr=ar*my0(ss), dm(i1,m1)=mr1;

   af=pass ax1, ax0=dm(i2,m2);
   ar=ax0+af, my0=pm(i6,m6);
   ar=ax0-af, dm(i0,m1)=ar;
   mr=ar*my0(ss), dm(i1,m1)=mr1;
   dm(i1,m1)=mr1;
   rts;                                    { end 1 16-way DIF }
.endmod;
```

Listing 7.2 DIF16 Subroutine

Discrete Cosine Transform   7

```
.module/ram do_DIF8;                    {2 8-way DIFs }
.external tmp;
.entry DIF8;

DIF8:
    i0=^tmp;
    i1=^tmp+4;
    i2=^tmp;
    i3=^tmp+4;
    m0=5;

    ax1=dm(i3,m1);
    af=pass ax1, ax0=dm(i2,m1);
    ar=ax0+af, ax1=dm(i3,m1), my0=pm(i6,m6);
    ar=ax0-af, dm(i0,m1)=ar;
    mr=ar*my0(ss);
    af=pass ax1, ax0=dm(i2,m1);
    ar=ax0+af, ax1=dm(i3,m1), my0=pm(i6,m6);
    ar=ax0-af, dm(i0,m1)=ar;
    mr=ar*my0(ss), dm(i1,m1)=mr1;
    af=pass ax1, ax0=dm(i2,m1);
    ar=ax0+af, ax1=dm(i3,m0), my0=pm(i6,m6);
    ar=ax0-af, dm(i0,m1)=ar;
    mr=ar*my0(ss), dm(i1,m1)=mr1;
    af=pass ax1, ax0=dm(i2,m0);
    ar=ax0+af,                      my0=pm(i6,m7);
    ar=ax0-af, dm(i0,m0)=ar;
    mr=ar*my0(ss), dm(i1,m1)=mr1;
    dm(i1,m0)=mr1;

    ax1=dm(i3,m1);
    af=pass ax1, ax0=dm(i2,m1);
    ar=ax0+af, ax1=dm(i3,m1), my0=pm(i6,m6);
    ar=ax0-af, dm(i0,m1)=ar;
    mr=ar*my0(ss);
    af=pass ax1, ax0=dm(i2,m1);
    ar=ax0+af, ax1=dm(i3,m1), my0=pm(i6,m6);
    ar=ax0-af, dm(i0,m1)=ar;
    mr=ar*my0(ss), dm(i1,m1)=mr1;
    af=pass ax1, ax0=dm(i2,m1);
    ar=ax0+af, ax1=dm(i3,m1), my0=pm(i6,m6);
    ar=ax0-af, dm(i0,m1)=ar;
    mr=ar*my0(ss), dm(i1,m1)=mr1;
    af=pass ax1, ax0=dm(i2,m1);
    ar=ax0+af,                      my0=pm(i6,m6);
    ar=ax0-af, dm(i0,m1)=ar;
    mr=ar*my0(ss), dm(i1,m1)=mr1;
    dm(i1,m1)=mr1;

    rts;                            { end 2 8-way DIFs }
.endmod;
```

Listing 7.3 DIF8 Subroutine

7 Discrete Cosine Transform

```
.module/ram do_DIF4;          ─  { 4 4-way DIFs }
.external tmp;
.entry DIF4;

DIF4:
    i0=^tmp;
    i1=^tmp+2;
    i2=^tmp;
    i3=^tmp+2;
    m2=3;

    ax1=dm(i3,m1);
    af=pass ax1, ax0=dm(i2,m1);
    ar=ax0+af, ax1=dm(i3,m2), my0=pm(i6,m6);
    ar=ax0-af, dm(i0,m1)=ar;
    mr=ar*my0(ss);
    af=pass ax1, ax0=dm(i2,m2);
    ar=ax0+af,                    my1=pm(i6,m6);
    ar=ax0-af, dm(i0,m2)=ar;
    mr=ar*my1(ss), dm(i1,m1)=mr1;
    dm(i1,m2)=mr1;

    ax1=dm(i3,m1);
    af=pass ax1, ax0=dm(i2,m1);
    ar=ax0+af, ax1=dm(i3,m2);
    ar=ax0-af, dm(i0,m1)=ar;
    mr=ar*my0(ss);
    af=pass ax1, ax0=dm(i2,m2);
    ar=ax0+af;
    ar=ax0-af, dm(i0,m2)=ar;
    mr=ar*my1(ss), dm(i1,m1)=mr1;
    dm(i1,m2)=mr1;

    ax1=dm(i3,m1);
    af=pass ax1, ax0=dm(i2,m1);
    ar=ax0+af, ax1=dm(i3,m1);
    ar=ax0-af, dm(i0,m1)=ar;
    mr=ar*my0(ss);
    af=pass ax1, ax0=dm(i2,m0);
    ar=ax0+af;
    ar=ax0-af, dm(i0,m0)=ar;
    mr=ar*my1(ss), dm(i1,m1)=mr1;
    dm(i1,m1)=mr1;
```

```
        ax1=dm(i2,m1);
        af=pass ax1, ax0=dm(i3,m1);
        ar=ax0+af, ax1=dm(i2,m1);
        ar=ax0-af, dm(i1,m1)=ar;
        mr=ar*my0(ss);
        af=pass ax1, ax0=dm(i3,m1);
        ar=ax0+af;
        ar=ax0-af, dm(i1,m1)=ar;
        mr=ar*my1(ss), dm(i0,m1)=mr1;
        dm(i0,m1)=mr1;
        rts;                        { end 4 4-way DIFs }
.endmod;
```

Listing 7.4 DIF4 Subroutine

7 Discrete Cosine Transform

```
    .module/ram do_DIF2;              { 8 2-way DIFs }
    .external tmp;
    .entry DIF2;

DIF2:
    i0=^tmp;
    i1=^tmp+1;
    i2=^tmp;
    i3=^tmp+1;
    m0=2;

    ax1=dm(i3,m0);
    af=pass ax1, ax0=dm(i2,m0);
    ar=ax0+af, ax1=dm(i3,m0), my0=pm(i6,m6);
    ar=ax0-af, dm(i0,m0)=ar;
    mr=ar*my0(ss);

    cntr=6;
    do difx2 until ce;
        af=pass ax1, ax0=dm(i2,m0);
        ar=ax0+af, ax1=dm(i3,m0);
        ar=ax0-af, dm(i0,m0)=ar;
difx2:  mr=ar*my0(ss), dm(i1,m0)=mr1;


    af=pass ax1, ax0=dm(i2,m0);
    ar=ax0+af;
    ar=ax0-af, dm(i0,m0)=ar;
    mr=ar*my0(ss), dm(i1,m0)=mr1;

    dm(i1,m1)=mr1;
    rts;                              { end 8 2-way DIFs }
.endmod;
```

Listing 7.5 DIF2 Subroutine

Discrete Cosine Transform 7

```
.module do_RLR4;
.external tmp;
.entry RLR4;

RLR4:
    i0=^tmp+3;
    i1=^tmp+2;
    i2=i0;
    m0=4;

    si=dm(i0,m0);
    ay0=dm(i1,m0);

    cntr=3;
    do rlrx4 until ce;
        sr=ashift si (hi),si=dm(i0,m0);
        ar=sr1-ay0,              ay0=dm(i1,m0);

    rlrx4: dm(i2,m0)=ar;
    sr=ashift si (hi);
    ar=sr1-ay0;
            dm(i2,m0)=ar; rts;
.endmod;
```

Listing 7.6 RLR4 Subroutine

7 Discrete Cosine Transform

```
.module do_RLR8;
.external tmp;
.entry RLR8;

RLR8:
    i0=^tmp+4;
    i1=^tmp+12;
    m0=1;
    m1=2;
    m2=-1;
    m3=-2;
                    ay0=dm(i0,m1);
                    si=dm(i0,m2);
    sr=ashift si (hi);
    ar=sr1-ay0,         si=dm(i0,m0);
    af=-ar,             dm(i0,m0)=ar;
    sr=ashift si (hi);
    ar=sr1+af,          si=dm(i0,m3);
    af=-ar,             dm(i0,m1)=ar;
    sr=ashift si (hi),  ay0=dm(i1,m1);
    ar=sr1+af,          si=dm(i1,m2);
    sr=ashift si (hi),  dm(i0,m0)=ar;
    ar=sr1-ay0,         si=dm(i1,m0);
    af=-ar,             dm(i1,m0)=ar;
    sr=ashift si (hi);
    ar=sr1+af,          si=dm(i1,m3);
    af=-ar,             dm(i1,m1)=ar;
    sr=ashift si (hi);
    ar=sr1+af;

                        dm(i1,m0)=ar;

    rts;
.endmod;
```

Listing 7.7 RLR8 Subroutine

Discrete Cosine Transform   7

```
.module do_RLR16;
.entry RLR16;

RLR16:
    i0=h#0407;      { h#0407 = bitrev(^tmp+8) when ^tmp=h#3800 }
    i1=i0;
    m0=2048;        { bitrev modifier = 16384/N = 2048 }
    ena bit_rev;
            ay0=dm(i0,m0);
            si=dm(i0,m0);
    sr=ashift si (hi),   dm(i1,m0)=ay0;
    ar=sr1-ay0,          si=dm(i0,m0);
    af=-ar,              dm(i1,m0)=ar;

    cntr=5;
    do rlrx16 until ce;
        sr=ashift si (hi);
        ar=sr1+af,   si=dm(i0,m0);
    rlrx16: af=-ar,      dm(i1,m0)=ar;

    sr=ashift si (hi);
    ar=sr1+af;
                         dm(i1,m0)=ar;

    dis bit_rev;
    rts;
.endmod;
```

Listing 7.8 RLR16 Subroutine

7 Discrete Cosine Transform

```
.module do_DC_and_brev;
.const sqrt2div2=h#5A82;
.external tmp;
.entry DC_AND_BREV;

DC_AND_BREV:        mx0=dm(tmp);
DCterm:      my0=sqrt2div2;
        mr=mx0*my0(rnd);         { calculate DC term using sqrt(2)/2 }
        dm(tmp)=mr1;
descramble:        i0=h#0007;  { h#0007 = bitrev(^tmp) when ^tmp=h#3800 }
        m0=1024;                 { bitrev modifier = 16384/N = 1024 }
        cntr=16;
        ena bit_rev;
        do unbrev until ce;
        ax0=dm(i0,m0);            { read from bit-reversed tmp buffer }
unbrev:          dm(i5,m5)=ax0;   { write to normal ordered x buffer }
        dis bit_rev;
        rts;
.endmod;
```

Listing 7.9 DC_AND_BREV Subroutine

466

Discrete Cosine Transform 7

```
{ TWO DIMENSIONAL, FAST, DISCRETE COSINE TRANSFORM, 16 x 16 POINTS

    Implementation:
        as described by Hsieh S. Hou in IEEE Transactions on Acoustics,
        Speech, and Signal Processing, Vol. ASSP-35, No.10, October 1987

    Target Processor:
        ADSP-2100 family of DSP processors from Analog Devices, Inc.

    Execution Benchmark:
        10046 instruction cycles - ADSP-2101 - 0.5023 ms at CLKOUT=20.00 MHz

    Memory Storage Requirement:
        304 PM = 291 program memory code, 15 program memory data (coefficients)
        274 DM = 18 data memory scratch pad, 256 data memory (16x16 image)
        Note: resulting transform coefficients written over original input data
Assumes: unsigned 8-bit input data, signed 16-bit output coefficients

    Release History: 27-March-1989, extensively revised: 17-July-1989
        Revised: 23-July-1989 Revised for ADSP-2101: 28-July-1993

    Analog Devices, Inc., DSP Division, P.O.Box 9106, Norwood, MA 02062, USA }

.module/ram/abs=0       fast_16x16_dct;
.var/pm/ram             cosvals[15];            { cosine coefficients }
.var/circ/abs=0x3800    tmp[16];                { temporary scratch memory }
.var            xadr, xadr2;
.var            x[256];                         { 16x16 block to transform }
.global  tmp;
.external   DIF16, DIF8, DIF4, DIF2, RLR4, RLR8, RLR16, DC_AND_BREV;
.init     x: <xx.dat>;
.init   cosvals[00]: h#7F6200, h#70E200, h#513300, h#252800,
          h#F37500, h#C3AA00, h#9D0E00, h#858300; .init  cosvals[08]: h#7D8A00,
h#471C00, h#E70800, h#959300; .init  cosvals[12]: h#764100, h#CF0500;
.init   cosvals[14]: h#5A8200;

        jump setup; rti; rti; rti;              { jump here at reset }
        rti; rti; rti; rti;                     { irq2 interrupt vector }
        rti; rti; rti; rti;                     { sport0 tx interrupt vector }
        rti; rti; rti; rti;                     { sport0 rx interrupt vector }
        rti; rti; rti; rti;                     { sport1 tx interrupt vector }
        rti; rti; rti; rti;                     { sport1 rx interrupt vector }
        rti; rti; rti; rti;                     { timer interrupt vector }
setup:  l0=0; l1=0; l2=0; l3=0; l5=0; l6=0; m6=1; m7=-3; se=1;
```

(listing continues on next page)

7   Discrete Cosine Transform

```
{ calculate the DCT values for the row addresses }
rows: si=^x;                          { cols: ^x }
    dm(xadr)=si;
    i2=si;
    si=^x+15;                         { cols: ^x+240 }
    dm(xadr2)=si;
    i3=si;
    m5=1;                             { cols: 16 }
    cntr=16;
    do rowdcts until ce;
        i6=^cosvals;
        m2=2;                         { cols: 32 }
        m3=-2;                        { cols: -32 }
        call DIF16;
            call DIF8;
            call DIF4;
        call DIF2;
        call RLR4;
        call RLR8;
        call RLR16;
        si=dm(xadr);
        i5=si;
        call DC_AND_BREV;

nextrow: ay0=16;                      { cols: 1 }
        ax0=dm(xadr);
        ar=ax0+ay0;
        dm(xadr)=ar;
        i2=ar;
        ax0=dm(xadr2);
        ar=ax0+ay0;
        dm(xadr2)=ar;

rowdcts: i3=ar;
```

```
{ calculate  DCT values for column addresses }
cols: si=^x;                         { cols: ^x }
   dm(xadr)=si;
   i2=si;
   si=^x+240;                        { cols: ^x+240 }
   dm(xadr2)=si;
   i3=si;
   m5=16;                            { cols: 16 }
   cntr=16;
   do coldcts until ce;
       i6=^cosvals;
       m2=32;                        { cols: 32 }
       m3=-32;                       { cols: -32 }
       call DIF16;
       call DIF8;
       call DIF4;
       call DIF2;
       call RLR4;
       call RLR8;
       call RLR16;
       si=dm(xadr);
       i5=si;
       call DC_AND_BREV;
nextcol: ay0=1;                      { cols: 1 }
       ax0=dm(xadr);
       ar=ax0+ay0;
       dm(xadr)=ar;
       i2=ar;
       ax0=dm(xadr2);
       ar=ax0+ay0;
       dm(xadr2)=ar;
coldcts: i3=ar;
wait1:  idle;
     jump wait1;
.endmod;
```

Listing 7.10 Two-Dimensional Fast Discrete Cosine Transform (16 X 16 Points) Routine

7   Discrete Cosine Transform

```
{ ONE DIMENSIONAL, FAST, DISCRETE COSINE TRANSFORM, 8 POINTS
    Implementation:
        as described by Hsieh S. Hou in IEEE Transactions on Acoustics,
        Speech, and Signal Processing, Vol. ASSP-35, No. 10, October 1987

    Target Processor:
    ADSP-2100 family of DSP processors from Analog Devices, Inc.

    Execution Benchmark:
        158 instruction cycles — ADSP-2101 — 7.90 us at CLKIN=20.00MHz

    Memory Storage Requirement:
        163 PM = 156 program memory code, 7 program memory data (coefficients)
        16 DM = 8 data memory scratch pad, 8 data memory (8-pt vector)
        Note: resulting transform coefficients written over original input data
        Assumes: unsigned 8-bit input data, signed 16-bit output coefficients

    Release History:
        27-March-1989, Revised: 23-July-1989, Revised for ADSP-2101 28-july-1993

    Analog Devices, Inc., DSP Division, P.O.Box 9106, Norwood, MA 02062, USA }

.module/ram/abs=0        fast_8pt_dct;
.var/pm/ram              cosvals[7];                  { cosine coefficients }
.var/circ/abs=0x3800     tmp[8];                      { temporary scratch memory }
.var             x[8];                                { 8-pt vector to transform }
.global          tmp;
.external    DIF8_8, DIF4_8, DIF2_8, RLR4_8, RLR8_8, DC_AND_BREV_8;
.init    x: <x.dat>;
.init   cosvals[0]: h#7D8A00, h#471C00, h#E70800, h#959300;
.init   cosvals[4]: h#764100, h#CF0500;
.init   cosvals[6]: h#5A8200;

        jump setup; rti; rti; rti;                    { jump here at reset }
        rti; rti; rti; rti;                           { irq2 interrupt vector }
        rti; rti; rti; rti;                           { sport0 tx interrupt vector }
        rti; rti; rti; rti;                           { sport0 rx interrupt vector }
        rti; rti; rti; rti;                           { sport1 tx interrupt vector }
        rti; rti; rti; rti;                           { sport1 rx interrupt vector }
        rti; rti; rti; rti;                           { timer interrupt vector }
```

470

Discrete Cosine Transform   7

```
setup:    l0=0; l1=0; l2=0; l3=0; l5=0; l6=0; m6=1; se=1;
dct8:     i2=^x;
          i3=^x+7;
          m5=1;
          i6=^cosvals;
          m2=2;
          m3=-2;
          call DIF8_8;
          call DIF4_8;
          call DIF2_8;
          call RLR4_8;
          call RLR8_8;
          i5=^x;
          call DC_AND_BREV_8;
wait2:    idle;
          jump wait2;
.endmod;
```

Listing 7.11 One-Dimensional Fast Discrete Cosine Transform (8 Points) Routine

7   Discrete Cosine Transform

```
.module/ram do_DIF8_8;          { 1 8-way DIFs }
.external tmp
.entry DIF8_8;

DIF8_8:
    i0=^tmp;
    i1=^tmp+4;
    m1=1;

    ax1=dm(i3,m3);
    af=pass ax1, ax0=dm(i2,m2);
    ar=ax0+af, ax1=dm(i3,m3), my0=pm(i6,m6);
    ar=ax0-af, dm(i0,m1)=ar;
    mr=ar*my0(ss);

    cntr=2;
    do d8x8 until ce;
        af=pass ax1, ax0=dm(i2,m2);
        ar=ax0+af, ax1=dm(i3,m3), my0=pm(i6,m6);
        ar=ax0-af, dm(i0,m1)=ar;
d8x8: mr=ar*my0(ss), dm(i1,m1)=mr1;

    af=pass ax1, ax0=dm(i2,m2);
    ar=ax0+af,      my0=pm(i6,m6);
    ar=ax0-af, dm(i0,m1)=ar;
    mr=ar*my0(ss), dm(i1,m1)=mr1;
    dm(i1,m1)=mr1;

    rts;                        { end 1 8-way DIFs }
.endmod;
```

Listing 7.12 DIF8_8 Subroutine

Discrete Cosine Transform   7

```
.module/ram do_DIF4_8;          { 2 4-way DIFs }
.external tmp;
.entry DIF4_8;

DIF4_8:
   i0=^tmp;
   i1=^tmp+2;
   i2=^tmp;
   i3=^tmp+2;
   m2=3;

   ax1=dm(i3,m1);
   af=pass ax1, ax0=dm(i2,m1);
   ar=ax0+af, ax1=dm(i3,m2),   my0=pm(i6,m6);
   ar=ax0-af, dm(i0,m1)=ar;
   mr=ar*my0(ss);
   af=pass ax1, ax0=dm(i2,m2);
   ar=ax0+af,                      my1=pm(i6,m6);
   ar=ax0-af, dm(i0,m2)=ar;
   mr=ar*my1(ss), dm(i1,m1)=mr1;
   dm(i1,m2)=mr1;

   ax1=dm(i3,m1);
   af=pass ax1, ax0=dm(i2,m1);
   ar=ax0+af, ax1=dm(i3,m2);
   ar=ax0-af, dm(i0,m1)=ar;
   mr=ar*my0(ss);
   af=pass ax1, ax0=dm(i2,m2);
   ar=ax0+af;
   ar=ax0-af, dm(i0,m2)=ar;
   mr=ar*my1(ss), dm(i1,m1)=mr1;
   dm(i1,m2)=mr1;

   rts;                         { end 2 4-way DIFs }
.endmod;
```

Listing 7.13 DIF4_8 Subroutine

7 Discrete Cosine Transform

```
.module/ram do_DIF2_8;          { 4 2-way DIFs }
.external tmp;
.entry DIF2_8;

DIF2_8:
    i0=^tmp;
    i1=^tmp+1;
    i2=^tmp;
    i3=^tmp+1;
    m0=2;

    ax1=dm(i3,m0);

    af=pass ax1, ax0=dm(i2,m0);
    ar=ax0+af, ax1=dm(i3,m0), my0=pm(i6,m6);
    ar=ax0-af, dm(i0,m0)=ar;
    mr=ar*my0(ss);

    af=pass ax1, ax0=dm(i2,m0);
    ar=ax0+af, ax1=dm(i3,m0);
    ar=ax0-af, dm(i0,m0)=ar;
    mr=ar*my0(ss), dm(i1,m0)=mr1;

    af=pass ax1, ax0=dm(i2,m0);
    ar=ax0+af, ax1=dm(i3,m0);
    ar=ax0-af, dm(i0,m0)=ar;
    mr=ar*my0(ss), dm(i1,m0)=mr1;

    af=pass ax1, ax0=dm(i2,m0);
    ar=ax0+af;
    ar=ax0-af, dm(i0,m0)=ar;
    mr=ar*my0(ss), dm(i1,m0)=mr1;

    dm(i1,m1)=mr1;
    rts;                         { end 4 2-way DIFs }
.endmod;
```

Listing 7.14 DIF2_8 Subroutine

```
.module do_RLR4_8;
.external tmp;
.entry RLR4_8;

RLR4_8:
   i0=^tmp+3;
   i1=^tmp+2;
   i2=i0;
   m0=4;
                  si=dm(i0,m0);
                  ay0=dm(i1,m0);
   sr=ashift si (hi),    si=dm(i0,m0);
   ar=sr1-ay0,                 ay0=dm(i1,m0);
            dm(i2,m0)=ar;
   sr=ashift si (hi);
   ar=sr1-ay0;
            dm(i2,m0)=ar;
   rts;
.endmod;
```

Listing 7.15 RLR4_8 Subroutine

7 Discrete Cosine Transform

```
.module do_RLR8_8;
.external tmp;
.entry RLR8_8;

RLR8_8:
    i0=^tmp+4;
    i1=^tmp+12;
    m0=1;
    m1=2;
    m2=-1;
    m3=-2;
            ay0=dm(i0,m1);
            si=dm(i0,m2);
    sr=ashift si (hi);
    ar=sr1-ay0,             si=dm(i0,m0);
    af=-ar,         dm(i0,m0)=ar;
    sr=ashift si (hi);
    ar=sr1+af,              si=dm(i0,m3);
    af=-ar,         dm(i0,m1)=ar;
    sr=ashift si (hi);
    ar=sr1+af;
            dm(i0,m0)=ar;

    rts;
.endmod;
```

Listing 7.16 RLR8_8 Subroutine

Discrete Cosine Transform 7

```
.module do_DC_and_brev_8;
.const sqrt2div2=h#5A82;
.external tmp;
.entry DC_AND_BREV_8;

DC_AND_BREV_8:     mx0=dm(tmp);
DCterm:       my0=sqrt2div2;
          mr=mx0*my0(rnd);              { calculate DC term using sqrt(2)/2 }
          dm(tmp)=mr1;
descramble:        i0=h#0007;           { h#0007 = bitrev(^tmp) when ^tmp=h#3800 }
          m0=2048;                      { bitrev modifier = 16384/N = 2048 }
          cntr=8;
          ena bit_rev;
          do unbrev until ce;
              ax0=dm(i0,m0);            { read from bit-reversed tmp buffer }
unbrev:            dm(i5,m5)=ax0;       { write to normal ordered x buffer }
          dis bit_rev;
          rts;
.endmod;
```

Listing 7.17 DC_AND_BREV_8 Subroutine

7 Discrete Cosine Transform

```
{ TWO DIMENSIONAL, FAST, DISCRETE COSINE TRANSFORM, 8 x 8 POINTS
    Implementation:
        as described by Hsieh S. Hou in IEEE Transactions on Acoustics,
        Speech, and Signal Processing, Vol. ASSP-35, No. 10, October 1987

    Target Processor:
        ADSP-2100 family of DSP processors from Analog Devices, Inc.

    Execution Benchmark:
        2492 instruction cycles — ADSP-2101 — 0.12460 ms at CLKOUT=20.00MHz

    Memory Storage Requirement:
        208 PM = 201 program memory code, 7 program memory data (coefficients)
        16 DM = 8 data memory scratch pad, 8 data memory (8-pt vector)
        Note: resulting transform coefficients written over original input data
        Assumes: unsigned 8-bit input data, signed 16-bit output coefficients

    Release History: 27-March-1989, Revised: 23-July-1989, Revised for ADSP-2101
            28-July-1993

    Analog Devices, Inc., DSP Division, P.O.Box 9106, Norwood, MA 02062, USA }

.module/ram/abs=0        fast_8x8_dct;
.var/pm/ram              cosvals[15];              { cosine coefficients }
.var/circ/abs=0x3800     tmp[8];                   { temporary scratch memory }
.var          xadr, xadr2;
.var          x[64];                               { 8x8 block to transform }
.global       tmp;
.external     DIF8_8, DIF4_8, DIF2_8, RLR4_8, RLR8_8, DC_AND_BREV_8;
.init    x: <xx.dat>;
.init    cosvals[0]: h#7D8A00, h#471C00, h#E70800, h#959300;
.init    cosvals[4]: h#764100, h#CF0500;
.init    cosvals[6]: h#5A8200;

    jump setup; rti; rti; rti;                     { start here on reset }
    rti; rti; rti; rti;                            { irq2 interrupt vector }
    rti; rti; rti; rti;                            { sport0 tx interrupt vector }
    rti; rti; rti; rti;                            { sport0 rx interrupt vector }
    rti; rti; rti; rti;                            { sport1 tx interrupt vector }
    rti; rti; rti; rti;                            { sport1 rx interrupt vector }
    rti; rti; rti; rti;                            { timer interrupt vector }
```

Discrete Cosine Transform   7

```
setup:  l0=0; l1=0; l2=0; l3=0; l5=0; l6=0; m6=1; se=1;
rows:   si=^x;                              { cols: ^x }
     dm(xadr)=si
     i2=si;
     si=^x+7;                               { cols: ^x+56 }
     dm(xadr2)=si;
     i3=si;
     m5=1;                                  { cols: 8 }
     cntr=8;
     do rowdcts until ce;
          i6=^cosvals;
          m2=2;                             { cols: 16 }
          m3=-2;                            { cols: -16 }
          call DIF8_8;
          call DIF4_8;
          call DIF2_8;
          call RLR4_8;
          call RLR8_8;
          si=dm(xadr);
          i5=si;
          call DC_AND_BREV_8;
nextrow: ay0=8;                             { cols: 1 }
          ax0=dm(xadr);
          ar=ax0+ay0;
          dm(xadr)=ar;
          i2=ar;
          ax0=dm(xadr2);
          ar=ax0+ay0;
          dm(xadr2)=ar;
rowdcts: i3=ar;
cols:   si=^x;                              { cols: ^x }
     dm(xadr)=si;
     i2=si;
     si=^x+56;                              { cols: ^x+56 }
     dm(xadr2)=si;
     i3=si;
     m5=8;                                  { cols: 8 }
     cntr=8;
     do coldcts until ce;
          i6=^cosvals;
          m2=16;                            { cols: 16 }
          m3=-16;                           { cols: -16 }
          call DIF8_8;
          call DIF4_8;
          call DIF2_8;
          call RLR4_8;
          call RLR8_8;
          si=dm(xadr);
          i5=si;
          call DC_AND_BREV_8;
```

(listing continues on next page)

```
nextcol: ay0=1;                                    { cols: 1 }
        ax0=dm(xadr);
        ar=ax0+ay0;
        dm(xadr)=ar;
        i2=ar;
        ax0=dm(xadr2);
        ar=ax0+ay0;
        dm(xadr2)=ar;
coldcts: i3=ar;
wait3: idle;
   jump wait3;
.endmod;
```

Listing 7.18 Two-Dimensional Fast Discrete Cosine Transform (8 X 8 Points) Routine

7.8   REFERENCES

Hou. H. 1986. "The Fast Recursive Algorithm for Computing the Discrete Cosine Transform," *SPIE Conference Proceedings*, vol. 697, pp. 18-25.

Kamanjar & Rao. 1982. "Fast Algorithms for the 2D DCT," *IEEE Trans on Computers*, pp. 899-906.

Lee, Byeong. 1984. "A New Algorithm to Compute the DCT," *IEEE ASSP*, vol. 32, No. 6, pp. 1243-1245.

Magal & Heiman. 1985. "Image Coding System–A Single Processor Implementation," *IEEE MilCom*, vol. 3, pp. 628-634.

Unknown. 1988. "A 1 Chip VLSI for Real Time Two Dimensional Discrete Cosine Transform," *ISACS Conference Proceedings*.

Digital Tone Detection 8

8.1   OVERVIEW

ADSP-2100 Family DSPs are well suited for applications that detect
sinusoidal tones. These applications include telephone signaling, remotely
controlled equipment, test instruments for tone based systems, and tone-
encoded data transmission.

One of the most common examples of tone detection is the touch-tone
signaling standard used in telephones. This standard is called DTMF, or
dual-tone, multi-frequency signaling. Since DTMF is an in-band signaling
system (superimposed on the voice channel), it rejects interference from
the simultaneously present voice frequencies. Telephones systems also use
other standards, for example, trunk switching circuits may use out-of-
band MF (multi-frequency) signaling, while other switching equipment
may use single-tone signaling.

Another common application for tone detection is remotely controlled
equipment, such as a remotely-piloted drone aircraft. These applications
pass servo instructions to the drone aircraft by radio control. These
instructions are binary numbers that are coded in frequency. Each binary
digit is assigned a frequency. The receiver reconstructs binary numbers by
detecting the presence (logic "1") or the absence (logic "0") of each
possible tone.

Digital tone detection applications usually have fast execution speeds and
require minimum memory storage. You can take advantage of these
features by coding tone detection as a sub task of a larger, single-chip DSP
application, or by using a single DSP to simultaneously handle tone
processing for many independent channels.

DTMF tone detection and generation is covered in Chapter 14, of *Digital
Signal Processing Applications Using the ADSP-2100 Family*, Volume 1. Refer
to that chapter for more details on the Goertzel method of tone detection
and validation, as well as precision sine wave generation using fast
polynomial expansions.

8 Digital Tone Detection

8.2 IMPLEMENTATION

This section outlines the steps you can use to implement the tone detection subroutines included at the end of this chapter.

8.2.1 Choosing A Sampling Frequency

Sometimes your application dictates the sampling frequency. This is true for telephone band applications where the local telephone administration specifies a sampling frequency (8000 Hz, for example). For applications where you choose the frequency, remember that the Nyquist theory states that the minimum sampling frequency must be at least twice the frequency of the highest frequency you want to process.

Once you have a list of frequencies, to help you select the best sampling frequency, factor each frequency into its prime factors. Listings 8.1 and 8.2 contain two C programs (FACTOR.C and PRIMES.C) to help you. Once each frequency is broken down into its constituent prime factors, pick out the prime factors that are most common to the greatest number of frequencies, then multiply the prime factors together. Call the resulting product "A". The best sampling frequency is an integer multiple of "A" that is greater than or equal to twice the highest input frequency of interest.

Listing 8.3 (BESTFS.C) is a C program that verifies if the chosen sampling frequency is the best fit. The program sweeps through the specified range of sampling frequencies and calculates the maximum squared mismatch error for the tone set. The mismatch error is how closely the tone of interest matches the integral subdivisions of the sampling frequency. Frequencies with many common prime factors will match more precisely.

For a given tone set, the mismatch error-squared is calculated for each individual tone. The largest mismatch in the tone set is chosen as the maximum error-squared value for the tone set at that sampling frequency. This is the term that should be minimized. The following equation describes the mismatch error-squared:

$$E^2(i) = \left[\left[\left(\frac{f_{sample}}{f_{tone}(i)}\right)\right] - n\left[\left(\frac{f_{sample}}{f_{tone}(i)}\right)\right]\right]^2$$

where $E^2(i)$ is the mismatch error-squared and n is the nearest integer.

BESTFS.C stores the resulting error values in a file called BESTFS.ERR and displays them on the terminal screen. To find the best sampling frequency,

Digital Tone Detection   8

sort the error file alphabetically using any sort utility. You may want to plot the error values before sorting them to graphically identify a minimum.

8.2.2   Picking The Best Value Of N For The Goertzel Iterations

The Goertzel algorithm operates on a sample-by-sample basis, like an IIR filter. After N iterations, or N samples received, the output value of this algorithm is the Goertzel output of interest. This output value is equivalent to what a single-frequency DFT calculates. You can think of the Goertzel algorithm as an IIR filter with an output that is sampled after every N samples. Consider the following parameters when you select a value for N:

- Leakage Loss

- Frequency Resolution

- Detection Time

8.2.2.1   *Leakage Loss*

The Goertzel algorithm has the same frequency characteristics as the DFT algorithm. In other words, if an N-point DFT is performed on a data sequence sampled at frequency F_S, the output frequency samples, sometimes called frequency bins, are equally spaced at F_S/N. If a tone is present that matches an integer multiple of F_S/N, it is completely contained in one of the output frequency bins. If however, the tone falls between the center of two adjacent frequency bins, the total energy is distributed among several neighboring frequency bins. This phenomenon is called spillage, or leakage. See Chapter 6, *One-Dimensional FFTs*, of *Digital Signal Processing Applications Using the ADSP-2100 Family*, Volume 1 for more information.

You can think of the frequency samples as if they were output frequency samples of an FFT or DFT calculation. Therefore, you can imagine that all frequency samples are present, even though the actual implementation only calculates the frequency samples of interest. As a result, spillage into neighboring bins appears to decrease the level present in the bin of interest. In a tone detection application, the missing energy (that, mathematically, is spilled into nearby bins) is never seen in the neighboring bins since the neighboring bin levels are not calculated.

8 Digital Tone Detection

Because of leakage, for a given sampling frequency and a given frequency to detect, some values of N show poor performance, some values have better performance, and some values perform optimally. In the optimal case, the tone to detect is an exact integer multiple of F_S/N, for example when $F_{tone}=k*F_S/N$. When decoding several frequencies, try to pick one value of N for all frequencies. This means that some frequencies will closely match $k*F_S/N$, and some will not. The reason you should try to use a single N value for multiple frequencies is because the valid Goertzel outputs are available after N samples are processed, so the valid output of all frequencies is available at the same time.

8.2.2.2 Frequency Resolution

Frequency resolution is the second consideration. Larger values for N provide better frequency resolution. If N is large, F_S/N (the individual frequency bin widths, or the spacing of the resulting frequency samples) is small. This means that the detector will reject more off-frequency tones, or resolve between two tones that are close together.

8.2.2.3 Detection Time

You must also consider detection time. When you choose a larger value for N, it takes longer for N samples to be received, and consequently, the time between valid Goertzel outputs is longer. This directly affects the speed at which the decoder detects the presence of a tone.

Listing 8.4, called BESTN.C, tests the "goodness of fit" for values of N within a specified range, given the sampling frequency. BESTN.C stores the results in a file called BESTN.ERR and displays them on the terminal screen. To find the best values for N, sort the error file alphabetically using any sort utility. The result is a list of the best values for N at the chosen sampling frequency in descending order of "goodness of fit."

8.2.2.4 Tone Detection Categories

Tone detection code falls into two categories:

- Symbol detection (applications such as DTMF)—More than one tone is detected, then tested for relative amplitudes, number of tones present, etc. to validate the presence of a symbol (made up tones).

- Independent, single, tone, presence detection—Any number of tones can be present, and the indication of a tone's presence is the only requirement. Tests, such as relative amplitudes and number of tones, are not necessary in this case.

Digital Tone Detection   8

Symbol detection follows the DTMF decoder described in Chapter 14 of Volume 1. That example only listens for two tones from a predetermined alphabet. By changing coefficients and post-testing thresholds, the symbol detector can be fine-tuned or reprogrammed for other tone standards, such as CCITT 2-of-6 Multi-Frequency (MF), call progress tones, US Air Force 412L, US Army TA-314/PT, etc.

Of the two categories of tone detection code, single tone presence detection is simpler to implement. Most of the post-testing can be eliminated and replaced by simpler energy presence (threshold comparison) tests. Section 8.2.3.5 contains an example.

Both categories use the basic Goertzel algorithm for each tone. Voice rejection requires the monitoring of energy content at the intended tone's second harmonic. Slight changes must be made when switching from single-channel decoding to multiple-channel decoding. When decoding several channels, the input samples of all channels are stored in a circular buffer. The buffer length is equal to the number of channels. When you decode a single channel, the replace the circular buffer with a single data memory variable.

8.2.2.5 Tone Detection Example

This example was designed for single tone detection on the frequencies shown in Table 8.1. Using FACTOR.C, the prime factors of the frequencies of interest are also shown in Table 8.1. To choose a good sampling frequency, pick the prime factors that are common to most of the frequencies of interest. In this example, most of the frequencies have the prime factors 3, 3, 5, 5, and 7. Multiply them together (the product is 1575), and select the integer multiple of that product that is the next one higher than twice the highest input frequency of interest.

Frequencies Of Interest	Prime Factors
11025 Hz	3, 3, 5, 5, 5, 7
12600 Hz	2, 2, 2, 3, 3, 5, 5, 7
14175 Hz	3, 3, 3, 3, 5, 5, 7
15750 Hz	2, 3, 3, 5, 5, 5, 7
17325 Hz	3, 3, 5, 5, 7, 11
18900 Hz	2, 2, 3, 3, 3, 5, 5, 7
20475 Hz	3, 3, 5, 5, 7, 13
23175 Hz	3, 3, 5, 5, 103

Table 8.1 Sample Frequencies & Prime Factors

8 Digital Tone Detection

The highest frequency of interest is 23175 Hz, therefore sampling must occur at a minimum frequency of 46350 Hz. The smallest integer multiple of 1575 that is greater than 46350 is 47250. This example uses BESTFS.C to verify the choice (47250 Hz). You can verify the sampling frequency by testing all integer frequencies between 46350 Hz and 49000 Hz with the following syntax:

```
c:> bestfs 46350 49000 1 | sort | more
```

Table 8.2 is an example of the resulting output. You can see that 47250 Hz is not the best choice; within the range 46350–49000 Hz, the best sampling frequency is 48600 Hz.

```
*** scanning f_sample from 46350.000000 Hz to 49000.000000 Hz,
        stepping 1.000000 Hz ***
0.183674 = maxerrsq      (at f_sample = 48600.000000)
0.183719 = maxerrsq      (at f_sample = 48599.000000)
0.183734 = maxerrsq      (at f_sample = 48601.000000)
0.183764 = maxerrsq      (at f_sample = 48598.000000)
0.183794 = maxerrsq      (at f_sample = 48602.000000)
0.183810 = maxerrsq      (at f_sample = 48597.000000)
0.183855 = maxerrsq      (at f_sample = 48596.000000)
0.183855 = maxerrsq      (at f_sample = 48603.000000)
0.183900 = maxerrsq      (at f_sample = 48595.000000)
0.183915 = maxerrsq      (at f_sample = 48604.000000)
0.183946 = maxerrsq      (at f_sample = 48594.000000)
0.183976 = maxerrsq      (at f_sample = 48605.000000)
0.183991 = maxerrsq      (at f_sample = 48593.000000)
0.184036 = maxerrsq      (at f_sample = 48592.000000)
0.184036 = maxerrsq      (at f_sample = 48606.000000)
0.184082 = maxerrsq      (at f_sample = 48591.000000)
0.184097 = maxerrsq      (at f_sample = 48607.000000)
0.184127 = maxerrsq      (at f_sample = 48590.000000)
0.184158 = maxerrsq      (at f_sample = 48608.000000)
     .
     .      etc,
     .
0.249947 = maxerrsq      (at f_sample = 47249.000000)
0.249947 = maxerrsq      (at f_sample = 47251.000000)
0.250000 = maxerrsq      (at f_sample = 47250.000000)
```

Table 8.2 Sorted Sampling Frequencies (BESTFS.ERR)

Next, look for the best value for N. For this example, assume that you must detect tones within 10 ms, and you want a frequency resolution of approximately 100 Hz. You can use the following syntax to select N:

```
c:> bestn 46350 0 600 | sort | more
```

Digital Tone Detection   8

This example evaluates all values for N from N=0 through N=600, at a sampling frequency of 48600 Hz. The program calculates the maximum square error of all tones of interest and their respective closest integer multiples of F_S/N. The program also displays the associated detection time in milliseconds and frequency resolution width in Hertz. Table 8.3 is a sample of the resulting output. The results show that several values for N are acceptable. If N=463, or N=493, it fits the requirements fairly well. If N=432, the tone set matches optimally with the sampling frequency of 48600 Hz, the detection time is less than 9 ms, but the frequency resolution is wider than the 100 Hz. For this example, let N=463.

```
*** scanning N from o to 600, where f_sample = 48600.000000 Hz ***
0.000000 = maxerrsq (N=      0) detect=    0.000 ms    resolu= Infinity Hz
0.000000 = maxerrsq (N=    216) detect=    4.444 ms    resolu=  225.000 Hz
0.000000 = maxerrsq (N=    432) detect=    8.889 ms    resolu=  112.500 Hz
0.047346 = maxerrsq (N=     31) detect=    0.638 ms    resolu= 1567.742 Hz
0.047346 = maxerrsq (N=    185) detect=    3.807 ms    resolu=  262.703 Hz
0.047346 = maxerrsq (N=    247) detect=    5.082 ms    resolu=  196.761 Hz
0.047346 = maxerrsq (N=    401) detect=    8.251 ms    resolu=  121.197 Hz
0.047346 = maxerrsq (N=    463) detect=    9.527 ms    resolu=  104.968 Hz
0.057955 = maxerrsq (N=    308) detect=    6.337 ms    resolu=  157.792 Hz
0.057955 = maxerrsq (N=    340) detect=    6.996 ms    resolu=  142.941 Hz
0.057955 = maxerrsq (N=    524) detect=   10.782 ms    resolu=   92.748 Hz
0.057955 = maxerrsq (N=    556) detect=   11.440 ms    resolu=   87.410 Hz
0.057957 = maxerrsq (N=     92) detect=    1.893 ms    resolu=  528.261 Hz
0.057957 = maxerrsq (N=    124) detect=    2.551 ms    resolu=  391.935 Hz
0.090552 = maxerrsq (N=    371) detect=    7.634 ms    resolu=  130.997 Hz
0.090552 = maxerrsq (N=    493) detect=   10.144 ms    resolu=   98.580 Hz
0.090552 = maxerrsq (N=    587) detect=   12.078 ms    resolu=   82.794 Hz
0.090557 = maxerrsq (N=     61) detect=    1.255 ms    resolu=  796.721 Hz
0.090557 = maxerrsq (N=    155) detect=    3.189 ms    resolu=  313.548 Hz
0.090557 = maxerrsq (N=    277) detect=    5.700 ms    resolu=  175.451 Hz
0.105024 = maxerrsq (N=     34) detect=    0.700 ms    resolu= 1429.412 Hz
0.105024 = maxerrsq (N=    182) detect=    3.745 ms    resolu=  267.033 Hz
0.105024 = maxerrsq (N=    250) detect=    5.144 ms    resolu=  194.400 Hz
0.105029 = maxerrsq (N=    398) detect=    8.189 ms    resolu=  122.111 Hz
0.105029 = maxerrsq (N=    466) detect=    9.588 ms    resolu=  104.292 Hz
0.111108 = maxerrsq (N=    354) detect=    7.284 ms    resolu=  137.288 Hz
0.111108 = maxerrsq (N=    510) detect=   10.484 ms    resolu=   95.294 Hz
    .
    .      etc,
    .
0.250000 = maxerrsq (N=    585) detect=   12.037 ms    resolu=   83.077 Hz
0.250000 = maxerrsq (N=    588) detect=   12.099 ms    resolu=   82.653 Hz
0.250000 = maxerrsq (N=    594) detect=   12.222 ms    resolu=   81.818 Hz
```

Table 8.3 Sorted Values For N (BESTN.ERR)

8 Digital Tone Detection

Next, the Goertzel coefficients must be calculated. Use the following syntax to start COEFGEN.C (Listing 8.5):

```
c:> coefgen
N > 463
f_sample > 48600
```

The resulting output is shown in Table 8.4. The columns show each tone to detect with its associated k(flt), k(int), and k(err) values. The k(flt) value is the floating point value of $N*(f_{tone}/f_{sample})$; k(int) is the closest integer to k(flt). This integer is the index of the frequency bin for the closest match. If k(flt) and k(int) are equal, they are perfectly matched and there is no leakage loss occurs. Discrepancies between k(flt) and k(int) lead to leakage losses, and this difference is measured in the k(err) variable. The "goodness of fit" of the N value is judged by the largest squared k(err) value of all the individual f_{tone} values.

```
N=463.000000
fs=48600.000000

f_tone[0]=   11025.00 Hz
             k(flt)=105.032410
             k(int)=105
             k(err)= +0.032410
             coef(flt)= +0.290734 coef(2.14 hex)=0x129B
f_tone[1]=   12600.00 Hz
             k(flt)=120.037041
             k(int)=120
             k(err)= +0.037041
             coef(flt)= -0.115286 coef(2.14 hex)=0xF89F
f_tone[2]=   14175.00 Hz
             k(flt)=135.041672
             k(int)=135
             k(err)= +0.041672
             coef(flt)= -0.516546 coef(2.14 hex)=0xDEF1
f_tone[3]=   15750.00 Hz
             k(flt)=150.046295
             k(int)=150
             k(err)= +0.046295
             coef(flt)= -0.896475 coef(2.14 hex)=0xC6A0
f_tone[4]=   17325.00 Hz
             k(flt)=165.050919
             k(int)=165
             k(err)= +0.050919
             coef(flt)= -1.239387 coef(2.14 hex)=0xB0AE
```

```
f_tone[5]=  18900.00 Hz
                k(flt)=180.055557
                k(int)=180
                k(err)= +0.055557
                coef(flt)= -1.531119 coef(2.14 hex)=0x9E02
f_tone[6]=  20475.00 Hz
                k(flt)=195.060181
                k(int)=195
                k(err)= +0.060181
                coef(flt)= -1.759627 coef(2.14 hex)=0x8F62
f_tone[7]=  23175.00 Hz
                k(flt)=220.782410
                k(int)=221
                k(err)= +0.217590
                coef(flt)= -1.979731 coef(2.14 hex)=0x814C
```

Table 8.4 Goertzel Coefficients

The Goertzel algorithm uses a single, real coefficient for each f_{tone} to detect. That coefficient is listed in a column with the hexadecimal equivalent for the ADSP-2100 family software. Since the equation for the coefficient is $2*\cos(2\pi*k(int)/N)$, the coefficient values are in the range -$2.0 \leq coef \leq 2.0$. For this reason, the coefficients are interpreted by the Goertzel algorithm is 2.14 format.

Listing 8.6 is an example of ADSP-2100 tone detection code. It establishes a routine that waits for interrupts. For each interrupt, the sample is counted and fed into the Goertzel feedback loop. When the sample count reaches N, the Goertzel feedforward instructions are executed, and a frequency-domain sample is calculated. Since N is the same for all tones being detected, all the results are available during the interrupt period. The following software checks the energy level found in each frequency bin of interest, and compares it to the predefined threshold. If the threshold is exceeded, a routine indicates the presence of that particular tone.

8.3   BENCHMARKS FOR THE EXAMPLE PROGRAM

The example detailed above detects the presence of energy in eight frequencies. The levels are threshold tested, and a binary number is output as a result every N input samples. Benchmarks will vary as specific applications deviate from this example, although this example is fundamental enough to demonstrate how you can evaluate your own benchmarks. Table 8.5 shows typical benchmark performance of the ADSP-2100A in this example application, as well as processor loading values for a similar example sampled at 8 kHz instead of the 48600 Hz.

8 Digital Tone Detection

Memory Usage:	PM RAM	DM RAM	
	102 Locations	302 Locations	

DSP	Processor Speed	Number of Cycles	Execution Time
ADSP-2101/2111	20 MHz	75	3.75 µs
ADSP-2171	33 MHz	75	2.25 µs

Table 8.5 Typical Benchmark Performance

Monitoring more frequencies requires more computation time. Having a faster sampling rate reduces the amount of time available for Goertzel feedback iterations between interrupts. The number of instructions available between interrupts is equal to the sampling period divided by the instruction cycle time.

The above example assumes that the ADSP-2100A is executing at a 12.5 MHz instruction rate. Dividing 12.5 MHz by the sampling frequency of 48600 Hz yields 257 available instructions between interrupts to maintain real-time processing.

Choosing different values of N has no impact on the computational benchmark. As long as the Goertzel feedback iterations can be performed between input samples, the Goertzel algorithm works. Large values of N mean that it takes the algorithm longer to generate an output value after N input iterations.

The example code uses very little memory. A major portion of the data memory storage (256 places out of 302 total) is taken by a lookup table for the µ-law PCM conversion. The ADSP-2101/2 has this function built into its serial ports and it does not usurp data memory storage.

Program memory is limited to 94 total instructions with 8 coefficients. All program memory and data memory requirements are easily fulfilled by the on-chip memory of the ADSP-2100 Family processors, leaving the remaining on-chip memory space for other DSP functions.

8.4   LISTINGS
This sections contains the listing for this chapter.

Digital Tone Detection   8

```c
#include <stdio.h>

int prime[2000], factor[2000];
FILE *fp;

int isaprime( a, howmany )
int a, howmany;
{
    int i;
    for (i=0; i<howmany; i++)
        if (a==prime[i]) return(1);
    return(0);
}

int findaprime( a, howmany )
int a, howmany;
{
    int i;
    for (i=1; i<howmany; i++)
        if ((a%prime[i])==0) return(prime[i]);
    return(0);
}

main(argc,argv)
int argc; char **argv;
{
    int i, j, orig, freq, num, maxprimes;

    if (argc!=2)
        {
        printf("number to factor> ");
        scanf("%d",&freq);
        }
    else
        sscanf(argv[1],"%d",&freq);

    fp=fopen("primes.dat","r");
    if (fp==NULL) {printf("\nerror opening primes.dat\n");return(-1);}
    i=0;
    while (!feof(fp)) fscanf(fp,"%d",&prime[i++]);
    maxprimes=i-1;
    printf("\n%d primes read\n", maxprimes);
    fclose(fp);
```

(listing continues on next page)

8 Digital Tone Detection

```
      orig=freq;
      i=0;
      while(1)
            {
            if (isaprime(freq,maxprimes)==1) { factor[i++]=freq; break; }
            num=findaprime(freq,maxprimes);
            freq=freq/num;
            factor[i++]=num;
            }
      printf("\n %d factored out = ", orig);
      for (j=0; j<i; j++) printf("%d ", factor[j]);
      printf("\n");
}
```

Listing 8.1 Prime Factors Routine (FACTOR.C)

```
#include <stdio.h>

int fact, num, k, prime[1000];
FILE *fp;

main()
{
   fp=fopen("primes.dat","w");
   fprintf(fp,"%4d\n",1);
   fprintf(fp,"%4d\n",2);
   num=2;
   k=0;
   while(num<=2000)
         {
         fact=num-1;
         while((num/fact*fact)!=num)
            {
            -fact;
            if (fact<=1)
                   {
                   prime[k] = num;
                   fprintf(fp,"%4d\n",prime[k]);
                   k++;
                   }
            }
         printf("\r%d",num);
         ++num;
         }
   printf("\nDONE.\n");
}
```

Listing 8.2 Prime Numbers Routine (PRIMES.C)

492

Digital Tone Detection   8

```
#include <string.h>
#include <stdio.h>
#include "tones.def"

int round( x )
float x;
{
    if (x>0) return ( (int) (x+0.5) );
    else if (x<0) return ( (int) (x-0.5) );
    else if (x==0) return ( 0 );
    else printf("\7bad data in round() function!");
    return(-1);
}

char f1name[]="bestfs.err";
char f2name[]="bestfs.fs";

main(argc,argv)
int argc;
char **argv;
{
    int i, kint;
    float f_min, f_max, f_incr, f_sample;
    float kflt, maxerrsqr, errsqr;
    FILE *f1, *f2;
    if (argc!=4)
            {
            printf("\n\7usage: %s <f_min> <f_max> <f_incr>\n",argv[0]);
            printf("\n(where f_min, f_max, f_incr are real numbers)\n");
            return(-1);
            }
    sscanf(argv[1],"%f",&f_min);
    sscanf(argv[2],"%f",&f_max);
    sscanf(argv[3],"%f",&f_incr);
    printf("*** scanning f_sample from %f Hz to %f Hz,
            stepping %f Hz ***\n",f_min,f_max,f_incr);
    if (f_min>=f_max)
            {
            printf("f_min>=f_max!");
            return(-1);
            }
    else if ((f_max<f_min+f_incr)||(f_incr<=0))
            {
            printf("bad f_incr value!");
            return(-1);
            }
```

(listing continues on next page)

8   Digital Tone Detection

```
else for (i=0; i<HOW_MANY_TONES; i++)
      {
      if (f_min<(2*f_tone[i]))
          {
          printf("\n\7Nyquist violation:\nf_tone=%f
                  at f_sampling=%f\n",f_tone[i],f_min);
          return(-1);
          }
      }
f1=fopen(f1name,"w"); if (f1==NULL) printf("\nerror opening %s\n",f1);
f2=fopen(f2name,"w"); if (f2==NULL) printf("\nerror opening %s\n",f2);
for (f_sample=f_min; f_sample<=f_max; f_sample=f_sample+f_incr)
      {
      maxerrsqr=0;
      for (i=0; i<HOW_MANY_TONES; i++)
          {
          kflt=f_sample/f_tone[i];
          kint=round(kflt);
          errsqr=(kflt-(float)(kint))*(kflt-(float)(kint));
          if (errsqr>maxerrsqr) maxerrsqr=errsqr;
          }
      printf("\n%f = maxerrsqr (at f_sample = %f)",
              maxerrsqr, f_sample);
      if ((f1)&&(f2))
          {
          fprintf(f1,"%f\n",maxerrsqr);
          fprintf(f2,"%f\n",f_sample);
          }
      }
printf("\nyou may want to pipe output to sort utility");
printf("\nor plot %s vs %s\n",f1name,f2name);
fclose(f1);  fclose(f2);
}
```

Listing 8.3 Best Sampling Frequency Routine (BESTFS.C)

Digital Tone Detection   8

```c
#include <string.h>
#include <stdio.h>
#include "tones.def"

int round( x )
float x;
{
   if (x>0) return ( (int) (x+0.5) );
   else if (x<0) return ( (int) (x-0.5) );
   else if (x==0) return ( 0 );
   else printf("\7bad data in round() function!");
   return(-1);
}

char f1name[]="bestn.err";
char f2name[]="bestn.N";

main(argc,argv)
int argc;
char **argv;
{
   int i, N, minN, maxN, kint;
   float f_sample, kflt, errsqr, maxerrsqr, detect, binwidth;
   FILE *f1, *f2;

   if (argc!=4)
        {
        printf("\nusage: %s <f_sample Hz [%%f] > <minN [%%d] > <maxN [%%d]
             >\n",argv[0]);
        return(-1);
        }
   sscanf(argv[1],"%f",&f_sample);
   sscanf(argv[2],"%d",&minN);
   sscanf(argv[3],"%d",&maxN);
   printf("*** scanning N from %d to %d, where f_sample
          = %f Hz ***\n",minN,maxN,f_sample);
   if (minN>=maxN)
        {
        printf("minN>=maxN!");
        return(-1);
        }
```

(listing continues on next page)

8   Digital Tone Detection

```
f1=fopen(f1name,"w"); if (f1==NULL) printf("\nerror opening %s\n",f1);
f2=fopen(f2name,"w"); if (f2==NULL) printf("\nerror opening %s\n",f2);
for (N=minN; N<=maxN; N++)
    {
    maxerrsqr=0;
    binwidth=f_sample/(float)N;
    for (i=0; i<HOW_MANY_TONES; i++)
        {
        kflt=((float)(N))*(f_tone[i]/f_sample);
        kint=round(kflt);
        errsqr=(kflt-(float)(kint))*(kflt-(float)(kint));
        if (errsqr>maxerrsqr) maxerrsqr=errsqr;
        }
    detect=((float)(N)*1000.0)/f_sample;
    printf("\n%f =maxerrsqr ",maxerrsqr);
    printf("(N=%6d) ",N);
    printf("detect= %10.3f ms ",detect);
    printf("resolu= %10.3f Hz",binwidth);
    if ((f1)&&(f2))
        {
        fprintf(f1,"%f\n",maxerrsqr);
        fprintf(f2,"%d\n",N);
        }
    }
printf("\nyou may want to pipe output to sort utility");
printf("\nsort according to incr maxerrsqr");
printf("\nor plot %s vs %s\n",f1name,f2name);
fclose(f1);
fclose(f2);
}
```

Listing 8.4 Best Number Of Samples Routine (BESTN.C)

Digital Tone Detection 8

```c
#include <math.h>
#include <stdio.h>
#include "tones.def"

int round( x )
float x;
{
    if (x>0) return ( (int) (x+0.5) );
    else if (x<0) return ( (int) (x-0.5) );
    else if (x==0) return ( 0 );
    else printf("\7bad data in round() function!");
    return(-1);
}

int flt_to_Q15( x, txt )
float x;
char txt[];
{
    int i, err=0;

    i = round(x*32768.0);
    if (x>=1.0)    { i=0x7FFF; err=1; }
    if (x<(-1.0)) { i=0x8000; err=(-1); }
    sprintf( txt, "%08X\n", i );
    txt[0]=txt[4]; txt[1]=txt[5]; txt[2]=txt[6];
    txt[3]=txt[7]; txt[4]='\000';
    return(err);
}

main(argc,argv)
int argc;
char **argv;
{
    int i, kint;
    float N, f_sample, kflt, kerr, coef;
    char Q15coef[255];

    switch(argc) /* get missing arguments */
        {
        case 1:  printf("N > ");          scanf("%f",&N);
        case 2:  printf("f_sample > "); scanf("%f",&f_sample);
        case 3:  break;
        default: printf("\n\7usage: %s <N [%%f]> <f_sample [%%f]>\n");
        return(-1);
        }
```

(listing continues on next page)

```
switch(argc) /* read the arguments */
     {
     case 3:   sscanf(argv[2],"%f",&f_sample);
     case 2:   sscanf(argv[1],"%f",&N);
     case 1:   break;
     }
printf("\nN=%f\nfs=%f\n",N,f_sample);
for (i=0; i<HOW_MANY_TONES; i++)
     {
     kflt=N*(f_tone[i]/f_sample);
     kint=round(kflt);
     kerr=kflt-(float)(kint);
     coef=2*cos((2*PI*(float)kint)/N);
     flt_to_Q15( coef/2, Q15coef );
     printf("\nf_tone[%2d]=%10.2f Hz",i,f_tone[i]);
     printf("\n\t\tk(flt)=%10.6f",kflt);
     printf("\n\t\tk(int)=%4d",kint);
     printf("\n\t\tk(err)=%+10.6f",kerr);
     printf("\n\t\tcoef(flt)=%+10.6f coef(2.14 hex)=0x%s",coef,Q15coef); }
     printf("\n");
}
```

Listing 8.5 Coefficient Generating Routine (COEFGEN.C)

498

Digital Tone Detection   8

```
.module/ram/abs=0    Tone_Detection;
{
    This example shows digital tone detection using the Goertzel algorithm.
    This example was designed to run on the ADSP-2100 Evaluation Board.
    Actual implementation in other systems would require modifications
    such as redefining the i/o ports and the data i/o handling.

    Analog Devices, Inc. — DSP Division — Norwood, MA 02062 — 4-April-1989
}

.const   f_sample  =48600;
.const   N         =463;
.const   tones     =8;
.const   tones_x_2 =16;

.var/circ   Q1Q2_buff[tones_x_2];    { Goertzel feedback loop storage elements  }
.var        outcode;
.var        in_sample;               { input samples (scaled down 8 bits)        }
.var        countN;                  { counts samples 1, 2, 3, ..., N            }
.var        mu_lookup_table[256];    { mu-law to linear tbl(scaled down 8 bits)  }
.var        min_tone_level[tones];   { min "tone-present" mnsqr level            }
.var        mnsqr[tones];            { 1.15 mnsqr Goertzel result values         }
.var        bits[tones];             { 1.15 mnsqr Goertzel result values         }
.var/pm/ram/circ
            coefs[tones];            { 2.14 Goertzel coefs: 2*cos(2*PI*k/N)      }
.var/pm  trashbin;                   { see release note about writes to pm(I4)   }

.port       codec;                   { telephone band speech i/o on Eval. Bd.    }
.port       cntl_port;               { part of above hardware                    }
.port       dac;                     { D/A converter used to monitor decode out  }

.init       coefs[00]: h#129B00, h#F89F00, h#DEF100, h#C6A000;
.init       coefs[04]: h#B0AE00, h#9E0200, h#8F6200, h#814C00;

.init mu_lookup_table:< mu255.q8 >;
.init min_tone_level: h#0003,h#0003,h#0003,h#0003,h#0003,h#0003,h#0003,h#0003;
.init bits:           h#0001,h#0002,h#0004,h#0008,h#0010,h#0020,h#0040,h#0080;
{——————————————————————————————————————————————————————————————————————————————}
{———————————— M A I N   C O D E ————————————————————————————————————————————————}
{——————————————————————————————————————————————————————————————————————————————}
IRQ0:   rti;
IRQ1:   rti;
IRQ2:   rti;
IRQ3:   jump sample;

    i4=^trashbin;         { see release note about writes to pm(I4)              }
    call setup;
    call restart;
    imask=b#1000;         { enable IRQ3 for samples                             }
here:       jump here;
```

(listing continues on next page)

8 Digital Tone Detection

```
{---------------------------------------------------------------------}
{----------- I N T E R R U P T   S E R V I C E   R O U T I N E --------}
{---------------------------------------------------------------------}
{                           )
{------ G E T   A   S A M P L E   T O   P R O C E S S ----------------}
{                                                                     }
sample:   ax0=h#00FF;
          ax1=^mu_lookup_table;
          ay0=dm(codec);                { read codec, mu-law data     }
          af=ax0 and ay0;
          ar=ax1+af;
          i6=ar;
          si=dm(i6,m4);                 { look-up scaled, linear value }
          dm(in_sample)=si;             { store input sample          }
          i0=^Q1Q2_buff;
          i5=^coefs;

{------ D E C R E M E N T   S A M P L E   C O U N T E R --------------}
{                                                                     }
decN:     ay0=dm(countN);
          ar=ay0-1;
          dm(countN)=ar;
          if lt jump skip_backs;

{------ G O E R T Z E L   F E E D B A C K   P H A S E ---------------}
{                                                                     }
feedback:
          ay1=dm(in_sample);            {get input sample   AY1=1.15  }
          cntr=tones;
          do backs until ce;
            mx0=dm(i0,m0), my0=pm(i5,m4);  {get Q1 and COEF Q1=1.15, COEF=2.14}
            mr=mx0*my0(rnd), ay0=dm(i0,m1);  {mult, get Q2  MR=2.30,  Q2=1.15  }
            sr=ashift mr1 by +1 (hi);   {change 2.30 to 1.15          }
            ar=sr1-ay0;                 {Q1*COEF - Q2            AR=1.15 }
            ar=ar+ay1;                  {Q1*COEF - Q2 + input    AR=1.15 }

            dm(i0,m0)=ar;               {result = new Q1              }
backs:      dm(i0,m0)=mx0;              {old Q1 = new Q2              }
          rti;

{------ W H E N   F E E D B A C K   P H A S E   I S   D O N E --------}
{                                                                     }
skip_backs:
          call feedforward;
          call test_and_output;
          call restart;
          rti;
```

Digital Tone Detection　　8

```
{------------------------------------------------------------------}
{----------- S U B R O U T I N E S ------------------------------}
{------------------------------------------------------------------}

{%%%%%%%%%%%%%%%%%%% O N E   T I M E   O N L Y   S E T U P %%%%%%%%%%%%%%%%%%}
{   initializes TP3051 codec control ports (ADSP-2100 Evaluation Board),   }
{   M and L registers in address generators, and sets ICNTL to edge-sens.  }
{                                                                          }
setup:  si=0;
        dm(cntl_port)=si;

        l0 = tones_x_2;
        l1 =      0;
        l2 =      0;
        l3 =      0;
        l4 =      0;
        l5 = tones_x_2;
        l6 =      0;

        m0 =      1;
        m1 = -1;
        m4 =      1;

        icntl=b#01111;
        rts;
{%%%%%%%%%%%%%%%%%%% E V E R Y   T I M E S E T U P %%%%%%%%%%%%%%%%%%%%%%%%%%%}
{       resets pointers to top of buffers, resets counter values,     }
{       clears Goertzel feedback buffers to zero, etc                 }
{                                                                     }
restart:
        i0=^Q1Q2_buff;
        i5=^coefs;
        cntr=tones_x_2;
        do zloop until ce;
zloop:  dm(i0,m0)=0;
        ax0=N;
        dm(countN)=ax0;
        rts;
```

(listing continues on next page)

8 Digital Tone Detection

```
{%%%%%%%%% G O E R T Z E L   F E E D F O R W A R D   P H A S E %%%%%%%%%%%%}
{                                                                          }
feedforward: cntr=tones;
    i2=^mnsqr;
    do forwards until ce;
        mx0=dm(i0,m0);          { get two copies of Q1       1.15      }
        my0=mx0;
        mx1=dm(i0,m0);          { get two copies of Q2       1.15      }
        my1=mx1;
        ar=pm(i5,m4);           { get COEF                   2.14      }
        mr=0;
        mf=mx0*my1(rnd);        {  Q1*Q2                      1.15      }
        mr=mr-ar*mf(rnd);       { -Q1*Q2*COEF                 2.14      }
        sr=ashift mr1 by +1 (hi); { 2.14 -> 1.15 format conv. 1.15      }
        mr=0;
        mr1=sr1;
        mr=mr+mx0*my0(ss);      { Q1*Q1 + -Q1*Q2*COEF         1.15      }
        mr=mr+mx1*my1(rnd);     { Q1*Q1 + Q2*Q2 + -Q1*Q2*COEF 1.15      }
forwards:   dm(i2,m0)=mr1;      { store in mnsqr buffer       1.15      }
    rts;

{%%%%% T E S T   T O N E   L E V E L S   A N D   O U T P U T   C O D E %%%%%}
{                                                                          }
test_and_output:
        i3=^bits;
        i1=^min_tone_level;
        i2=^mnsqr;
        cntr=tones;
        af=pass 0;
        do thresholds until ce;
            ax1=dm(i3,m0);      { get bit position to set/clear        }
            ax0=dm(i2,m0);      { get tone mnsqr calculated value      }
            ay0=dm(i1,m0);      { get min tone level threshold value   }
            ar=ax0-ay0;         { mnsqr - min_tone_level               }
thresholds:
        if gt af=ax1 or af;
        ar=pass af;
        dm(outcode)=ar;         { write bit-coded result to output     }
        rts;
.endmod;
```

Listing 8.6 Tone Detection Routine (EXAMPLE.DSP)

Digital Control System ◼ 9
Design

9.1 OVERVIEW

The ADSP-2100 family of Digital Signal Processors is well suited for implementing complex measurement and control algorithms in embedded control systems with high sampling rates. This is mainly due to their computing speed, which is much greater than that of conventional microcontrollers and microprocessors. Typical application areas include servo motor control, process control, robot arm control, disk drive head control, flight control and general servomechanisms.

This chapter presents the implementation of several common control algorithms on the ADSP-2100 family of DSP processors, and presents software and hardware design methods as well as guidelines for designing high speed digital control systems with the ADSP-2100 family. A table of representative benchmarks for common digital control algorithms can be found at the end of this chapter.

9.2 DIGITAL CONTROL SYSTEMS OVERVIEW

A controller is a system used to control closed-loop feedback systems. It implements algebraic algorithms, such as filters and compensators, in order to regulate, correct, or change the behavior of a controlled system. Controllers can be implemented using analog or digital circuitry. A digital control system is comprised of a digital controller, the controlled plant (or system), and the necessary input/output devices. A general digital control system is shown in Figure 9.1. Note the analog-to-digital (A/D) and digital-to-analog (D/A) converters that are used to interface the digital controller with the plant (which is a continuous time system). There are several advantages to using a digital controller implementation instead of an analog one. In the case of digital controllers, complex control algorithms can be implemented in software or firmware rather than in special hardware. Digital controller designs and parameters can be changed without affecting the hardware. In digital control systems increased noise immunity is guaranteed and parameter drift is eliminated. Such systems are more reliable, maintainable, and testable. Finally, digital control systems feature reduced size, power, weight and costs.

9 Digital Control System Design

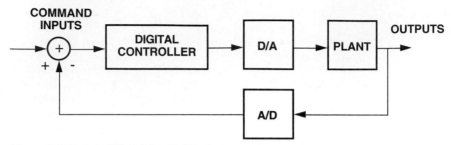

Figure 9.1 General Digital Control System

Analog Devices' ADSP-2100 family of Digital Signal Processors has several beneficial features for implementing digital controllers. These features include the following:

- Single-cycle instruction execution
- Three arithmetic function units arranged in parallel
- Single-cycle 16x16-bit multiplications
- Single-cycle ([16x16]+40 bit) multiply-accumulate operations with 40-bit results
- Single-cycle 16-bit additions, subtractions and logical operations
- Single-cycle bit shifts (up to 32 bits at a time)
- Efficient execution of 32-bit (or higher) arithmetic operations
- Efficient modulo addressing for data and coefficient arrays in memory
- No cycle penalty for looped code execution
- Single-cycle access of internal and external memory
- Single or multi-cycle parallel accesses of external peripherals (i.e., A/D, D/A)
- Up to four levels of nested external interrupts
- On-chip interval timer and serial ports
- Low power consumption (CMOS) and power down "idle" mode
- Easy-to-read algebraic assembly language syntax
- Complete set of hardware and software development tools

9.3   DIGITAL CONTROL SYSTEM MODEL

Most practical control systems use feedback in their operation. Figure 9.2 shows a model for a typical closed-loop digital control system. R(z), E(z), U(z) and Y(z) are the z-transforms of the reference input, the error signal, the control signal, and the plant output respectively. G(z) is the transfer function corresponding to the digital controller, while P(z) is the transfer function describing the input-output behavior of the object to be

Digital Control System   9
Design

controlled (e.g., plant). This does not imply that the object to be controlled (e.g., a plant) must be a discrete system, but rather that it must be modeled as one. P(z) is also assumed to contain the transfer characteristics of the A/D and the D/A converters that are needed to implement a real system.

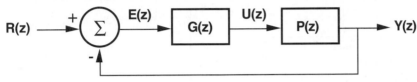

Figure 9.2 Digital Control System Model

9.4   DIGITAL CONTROL SYSTEM HARDWARE IMPLEMENTATION

A digital controller G(z), as shown in Figure 9.2, must be able to observe and alter certain characteristics of the controlled system. For example, an ADSP-2100 family-based digital controller can be used to control fast and accurate positioning of an actuator shaft upon an external command R(z). In this case, the output of the controller can be used to alter the amount of current U(z) that is fed into the actuator windings which in turn would move the actuator shaft. In a closed-loop system, the same controller would also need to observe the position of the actuator at all times. This can be achieved by recording the position Y(z) of the shaft at specific intervals and feeding it back to the controller. This would allow the controller to compare the desired shaft position to the actual measured position and make the necessary adjustments in the actuator current. This simple controller example can serve as a starting point for constructing an actual hardware implementation.

The block diagram for a digital control system based on the ADSP-2102 is shown in Figure 9.3. The ADSP-2102 performs the digital control algorithms by executing instructions from its on-chip program memory ROM. The ROM is also used to store fixed coefficients and scale factors. The processor uses its on-chip data memory RAM and program memory RAM to store incoming data values and other intermediate variables.

The ADSP-2102 accepts up to three external hardware interrupts. In a typical digital control system, the processor operation is interrupt-driven. In the system shown in Figure 9.3, an external clock (sample clock) drives one of the ADSP-2102 interrupts. The same clock is typically used to initiate A/D conversions at regular intervals. Other interrupts can be set by the host to notify the ADSP-2102 of new commands, expiration of a watchdog timer, etc.

9 Digital Control System Design

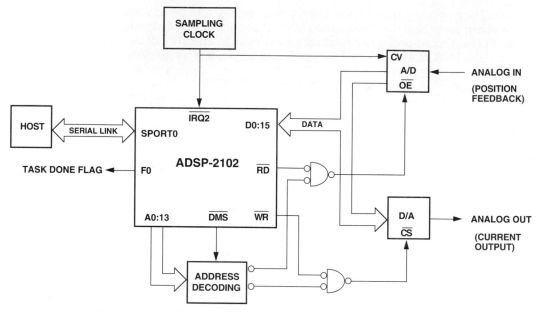

Figure 9.3 ADSP-2101-Based Actuator Controller

The ADSP-2102 outputs its control current via a D/A converter, whose output is amplified before it is fed into the motor. The processor receives its feedback from a position encoder which can be an optical shaft encoder, a synchro-to-digital converter, a resolver-to-digital converter, or some other circuitry with an A/D converter. The A/D label is used in the figure since feedback essentially involves an analog-to-digital conversion. Data transfers between the processor and the converters are done over the 16-bit data bus. The converters are mapped into the ADSP-2102's external data memory space. This allows the processor to access them as memory locations. The address decoding circuitry shown in Figure 9.3 is used to map every converter to a separate data memory location. If the converters have slow data bus interfaces, the processor can extend the duration of the converter access cycles by inserting wait states. The data converters can also be tied to the ADSP-2102's serial ports if it is more convenient to do so. Several serial input and serial output converters are available from Analog Devices, such as the AD766, AD7772, AD7868, and AD7878.

Digital Control System   9
Design

The ADSP-2102 receives its reference position command R(z) from a host processor or an internal software routine that is running concurrently with the shaft positioning program. Figure 9.3 depicts the case where serial port 0 on the ADSP-2102 is used to exchange commands and results with a host processor. The flag out pin on the ADSP-2102 can be used to notify the host of the completion of a specific task.

The ADSP-2100 family processors can interface with multiple A/D and D/A converters in order to monitor and control several motors, actuators, or processes. These converters can simply be added as memory-mapped peripherals like the ones shown in Figure 9.3. Several general purpose and special purpose data converters for digital control applications are available from Analog Devices.

9.5   DIGITAL CONTROL SYSTEM SOFTWARE IMPLEMENTATION

The software running in a digital controller system is responsible for executing the control algorithms which are represented by G(z) in the model on Figure 9.2. Typically, G(z) can be broken into smaller sub-tasks. For example it may be necessary to execute a state estimator, several notch filters and some PID (Proportional, Integral, Derivative) control as a whole function. Generally, a separate portion of the software must manage the input/output operations of the controller with the host and other peripherals. A diagnostic error checking and handling routine is also usually developed, to be run at powerup or at specified intervals during program execution. Finally there is a main manager routine that is responsible for the orchestration of these different subroutines.

The software must be organized in a modular manner in order for the main managing program to call every sub-task as a subroutine. The ADSP-2100 Family Development Software tools encourage modular programming. The subroutines can be written, assembled, and debugged as independent modules which can later be linked with the main manager program. Parameter passing and symbolic coding is supported on the assembler, linker and simulator. An example of a fully coded notch filter algorithm is shown in a later section of this application note. The ADSP-2100 Assembler and Simulator manuals describe the software tools.

Memory management is very straightforward in the ADSP-2100 family processors. The Data memory (DM) space is typically used for variables and data storage. The incoming A/D samples can be stored in data memory buffers. A large number of variables and intermediate values can

9 Digital Control System Design

also be stored in DM space. The Program memory (PM) space is divided into two sections: the PM instruction space and the PM data space. The instruction space is used to store the programs to be executed. The PM space can also be freely used for additional data and variable storage. This data space is usually used to store filter coefficients and various other tables that may need to be present during program execution. The ADSP-2100 family processors can read or write to both DM and PM locations in a single instruction cycle and execute an arithmetic operation at the same time. This not only allows classical control algorithms to execute at very high speeds but also allows very efficient implementation of adaptive control algorithms. This is due to the fact that in adaptive control, filter coefficients must be updated periodically with every new incoming sample. These coefficient values can be updated easily in the PM space and can be readily available on the next processing cycle.

The following sections discuss the implementation of first, second, and higher-order control algorithms with the ADSP-2100 family processors.

9.6   DIGITAL PID CONTROLLER DESIGN

The controller G(z) shown in Figure 9.2 can be designed to vary its output U(z) in relation to the error feedback E(z). A PID (Proportional, Integral, Derivative) controller derives its name from the fact that its output U(z) is a weighted sum of the error signal, its integral, and its derivative. PID controllers are widely-used building blocks in a large variety of servo control applications.

Since analog PID controllers are well understood, it is often desirable to start a digital PID controller design in the continuous domain and then create discrete equivalents. In the continuous time case if E(t) is the error feedback, the PID output U(t) can be expressed as:

$$U(t) = K_p \bullet E(t) + K_d \bullet dE(t)/dt + K_i \bullet \int_o{}^t E(\pi)d\pi \quad (1)$$

where K_p, K_d, and K_i are the gains associated with the proportional, derivative, and integral terms, respectively. Equation (1) can be represented in the frequency domain by using Laplace transforms:

$$U(s) = K_p \bullet E(s) + K_d \bullet s \bullet E(s) + (K_i/s) E(s) \quad (2)$$

where it is assumed that the initial conditions are 0. The equations (1) and (2) are graphically represented in Figure 9.4.

508

Digital Control System 9
Design

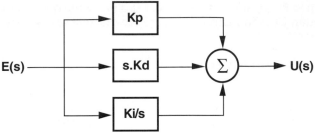

Figure 9.4 PID Block Diagram

The coefficients Kp, K_d and K_i must be determined during the design process. These coefficients will depend on the desired controller characteristics and will be varied in different systems.

Other types such as PD (Proportional, Derivative) and PI (Proportional, Integral) controllers can also be expressed in a similar manner to the PID relationships in (1) and (2). These controllers lack either the differential or the integral term that exists in the relationships above.

The next step is to derive the discrete equivalent for the controller described by equations (1) and (2). The backward difference is defined as the discrete-time equivalent for the continuous-time derivative of a function. It is obtained by:

$$\Delta f(t) = [f(t) - f(t-T)] / T \qquad\qquad (3)$$

where T is the sample period.

The definite sum is defined as the discrete time equivalent for the continuous time integral of a function. It is obtained by:

$$_a\int^b f(\tau) = T\,[\,f(a+T) + f(a+2T) + \cdots\cdots + f(b)] \qquad\qquad (4)$$

where T is the sample period.

9 Digital Control System Design

By applying the relationships shown in (3) and (4) to the ones shown in (1) and (2) we obtain U(n) and U(z), which are the discrete equivalents of the PID output U(t):

$$U(n) = U(n-1) + A_1 \cdot E(n) + A_2 \cdot E(n-1) + A_3 \cdot E(n-2) \qquad (5)$$

and

$$U(z) / E(z) = [A_1 \cdot z^2 + A_2 \cdot z + A_3] / [z^2 - z] \qquad (6)$$

with

$$A_1 = K_p + K_i \cdot T + K_d / T$$
$$A_2 = - [K_p + 2K_d / T]$$
$$A_3 = K_d / T$$

Transfer functions and difference equations for PD and PI controllers can also be derived in a similar manner. Figure 9.5 shows the results for these as well as for the PID controllers.

	PD Controller	PI Controller	PID Controller
Definitions	$A_1 = K_P + \dfrac{K_D}{T}$ $A_2 = - \dfrac{K_D}{T}$	$A_1 = K_P + K_I T$ $A_2 = - K_P$	$A_1 = K_P + K_I T + \dfrac{K_D}{T}$ $A_2 = -[K_P + 2\dfrac{K_D}{T}]$ $A_3 = \dfrac{K_D}{T}$
Transfer Function, $\dfrac{C(z)}{E(z)}$	$\dfrac{A_1 z + A_2}{z}$	$\dfrac{A_1 z + A_2}{z - 1}$	$\dfrac{A_1 z^2 + A_2 z + A_3}{z^2 - z}$
Difference Equation	$U(n) = A_1 E(n) + A_2 E(n-1)$	$U(n) = U(n-1) + A_1 E(n)$ $+ A_2 E(n-1)$	$U(n) = U(n-1) + A_1 E(n)$ $+ A_2 E(n-1) + A_3 E(n-2)$

Figure 9.5 PD, PI, & PID Controllers

Digital Control System   9
Design

9.7   PID CONTROLLER IMPLEMENTATION

An ADSP-2100 family assembly language subroutine that implements the PID algorithm is shown in Listing 9.1. There are a number of registers that need to be initialized in order to execute this subroutine. It may be sufficient to do this initialization only once (e.g. on powerup) if other algorithms that are being executed do not need to use these registers. In most typical cases, however, some of these registers may need to be set every time the PID subroutine is called.

The PID routine in Listing 9.1 takes its input from the AR register. This register must contain the 16-bit error input $E(n)$. $E(n)$ is assumed to be already computed before the PID subroutine is called. The output of the PID algorithm, $U(n)$, is made available in the SR1 register.

After the initial design of a digital PID controller, all coefficients must be scaled down by the same factor. This is necessary in order to conform to the 16-bit fixed-point fractional number format as well as to insure that overflows won't occur in the final stage of the multiply-accumulate operations. The scaled down coefficients are the ones that get stored in the processor's memory. The result of the multiply and accumulate operations is eventually scaled up before being output to the controlled system. The choice of a proper scaling factor depends greatly on the design objectives and in some cases it may even be unnecessary. The PID controller coefficients are usually designed with a commercial software package in higher precision arithmetic than 16 bits. System performance deviates from ideal when such high precision PID coefficients are quantized to 16 bits and further scaled down. In systems that require stringent PID specifications, careful simulations of quantization and scaling effects must be performed.

During the initialization for the PID routine, the scaling factor for the coefficients must be stored in the SE register. The index register I0 points to the circular data memory buffer that contains the previous error inputs and the previous PID output. This buffer must be initialized to zero at powerup unless some non-zero initial condition is desired. The index register I4, on the other hand, points to the circular program memory buffer that contains the scaled PID coefficients. These coefficients include a term "B" (for $U(n-1)$), which is equal to the value " 1/scaling factor ". This value is derived from the fact that the real coefficient for $U(n-1)$ is " 1 " and that it must be scaled down along with the other coefficients. The order that these scaled coefficients are stored in program memory is: A2, A1, A0, B.

9 Digital Control System Design

```
.MODULE      PID_CONTROLLER;

{  This is a PID controller subroutine that executes the following equation:

                 U(n) = B•U(n-1) + A0•E(n) + A1•E(n-1) + A2•E(n-2)

          Calling Parameters:

              AR=      error input E(n), [E(n) = Y(n) - R(n)]
              I0 —>    circular delay line buffer for E(n-2), E(n-1) and U(n-1)
                       this delay line buffer must be initialized to zero at powerup
              I4 —>    circular buffer for the scaled coefficients A2, A1, A0, B
              M0,M4=   1
              L0 =     3
              L4 =     4
              SE=      scaling factor for the coefficients

          Return Value:
              SR1= output sample U(n)

          Altered Registers:
              MX0, MX1, MY0, MR, SR

          Computation Time:
              ADSP2101  :    8 Instruction Cycles
              ADSP2102  :    8 Instruction Cycles
              ADSP2100  :   12 Instruction Cycles
              ADSP2100A :   12 Instruction Cycles

          All coefficients and data values are assumed to be in 1.15 format
}

.ENTRY    PID;

PID:      MX0 = DM(I0,M0), MY0 = PM(I4,M4);
          MR = MX0*MY0 (SS), MX1 = DM(I0,M0), MY0 = PM(I4,M4);
          MR = MR+MX1*MY0 (SS), MY0 = PM(I4,M4);
          MR = MR+AR*MY0(SS), MX0 = DM(I0,M0), MY0 = PM(I4,M4);
          MR = MR+MX0*MY0 (RND), DM(I0,M0)= MX1;
          SR = ASHIFT MR1 (HI), DM(I0,M0) = AR;
          DM(I0,M0) = SR1;
          RTI;

.ENDMOD;
```

Listing 9.1 PID_CONTROLLER Routine

512

Digital Control System 9
Design

The PID core routine fetches the coefficients and data values from memory following the sequence that they have been stored. These values are multiplied and accumulated until all of them are accessed. Note that both of the address generators are used in parallel with the multiply-accumulator throughout these operations. Finally, the data memory buffer is updated with the new samples and the output is obtained by scaling up the result of the multiplication and accumulation operations.

The PID routine executes in 8 instruction cycles on the ADSP-2101 and ADSP-2105. It executes in 12 instruction cycles on the ADSP-2100 and ADSP2100A. In the case that the initialization registers have been modified by other routines, it may be necessary to execute up to 7 overhead setup cycles before calling the core PID routine.

9.8   N'TH ORDER DIGITAL CONTROLLER DESIGN
There are several methods to design high order digital controllers. This section briefly outlines three approaches and cites some references on this topic. The three methodologies are "analog-controller-based digital design", "direct digital design" and "state-space design".

9.8.1   Analog-Controller-Based Digital Design
This is a very common way of designing digital controllers. In this method, an analog controller that satisfies the desired requirements is first created using well established analog design procedures. This controller is then transformed into the digital domain and implemented.

The analog controller design may be performed in the s-plane using common design techniques such as root-locus methods, Bode plots, the Routh-Hurwitz criterion, state variable techniques and other methods. The resulting analog transfer function is then transformed into a digital transfer function in the z-domain. Finally, the z-domain transfer function is inverse-z transformed into a difference equation that can be implemented on a digital processor.

The transformation from the s-domain to the z-domain can be accomplished using various techniques such as the matched pole-zero method, the bilinear (Tustin) transformation the method of mapping differentials, the impulse-invariance method, the step-invariance method, and the zero-order hold technique.

9 Digital Control System Design

The most commonly used of these methods is the bilinear transformation. This transformation approximates the s-domain transfer function with a z-domain transfer function by use of the substitution:

s = (2/T) (z-1/z+1)   (7)

Analog controllers that are in parallel or cascade maintain their respective structures after going through this transformation. The Tustin transformation maps the stable region of the s-plane exactly into the stable region of the z-plane although the entire jω-axis of the s-plane is stuffed into the 2π-length of the unit circle. Obviously a great deal of distortion takes place in the mapping in spite of the consistency of the stability regions. This distortion can be corrected by using a frequency pre-warping scheme. The pre-warping matches the single most important critical frequency in the analog domain and the digital domain. This method replaces each "s" in the analog transfer function with

(ω1 / ω_2) s

where w_1 is the frequency to be matched in the digital transfer function and with

ω_2 = (2/T) tan (ω_1T/2)   (8)

Bilinear transformation with frequency pre-warping is one of the most commonly used analog based design techniques. The most significant drawback of this type of design is that the digital controller that results from it is only an approximation to the analog one. The analog controller puts an implicit upper bound on the digital controller's performance. More information on the other analog based controller design methods can be found in References 1 and 3.

9.8.2   Direct Digital Design
This method allows us to perform the control system design directly in the digital domain. Thus, the design can be carried out in the z-plane. The approximations and limitations that arise from starting in the s-domain and transforming into the z-domain are eliminated. Conventional design techniques can be used to place the closed-loop poles and zeros exactly where appropriate. Some of these z-domain techniques include the root-locus method, the pole-zero cancellation method and the w-transform. More detailed information on these methods can be found in References 1 and 3.

Digital Control System 9
Design

9.8.3   State-Space Design

The digital controller design methods discussed above are designated as classical design methods. Same design tasks can be accomplished by using a different set of techniques based on the state-space or modern control formulation. Modern control design methodology is especially advantageous when designing controllers for multi-input and multi-output systems. However, single-input and single-output systems that are discussed in this application note can also be efficiently designed using state-space methods. More detailed information on this topic can be found in References 1 and 3.

9.9   N'TH ORDER DIGITAL CONTROLLER STRUCTURES

Standard second-order (N=2) digital controller implementation is directly analogous to IIR (Infinite Impulse Response) filter implementations. These second-order controller blocks can be implemented as biquad second order IIR filter sections. A second-order biquad section is shown on Figure 9.6 and its corresponding transfer function in the z-domain is given by:

$$G(z) = U(z)/E(z) = (B_0 + B_1 \bullet z^{-1} + B_2 \bullet z^{-2})/(1 + A_1 \bullet z^{-1} + A_2 \bullet z^{-2}) \qquad (9)$$

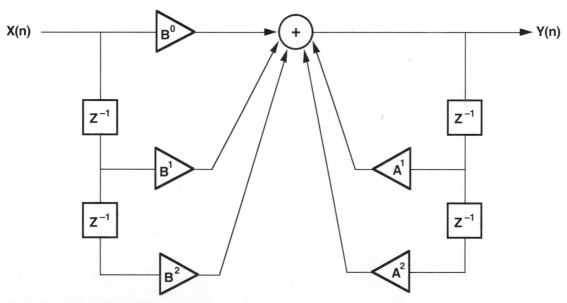

Figure 9.6 Second-Order Biquad Structure

9 Digital Control System Design

where A_1, A_2, B_0, B_1 and B_2 are coefficients that determine the desired impulse response of the system $G(z)$. Furthermore, the corresponding difference equation for a biquad section is given by:

$$U(n) = B_0 \bullet E(n) + B_1 \bullet E(n-1) + B_2 \bullet E(n-2) - A_1 \bullet U(n-1) - A_2 \bullet U(n-2) \qquad (10)$$

Higher order (N'th order) controllers can be obtained by cascading several biquad sections with appropriate coefficients. An example is shown on Figure 9.7 where three biquad sections are cascaded to construct the overall $G(z)$ transfer function. Another way to design higher order controllers is to use only one complicated single section. This approach is also called the direct form implementation. The block diagram of a direct form fourth-order controller is shown on Figure 9.8 as an example.

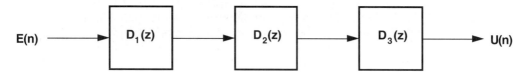

$$G(z) = \frac{U(z)}{E(z)} = D_1(z) \bullet D_2(z) \bullet D_3(z)$$

Figure 9.7 Cascaded Biquad Sections

The direct form implementation executes faster but generates larger numerical errors than the biquad implementation. The biquads can be scaled separately and then cascaded in order to minimize the coefficient quantization and the recursive accumulation errors. The coefficients and data in the direct form implementation must be scaled all at once, which gives rise to larger errors. Another disadvantage of the direct form implementation is that the poles of such single stage high order polynomials get increasingly sensitive to quantization errors. The second-order polynomial sections (i.e., biquads) are less sensitive to quantization effects.

Digital Control System   9
Design

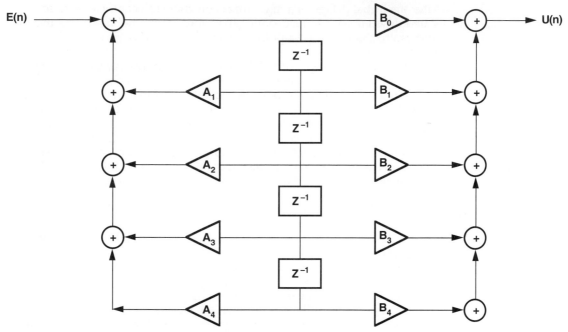

Figure 9.8 Fourth-Order Direct Form Controller

9.10   N'TH ORDER CONTROLLER IMPLEMENTATION

An ADSP-2100 family assembly language subroutine that implements a
high order controller is shown in Listing 9.2. The subroutine is arranged as
a module and is labeled BIQUAD_CONTROLLER. There are a number of
registers that need to be initialized in order to execute this subroutine. It
may be sufficient to do this initialization only once (e.g. on powerup) if
other executed algorithms do not need these registers. In most typical
cases, however, some of these registers may need to be set every time the
BIQUAD_CONTROLLER subroutine is called. It may sometimes be
beneficial, from a modular software point of view, to always initialize all
the setup registers as a part of this subroutine.

The BIQUAD_CONTROLLER routine in Listing 9.2 takes its input from
the SR1 register. This register must contain the 16 bit error input E(n). E(n)
is assumed to be already computed before this subroutine was called. The
output of the controller is also made available in the SR1 register.

517

9 Digital Control System Design

After the initial design of a high order controller, all coefficients must be scaled down in each biquad stage separately. This is necessary in order to conform to the 16-bit fixed-point fractional number format as well as to insure that overflows won't occur in the final multiply-accumulate operations in each stage. The scaled down coefficients are the ones that get stored in the processor's memory. The operations in each biquad are performed with scaled data and coefficients and are eventually scaled up before being output to the next one. The choice of a proper scaling factor depends greatly on the design objectives and in some cases it may even be unnecessary. The controller coefficients are usually designed with a commercial software package in higher precision arithmetic than 16 bits. System performance deviates from ideal when such high precision coefficients are quantized to 16 bits and further scaled down. In systems that require stringent specifications, careful simulations of quantization and scaling effects must be performed.

During the initialization of the BIQUAD_CONTROLLER routine, the index register I0 points to the data memory buffer that contains the previous error inputs and the previous biquad section outputs. This buffer must be initialized to zero at powerup unless some non-zero initial condition is desired. The index register I1 points to another buffer in data memory that contains the individual scale factors for each biquad. The buffer length register L1 is set to zero if the controller has only one biquad section. L1 is initialized with the number of biquad sections in the case of multiple biquads. The index register I4, on the other hand, points to the circular program memory buffer that contains the scaled biquad coefficients. These coefficients are stored in the order : B2, B1, B0, A2, A1 for each biquad. All of the individual biquad coefficient groups must be stored in the same order that the biquads were cascaded in such as: B2, B_1, B_0, A_2, A_1, B_2^*, B_1^*, B_0^*, A_2^*, A_1^*, B_2^{**}...etc The buffer length register L4 must be set to the value given by: (5 x number of biquad sections). Finally, the loop counter register "CNTR" must be set to the number of biquad sections since the controller code will be executed as a loop.

The core of the BIQUAD_CONTROLLER routine starts its execution at the "biquad" label. The routine is organized in a looped fashion where the end of the loop is the instruction labeled "sections". Each iteration of the loop executes the computations for one biquad. The number of loops to be executed is determined by the CNTR register contents. The SE register is loaded with the appropriate scaling factor for the particular biquad at the beginning of each loop iteration. After this operation, the coefficients and the data values are fetched from memory in the sequence that they have

Digital Control System 9
Design

```
.MODULE              BIQUAD_CONTROLLER;

{   This is an Nth order cascaded biquad controller subroutine

            Calling Parameters:

                    SR1= error input E(n),  [ E(n) = Y(n) - R(n) ]
                    I0 -> delay line buffer for E(n-2), E(n-1), Y(n-2), Y(n-1)
                    L0 = 0
                    I1 -> scaling factors for each biquad section
                    L1 = 0  (in the case of a single biquad)
                    L1 = number of biquad sections (for multiple biquads)
                    I4 -> scaled biquad coefficients
                    L4 = 5 x [number of biquads]
                    M0,M4 = 1
                    M1 = -3
                    M2 = 1 (in the case of multiple biquads)
                    M2 = 0 (in the case of a single biquad)
                    M3 = (1 - length of delay line buffer)

            Return Value:
                    SR1 = output sample U(n)

            Altered Registers:
                    SE, MX0, MX1, MY0, MR, SR

            Computation Time (with N even):
                    ADSP-2101/2102:    (8 x N/2) + 5 instruction cycles
                    ADSP-2100/2100A:   (8 x N/2) + 5 + 5 instruction cycles

            All coefficients and data values are assumed to be in 1.15 format
}

.ENTRY               CONTROLLER;

BIQUAD:              CNTR = number of biquads
                     DO SECTIONS UNTIL CE;
                         SE = DM(I1,M2);
                         MX0 = DM(I0,M0), MY0 = PM(I4,M4);
                         MR = MX0*MY0(SS), MX1 = DM(I0,M0), MY0 = PM(I4,M4);
                         MR = MR+MX1*MY0(SS), MY0=PM(I4,M4);
                         MR = MR+SR1*MY0(SS), MX0 = DM(I0,M0), MY0 = PM(I4,M4);
                         MR = MR+MX0*MY0(SS), MX0 = DM(I0,M1), MY0 = PM(I4,M4);
                         DM(I0,M0) = MX1, MR = MR+MX0*MY0(RND);
SECTIONS:                DM(I0,M0) = SR1, SR = ASHIFT MR1 (HI);
                     DM(I0,M0) = MX0;
                     DM(I0,M3) = SR1;
                     RTS;
.ENDMOD;
```

Listing 9.2 BIQUAD_CONTROLLER Routine

9 Digital Control System Design

been stored. These numbers are multiplied and accumulated until all of the values for a particular biquad have been accessed. The result of the last multiply accumulate is rounded to 16 bits and upshifted by the scaling value. At this point the "biquad" loop is executed again or the controller computations are completed by doing the final update to the delay line. The delay lines for data values are always being updated within the biquad loop as well as outside of it.

The controller coefficients must be scaled appropriately so that no overflows occur after the upshifting operation between the biquads. If this is not insured by design, it may be necessary to include some overflow checking between the biquads.

The execution time for an N'th order BIQUAD_CONTROLLER routine can be calculated as follows (assuming that the appropriate registers have been initialized and N is a power of 2):

ADSP2101/2105   : $(8 \times N/2) + 4$ processor cycles
ADSP2100/2100A   : $(8 \times N/2) + 4 + 5$ processor cycles

It may take up to a maximum of 11 cycles to initialize the appropriate registers every time the controller is called. But typically this number will be lower in most applications.

9.11   NOTCH FILTER EXAMPLE FOR THE ADSP-2100A

A fully coded ADSP-2100A notch filter program is presented in this section. The program executes two cascaded biquad sections and is designed to run standalone on an ADSP-2100A. The processor reads the input samples from a data memory mapped A/D converter and sends the filtered output to a data memory mapped D/A converter. The operation of the ADSP-2100A is interrupt driven. The occurrence the IRQ0 interrupt notifies the processor that there is a new sample ready at the A/D converter output. The ADSP-2100A normally waits in a wait loop, processes the incoming samples upon an interrupt and returns to the wait loop again. This process starts with a reset or powerup and repeats until a powerdown or another reset occurs. The full code of the notch filter is shown in Listing 9.3.

The program starts with some variable, constant and port declarations. These declarations allow the program to refer to specific memory addresses symbolically. This greatly eases the software maintenance and debugging tasks at the assembly level. In order to run this program, the NOTCH_FILTER assembly module shown in Listing 9.3 must be first assembled using the ADSP-2100 assembler. It must then be linked with an architecture description file that was built using the ADSP-2100 system builder. This architecture file should be specific to the particular hardware configuration that the ADSP-2100A is being built in. The assembly module also needs to be linked with three data files along with the architecture file. The first data file must contain the scale factors, the second one must contain the scaled coefficients and the third one must contain the initial values for the delay taps of the filter.

```
{THIS IS AN ADSP2100A ASSEMBLY PROGRAM THAT EXECUTES A FOURTH}
{ORDER NOTCH FILTER IN A CASCADED BIQUAD IMPLEMENTATION}

.MODULE/RAM           NOTCH_FILTER;          {The name of the module}

.VAR/PM/CIRC          COEFFICIENTS[10];  {These are the declarations for}
.VAR/DM               DATA_BUFFER[6];    {data and program memory buffers}
.VAR/DM/CIRC          SCALE_FACTORS[2];

.PORT                 AD_CONVERTER;          {There is one input port and one}
.PORT                 DA_CONVERTER;          {output port in the system}

.INIT                 COEFFICIENTS: <COEFF.DAT>;       {The memory buffers}
.INIT                 DATA_BUFFER: <INITIAL.DAT>;      {are initialized here}
.INIT                 SCALE_FACTORS: <SCALE.DAT>;

              JUMP BIQUAD;                {Interrupt vector for IRQ0}
              RTI;
              RTI;
              RTI;

              I0 = ^DATA_BUFFER;          {These are the proper initializations}
              L0 = 0;                     {for the index, length and modify}
              I1 = ^SCALE_FACTORS;        {registers to be used}
              L1 = %SCALE_FACTORS;
              I4 = ^COEFFICIENTS;
              L4 = %COEFFICIENTS;
              M0 = 1;
              M1 = -3;
              M2 = 1;
              M3 = -5;
              M4 = 1;
              ICNTL = B#00001;            {Set IRQ0 to be edge sensitive}
              IMASK = B#0001;             {Enable IRQ0 interrupt only}
WAIT:         JUMP WAIT;                  {Wait for IRQ0 interrupt to occur}
```

(listing continues on next page)

9 Digital Control System Design

{The interrupt service routine below executes the two biquad sections of the filter}

```
BIQUAD:             SR1=DM(AD_CONVERTER);    {Read the A/D converter}
                    CNTR = %SCALE_FACTORS;
                    DO SECTIONS UNTIL CE;
                        SE = DM(I1,M2);
                        MX0 = DM(I0,M0), MY0 = PM(I4,M4);
                        MR = MX0*MY0(SS), MX1 = DM(I0,M0), MY0 = PM(I4,M4);
                        MR = MR+MX1*MY0(SS), MY0=PM(I4,M4);
                        MR = MR+SR1*MY0(SS), MX0 = DM(I0,M0), MY0 = PM(I4,M4);
                        MR = MR+MX0*MY0(SS), MX0 = DM(I0,M1), MY0 = PM(I4,M4);
                        DM(I0,M0) = MX1, MR = MR+MX0*MY0(RND);
SECTIONS:               DM(I0,M0) = SR1, SR = ASHIFT MR1 (HI);
                    DM(I0,M0) = MX0;
                    DM(I0,M3) = SR1;
                    DM(DA_CONVERTER) = SR1; {Send the filtered output to the D/A}
                    RTI;                    {Return to the wait loop}
.ENDMOD;
```

Listing 9.3 NOTCH_FILTER Routine

ADSP-2100 Family Benchmarks For Digital Control Applications		
	ADSP-2101 (60 ns)	ADSP-2105 (100 ns)
PID Loop	0.48 µs	0.8 µs
FIR Filter	60 ns/tap	100 ns/tap
IIR Biquad – 16 Bit (2nd Order)	0.72 µs	1.2µs
IIR Biquad – 16 Bit (nth Order)	4N + 4 cycles	4N + 9 cycles
IIR Biquad – 32 Bit (2nd Order)	2.64 µs	4.4 µs
IIR Biquad – 32 Bit (nth Order)	16N + 12 cycles	16N + 22 cycles
256-Point FFT (Complex)	0.405 ms	0.675 ms
Matrix Multiply (3x3 * 3x1)	3.12 µs	5.2 µs
Stochastic Gradient (LMS) N Tap Coefficient Update	2N + 9 cycles	2N + 9 cycles

Digital Control System Design   9

9.12   REFERENCES

Borrie, J. A. 1986. "Modern Control Systems". Englewood Cliffs, NJ: Prentice-Hall Inc.

Franklin, G.F., J.D. Powell and M.L. Workman 1990. "Digital Control of Dynamic Systems". Reading, MA: Addison-Wesley Publishing Company.

Kazanzides, P. 1985. "A Microprocessor Based Control System with Robotics Applications". Brown University, LEMS, Providence, RI.

Oppenheim, A. V. and A. Willsky 1983. "Signals and Systems". Englewood Cliffs, NJ: Prentice-Hall, Inc.

Variations On IIR Biquad ■ 10
Filters

10.1 OVERVIEW

Digital Signal Processing Applications Using the ADSP-2100 Family, Volume 1, contains a chapter about digital filters. That chapter (Chapter 5) includes information about second-order sections of Infinite Impulse Response, or IIR, filters. The particular second-order sections discussed in Volume 1, are commonly referred to as biquads.

This chapter includes the following variations on the basic IIR biquad filter and the filter subroutines described in Volume 1:

- Multiprecision filters
- Optimized filter subroutines

10.1.1 IIR Biquad Filter

Figure 10.1 shows the structure of a second-order biquad IIR filter section. You can design IIR filters of an order greater than two by cascading multiple second-order biquad IIR filter sections. The filter sums the products of the current input, x(n), the previous two inputs, x(n-1) and x(n-2), the past two results, y(n-1) and y(n-2), and their respective coefficients, b0, b1, b2, a1 and a2. The biquad has a necessary coefficient scaling factor when one or more of the coefficients are greater than 1.0.

10.1.2 Biquad Filter Subroutine

Listing 10.1 is the subroutine from *Digital Signal Processing Applications Using the ADSP-2100 Family*, Volume 1, for a basic biquad filter. This filter has a 16-bit input, 16-bit output, and 16-bit coefficients. This code lets the ADSP-2100 Family DSP perform a Nth-order IIR filter by performing N/2 biquads. The execution time is [8*(N/2) + 10] instruction cycles. For example, a tenth-order filter executes in 50 instruction cycles, or in 3 µs using a DSP with 60 ns cycle time. The DSP can perform a tenth-order IIR filter on a signal sampled at more than 300 kSa/s.

10 Variations On IIR Biquad Filters

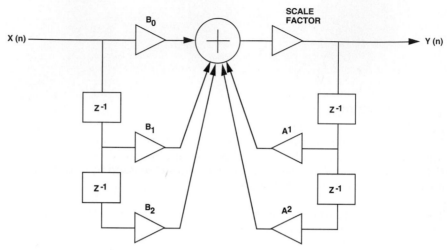

Figure 10.1 Second-Order Biquad IIR Filter Section

```
.MODULE biquad_sub;
{      Cascaded Biquad IIR Filter Subroutine
       Calling Parameters
              SR1 = input sample
              I0 -> delay line buffer
              L0 = 0
              I1 -> list of scale factors for each biquad section
              L1 = 0 (in the case of a single biquad)
              L1 = N/2 where N is the filter order
              I4 -> scaled coefficients b2,b1,b0,a2,a1, b2,b1,b0,a2,a1,
              L4 = 5 * N/2
              M0,M4 = 1
              M1 = -3
              M2 = 1 (in the case of multiple biquads)
              M2 = 0 (in the case of a single biquad)
              M3 = (1 - length of delay line buffer)
              CNTR = number of biquad sections
       Return Values
              SR1 = output sample
              I0 -> inside delay line buffer
              I1 -> top of scale factor list
              I4 -> top of coefficients
```

Variations On IIR Biquad 10
Filters

```
        Altered Registers
            MX0,MX1,MY0,MR,SE,SR
        Computation Time
            8 * number of biquad sections + 10 cycles
        All coefficients and data values are assumed to be in 1.15 format. }

.CONST N = 3;                      {number of biquad sections, example: 3}
.CONST N_x_5 = 15;                 {number of biquad sections times five}

.VAR/DM delayline[4];              {this is scratchpad memory}
.VAR/DM scalelist[N];              {initialize scale factor for each biquad}
.VAR/PM coefflist[N_x_5];          {init with filter coefficients for each biquad}

.ENTRY biquad;
biquad: I0=^delayline;
     DO sections UNTIL CE;
            SE=DM(I1,M2);
            MX0=DM(I0,M0), MY0=PM(I4,M4);                       {get x(n-2), b2}
            MR=MX0*MY0(SS), MX1=DM(I0,M0), MY0=PM(I4,M4);   {get x(n-1), b1}
            MR=MR+MX1*MY0(SS), MY0=PM(I4,M4);                {get b0}
            MR=MR+SR1*MY0(SS), MX0=DM(I0,M0), MY0=PM(I4,M4);{get y(n-2), a2}
            MR=MR+MX0*MY0(SS), MX0=DM(I0,M1), MY0=PM(I4,M4);{get y(n-1), a1}
            DM(I0,M0)=MX1, MR=MR+MX0*MY0(RND);        {store x(n-1) as new x(n-2)}
sections: DM(I0,M0)=SR1, SR=ASHIFT MR1 (HI);          {store x(n) as new x(n-1)}
     DM(I0,M0)=MX0;
     DM(I0,M3)=SR1;
     RTS;
.ENDMOD;
```

Listing 10.1 Basic Biquad Filter Subroutine

10.2   MULTIPRECISION FILTERS

When your calculations require more precision than 16-bit arithmetic
provides, you can still use ADSP-2100 family 16-bit DSPs. These DSPs
have architectural features that make multiprecision calculations possible.
Unlike other DSPs, the ADSP-2100 family processors let you multiply
mixed-mode numbers (signed numbers x unsigned numbers). This
chapter shows you how to take advantage of the multiprecision features of
the ADSP-2100 Family.

10 Variations On IIR Biquad Filters

Multiprecision filters can have input data, output data, delay line data, or coefficients greater than 16-bits. While a 16-bit IIR filter is adequate for most applications, use multiprecision filters under the following conditions:

- The filter has a small passband or stopband relative to the sample rate
- You require more than 16-bit precision on the delay line, or the delay line and the coefficients

For example, consider a graphic equalizer where the sample rate is 44.1 kHz. The lowest equalizer band is centered at 31 Hz with stop bands at 0 Hz and 62 Hz. Although this filter is second-order, both the delay line and the coefficients require 32-bit arithmetic for the desired accuracy.

ADC converters, like the ones in the example graphic equalizer, are now available with 18-bit and 20-bit resolution. A digital filter with 32-bit accuracy preserves the arithmetic precision of the filter algorithm. This lets the DSP programmer maintain the signal-to-noise ratio delivered by the ADC.

10.2.1 Multiprecision Multiplication On ADSP-2100 Family DSPs

Multiprecision filters require multiprecision multiplication. ADSP-2100 Family DSPs include a 16-bit multiplier/accumulator with a 40-bit result register that is most efficient when working with 16-bit inputs. When you multiply 32-bit or 48-bit inputs, multiplication is accomplished by breaking the inputs into 16-bit components.

The code segment in Listing 10.2 is an example of multiplying two 32-bit inputs. Normally, when you multiply two 32-bit values, you get a 64-bit result. In this example, only 32 bits are required. To save instruction cycles and properly scale the partial products of each multiplication, the code shifts the contents of the multiply results registers "on-the-fly" to overwrite the lower 16 bits. This is more efficient than using processor cycles to write the contents to the barrel shifter.

Registers MX0 and MX1 contain the least significant word (lsw) and most significant word (msw) of one input. Registers MY0 and MY1 contain the least and most significant words of the second input.

```
MR=MX0*MY0(UU);     {multiply unsigned lsws}
MR0=MR1;            {shift product 16-bits right}
MR1=MR2;
MR=MR+MX1*MY0(SU);  {multiply signed msws times unsigned lsws}
MR=MR+MX0*MY1(US);  {and accumulate with shifted lsw product}
MR0=MR1;            {shift product 16-bits right}
MR1=MR2;
MR=MR+MX1*MY1(SS);  {multiply signed msws and accumulate with}
                    {shifted intermediate product}
```

Listing 10.2 Double-Precision Multiply Routine

To generate the product, the code in Listing 10.2 performs the following operations, which are also shown in Figure 10.2.

1. It multiplies the unsigned contents of registers MX0 (lower 16 bits) and MY0 (lower 16 bits). The 32-bit product is put in the multiplier/ accumulator results register MR (MR0, MR1, and MR2).

2. The contents of MR1 are shifted right 16 bits to MR0, this overwrites, or eliminates, the lower 16 bits. The contents of MR2 are shifted right 16 bits to MR1.

3. Next, it multiplies the signed contents of register MX1 (upper 16 bits) and unsigned contents of register MY0 (lower 16 bits), and accumulates this product with the contents of MR.

4. It then multiplies the unsigned contents of register MX0 (lower 16 bits) and the signed contents of register MY1 (upper 16 bits), and accumulates this product with the contents of MR.

5. This product is shifted 16 bits right, as described in step 2.

6. Finally, it multiplies the signed contents of registers MX1 (upper 16 bits) and MY1 (upper 16 bits), and accumulates this 32-bit product with the contents of MR.

This method assumes that the inputs to the filter are twos complement fractional where the most significant bit weighting is -2^0, or -1. Multiplying two 32-bit numbers generates a 64-bit product. With fractional products, the most significant 32 bits are saved; the lower 32 bits are used to properly scale the partial products and then are overwritten by MR1 (MR0=MR1) during the 16-bit right shifts. The 32-bit fractional product is in the accumulator with the msw in register MR1 and the lsw in register MR0.

529

10 Variations On IIR Biquad Filters

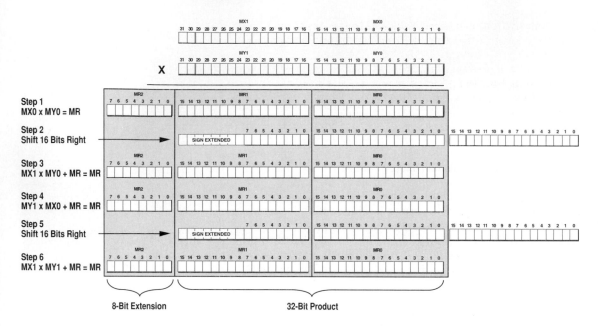

Figure 10.2 Multiprecision Multiplication Of 32-Bit Numbers

The technique used in Listing 10.2 can be applied to multiprecision IIR biquads. For example, an IIR biquad has five product terms that are accumulated:

```
y(n) = b0*x(n) + b1*x(n-1) + b2*x(n-2) + a1*y(n-1) + a2*y(n-2)
```

Instead of multiplying each 32-bit delay line input and coefficient pair one at a time, follow these steps to calculate a 32-bit delay line/coefficient biquad:

1. First, accumulate the lsw products of the five delay line/coefficient pairs.

2. Then, shift the accumulator 16-bits right.

3. Next, accumulate the msw/lsw and lsw/msw products.

4. Again, shift the accumulator 16-bits right.

5. Finally, accumulate the msw products.

530

Variations On IIR Biquad   10
Filters

In this method, the accumulator is shifted only twice for each biquad. As a result, this technique is more efficient than generating the full product for each pair and accumulating the full products.

10.2.2 Double-Precision Biquad

Listing 10.3 shows a double-precision IIR biquad subroutine with a 32-bit delay line and 32-bit coefficients. The calling routine initializes registers with the following items:

- 32-bit input sample
- Start addresses of the delay line
- Coefficient
- Scaling factor buffers
- Number of sections

Also, the calling routine sets the data address generator (DAG) length and modify registers to support modulo addressing. When you initialize registers in the calling routine, the biquad subroutine is reusable.

The code only clears the delay line buffer once before the first filter call. After the filter subroutine is called, the code stores delay line data in the buffer (lsw first, then msw). The filter coefficients are stored in the reverse order of the delay line data (msw first). After completing the calculations, the routine returns a 32-bit result in the shifter result registers SR1 and SR0.

This routine uses the circular addressing mode to retrieve and store data efficiently into the delay line. The starting address of the buffer in register I0 changes each time the filter is called, therefore, this address must be saved to memory if other routines use I0.

The core loop of Listing 10.3, starting at "DO biq UNTIL CE: ", groups line of code according to function. The comment above each group describes its function.

The execution time for this code is $[28*(N/2) + 10]$ instruction cycles. A tenth-order filter executes in 150 instruction cycles, or in 9 μs using a DSP with 60 ns cycle time. The DSP can perform a tenth-order IIR filter on a signal sampled at more than 100 kSa/s.

10 Variations On IIR Biquad Filters

```
.MODULE/RAM/BOOT=0                 dpiir_2p2z;
{
     Nth Order IIR Filter Constructed From N/2 Biquads of the Form:
          y(n)=b0*x(n)+b1*x(n-1)+b2*x(n-2)+a1*y(n-1)+a2*x(n-2)

     Where:
          x(n), x(n-1), x(n-2), y(n-1), y(n-2) are 32-bits
          b0, b1, b2, a1, a2 are 32-bits

     Calling Parameters:
          AX0=least significant word (LSW) of input
          AX1=most significant word (MSW) of input
          I0=start address of delay line in data memory stored in
               LSW,MSW order [organized x(n-1),x(n-2),y(n-1),y(n-2)]
               buffer length = (4*N/2)+4 where N is the order of the
               filter - must be declared circular
          I4=start address of coefficients in program memory stored
               in MSW,LSW order [organized b0,b1,b2,a1,a2]
               buffer length = 10*N/2 - circular declaration not required
          I5=start address of coefficient scaling factors in program
               memory (one/section)
               buffer length = N/2 - circular declaration not required
          L0=(4*N/2)+4
          L4=0
          L5=0
          M0=-1, M1=1, M2=2, M3=-6
          M5=1, M6=2, M7=-9
          CNTR=N/2

     Return Values:
          I0=new start address of delay line must be saved to memory
          SR1=MSW result
          SR0=LSW result

     Altered Registers:
          MX0,MY0,MR,SE,SR

     Computation Time:
          10 + (28 * N/2)


     Coefficients, delay line, input sample in 1.31 (Q.31) format
}
```

Variations On IIR Biquad 10 Filters

```
.ENTRY                  biq32;

biq32: SR1=AX1;                         {transfer input to SR}
       SR0=AX0;
       MODIFY(I4,M5);
       MY0=PM(I4,M6);                   {read first coefficient}
       SE=PM(I5,M5);                    {read first scaling factor}

       DO biq UNTIL CE;                 {set up biquad loop}

              {multiply/accumulate LSW*LSW}

          MR=SR0*MY0(UU),     MX0=DM(I0,M2), MY0=PM(I4,M6);
          MR=MR+MX0*MY0(UU),  MX0=DM(I0,M2), MY0=PM(I4,M6);
          MR=MR+MX0*MY0(UU),  MX0=DM(I0,M2), MY0=PM(I4,M6);
          MR=MR+MX0*MY0(UU),  MX0=DM(I0,M3), MY0=PM(I4,M7);
          MR=MR+MX0*MY0(UU),  MY0=PM(I4,M5);

          MR0=MR1;            {16-bit right shift}
          MR1=MR2;

              {multiply/accumulate LSW*MSW, MSW*LSW}

          MR=MR+SR0*MY0(US),  MY0=PM(I4,M5);
          MR=MR+SR1*MY0(SU),  MX0=DM(I0,M1), MY0=PM(I4,M5);
          MR=MR+MX0*MY0(US),  MX0=DM(I0,M1), MY0=PM(I4,M5);
          MR=MR+MX0*MY0(SU),  MX0=DM(I0,M1), MY0=PM(I4,M5);
          MR=MR+MX0*MY0(US),  MX0=DM(I0,M1), MY0=PM(I4,M5);
          MR=MR+MX0*MY0(SU),  MX0=DM(I0,M1), MY0=PM(I4,M5);
          MR=MR+MX0*MY0(US),  MX0=DM(I0,M1), MY0=PM(I4,M5);
          MR=MR+MX0*MY0(SU),  MX0=DM(I0,M1), MY0=PM(I4,M5);
          MR=MR+MX0*MY0(US),  MX0=DM(I0,M3), MY0=PM(I4,M7);
          MR=MR+MX0*MY0(SU),  MY0=PM(I4,M6);

          MR0=MR1;            {16-bit right shift}
          MR1=MR2;
```

(listing continues on next page)

10 Variations On IIR Biquad Filters

```
                    {multiply/accumulate MSW*MSW}

         MR=MR+SR1*MY0(SS),  MX0=DM(I0,M2),  MY0=PM(I4,M6);
         MR=MR+MX0*MY0(SS),  MX0=DM(I0,M2),  MY0=PM(I4,M6);
         MR=MR+MX0*MY0(SS),  MX0=DM(I0,M2),  MY0=PM(I4,M6);
         MR=MR+MX0*MY0(SS),  MX0=DM(I0,M3),  MY0=PM(I4,M6);
         MR=MR+MX0*MY0(SS),  MX0=DM(I0,M1),  MY0=PM(I4,M5);

              {apply scale factor, read next scale factor}

         SR=ASHIFT MR1 (HI),  MY0=PM(I4,M6);
         SR=SR OR LSHIFT MR0 (LO),  SE=PM(I5,M5);

              {store new y(n) to delay line}

         DM(I0,M1)=SR0;
biq:     DM(I0,M1)=SR1;

         MODIFY(I0,M2);
         DM(I0,M1)=AX0;       {store original input to delay line}
         DM(I0,M0)=AX1;

         RTS;

.ENDMOD;
```

Listing 10.3 Double-Precision IIR Biquad Subroutine

Listing 10.4 is an optimized version of Listing 10.3. In Listing 10.4, the lsw multiply/accumulates are eliminated, reducing the feedback accuracy from 32-bits to 31-bits. In most cases, the code in listing 10.4 provides adequate performance.

The execution time is $[20*(N/2) + 9]$ instruction cycles. A tenth-order filter executes in 109 instruction cycles, or in 6.54 µs using a DSP with 60 ns cycle time. The DSP can perform a tenth-order IIR filter on a signal sampled at more than 150 kSa/s.

534

Variations On IIR Biquad 10 Filters

```
.MODULE/RAM/BOOT=0              dpiir_2p2z_optimized;
{
      IIR Filter of the Form:

            y(n)=b0*x(n)+b1*x(n-1)+b2*x(n-2)+a1*y(n-1)+a2*x(n-2)

      Where:
            x(n), x(n-1), x(n-2), y(n-1), y(n-2) are 32-bits
            b0, b1, b2, a1, a2 are 32-bits

      Calling Parameters:
            AX0=least significant word (LSW) of input
            AX1=most significant word (MSW) of input
            I0=start address of delay line in data memory stored in
                  LSW,MSW order [organized x(n-1),x(n-2),y(n-1),y(n-2)]
                  buffer length = (4*N/2)+4 where N is the order of the
                  filter - must be declared circular
            I4=start address of coefficients in program memory stored
                  in MSW,LSW order [organized b0,b1,b2,a1,a2]
                  buffer length = 10*N/2 - circular declaration not required
            I5=start address of coefficient scaling factors (one/section)
                  buffer length = N/2 - circular declaration not required
            L0=(4*N/2)+4
            L4=0
            L5=0
            M0=-1, M1=1, M2=2, M3=-6
            M5=1, M6=2, M7=-9
            CNTR=N/2

      Return Values:
            I0=new start address of delay line must be saved to memory
            SR1=MSW result
            SR0=LSW result

      Altered Registers:
            MX0,MY0,MY1,MR,SE,SR

      Computation Time:
            9 + (20 * N/2)

      Coefficients, delay line, input sample in 1.31 (Q.31) format
}
```

(listing continues on next page)

10 Variations On IIR Biquad Filters

```
ENTRY          opt_biq32;

opt_biq32:

    SR1=AX1;                    {transfer input to SR}
    SR0=0;
    MY0=PM(I4,M5);              {read first coefficient}
    MY1=PM(I4,M5);              {read second coefficient}
    DO biq UNTIL CE;           {set up biquad loop}


            {multiply/accumulate LSW*LSW, read scale factor}

        MR=SR0*MY0(US),    SE=PM(I5,M5);
        MR=MR+SR1*MY1(SU), MX0=DM(I0,M1), MY0=PM(I4,M5);
        MR=MR+MX0*MY0(US), MX0=DM(I0,M1), MY0=PM(I4,M5);
        MR=MR+MX0*MY0(SU), MX0=DM(I0,M1), MY0=PM(I4,M5);
        MR=MR+MX0*MY0(US), MX0=DM(I0,M1), MY0=PM(I4,M5);
        MR=MR+MX0*MY0(SU), MX0=DM(I0,M1), MY0=PM(I4,M5);
        MR=MR+MX0*MY0(US), MX0=DM(I0,M1), MY0=PM(I4,M5);
        MR=MR+MX0*MY0(SU), MX0=DM(I0,M1), MY0=PM(I4,M5);
        MR=MR+MX0*MY0(US), MX0=DM(I0,M3), MY0=PM(I4,M7);
        MR=MR+MX0*MY0(SU), MY0=PM(I4,M6);

        MR0=MR1;              {16-bit right shift}
        MR1=MR2;

            {multiply/accumulate LSW*MSW, MSW*LSW}

        MR=MR+SR1*MY0(SS), MX0=DM(I0,M2), MY0=PM(I4,M6);
        MR=MR+MX0*MY0(SS), MX0=DM(I0,M2), MY0=PM(I4,M6);
        MR=MR+MX0*MY0(SS), MX0=DM(I0,M2), MY0=PM(I4,M6);
        MR=MR+MX0*MY0(SS), MX0=DM(I0,M3), MY0=PM(I4,M6);
        MR=MR+MX0*MY0(SS), MX0=DM(I0,M1), MY0=PM(I4,M5);

            {apply scale factor, store y(n) to delay line}
```

```
              SR=LSHIFT MR0 (LO), MY1=PM(I4,M5);
              DM(I0,M1)=SR0, SR=SR OR LSHIFT MR1 (HI);
biq:       DM(I0,M1)=SR1;

           MODIFY(I0,M2);
           DM(I0,M1)=AX0;                {store inputs to delay line}
           DM(I0,M0)=AX1;

           RTS;

.ENDMOD;
```

Listing 10.4 Optimized Double-Precision IIR Biquad Subroutine

10.2.3   Half, Double-Precision Biquad

The half, double-precision IIR Biquad subroutine in Listing 10.5 maintains a 32-bit delay line, but reduces the coefficients to 16-bit precision. This routine is useful when a 16-bit biquad (as shown in Listing 10.1) has the desired accuracy, but lacks the necessary stability.

The execution time is $[16*(N/2) + 9]$ instruction cycles. A tenth-order filter executes in 89 instruction cycles, or in 5.34 µs using a DSP with 60 ns cycle time. The DSP can perform a tenth order IIR filter on a signal sampled at more than 180 kSa/s.

```
.MODULE/RAM/BOOT=0          dpiir_2p2z_32_16;

{
     IIR Filter of the Form:
          y(n)=b0*x(n)+b1*x(n-1)+b2*x(n-2)+a1*y(n-1)+a2*x(n-2)

     Where:
          x(n), x(n-1), x(n-2), y(n-1), y(n-2) are 32-bits
          b0, b1, b2, a1, a2 are 16-bits

     Calling Parameters:
          AX0=least significant word (LSW) of input
          AX1=most significant word (MSW) of input
          I0=start address of delay line in data memory stored in
               LSW,MSW order [organized x(n-1),x(n-2),y(n-1),y(n-2)]
               buffer length = (4*N/2)+4 where N is the filter order
               buffer must be declared circular
```

(listing continues on next page)

10 Variations On IIR Biquad Filters

```
            I4=start address of coefficients in program memory
                    [organized b0,b1,b2,a1,a2] - buffer length = 5*N/2
                    does not require circular declaration
            I5=start address of coefficient scaling factors in program
                    memory (one/section) - buffer length = N/2
                    does not require circular declaration
            L0=(4*N/2)+4
            L4=0
            L5=0
            M0=-1, M1=1, M2=2, M3=-5
            M5=1, M7=-4
            CNTR=N/2

    Return Values:
            I0=new start address of delay line must be saved to memory
            SR1=MSW result
            SR0=LSW result

    Altered Registers:
            MX0,MY0,MR,SE,SR

    Computation Time:
            9 + (16 * number of sections)

    Coefficients in 1.15 (Q.15) format
    Delay line, input sample in 1.31 (Q.31) format
}


.ENTRY                  biq_3216;

biq_3216:   SR1=AX1;                {copy input to SR}
                SR0=AX0;
                MY0=PM(I4,M5);      {read first coefficient}
                SE=PM(I5,M5);       {read first scaling factor}

                DO biq UNTIL CE;    {set up biquad loop}

                        {multiply/accumulate LSW delay line * coef}
```

```
        MR=SR0*MY0(US),      MX0=DM(I0,M2), MY0=PM(I4,M5);
        MR=MR+MX0*MY0(US),   MX0=DM(I0,M2), MY0=PM(I4,M5);
        MR=MR+MX0*MY0(US),   MX0=DM(I0,M2), MY0=PM(I4,M5);
        MR=MR+MX0*MY0(US),   MX0=DM(I0,M3), MY0=PM(I4,M7);
        MR=MR+MX0*MY0(US),   MY0=PM(I4,M5);

        MR0=MR1;             {16-bit right shift}
        MR1=MR2;


             {multiply/accumulate MSW delay line * coef}

        MR=MR+SR1*MY0(SS),   MX0=DM(I0,M2), MY0=PM(I4,M5);
        MR=MR+MX0*MY0(SS),   MX0=DM(I0,M2), MY0=PM(I4,M5);
        MR=MR+MX0*MY0(SS),   MX0=DM(I0,M2), MY0=PM(I4,M5);
        MR=MR+MX0*MY0(SS),   MX0=DM(I0,M3), MY0=PM(I4,M5);
        MR=MR+MX0*MY0(SS);


             {scale factor correction}

        SR=ASHIFT MR1 (HI),  MY0=PM(I4,M5);
        SR=SR OR LSHIFT MR0 (LO),  SE=PM(I5,M5);

        DM(I0,M1)=SR0;    {store y(n) to delay line}
biq:    DM(I0,M1)=SR1;

        MODIFY(I0,M2);
        DM(I0,M1)=AX0;             {store input to delay line}
        DM(I0,M0)=AX1;

        RTS;

.ENDMOD;
```

Listing 10.5 Half, Double-Precision IIR Biquad Subroutine

10 Variations On IIR Biquad Filters

10.2.4   Half, Triple-Precision Biquad

The half, triple-precision IIR Biquad subroutine in Listing 10.6 has a 48-bit delay line and increases the coefficients to 32-bit precision. This subroutine provides a filter resolution that is usually reserved for floating-point arithmetic. The code in Listing 10.6 gives you 32-bit, floating-point precision on a 16-bit fixed-point DSP. Filters with extremely narrow pass or reject bands may require this precision.

The execution time is [45*(N/2) + 12] instruction cycles. A tenth-order filter executes in 237 instruction cycles, or in 14.22 µs using a DSP with 60 ns cycle time. The DSP can perform a tenth-order IIR filter on a signal sampled at more than 70 kSa/s.

```
.MODULE/RAM/BOOT=0                  half_triple_precision_iir_biquad;

{

    IIR Filter of the Form:
        y(n)=b0*x(n)+b1*x(n-1)+b2*x(n-2)+a1*y(n-1)+a2*x(n-2)

Where:
        x(n), x(n-1), x(n-2), y(n-1), y(n-2) are 48-bits
        b0, b1, b2, a1, a2 are 32-bits

Calling Parameters:
        AR=lower word of input
        AX0=middle word of input
        AX1=upper word of input
        I0=start address of delay line lower word in data memory stored in
            [x(n-1),x(n-2),y(n-1),y(n-2)] order buffer length = (2*N/2)+2
            where N is filter order must be declared circular
        I1=start address of delay line middle word in data memory stored in
            [x(n-1),x(n-2),y(n-1),y(n-2)] order buffer length = (2*N/2)+2
            where N is filter order must be declared circular
        I2=start address of delay line upper word in data memory stored in
            [x(n-1),x(n-2),y(n-1),y(n-2)] order buffer length = (2*N/2)+2
            where N is filter order must be declared circular
        I4=start address of coefficients in program memory stored in MSW,LSW
            order [organized b0,b1,b2,a1,a2] buffer length = 10*N/2
            does not require circular declaration
        I5=start address of coefficient scaling factors in program
            memory (one/section) - buffer length = N/2
            does not require circular declaration
```

540

Variations On IIR Biquad 10
Filters

```
            L0=(2*N/2)+2
            L1=(2*N/2)+2
            L2=(2*N/2)+2
            L4=0
            L5=0
            M0=0,  M1=1,  M3=-3
            M5=1,  M6=2,  M7=-9
            CNTR=N/2

    Return Values:
            I0,I1,I2 -> new start address of delay line must be saved to memory
            SR1=upper result
            SR0=middle result
            AR=lower result

    Altered Registers:
            MX0,MY0,MR,SE,SR,SI,AF

    Computation Time:
            12 + (45 * number of sections)

    Coefficients in 1.31 (Q.31) format
    Delay line, input sample, result in 1.47 (Q.47) format
}

.ENTRY                  biq_3P;

biq_3P:

            SR1=AX1;                 {SR = upper/middle word}
            AF=PASS AR, SR0=AX0;     {AF = lower word}
            MODIFY(I4,M5);
            MY0=PM(I4,M6);           {read first coefficient}

            DO biq UNTIL CE;         {set up biquad loop}

                         {multiply/accumulate - lower delay*lower coef}
```

(listing continues on next page)

10 Variations On IIR Biquad Filters

```
        MR=AR*MY0(UU),      MX0=DM(I0,M1), MY0=PM(I4,M6);
        MR=MR+MX0*MY0(UU), MX0=DM(I0,M1), MY0=PM(I4,M6);
        MR=MR+MX0*MY0(UU), MX0=DM(I0,M1), MY0=PM(I4,M6);
        MR=MR+MX0*MY0(UU), MX0=DM(I0,M3), MY0=PM(I4,M7);
        MR=MR+MX0*MY0(UU), MY0=PM(I4,M5);
        MR=MR(RND), SE=PM(I5,M5);

                {shift accumulator 16-bits right}

        MR0=MR1;
        MR1=MR2;


{multiply/accumulate - middle delay*lower coef, lower delay*upper coef}

        MR=MR+AR*MY0(US), MY0=PM(I4,M5);
        MR=MR+SR0*MY0(UU), MX0=DM(I0,M1), MY0=PM(I4,M5);
        MR=MR+MX0*MY0(US), MX0=DM(I1,M1), MY0=PM(I4,M5);
        MR=MR+MX0*MY0(UU), MX0=DM(I0,M1), MY0=PM(I4,M5);
        MR=MR+MX0*MY0(US), MX0=DM(I1,M1), MY0=PM(I4,M5);
        MR=MR+MX0*MY0(UU), MX0=DM(I0,M1), MY0=PM(I4,M5);
        MR=MR+MX0*MY0(US), MX0=DM(I1,M1), MY0=PM(I4,M5);
        MR=MR+MX0*MY0(UU), MX0=DM(I0,M3), MY0=PM(I4,M5);
        MR=MR+MX0*MY0(US), MX0=DM(I1,M3), MY0=PM(I4,M7);
        MR=MR+MX0*MY0(UU), MX0=DM(I0,M1), MY0=PM(I4,M5);
        MR=MR(RND);

                {shift accumulator 16-bits right}

        MR0=MR1;
        MR1=MR2;

{multiply/accumulate - upper delay*lower coef, middle delay*upper coef}

        MR=MR+SR0*MY0(US), MY0=PM(I4,M5);
        MR=MR+SR1*MY0(SU), MX0=DM(I1,M1), MY0=PM(I4,M5);
        MR=MR+MX0*MY0(US), MX0=DM(I2,M1), MY0=PM(I4,M5);
        MR=MR+MX0*MY0(SU), MX0=DM(I1,M1), MY0=PM(I4,M5);
        MR=MR+MX0*MY0(US), MX0=DM(I2,M1), MY0=PM(I4,M5);
        MR=MR+MX0*MY0(SU), MX0=DM(I1,M1), MY0=PM(I4,M5);
        MR=MR+MX0*MY0(US), MX0=DM(I2,M1), MY0=PM(I4,M5);
        MR=MR+MX0*MY0(SU), MX0=DM(I1,M3), MY0=PM(I4,M5);
        MR=MR+MX0*MY0(US), MX0=DM(I2,M3), MY0=PM(I4,M7);
        MR=MR+MX0*MY0(SU), MX0=DM(I1,M1), MY0=PM(I4,M6);
```

Variations On IIR Biquad Filters

```
                    {shift accumulator 16-bits right}

            SR0=MR0;
            MR0=MR1;
            MR1=MR2;

                    {multiply/accumulate - upper delay*upper coef}

            MR=MR+SR1*MY0(SS), MX0=DM(I2,M1), MY0=PM(I4,M6);
            MR=MR+MX0*MY0(SS), MX0=DM(I2,M1), MY0=PM(I4,M6);
            MR=MR+MX0*MY0(SS), MX0=DM(I2,M1), MY0=PM(I4,M6);
            MR=MR+MX0*MY0(SS), MX0=DM(I2,M3), MY0=PM(I4,M6);
            MR=MR+MX0*MY0(SS), MX0=DM(I2,M1), MY0=PM(I4,M5);


        {apply scale factor to biquad result, write scaled result to delay line}

scale_it:   SR=LSHIFT SR0 (LO), MY0=PM(I4,M6);
            AR=SR0;
            SR=LSHIFT SR1 BY 0 (LO);
            DM(I0,M1)=AR, SR=SR OR LSHIFT MR0 (LO);
            DM(I1,M1)=SR0, SR=SR OR ASHIFT MR1 (HI);
biq:        DM(I2,M1)=SR1;

                    {store original inputs to delay line}

            MODIFY(I0,M1);
            MODIFY(I1,M1);
            MODIFY(I2,M1);
            DM(I2,M0)=AX1;
            DM(I1,M0)=AX0, AR=PASS AF;
            DM(I0,M0)=AR;
            RTS;        {return}

.ENDMOD;
```

Listing 10.6 Half, Triple-Precision IIR Biquad Subroutine

10 Variations On IIR Biquad Filters

10.3   OPTIMIZED 16-BIT BIQUADS

If 16-bit accuracy is adequate for your application, you can use the two subroutines included in this section. While these routines are similar to the program in Listing 10.1, they are optimized to require fewer instruction cycles to execute.

The program in Listing 10.7 provides the identical results to the program in Listing 10.1, but, in Listing 10.7, it executes the biquad loop in six instruction cycles rather the seven. This decrease results from the ADSP-2100 Family's modulo addressing capability.

To optimize the filter, the program uses a circular buffer that contains input data and output data. If you carefully arrange the data in the delay line and use the modulo addressing of the data address generators, you will have an efficient addressing scheme that lets you use any address in the circular buffer as the starting address. Modulo addressing only applies to the delay line; it is not required for the coefficient and scale factor buffers. Modulo addressing is also used in the multiprecision IIR filters in this chapter. You must save the delay line address pointer, index register I0, to memory after each call if I0 is used elsewhere in the program.

Figure 10.3 is a memory map that illustrates modulo addressing. This figure shows three time intervals. In the first interval (t), addresses 1, 3, and, 5 contain the oldest data. Because the oldest data is not needed during the second interval (t + 1), new data is written into those locations. Data in addresses 0, 2, and 4 is preserved for the second interval, but it becomes the oldest data. For example, $X_0(n-1)$ from the first interval becomes $X_0(n-2)$ in the second interval. In the third sample interval, the oldest data from the second interval is overwritten. The cycle continues for every new sample interval. This method lets save instruction cycles because you move the pointers to the circular buffer, rather than move the buffer contents. You can also use the modulo addressing advantages for feedback values in the delay line or past results. For example, past result $y(n-1)$ is used as $y(n-2)$ during the next sample interval without moving to another memory location.

Variations On IIR Biquad Filters 10

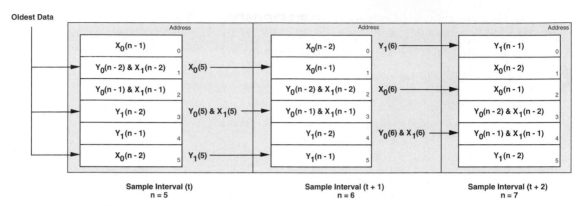

Oldest Data

Sample Interval (t)
n = 5

Sample Interval (t + 1)
n = 6

Sample Interval (t + 2)
n = 7

Figure 10.3 Modulo Addressing & Delay Line Data

The execution time is $[7*(N/2) + 6]$ instruction cycles. A tenth-order filter executes in 41 instruction cycles, or in 2.46 µs using a DSP with 60 ns cycle time. The DSP can perform a tenth-order IIR filter on a signal sampled at more than 400 kSa/s.

```
.MODULE/boot=0 optimized_biquad_sub;

{
     Optimized Cascaded Biquad IIR Filter Subroutine (Direct Form I)
     Calling Parameters
          SR1 = input sample
          I0 —> delay line buffer in data memory
               x(n-2),x(n-1),y(n-2),y(n-1) order
               buffer length = 2N+2 where N is the filter order
          L0 = 2N+2 - circular buffer declaration required
          I4 —> scaled coefficients in program memory b2,b1,b0,a2,a1 order
               buffer length = 5*N/2 - circular buffer not required
          L4 = 0
          I5 —> coefficient scale factors in program memory
               buffer length = N - circular buffer not required
          L5 = 0
           M0 = 0
          M1,M4 = 1
          M3,M5 = -1
          CNTR = N/2
```

(listing continues on next page)

545

10 Variations On IIR Biquad Filters

```
Return Values
      SR1 = output sample
      I0 -> new delay line buffer start address must be saved to memory

Altered Registers

      MX0,MY0,MR,SR,SE

Computation Time

      7 * number of biquad sections + 6 cycles

All coefficients and data values are assumed to be in 1.15 format.
}

.ENTRY      optbiq;

optbiq:

            SE=PM(I5,M4);                     {read coefficient sclaing factor}
            MX0=DM(I0,M1), MY0=PM(I4,M4); {x=x(n-2), y=b2}
            DO sections UNTIL CE;
                  MR=MX0*MY0(SS), MX0=DM(I0,M1), MY0=PM(I4,M4);
                                             {mult, x=x(n-1), y=b1}
                  MR=MR+MX0*MY0(SS), MY0=PM(I4,M4);
                                             {mac, y=b0}
                  MR=MR+SR1*MY0(SS), MX0=DM(I0,M1), MY0=PM(I4,M4);
                                             {mac, x=y(n-2), y=a2}
                  MR=MR+MX0*MY0(SS), MX0=DM(I0,M3), MY0=PM(I4,M4);
                                             {mac, x=y(n-1), y=a1}
                  MR=MR+MX0*MY0(RND), MX0=DM(I0,M0), MY0=PM(I4,M4);
                                             {mac, x=next x(n-2), y=next b2}
                  DM(I0,M1)=SR1, SR=ASHIFT MR1 (HI);
                                             {x(n)->new x(n-1), scale result}
sections:   SR=SR OR LSHIFT MR0 (LO), SE=PM(I5,M4);
                                             {continue scaling, new scale factor}
            MX0=DM(I0,M1);                   {dummy read to modify pointer}
            DM(I0,M1)=SR1;                   {store last result into delay line}
                                             {must return new I0}

            RTS;

.ENDMOD;
```

Listing 10.7 Optimized Basic Biquad Filter Subroutine

546

Listing 10.8 further optimizes the filter because the coefficient scaling factor for each biquad is the same. This eliminates the need to read the scaling factor from program memory.

The biquad loop in Listing 10.7 includes a double-precision shift for scaling factor correction. Listing 10.8 performs a single-precision shift. Since the scaling factor typically implies a one-bit left shift, a single precision shift yields a zero as the least significant bit. Therefore, the result from Listing 10.8 only has 15-bit accuracy. This may change filter performance if the filter was designed for 90 dB or greater stopband attenuation, which may effect the filter's stability.

The execution time is $[6*(N/2) + 5]$ instruction cycles. A tenth-order filter executes in 35 instruction cycles, or in 2.1 μs using a DSP with 60 ns cycle time. The DSP can perform a tenth-order IIR filter on a signal sampled at more than 470 kSa/s.

```
.MODULE/boot=0 optimized_biquad_sub;
{
     Optimized Cascaded Biquad IIR Filter Subroutine (Direct Form I)

     Calling Parameters
          SR1 = input sample
          I0 —> delay line buffer in data memory
               x(n-2),x(n-1),y(n-2),y(n-1) order
               buffer length = 2N+2 where N is the filter order
               buffer must be declared circular
          L0 = 2N+2
          I4 —> scaled coefficients in program memory
               b2,b1,b0,a2,a1 order
               buffer length = 5*N/2 - circular buffer not required
          L4 = 2.5 * filter order —or— 5 * number of biquad sections
          M0 = 0
          M1,M4 = 1
          M3,M5 = -1
          CNTR = number of biquad sections
          SE = shift count (must be same for all biquad sections)

     Return Values
          SR1 = output sample
          I0 —> new delay line buffer start address must be stored to memory
```

(listing continues on next page)

10 Variations On IIR Biquad Filters

```
        Altered Registers
            MX0,MY0,MR,SR

        Computation Time
            6 * number of biquad sections + 5 cycles

        All coefficients and data values are assumed to be in 1.15 format.
}


.ENTRY optbiq;


optbiq:

            MX0=DM(I0,M1), MY0=PM(I4,M4); {x=x(n-2), y=b2}
            DO sections UNTIL CE;
                MR=MX0*MY0(SS), MX0=DM(I0,M1), MY0=PM(I4,M4);
                                              {mult, x=x(n-1), y=b1}

                MR=MR+MX0*MY0(SS), MY0=PM(I4,M4);
                                              {mac, y=b0}
                MR=MR+SR1*MY0(SS), MX0=DM(I0,M1), MY0=PM(I4,M4);
                                              {mac, x=y(n-2), y=a2}
                MR=MR+MX0*MY0(SS), MX0=DM(I0,M3), MY0=PM(I4,M4);
                                              {mac, x=y(n-1), y=a1}
                MR=MR+MX0*MY0(RND), MX0=DM(I0,M0), MY0=PM(I4,M4);
                                              {mac, x=next x(n-2), y=next b2}
sections:   DM(I0,M1)=SR1, SR=ASHIFT MR1 (HI);
                                              {x(n)->new x(n-1), scale result}
            MX0=DM(I0,M1), MY0=PM(I4,M5);
            DM(I0,M1)=SR1;                    {store last result into delay line}
                                              {must return new I0}

            RTS;

.ENDMOD;
```

Listing 10.8 Second-Level Optimization Of Basic Biquad Filter

Variations On IIR Biquad 10 Filters

10.4   CONCLUSION

This chapter provides you with precision-related options when designing digital IIR filters. Depending on the filter characteristics, 15-bit, 16-bit, 32-bit or 48-bit, precision may be required in the delay line for an IIR biquad to function correctly. Filter coefficients may also require 32-bit precision to ensure filter stability. Higher precision IIR filters are available at the expense of instruction cycles. Table 10.1 lists the subroutines listed in this chapter, their characteristics, and performance.

Listing #	Filename	delay line # bits	coefficients # bits	performance # cycles
8	iir1516.dsp	15-bits	16-bits	$6*(N/2) + 5$
7	iir1616.dsp	16-bits	16-bits	$7*(N/2) + 6$
5	iir3216.dsp	32-bits	16-bits	$16*(N/2) + 9$
4	iir3132.dsp	31-bits	32-bits	$20*(N/2) + 9$
3	iir3232.dsp	32-bits	32-bits	$28*(N/2) + 10$
6	iir4832.dsp	48-bits	32-bits	$45*(N/2) + 12$

where N = filter order

Table 10.1 Filter Routine Characteristics Summary

Other variations of IIR biquad sections are possible. You can achieve more precision by increasing the feedback variables or coefficients to 64-bits. Existing 16-bit or 32-bit subroutines can be optimized to exclude those multiplications when the coefficients are known to be zero.

Coefficients for these filters can be generated by many digital filter design software tools. These packages, however, do not determine if a 16-bit or 32-bit filter is required for proper convergence. A few of these companies are listed below.

Momentum Data Systems	(714) 557-6884
Hyperception	(214) 343-8525
The Athena Group	(904) 371-2567
Signalogic	(214) 343-0069

Software UART ■ 11

11.1   OVERVIEW

This chapter describes a software implementation of a Universal Asynchronous Receiver/Transmitter (UART). The UART is implemented as a program running on an ADSP-2101, with the Flag In (FI) and Flag Out (FO) signals used as asynchronous receive and transmit lines. The UART software was developed to provide the following features:

- Full-duplex operation—independent receiver and transmitter
- Double-buffered receive and transmit registers, to allow continuous data flow
- Asynchronous operation—no synchronization required between transmitted and received bit streams
- Programmability—to provide a variety of baud rates and flexible data formats (7 or 8 data bits, 1 or 2 stop bits)

11.2   HARDWARE

A general system configuration is shown in Figure 11.1. The ADSP-21xx is interfaced to an RS-232 line driver chip which is in turn connected to any RS-232-compatible device. The RS-232 line driver is needed to convert the 5-volt logic level of the ADSP-21xx to the proper RS-232 line voltages, and vice versa.

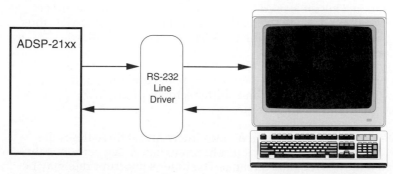

Figure 11.1 General System Configuration

11 Software UART

Figure 11.2 shows a specific example of a hardware implementation for the UART in which an ADSP-2101 processor and AD233 RS-232 Driver/Receiver are used. The Flag In (FI) and Flag Out (FO) pins of the ADSP-21xx are used as independent receive and transmit lines. The AD233 is an ideal choice for the line driver because it requires no external capacitors and is powered by a single 5V supply.

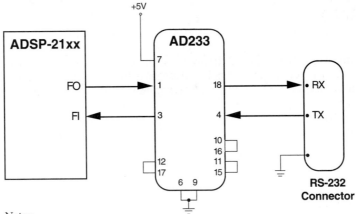

Notes:
1. Pins 2, 5, 8, 13, 14, 19, and 20 on the AD233 have no connection.
2. For autobaud operation, connect the ADSP-21xx $\overline{IRQ2}$ pin to the FI pin.

Figure 11.2 Example System Configuration

11.3   SOFTWARE

Two separate sets of receive and transmit registers are used in this UART implementation. One set is used by the UART to to clock data words in and out. The other set is used to read from (and write to) the UART, thus providing access to the UART while it is in operation. This allows a continuous data flow.

The UART program can handle a wide variety of baud rates and data formats by modifying the timer and shifter settings of the ADSP-21xx. Autobaud operation is also possible.

The ADSP-21xx's on-chip timer generates interrupts at three times the baud rate, providing enough clock resolution to handle asynchronously transmitted and received data streams. The timer's interrupt rate may be increased to provide additional clock resolution, if desired—this is described below under "Initialization & Timer Interrupt Routines."

11.3.1 Program Flow

The UART program consists of six subroutines:

- Initialization routine
- Timer interrupt routine
- Transmit character routine
- Receive character routine
- Enable receive routine
- Disable receive routine

The *initialization routine* must be called after a system reset. The *timer interrupt routine* is the heart of the UART program. It transmits and receives bits when necessary. The receive portion of this routine can be disabled by executing the disable receive routine. The timer interrupt routine prepares the UART for use by: 1) setting up the timer to generate its interrupt at the proper rate, 2) configuring SPORT1 (Serial Port 1) as the FI/FO pins (Flag In, Flag Out), 3) setting flags to indicate that the UART is not busy, and 4) clearing any pending interrupts and enabling the timer.

The *transmit character routine* waits for any previously transmitted character to be completely sent, and then sends the next character.

The *receive character routine* waits for a character to be completely received, and then gets the character and returns it to the calling program.

The *enable receive routine* enables the UART receive portion of the timer interrupt routine.

The *disable receive routine* disables the UART receive portion of the timer interrupt routine.

11   Software UART

11.3.2   Initialization & Timer Interrupt Routines

For the following discussion of each subroutine, refer to the code shown in Listing 11.1.

```
{**************************************************************
   ADSP-2101 Software UART                         UART.DSP

This program uses FLAG_IN, FLAG_OUT and the TIMER of the ADSP-2101 to
interface to an RS-232 asynchronous serial device such as a VT100 terminal.

for example:

  ADSP-2101 FLAG_OUT ——> AD233 ——> RS232 RX

  ADSP-2101 FLAG_IN  <—— AD233 <—— RS232 TX

                  (TIMER maintains baudrate)

Parameters bits/word, baudrate, stopbits & parity are user-programmable.
An RS-232 line driver chip (such as the AD233) can be used to electrically
interface +5 VDC to the RS-232 line voltage levels.

The operation of the transmitter setup routine is completely independent on
the receiver setup routine operation. Although both tx and rx use the same
timer as a master clock source, the transmitted bits need not be in sync
with the received bits. The default state of the reciever is OFF, so the
"turn_rx_on" subroutine must be used to enable RX.

Calling Argument:
  For autobaud load the baud constant:
    dm(baud_period)=(Proc_frequency/(3*Baudrate))-1

Useful Subroutines:
  init_uart        Must be called after system reset.
  get_char_ax1     Waits for RX input and returns with it in AX1.
  out_char_ax1     Waits for last TX output and transmits data from AX1.
  turn_rx_on       Must be called to enable the receipt of RX data.
  turn_rx_off      Can be used to ignore input RX data.

Useful Flag:
  DM(flag_rx_ready)      If this DM location is all ones it indicates that
                         the UART is ready to receive a new word. If it is
                         zero then data is being received. Can be used for
                         xon/xoff flow control.

**************************************************************}
```

Software UART 11

```
.module/boot=1    UART;

{The constants below must be changed to modify the UART parameters}

.const tx_num_of_bits = 10;        {start bits + tx data bits + stop bits}
.const rx_num_of_bits = 8;         {rx data bits (start&stop bits not counted)}
.const RX_BIT_ADD = 0x0100;        {= 1<<rx_num_of_bits}
.const TX_BIT_ADD = 0xfe00;        {= 0xffff<<(tx data bits+1)}

{___These constants can be used if autobaud is not needed___}

.const PERIOD=74;         {13 & 57600}    {PERIOD=(Proc_freq/(3*Baudrate))-1}
{.const PERIOD=112;}      {13 & 38400}    {PERIOD=(Proc_freq/(3*Baudrate))-1}
{.const PERIOD=225;}      {13 & 19200}    {PERIOD=(Proc_freq/(3*Baudrate))-1}
{.const PERIOD=450;}      {13 & 9600}     {PERIOD=(Proc_freq/(3*Baudrate))-1}

{_____Definitions for memory-mapped control registers_____}

.const TSCALE = 0x3ffb;
.const TCOUNT = 0x3ffc;
.const TPERIOD = 0x3ffd;
.const System_Control_Reg = 0x3fff;

{_____}

.entry init_uart;        {UART initialize baudrate etc.}
.entry out_char_ax1;     {UART output a character}
.entry get_char_ax1;     {UART wait & get input character}
.entry turn_rx_on;       {UART enable the rx section}
.entry turn_rx_off;      {UART disable the rx section}
.entry process_a_bit;    {UART timer interrupt routine for RX and TX}

.global flag_rx_ready;
.global baud_period;

.var flag_tx_ready;         {flag indicating UART is ready for new tx word}
.var flag_rx_ready;         {flag indicating UART is ready to rx new word}
.var flag_rx_stop_yet;      {flag tells that a rx stop bit is not pending}
.var flag_rx_no_word;       {indicates a word is not in the user_rx_buffer}
.var flag_rx_off;           {indicates a that the receiver is turned off}
.var timer_tx_ctr;          {divide by 3 ctr, timer is running @ 3x baudrate}
.var timer_rx_ctr;          {divide by 3 ctr, timer is running @ 3x baudrate}
.var user_tx_buffer;        {UART tx reg loaded by user before UART xmit}
.var user_rx_buffer;        {UART rx reg read by user after word is rcvd}
.var internal_tx_buffer;    {formatted for serial word, adds start&stop bits}
                            {'user_tx_buffer' is copied here before xmission}
.var internal_rx_buffer;
.var bits_left_in_tx;       {number of bits left in tx buffer (not yet clkd out) }
.var bits_left_in_rx;       {number of bits left to be rcvd (not yet clkd in) }
.var baud_period;           {loaded by autobaud routine}
```

(listing continues on next page)

11 Software UART

```
{_____Initializing subroutine_____}

init_uart:

  ax0=0;
  dm(TSCALE)=ax0;              {decrement TCOUNT every instruction cycle}

  ax0=dm(baud_period);         {from autobaud or use constant: ax0=PERIOD;}
                               {...and comment in the appropriate constant}
  dm(TCOUNT)=ax0;
  dm(TPERIOD)=ax0;             {interrupts generated at 3x baudrate}
  ax0=0;
  dm(System_Control_Reg)=ax0;  {no bmwait,pmwait states, SPORT1=FI/FO}

  ax0=1;
  dm(flag_tx_ready)=ax0;       {set the flags showing that UART is not busy}
  dm(flag_rx_ready)=ax0;
  dm(flag_rx_stop_yet)=ax0;
  dm(flag_rx_no_word)=ax0;
  dm(flag_rx_off)=ax0;         {rx section off}

  set flag_out;                {UART tx output is initialized to high}
  ifc=0x003f;                  {clear all pending interrupts}
  nop;                         {wait for ifc latency }
  imask=b#000001;              {enable TIMER interrupt handling}
  ena timer;                   {start timer now}
  rts;

{_____process_a_bit_____
                (TIMER interrupt routine)
```

This routine is the heart of the UART. It is called every timer
interrupt (i.e. 3x baudrate). This routine will xmit one bit at a time
by setting/clearing the FLAG_OUT pin of the ADSP-2101. This routine
will then test if the UART is already receiving. If not it will test
flagin (rx) for a start bit and place the UART in receive mode if true.
If already in receive mode it will shift in one bit at a time by
reading the FLAG_IN pin. Since the internal timer is running at 3x
baudrate, bits need only be transmitted/received once every 3 timer
interrupts.

```
_____}

process_a_bit:

  ena sec_reg;                       {Switch to background register set}
  ax0=dm(flag_tx_ready);             {if not in "transmit", go right to
                                        "receive"}
  ar=pass ax0;
  if ne jump receiver;
```

556

Software UART   11

```
{_____Transmitter Section_____}

   ay0=dm(timer_tx_ctr);              {test timer ctr to see if a bit}
   ar=ay0-1;                          {is to be sent this time around}
   dm(timer_tx_ctr)=ar;              {if no bit is to be sent}
   if ne jump receiver;               {then decrement ctr and return}

   sr1=dm(internal_tx_buffer);        {shift out LSB of internal_tx_buffer}
   sr=lshift sr1 by -1 (hi);          {into SR1.  Test the sign of this bit}
   dm(internal_tx_buffer)=sr1;        {set or reset FLAG_OUT accordingly}
   ar=pass sr0;                       {this effectively clocks out the}
   if ge reset flag_out;              {word being xmitted one bit at a time}
   if lt set flag_out;                {LSB out first at FLAG_OUT.}

   ay0=3;                             {reset timer ctr to 3, i.e. next bit}
   dm(timer_tx_ctr)=ay0;              {will be sent after 3 timer interrupts}

   ay0=dm(bits_left_in_tx);           {number of bits left to be xmitted}
   ar=ay0-1;                          {is now decremented by one,}
   dm(bits_left_in_tx)=ar;            {indicating that one is now xmitted}
   if gt jump receiver;               {if no more bits left, then ready}

   ax0=1;                             {flag is set to true indicating}
   dm(flag_tx_ready)=ax0;             {a new word can now be xmitted}

{_____Receiver Section_____}

receiver:
   ax0=dm(flag_rx_off);               {Test if receiver is turned on}
   ar=pass ax0;
   if ne rti;

   ax0=dm(flag_rx_stop_yet);          {Test if finished with stop bit of}
   ar=pass ax0;                       {last word or not. if finished then}
   if ne jump rx_test_busy;           {continue with check for receive.}

   ay0=dm(timer_rx_ctr);              {decrement timer ctr and test to see}
   ar=ay0-1;                          {if stop bit period has been reached}
   dm(timer_rx_ctr)=ar;               {if not return and wait}
   if ne rti;

   ax0=1;                             {if stop bit is reached then reset}
   dm(flag_rx_stop_yet)=ax0;          {to wait for next word}
   dm(flag_rx_ready)=ax0;

   ax0=dm(internal_rx_buffer);        {copy internal rx buffer}
   dm(user_rx_buffer)=ax0;            {to the user_rx_buffer}
```

(listing continues on next page)

11 Software UART

```
    ax0=0;                              {indicated that a word is ready in}
    dm(flag_rx_no_word)=ax0;            {the user_rx_buffer}
    rti;

rx_test_busy:
    ax0=dm(flag_rx_ready);              {test rx flag, if rcvr is not busy}
    ar=pass ax0;                        {receiving bits then test for start.If it}
    if eq jump rx_busy;                 {is busy, then clk in one bit at a time}

    if flag_in jump rx_exit;            {Test for start bit and return if none}

    ax0=0;
    dm(flag_rx_ready)=ax0;              {otherwise, indicate rcvr is now busy}
    dm(internal_rx_buffer)=ax0;         {clear out rcv register}

    ax0=4;                              {Timer runs @ 3x baud rate, so rcvr}
    dm(timer_rx_ctr)=ax0;               {will only rcv on every 3rd interrupt.}
                                        {Initially this ctr is set to 4.}
                                        {This will skip the start bit and will}
    ax0=rx_num_of_bits;                 {allow us to check FLAG_IN at the center}
    dm(bits_left_in_rx)=ax0;            {of the received data bit.}

rx_exit:
    rti;

rx_busy:
    ay0=dm(timer_rx_ctr);               {decrement timer ctr and test to see}
    ar=ay0-1;                           {if bit is to be rcvd this time around}
    dm(timer_rx_ctr)=ar;                {if not return, else receive a bit}
    if ne rti;

rcv:                                    {Shift in rx bit}
    ax0=3;                              {reset the timer ctr to 3 indicating}
    dm(timer_rx_ctr)=ax0;               {next bit is 3 timer interrupts later}

    ay0=RX_BIT_ADD;
    ar=dm(internal_rx_buffer);
    if not flag_in jump pad_zero;       {Test RX input bit and}
    ar=ar+ay0;                          {add in a 1 if hi}

pad_zero:
    sr=lshift ar by -1 (lo);            {Shift down to ready for next bit}
    dm(internal_rx_buffer)=sr0;

    ay0=dm(bits_left_in_rx);            {if there are more bits left to be rcvd}
    ar=ay0-1;                           {then keep UART in rcv mode}
    dm(bits_left_in_rx)=ar;             {and return}
    if gt rti;                          {if there are no more bits then ...}
                                        {...that was the last bit }
```

```
    ax0=3;                              {set timer to wait for middle of the}
    dm(timer_rx_ctr)=ax0;              {stop bit}
    ax0=0;                              {flag indicated that uart is waiting}
    dm(flag_rx_stop_yet)=ax0;          {for the stop bit to arrive}
    rti;
```

```
{_____invoke_UART_transmit subroutine_____
```

This is the first step in the transmit process. The user has now loaded
'user_tx_buffer' with the ascii code and has also invoked this routine.
_____}

```
invoke_UART_transmit:

    ax0=3;                              {initialize the timer decimator ctr}
    dm(timer_tx_ctr)=ax0;              {this divide by three ctr is needed}
                                        {since timer runs @ 3x baud rate}

    ax0=tx_num_of_bits;                 {this constant is defined by the}
    dm(bits_left_in_tx)=ax0;           {user and represents total number of}
                                        {bits including stop and parity}
                                        {ctr is initialized here indicating}
                                        {none of the bits have been xmitted}

    sr1=0;
    sr0=TX_BIT_ADD;                     {upper bits are hi to end txmit with hi}
    ar=dm(user_tx_buffer);             {transmit register is copied into }
    sr=sr or lshift ar by 1 (lo);      {the internal tx reg & left justified}
    dm(internal_tx_buffer)=sr0;        {before it gets xmitted}

    ax0=0;                              {indicate that the UART is busy}
    dm(flag_tx_ready)=ax0;
    rts;
```

```
{_____get an input character_____

    output:    ax1
    modifies:  ax0
    _____}
```

```
get_char_ax1:
    ax0=dm(flag_rx_no_word);
    ar=pass ax0;
    if ne jump get_char_ax1;           {if no rx word input, then wait}

    ax1=dm(user_rx_buffer);            {get received ascii character}
    ax0=1;
    dm(flag_rx_no_word)=ax0;           {word was read}
    rts;
```

(listing continues on next page)

11 Software UART

```
{_____output a character_____

   input:      ax1
   modifies:   ax0, sr1, sr0, ar
_____}

out_char_ax1:
  ax0=dm(flag_tx_ready);
  ar=pass ax0;
  if eq jump out_char_ax1;          {if tx word out still pending, then wait}
  dm(user_tx_buffer)=ax1;
  call invoke_UART_transmit;        {send it out}
  rts;

{_____enable the RX section_____

   modifies:  ax0
_____}

turn_rx_on:
  ax0=0;
  dm(flag_rx_off)=ax0;
  rts;

{_____disable the RX section_____

   modifies:  ax0
_____}

turn_rx_off:
  ax0=1;
  dm(flag_rx_off)=ax0;
  rts;

.endmod;
```

Listing 11.1 UART.DSP Code

Software UART 11

The *initialization routine* (init_uart) first sets the timer to generate interrupts at three times the baud rate. (**Note:** autobaud is supported.) This provides sufficient clock resolution to handle the asynchronous data stream in most cases. For applications requiring greater resolution, the interrupt rate may be increased to higher *odd* multiples of the baud rate. The reason only odd multiples of the baud rate are used is to locate the sampling of the bitstream as close to the bit center as possible.

The initialization routine next configures Serial Port 1 as the Flag In and Flag Out pins. It also configures the system for no boot memory or program memory wait states. (The number of wait states may be changed as needed for a particular application.) The routine then sets the following flags: transmit ready (tx_ready), receive ready (rx_ready), receive stop yet (rx_stop_yet), receive no word (rx_no_word), and receive section off/disable (rx_off). This indicates that the UART is ready to both transmit and receive. However, the UART receive function is initially disabled, and must be enabled by the *enable receive routine* before the UART can receive data. The initialization routine then sets Flag Out (FO) to initialize the UART transmit output high, clears all pending interrupts, and enables the timer interrupt handling. Lastly, the initialization routine enables the timer.

The *timer interrupt routine* (process_a_bit) is the central routine of the UART program. It processes the individual bits received and transmitted. The routine is divided into two main sections: a transmitter section and a receive section.

The timer interrupt routine begins by switching to the secondary data register set of the ADSP-21xx. Next, it checks the transmit ready flag to see if the UART is in transmit mode. If transmit mode is active, the transmit section is executed; if not, the routine skips to the receiver section.

The transmitter section decrements the timer transmit counter to determine whether or not a bit must be sent during the current interrupt. (A bit will be sent once every three interrupts in transmit mode, since the timer is running at three times the baud rate.) If no bit is to be sent, the routine jumps to the receiver section. If a bit is to be sent, it is shifted out of the internal transmit buffer, LSB first. This bit will determine whether Flag Out is set high or low. All necessary counters and flags are then updated and the receiver section is executed.

11 Software UART

The receiver section of the timer interrupt routine first checks to see if the receiver is enabled, via the `rx_off` flag. The routine then checks to see if the stop bit of the last word has been received, via the stop bit flag.

If the stop bit has been received, the timer receive counter is decremented and checked to see if the stop bit period has been reached. This must be done because the timer runs at three times the baud rate. If the stop bit of the last word has been received, the routine sets the stop bit flag and the receiver ready flag, copies the internal RX buffer to the user RX buffer, and clears the `rx_no_word` flag to indicate that a word is ready in the user RX buffer.

The routine then returns from the timer interrupt. If the stop bit has not been reached, the routine checks to see if the receiver is either: 1) in the middle of receiving a word, or 2) is waiting for a start bit (via the `rx_ready` flag). If it is waiting for a start bit, the routine checks for one. If one is not found, it exits from the interrupt. If one is found, it sets the appropriate flag to indicate that the receiver is busy, clears out the internal receive register, sets the timer counter to receive on every third interrupt, and sets the "number of bits left to be received" counter.

Initially the timer counter is set to four, not three. This is done to skip the start bit and align the Flag In check at the middle of the received data bit, as shown in Figure 11.3.

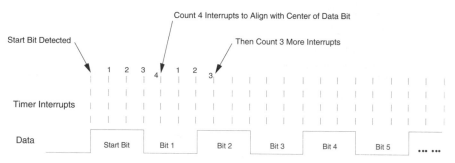

Figure 11.3 Receive Data Timing

The timer routine then exits from the interrupt. If the receiver is in the middle of receiving a word, it decrements and checks the timer RX counter to determine if a bit is to be received during the current interrupt. (A bit will be received every three timer interrupts when receiving, since the timer is running at three times the baud rate.)

If a data bit is not due to be received, it returns from the interrupt; otherwise, it shifts in the bit by checking if Flag In is high or low and shifting in a one or zero to the internal RX buffer. The number of bits left in the RX counter is then decremented, and, if it has expired, a flag is set to indicate that the UART is waiting for a stop bit. The routine then returns from the interrupt.

11.3.3   Transmit & Receive Subroutines

The *transmit character routine* (`out_char_ax1`) first checks the transmit ready flag to determine if a character transmit is still pending. If one is, the routine waits until it is complete. Then the routine copies the character stored in AX1 (by the calling routine) to the user transmit buffer; the invoke_uart_transmit (IUT) subroutine is then invoked.

The IUT subroutine initializes the timer transmit counter to set up operation at the proper baud rate, in this case one-third of the timer interrupt rate since the timer runs at three times the baud rate. The subroutine then initializes the "bits left to be transmitted" counter. Next, the subroutine copies the user transmit register to the internal transmit register and sets a flag indicating that the UART is busy. The IUT subroutine then returns to the transmit character routine, which in turn returns to the calling routine. As described later, the timer interrupt routine handles the actual transmission of this character.

The *receive character routine* (`get_char_ax1`) first checks to see if a character has been received. If not, it waits until one has been received, then returns to the calling routine with the character's ASCII code in AX1.

The *enable receive routine* (`turn_rx_on`) enables the UART to receive by clearing the `rx_off` flag.

(**Note:** The UART program starts with its receive function disabled—it must be activated by calling the enable receive routine in order to receive data. Otherwise, the receive character routine will loop indefinitely, waiting for a character to be received.)

The *disable receive routine* (`turn_rx_off`) disables the UART receive function by setting the `rx_off` flag.

11 Software UART

11.4 BAUD RATES

The worst-case time required for the timer interrupt routine to execute is 55 processor cycles. Since this routine is executed for every timer interrupt, the maximum allowable baud rate can be determined by the formula:

$$\text{Baudrate} = (\text{CLK frequency}) \left(\frac{1}{55 \text{ cycles/interrupt}} \right) \left(\frac{1}{3 \text{ interrupts/baud}} \right)$$

The following table can thus be derived, assuming a timer interrupt rate of three times the baud rate:

Maximum Baud Rate (approx.)	ADSP-21xx Clock Frequency
60600	10 MHz
66600	11 MHz
72700	12 MHz
75700	12.5 MHz
78700	13 MHz
101000	16.67 MHz

11.5 AUTOBAUD FEATURE

Listing 11.2 shows `AUTOTEST.DSP`, an example program that automatically selects the proper baud rate for the UART. As noted in the program comments, the interrupt vectors used are for the ADSP-2101, but can easily be modified for other ADSP-21xx processors. Also, the code can be modified for wakeup characters other than the ASCII Bell and for other processor clock frequencies.

It is important to note that in order to use the autobaud feature of the UART, the processor's $\overline{\text{IRQ2}}$ line, as well as Flag In (FI), must be tied to the serial receive input (RX).

The `AUTOTEST.DSP` program begins by booting page zero and invoking the autobaud program. Autobauding works as follows:

1. The program counts (using the timer) three high baud periods after being awakened.

2. It then compares this amount of time to known amounts of time for different baud rates, and computes the baud rate constant. This constant is left in the AR register and can be used to initialize the UART routine's `baud_period` variable.

The program then boots page one, where the UART is initialized, turned on, and then echoes the characters input.

```
{***************************************************************
  ADSP-2101 UART Autobaud Example                 AUTOTEST.DSP

This example program waits for a bell character to interrupt IRQ2, then performs an
autobaud on it, loads the autobaud constant, and boots the next page.

***************************************************************}

.module/boot=0/abs=0     IO_shell;

{_____ADSP-2101 Interrupt Vector Table_____}

chip_reset:     call init_irq; jump main; rti; rti; {Reset Vector}
ext_IRQ2:       jump boot_pg1; rti; rti; rti;       {external IRQ2}
sport0_tx:      rti; rti; rti; rti;                 {sport0 TX}
sport0_rx:      rti; rti; rti; rti;                 {sport1 RX}
s1_tx_IRQ1:     rti; rti; rti; rti;                 {sport1 TX or IRQ1}
s1_rx_IRQ0:     rti; rti; rti; rti;                 {sport1 RX or IRQ0}
timer_done:     rti; rti; rti; rti;                 {Timer not used}

{_____main idle loop_____}

main:   idle;
        jump main;

{_____initialize interrupt for UART monitor_____}

init_irq:       icntl=0x0007;
                ifc=0x003f;         {make sure that interrupts are clear}
                nop;                {nop is for latency in loading IFC}

                ay0=0x0020;         {enable irq2}
                ar=imask;
                ar=ar or ay0;
                imask=ar;

                mr1=0x0000;         {set flag in and out on sport}
                dm(0x3fff)=mr1;

                set flag_out;       {default state for UART TX}
                rts;
```

(listing continues on next page)

11 Software UART

```
{_____IRQ2 interrupt from PC, Autobaud, Boot page 1_____

After the PC monitor sends a Bell character to wakeup the 2101, it waits
for a "1", counts cycles and waits for a "0", then selects the
appropriate baud rate. Exit the autoboot with AR = the baud period value
to the uart routine. The maximum baudrate supported is 57600. This
routine can be modified for other wakeup characters and other processor
clock frequencies. It can also be collapsed into one boot page if needed.
The IRQ2 interrupt is used to wakeup the processor; the FLAGIN pin is
used to test the level of rx.

The IRQ2 pin and the FLAG IN pin must be tied to the serial rx input.

Ascii Bell,Start bit,Stop bit = 00000111 0 1

Determine baud constant for the UART = (proc_freq/(3*Baudrate))-1

   9600 = 4062 cycles at 13MHz        if>3046              ar=450
  19200 = 2031 cycles at 13MHz        else if>1523         ar=225
  38400 = 1016 cycles at 13MHz        else if>876          ar=112
  57600 =  676 cycles at 13MHz        else                 ar= 74
  _____}

boot_pg1:   dis timer;                 {be sure timer is off}

            ar=0;
            dm(0x3ffb)=ar;             {TSCALE=0, decrement every cycle}
            ar=10000;                  {TPERIOD=TCOUNT=10000 to start}
            dm(0x3ffc)=ar;
            dm(0x3ffd)=ar;

wait_1: if not flag_in jump wait_1; {wait for first high bit}
wait_2: if not flag_in jump wait_2; {be sure it's not a glitch}
wait_3: if not flag_in jump wait_3; {be sure it stays high}

            ena timer;

count:      if flag_in jump count; {determine # cycles for 3 bauds}
            if flag_in jump count; {be sure it's not a noise spike}

            dis timer;

            ay0=dm(0x3ffc);            {get TCOUNT value}
            ax0=10000;
            ar=ax0-ay0;                {calc elapsed cycles}
            ax0=ar;                    {ax0=3 baud periods}
```

```
            ay0=3046;                    {what's the baud rate?}
            ar=ax0-ay0;
            if ge jump b_9600;

            ay0=1523;
            ar=ax0-ay0;
            if ge jump b_19200;

            ay0=876;
            ar=ax0-ay0;
            if ge jump b_38400;

b_57600:    ar= 74; jump baud_done;
b_38400:    ar=112; jump baud_done;
b_19200:    ar=225; jump baud_done;
b_9600:     ar=450;

baud_done:  MSTAT=0;                     {clear modes for boot}

            ax0=0x0A58;                  {sport 1 in flag/interrupt mode}
            dm(0x3fff)=ax0;              {force a boot to page 1}
.endmod;
```

Listing 11.2 Autobaud Example Program

11.6   CHARACTER ECHO EXAMPLE

Listing 11.3 shows a program called AUTOECHO.DSP that uses the software UART routines to provide a simple example of how to use the UART monitor. The program reads in a character and writes ("echoes") it back out.

The program also uses the autobaud capability of the UART, modified slightly for this example. (The modified autobaud code is included in the AUTOECHO.DSP file, provided on the disk accompanying this handbook.)

The only hardware needed to run this example is an ADSP-2101 EZ-LAB board and an interface board with an RS-232 line driver chip connected to the Flag In (FI) and Flag Out (FO) pins on the EZ-LAB's J2 SPORT Connector. (**Note:** You must supply the input signals to the line driver chip.)

11 Software UART

```
{*************************************************************
   ADSP-2101 EZLAB        UART Example          AUTOECHO.DSP

This program uses the software UART routines to provide a simple example of
how to use the UART monitor. The program reads in a character and writes
("echoes") it back out. The program also uses the autobaud capability of
the UART, modified slightly for this example.
 *************************************************************}

.module/boot=1/abs=0     AUTOEcho;

.external init_uart;            {UART initialize baudrate etc.}
.external turn_rx_on;           {UART enable the rx section of the uart}
.external turn_rx_off;          {UART disable the rx section of the uart}
.external out_char_ax1;         {UART output a character}
.external get_char_ax1;         {UART wait & get input character}
.external process_a_bit;        {UART timer interrupt routine for RX and TX}
.external baud_period;          {UART load with period from autobaud}

{_____ADSP-2101 Interrupt Vector Table_____}

jump start; rti; nop; nop;                 {Reset Vector}
rti; nop; nop; nop;                        {IRQ2 Interrupt}
rti; nop; nop; nop;                        {SPORT0 Transmit Interrupt}
rti; nop; nop; nop;                        {SPORT0 Receive Interrupt}
rti; nop; nop; nop;                        {SPORT1 Transmit Interrupt}
rti; nop; nop; nop;                        {SPORT1 Receive Interrupt}
jump process_a_bit; rti; nop; nop;         {Timer Interrupt}

{_____Initialization Routine_____}

start:    DM(Baud_Period)=AR;            {UART Autobaud}
          CALL Init_UART;                {Initialize UART}
          CNTR=15000;                    {Wait approximately one}
          DO xloop UNTIL CE;             { character to insure last}
xloop:      NOP;                         { one made it through}
          CNTR=15000;
          DO yloop UNTIL CE;
yloop:      NOP;

          CALL Turn_RX_On;               {Enable UART Receive}

{_____Main System Loop_____}

          DO mloop UNTIL FOREVER;
            CALL Get_Char_AX1;           {Read in character}
            CALL Out_Char_AX1;           {and Echo it back out}
mloop:      NOP;

.endmod;
```

Listing 11.3 Character Echo Program

568

Software UART 11

11.7 PROGRAM FILES

The disk included with this handbook contains the following files for the software UART:

UART.DSP	UART program subroutines
AUTOTEST.DSP	UART example that echoes received characters
AUTOBAUD.DSP	Section of program that automatically selects the correct baud rate
210X.SYS	System Builder source file used to generate the .ACH architecture file
MAKEIT.BAT	Batch file that runs the assembler, linker, and PROM splitter with the files described above

After building the 210x.SYS file, the MAKEIT.BAT file can be used to create the .BNM version of the UART program (called 2101.BNM). The MAKEIT batch program creates this by assembling the UART.DSP, AUTOTEST.DSP, and AUTOECHO.DSP programs.

It then links the assembled versions with the architecture file 210x.ACH to create the executable file 2101.EXE. After this, it runs the PROM splitter on the executable file to generate the file that will be downloaded (2101.BNM).

The UART program and 210X.SYS system file are designed to run on an ADSP-2101 EZ-LAB board connected to an RS-232 line driver chip. The files can be easily modified for other hardware configurations.

Hardware Interfacing ◾ 12

12.1   OVERVIEW

This chapter describes several hardware interface solutions for connecting ADSP-2100 Family digital signal processors to peripheral devices, such as codecs. Because the peripheral devices are also programmable, this chapter includes the code for configuring these devices. The corresponding subroutines are listed at the end of each section.

This chapter includes the following hardware interface solutions:

- ADSP-2105/AD1849 SoundPort®

- ADSP-2111/AD1849 SoundPort

- ADSP-2101/AD1847 SoundPort

- ADSP-2100 Family/DRAM Interfacing

- Loading an ADSP-2101 Program Through the Serial Port

- Memory Interfacing with the ADSP-2105

Although most sections of this chapter contain solutions for specific ADSP-2100 Family processor hardware interfaces, the code listings can be modified to accommodate additional solutions. Several sections also include suggestions for modifying the code.

When you change the code to accommodate ADSP-2100 Family DSPs other than the ones specified, you must consider differences in interrupt vectors, chip architecture, peripheral devices, and memory configurations.

12.2   SOUNDPORT INTERFACES

This section contains three solutions for interfacing the ADSP-2100 Family DSPs with AD1849 and AD1847 SoundPorts.

571

12 Hardware Interfacing

12.2.1 ADSP-2111/AD1849 SoundPort Interface

This code in this section provides the initialization of the control functions needed to connect the AD1849 SoundPort Stereo Codec to the ADSP-2111. Listing 12.1 performs an ADSP-2111/AD1849 talk-through routine.

You should consider the following points when using this program:

- The AD1849 works on 64-bit words. The DSP sees this as four 16-bit words. Be careful when you set up control words so that the appropriate control bits are in the correct locations.

- The source of SPORT signals is different between data and control modes. In *control mode*, the DSP is expected to provide all signals. In *data mode*, the AD1849 provides the signals. You must reset the DSP SPORT when you change from control mode to data mode.

- When you change sampling rates (or other system parameters), you must change the appropriate control field in the command word *and* in the mask word used to check DCB status. Also, remember that the AD1849 initiates its auto calibration sequence once the sampling rate is changed.

- If you retain indirect addressing into the receive and transmit autobuffers, the I3 pointer must be set immediately after the code enters the interrupt routine. A delay can cause pointer misalignment and render the code inoperable.

The following suggestions are included for modifying the code:

- Direct Addressing: Instead of resetting an indirect pointer into the receive and transmit buffers, try using direct addressing. For example, *dm(datain)* is the left channel data, *dm(datain+1)* is the right channel data.

- Transmit Interrupt: Because of interrupt timing and other system constraints, it may be advantageous to time your system from the transmit interrupt instead of the receive interrupt. Refer to the *ADSP-2100 Family User's Manual* for more information about the different interrupt latencies that are inherent with the different approaches.

Hardware Interfacing 12

```
{**********************************************************************
ADSP-2111 - AD1849(SOUNDPORT)  INTERFACE PROGRAM
Analog Devices  Sept, 1991

This program is written to run on an ADSP-2111. However, it can be easily
changed to work with an ADSP-2101 by modifying the interrupt vector table.

For the hardware connections between the ADSP-2111 and AD1849 please refer to
the diagrams shown on the AD1849 data sheet.

The AD1849 will be set up as follows:
(NC indicates a no care state, all control reg are 1 bit unless indicated
 by [#bits])

Control time slot control bits:
DCB=0          followed by 1
AC=1           autocalibrate
DFR=5          Data conversion frequency,44.1kH [3]
ST=1           stereo mode
DF=0           dataformat is 16 bit twos complement [2]
MCK=2          16.9344 MHz is the master clock [2]
FSEL=0         frame size, 64 bits [2]
MS=1           master mode (i.e. receive external serial clock)
TXDIS=0        enable serial output
ENL=0          disable loopback testing
ADL=NC         loopback mode analog/digital (disabled)
PIO=NC         paralell I/O bits not used [2]
REVID=NC


Data Timeslot Control Bits:
OM0=1          enable line 0 output
OM1=1          enable line 1 output
LO=0           no attenuation of left channel output [6]
SM=0           mono output is muted
RO=0           no attenuation of right channel output [6]
PIO=NC         paralell I/O bits not used [2]
OVR=NC         overrange INPUT
IS=0           line-level stereo input selected
LG=0           no gain for left channel [4]
MA=15          no monitor mix (i.e. ADC output is not mixed with DAC input
RG=0           no gain for right channel [4]

**********************************************************************

This program makes use of the multi-channel mode that is available on the
SPORT0 and it also uses autobuffering to reduce interrupt service overhead.
For a description of the multi-channel and autobuffering features, please
refer to the ADSP-2111 or ADSP-2101 architecture user's manuals.
```

(listing continues on next page)

12 Hardware Interfacing

In its current condition, this program can be used "as is" to perform straight talkthrough (at 44.1 KHz sample rate) on both left and right channels of the AD1849. The incoming audio data from the 16bit ADCs is placed in a short buffer and immediately sent out to the 16bit DACs.

The initial setup and handshaking with the AD1849 starts with the "START" routine. A state machine to perform the handshaking is executed in software. Once the AD1849 is configured properly, the processor enters the WAIT_DATA loop and waits for serial port interrupt requests.

This program is booted into the ADSP-2111 on power-up.

```
*********************************************************************}

.MODULE/ABS=0/BOOT=0      AD1849;
.VAR/CIRC   CTRLIN[4];                   {circular buffers for data input and}
.VAR/CIRC   CTRLOUT[4];                  {output for data mode and control mode}
.VAR/CIRC   DATAIN[4];
.VAR/CIRC   DATAOUT[8];
.VAR        FIRST_FLG;
.VAR        DCB_FLG;
.VAR        DMODE_FLG;

            JUMP START;nop;nop;nop;               {restart interrupt}
            RTI;nop;nop;nop;                      {IRQ2 int, not used}
            RTI;nop;nop;nop;                      {HIP write int, not used}
            RTI;nop;nop;nop;                      {HIP read int, not used}
            JUMP SETUPCONTROL;nop;nop;nop;        {SPORT0 transmit int}
            JUMP NEWDATA;nop;nop;nop;             {SPORT0 receive int}
            RTI;nop;nop;nop;                      {SPORT1 tx or IRQ1 int, not used}
            RTI;nop;nop;nop;                      {SPORT1 rx or IRQ0 int, not used}
            nop;nop;nop;nop;                      {timer int,not used}

START:      RESET FL0;                    {set the AD1849 D/C pin low}
            L0=%CTRLIN;
            M0=1;
            I0=^CTRLIN;
            L1=%CTRLOUT;
            M1=1;
            I1=^CTRLOUT;
            AX0=1;
            DM(FIRST_FLG)=AX0;
            DM(DCB_FLG)=AX0;
            AX0=0;
            DM(DMODE_FLG)=AX0;
            AY0=B#0010000100101100;
```

```
{Initialize contol mode output buffer}

            DM(I1,M1)= B#0010000100101100;    {DCB=0,AC=1,DFR=05,ST=1,DF=00}
            DM(I1,M1)= B#0010001000000000;    {MCK=02,FSEL=0,MS=0,TXDIS=0}
                                              {ENL=0,ADL=0}
            DM(I1,M1)= B#0000000000000000;    {PIO = 00}
            DM(I1,M1)= B#0000000000000000;    {REVID=NC}
            L2 = 0;                           {linear addressing for register }
            I2 = 0x3fef;                      {pt. to last DM cntrl reg}
{3FEF}      DM(I2,M1) = 0x0000;               {sport1 not used}
{3FF0}      DM(I2,M1) = 0x0000;
{3FF1}      DM(I2,M1) = 0x0000;
{3FF2}      DM(I2,M1) = 0x0000;
{3FF3}      DM(I2,M1) = 0X0283;               {autobuf. rx:i0, m0 tx:i1,m1}
{3FF4}      DM(I2,M1) = 383;                  {sport rfsdiv, sets up FRAME sync freq}
{3FF5}      DM(I2,M1) = 849;                  {sport0 sclkdiv, SCLK will be less than}
                                              {8KHz from control mode}
{3FF6}      DM(I2,M1) = B#1100010100011111;
                                              {sport0 control register:
                                              multi-channel mode, 24 channels
                                              internal sclk & rfs
                                              normal framing mode
                                              frame sync not inverted
                                              16 bit word length}
{3FF7}      DM(I2,M1) = 0x000F;               {first 4 xmit multichannels used}
{3FF8}      DM(I2,M1) = 0x0000;
{3FF9}      DM(I2,M1) = 0x000F;               {first 4 rx multichannels used}
{3FFA}      DM(I2,M1) = 0x0000;
{3FFB}      DM(I2,M1) = 0x0000;               {TSCALE register}
{3FFC}      DM(I2,M1) = 0x0000;               {TCOUNT register}
{3FFD}      DM(I2,M1) = 0x0000;               {TPERIOD register,initializing value
                                              for TCOUNT after every interrupt}
{3FFE}      DM(I2,M1) = 0x0000;               {external data memory waits=0}

            ICNTL=0x00;
            IMASK=B#00010000;                 {only SPORT transmit intrpt enabled
                                              initially while in control mode}

            AX0=DM(I1,M1);                    {send first 16bits of ctrl word}
            TX0=AX0;
{3FFF}      DM(I2,M1) = 0x1418;               {system control reg: sport0 enabled}
```

(listing continues on next page)

12 Hardware Interfacing

```
{Wait for an interrupt indicating that transmit register is ready for
 new data and that the 2111 has received a 16bit word}

WAIT1:     AX1=DM(DMODE_FLG);
           AR=PASS AX1;
           IF GT JUMP GO_DMODE;
           JUMP WAIT1;

SETUPCONTROL:
           AX0=DM(FIRST_FLG);
           AF=PASS AX0;
           IF NE JUMP DECR_FIRST;
           AX0=DM(DCB_FLG);
           AR=PASS AX0;
           IF EQ JUMP DCBFLG_SET;
           AX0=DM(CTRLIN);            {DCB_FLG has not been set yet}
           AR=AX0 XOR AY0;            {check all incoming bits including DCB bit}
           IF EQ JUMP SET_DCB;        {set flag if DCB was 0}
           RTI;

DCBFLG_SET:
           AX0=DM(CTRLIN);            {DCB_FLG was set}
           AR=AX0 AND AY0;            {only check for DCB bit}
           IF NE JUMP SETDMODE;       {if DBC=1 ready for datamode}
           RTI;

SET_DCB:   AX0=0;
           DM(DCB_FLG)=AX0;
           AX0=B#0010010100101100;    {DCB was 0, prepare to send DCB=1}
           DM(CTRLOUT)=AX0;
           AY0=B#0000010000000000;
           RTI;

DECR_FIRST:
           AX0=0;
           DM(FIRST_FLG)=AX0;
           RTI;

SETDMODE:
           IMASK=0;
           AX0=0X0818;                {disable sport0}
           DM(0X3FFF)=AX0;

{ At this point we could boot page#1 and do the following data mode setup
  after the reboot occurs. This would free up some PM RAM for other uses}
           I1 = ^DATAOUT;
           L1=0;
           DM(I1,M1) = 0x0000;        {reset output & input data buffers}
           DM(I1,M1) = 0x0000;        {initialize embedded control bits}
           DM(I1,M1) = B#1100000000000000;  {OM1=1,OM1=1,LO=0,SM=1,RO=0}
           DM(I1,M1) = B#0000000011110000;
                                      {PIO=00,OVR=1,IS=0,LG=0,MA=15,RG=0}
```

576

```
        DM(I1,M1) = 0x0000;          {reset output & input data buffers}
        DM(I1,M1) = 0x0000;          {initialize embedded control bits}
        DM(I1,M1) = B#1100000000000000;   {OM1=1,OM1=1,LO=0,SM=1,RO=0}
        DM(I1,M1) = B#0000000011110000;   {PIO=00,OVR=1,IS=0,LG=0,MA=15,RG=0}
        AX0=0X861F;
        DM(0X3FF6)=AX0;              {sport0 control:
                                     multi-channel mode
                                     external sclk & rfs
                                     32 word frames, 16bit words}
        SET FL0;                     {set D/C high}
        AX0=1;
        DM(DMODE_FLG)=AX0;
        RTI;

GO_DMODE:
        L0=%DATAIN;
        I0=^DATAIN;
        L1=%DATAOUT;
        I1=^DATAOUT;
        M2=3;
        L3=L1;
        AX0=0XFFFF;
        DM(0X3FF7)=AX0;              {enable all multi-channel words}
        DM(0X3FF8)=AX0;
        DM(0X3FF9)=AX0;
        DM(0X3FFA)=AX0;
        AX0=DM(I1,M1);               {send first 16bits of data}
        TX0=AX0;
        AX0=0X1418;
        DM(0X3FFF)=AX0;              {turn on sport0}
        IFC=B#000000111111;          {clear all pending interrupts}
        IMASK=B#00001000;            {sport0 rx interrupt on}

WAIT_DATA:
        JUMP WAIT_DATA;              {wait for sport0 rx autobuffer interrupt}

NEWDATA:
        I3=I1;
        MODIFY(I3,M2);
        AX0=DM(DATAIN);              {this routine sends the incoming}
        DM(I3,M1)=AX0;              {a/d data straight to the d/a by}
        AX0=DM(DATAIN+1);            {copying newly arrived data words from}
        DM(I3,M1)=AX0;              {DATAIN into the DATAOUT buffer}
        RTI;

{The DATAOUT buffer is twice as long as the DATAIN buffer, since it contains}
{control words, as well as output data. This program ignores the control}
{words arriving back from the AD1849 during the data mode}

.ENDMOD;
```

Listing 12.1 ADSP-2111/AD1849 Talk-Through Routine (18492111.DSP)

577

12 Hardware Interfacing

12.2.2 ADSP-2105/AD1849 SoundPort Interface

This section contains a program that provides the initialization of control functions needed to interface the AD1849 SoundPort Stereo Codec to the ADSP-2105. Listing 12.2 is an ADSP-2105/AD1849 talk-through routine: incoming data is echoed immediately to the output.

You should consider the following points when using this program:

- Multichannel mode is not available on SPORT1, so SPORT1 is run in unframed mode. Any disruption in the serial data stream can cause the AD1849 to lose synchronization. The interface code checks incoming control words, and if there is a discrepancy, the code resets SPORT1. This could cause a loss of data if the data stream is corrupted. In a stable, electrically clean environment you should not have any significant problems. The SPORT receive interrupt routine checks for errors, and this routine can be removed if your environment does not require it.

- Control for the AD1849's D/C line is derived from a latched external memory-mapped port (the *mode_sel* port definition). The program uses data bit D8 (the LSB) as the mode select flag.

- RFS1 and TFS1 must be tied together for codec initialization and operation.

Hardware Interfacing 12

```
{*************************************************************************

ADSP-2105 - AD1849(SOUNDPORT)  INTERFACE PROGRAM
Analog Devices Jan, 1992

Revised 12/10/92: Equal length Tx and Rx buffers. Direct addressing
                  for Tx autobuffer writes (NEWDATA routine)

This program is written to run on an ADSP-2105.

For the hardware connections between the ADSP-2105 and AD1849, please refer to
the diagrams shown on the AD1849 data sheet.

Note:RFS1 and TFS1 MUST be tied together in order to initialize and operate
     the codec.

The AD1849 will be set up as follows:
(NC indicates a no care state, all control reg are 1 bit unless indicated
by [#bits])

Control time slot control bits:
DCB=0       followed by 1
AC=1        autocalibrate
DFR=5       Data conversion frequency,44.1kH [3]
ST=1        stereo mode
DF=0        dataformat is 16 bit twos complement [2]
MCK=2       16.9344 MHz is the master clock [2]
FSEL=0      frame size, 64 bits [2]
MS=1        master mode (i.e. receive external serial clock)
TXDIS=0     enable serial output
ENL=0       disable loopback testing
ADL=NC      loopback mode analog/digital (disabled)
PIO=NC      parallel I/O bits not used [2]
REVID=NC


Data Timeslot Control Bits:
OM0=1       enable line 0 output
OM1=1       enable line 1 output
LO=0        no attenuation of left channel output [6]
SM=0        mono output is muted
RO=0        no attenuation of right channel output [6]
PIO=NC      parallel I/O bits not used [2]
OVR=NC      overrange INPUT
IS=0        line-level stereo input selected
LG=0        no gain for left channel [4]
MA=15       no monitor mix (i.e. ADC output is not mixed with DAC input)
RG=0        no gain for right channel [4]

*************************************************************************
```

(listing continues on next page)

12 Hardware Interfacing

In its current condition, this program can be used "as is" to perform straight talkthrough (at 44.1 KHz sample rate) on both left and right channels of the AD1849. The incoming audio data from the 16bit ADCs is placed in a short buffer and immediately sent out to the 16bit DACs.

The initial setup and handshaking with the AD1849 starts with the "START" routine. A state machine to perform the handshaking is executed in software. Once the AD1849 is configured properly, the processor enters the WAIT_DATA loop and waits for serial port interrupt requests.
This program is booted into the ADSP-2105 on power-up.

```
********************************************************************************}
.
MODULE/ABS=0/BOOT=0  AD1849;
.VAR/CIRC    CTRLIN[4];                  {circular buffers for data input and}
.VAR/CIRC    CTRLOUT[4];                 {output for data mode and control mode}
.VAR/CIRC    DATAIN[4];
.VAR/CIRC    DATAOUT[4];
.VAR         FIRST_FLG;
.VAR         DCB_FLG;
.VAR         DMODE_FLG;
.var         sync_flag;

.port mode_sel;                          {latched control for Control/Data line}

        JUMP START;nop;nop;nop;              {restart interrupt}
        RTI;nop;nop;nop;                     {IRQ2 int, not used}
        RTI;nop;nop;nop;                     {SPORT0 tx not used}
        RTI;nop;nop;nop;                     {SPORT0 rx not used}
        JUMP SETUPCONTROL;nop;nop;nop;       {SPORT1 transmit int}
        JUMP NEWDATA;nop;nop;nop;            {SPORT1 receive int}
        RTI;nop;nop;nop;                     {timer int,not used}

START:  AX0=0;
        dm(mode_sel)=AX0;                    {set AD1849 D/C pin low}
        L0=%CTRLIN;
        M0=1;
        I0=^CTRLIN;
        L1=%CTRLOUT;
        M1=1;
        I1=^CTRLOUT;
        AX0=1;
        DM(FIRST_FLG)=AX0;
        DM(DCB_FLG)=AX0;
        AX0=0;
        DM(DMODE_FLG)=AX0;                   {in control mode}
        AY0=B#0010000100101100;
```

```
{Initialize contol mode output buffer}

          DM(I1,M1)= B#0010000100101100;    {DCB=0,AC=1,DFR=05,ST=1,DF=00}
          DM(I1,M1)= B#0010001000000000;    {MCK=02,FSEL=0,MS=1,TXDIS=0}
                                            {ENL=0,ADL=0}
          DM(I1,M1)= B#0000000000000000;    {PIO = 00}
          DM(I1,M1)= B#0000000000000000;    {REVID=NC}
          L2 = 0;                           {linear addressing for register}
          I2 = 0x3fef;                      {point to last DM cntrl reg}
{3FEF}    DM(I2,M1) = 0x0283;               {sport1 autobuffer register}
{3FF0}    DM(I2,M1) = 383;                  {rfsdiv1, not really used}
{3FF1}    DM(I2,M1) = 849;                  {sclkdiv1}
{3FF2}    DM(I2,M1) = B#0100000100011111;   {sport1 control register:
                                            internal sclk & rfs
                                            normal framing mode
                                            frame sync not inverted
                                            16-bit word length      }
          I2=0x3ffb;
{3FFB}    DM(I2,M1) = 0x0000;               {TSCALE register}
{3FFC}    DM(I2,M1) = 0x0000;               {TCOUNT register}
{3FFD}    DM(I2,M1) = 0x0000;               {TPERIOD register,initializing value
                                            for TCOUNT after every interrupt }
{3FFE}    DM(I2,M1) = 0x0000;               {external data memory waits=0}

          ICNTL=0x00 ;
          IMASK=B#000100;                   {only SPORT1 tx interurpt enabled
                                            initially while in control mode }

{.... Set bit test mask for DCB bit, used in tx interrupt state machine ....}
          AY0=B#0010000100101100;           {test mask for DCB bit}

{....     send first control word to switch codec to data mode      ....}
          AX0=DM(I1,M1);                    {send first 16bits of ctrl word}
          TX1=AX0;
{3FFF}    DM(I2,M1) = 0x0c18;               {system control reg: sport1 enabled}

{.... Wait for an interrupt indicating that transmit register is ready for
   new data and that the 2105 has received a 16bit word              ....}

WAIT1:    AX1=DM(DMODE_FLG);                {check dmode flag}
          AR=PASS AX1;
          IF GT JUMP GO_DMODE;             {if set, in data mode}
          JUMP WAIT1;                      {else, wait for initialization to
                                            be completed from tx interrupt
                                            routine          }

GO_DMODE: L0=%DATAIN;                       {init I0, L0 for rx autobuffer}
          I0=^DATAIN;
          L1=%DATAOUT;                      {init I1, L1 for tx autobuffer}
          I1=^DATAOUT;
```

(listing continues on next page)

12 Hardware Interfacing

```
        M2=3;
        AX0=DM(I1,M1);                  {send first 16bits of data}
        TX1=AX0;
        AX0=0X0c18;
        DM(0X3FFF)=AX0;                 {turn on sport1}
        IFC=B#000000111111;             {clear all pending interrupts}
        nop;                            {cycle for IFC latency}
        IMASK=B#000010;                 {sport1 rx interrupt on}

WAIT_DATA:
        JUMP WAIT_DATA;                 {wait for sport1 rx autobuffer interrupt}

{sport1 tx interrupt routine}

{This routine initializes the AD1849 control mode and then waits to    }
{see if the codec is ready to be switched to data mode. The routine    }
{also initializes the transmit autobuffer with the appropriate         }
{data-mode control words. See the AD1849 data sheet for a complete}
{explanation of control word bits.                                     }
{                                                                 }
{Note: AY0 contains a bit mask and must NOT be modified elsewhere      }
{                                                                 }
{                                                                 }

SETUPCONTROL:
        AX0=DM(FIRST_FLG);             {first time through?}
        AF=PASS AX0;
        IF NE JUMP DECR_FIRST;         {if so, wait until next word transmitted}
        AX0=DM(DCB_FLG);
        AR=PASS AX0;
        IF EQ JUMP DCBFLG_SET;

        AX0=DM(CTRLIN);                {DCB_FLG has not been set yet}
        AR=AX0 XOR AY0;                {check all incoming bits including DCB bit}
        IF EQ JUMP SET_DCB;            {set flag if DCB was 0}
        RTI;

DCBFLG_SET:
        AX0=DM(CTRLIN);                {DCB_FLG was set}
        AR=AX0 AND AY0;                {only check for DCB bit}
        IF NE JUMP SETDMODE;           {if DBC=1 ready for datamode}
        RTI

SET_DCB:
        AX0=0;
        DM(DCB_FLG)=AX0;
        AX0=B#0010010100101100;        {DCB was 0, prepare to send DCB=1}
        DM(CTRLOUT)=AX0;
        AY0=B#0000010000000000;
        RTI;
```

```
DECR_FIRST: AX0=0;
            DM(FIRST_FLG)=AX0;               {if first time, set flag=0}
            RTI;

SETDMODE:   IMASK=0;
            AX0=0X0418;                      {disable sport1}
            DM(0X3FFF)=AX0;
```

{ At this point we could boot page#1 and do the following data mode setup
   after the reboot occurs. This would free up some PM RAM for other uses.

   The data mode setup initializes the transmit autobuffer control word
   elements. These values are not changed in this talkthrough application.   }

```
            I1 = ^DATAOUT;
            L1=0;
            DM(I1,M1) = 0x0000;              {reset output & input data buffers}
            DM(I1,M1) = 0x0000;              {initialize embedded control bits}
            DM(I1,M1) = B#1100000000000000;  {OM1=1,OM1=1,LO=0,SM=0,RO=0}
            DM(I1,M1) = B#0000000011110000;  {PIO=00,OVR=0,IS=0,LG=0,MA=15,RG=0}
            DM(I1,M1) = 0x0000;              {reset output & input data buffers}
            DM(I1,M1) = 0x0000;              {initialize embedded control bits}
            DM(I1,M1) = B#1100000000000000;  {OM1=1,OM1=1,LO=0,SM=0,RO=0}
            DM(I1,M1) = B#0000000011110000;  {PIO=00,OVR=0,IS=0,LG=0,MA=15,RG=0}
            AX0=0X001F;
            DM(0X3FF2)=AX0;                  {   sport1 control:}
                                             internal tfs
                                             external sclk & rfs
                                             16bit words   }

            AX0=1;
            dm(mode_sel)=AX0;                {set D/C high}
            DM(DMODE_FLG)=AX0;               {set data mode flag high}
            RTI;
```

```
{                                                                              }
{                                                                              }
{ sport1 rx interrupt routine                                                  }
{                                                                              }
{ This routine sends the incoming a/d data straight to the d/a by copying      }
{ newly arrived data words from the DATAIN buffer into the DATAOUT buffer      }
{                                                                              }
{                                                                              }
```

```
NEWDATA: AX0=DM(DATAIN);                     {get LEFT channel data}
         DM(dataout)=AX0;                    {output LEFT channel data}
         AX0=DM(DATAIN+1);                   {get RIGHT channel data}
         DM(dataout+1)=AX0;                  {output RIGHT channel data}
```

(listing continues on next page)

```
{                                                                   }
{ Error checking section (optional). Incoming control words are compared}
{ to the outgoing control words.                                    }
{                                                                   }
        ax0=dm(dataout+2);         {read known output control word}
        ay0=dm(datain+2);          {read newly received control word}
        ar=ax0-ay0;                {compare, they should be the same}
        if eq jump no_error;       {if same, no error}
        ax0=dm(sync_flag);         {if not, read SYNC_FLAG}
        ar=pass ax0;               {test for 0 or 1}
        if ne jump reset_sport;    {if 1, second failure, reset SPORT1}
        ar=pass 1;                 {else, first failure, set SYNC_FLAG}
        dm(sync_flag)=ar;
        rti;                       {return}

reset_sport:
        si=0x418;                  {disable SPORT1}
        dm(0x3fff)=si;
        si=0xc18;                  {re-enable SPORT1}
        dm(0x3fff)=si;
no_error:
        ar=pass 0;
        dm(sync_flag)=ar;          {reset SYNC_FLAG}
        rti;                       {return}
{                                                                   }
{ The DATAOUT buffer is twice as long as the DATAIN buffer, since it }
{ contains control words, as well as output data. This program ignores the }
{ control words arriving back from the AD1849 during the data mode, except }
{ for the purpose of error checking.                                }
{                                                                   }

next:   ax0=dm(0x3fff);
        ay0=b#0000001001000000;
        ar=ax0 OR ay0;
        dm(0x3fff)=ar;
        rti;
.ENDMOD;
```

Listing 12.2 ADSP-2105/AD1849 Talk-Through Routine (18492105.DSP)

Hardware Interfacing   12

12.2.3   ADSP-2101/AD1847 SoundPort Interface

This section contains a program that provides the initialization of control functions needed to interface the AD1847 SoundPort Stereo Codec to the ADSP-2101.

The AD1847 has a TDM serial interface. The code provided should work without modifications for the ADSP-2101, ADSP-2103, and ADSP-2115. The code requires minor modifications to the interrupt vector table to use it with the ADSP-2111, ADSP-2171, and ADSP-21msp5x processors. Also, you can probably use the additional flags available on these processors to optimize the code.

Analog Devices does not recommend using this code on the ADSP-2105 because the ADSP-2105 does not have a TDM serial port. For additional information about AD1847 interfacing, refer to the "README" file included with the other files for this chapter.

Listing 12.3 is an ADSP-2101/AD1847 talk-through routine.

You should consider the following points when using this program:

- Although the AD1847 and the ADSP-2101 are set for 32-word blocks, the AD1847 works on 16-word cycles. Therefore, control words are sent on time slots 0 and 16, left channel data is sent on slots 1 and 17, and right channel data is sent on slots 2 and 18. Receive data follows a similar pattern.

- The AD1847 has a two-sample buffer to allow for slower sampling rates. Since the AD1847's serial bit clock rate is fixed, the interval between time slot 0 and time slot 16 is less than the sample period. For example, for an 8 kHz sampling rate, data is expected every 125 μsec. At that sampling rate, the serial bit clock produced by the AD1847 is 12.288 MHz, yielding a time span between data samples of 20.83 μsec (81.38 ns x 16 bits/word x 16 words). The actual data received, however, is sampled at the correct time interval and stored in the AD1847's output buffer. Remember that time is needed to convert the data, and that the AD1847 can wait between generating frame syncs to insure proper sample timing.

12 Hardware Interfacing

```
.module/ram/abs=0          ad1847;

/****************************************************************************
ADSP-2101 -> AD1847: Talkthru Interface

This code provides the necessary initializations and setup to allow for
communication between the AD1847 SoundPort codec and the ADSP-2100 family
Serial Port 0.

This interface code was written around an AD1847 connected to an ADSP-2101
EZ-Lab board through the J2 SPORT Connector.

When the board is RESET, the codec is initialized for an
8kHz sampling rate of stereo PCM data.

This module can be used without modification for the ADSP-2101, ADSP-2103,
and ADSP-2115. Use with the ADSP-2111, ADSP-2171, or ADSP-21msp5x processors
would require modification of the interrupt vector table and any instructions
relating to IFC, IMASK, or any other interrupt-related structure.

This interface is not recommended for the ADSP-2105.
****************************************************************************/


.var/dm/ram/circ       rx_buf[3];              /* Status + L data + R data */
.var/dm/ram/circ       tx_buf[3];              /* Cmd + L data + R data    */
.init tx_buf:          0xc000, 0x0000, 0x0000; /* Initially set MCE        */

.var/dm/ram/circ       init_cmds[13];
/***********************************************************/
/*         Initial codec setup:                          */
/* 0:  Left Input Control: 0 gain, Line 1 input          */
/* 1:  Right Input Control: 0 gain, Line 1 input         */
/* 2:  Left Aux #1 Input Control: Muted                  */
/* 3:  Right Aux #1 Input Control: Muted                 */
/* 4:  Left Aux #2 Input Control: Muted                  */
/* 5:  Right Aux #2 Input Control: Muted                 */
/* 6:  Left DAC Control: 0 attenuation, muted            */
/* 7:  Right DAC Control: 0 attenuation, muted           */
/* 8:  Data Format: XTAL1, 8kHz sampling, stereo,        */
/*          16-bit linear PCM                            */
/* 9:  Interface Config: Playback enabled, ACAL allowed  */
/* 10: Pin Control: CLKOUT active, XCTL1/0 LO            */
/* 12: Misc. Info: Transmit on 0,1,2, 32-word frame      */
/* 13: Digital Mix Control: DME Disabled, 0 attenuation  */
/*                                                       */
/* To start-up the codec with any other properties,     */
/* change the appropriate initialization value.          */
/* Refer to the AD1847 data sheet for detailed register  */
/* descriptions.                                         */
/***********************************************************/
```

Hardware Interfacing 12

```
.init init_cmds:
            0xc000,
            0xc100,
            0xc280,
            0xc380,
            0xc480,
            0xc580,
            0xc680,
            0xc780,
            0xc850,
            0xc909,
            0xca00,
            0xcc40,
            0xcd00;

.var/dm     stat_flag;

reset_vect: jump sys_init; nop; nop; nop;

irq2_svc:   rti; nop; nop; nop;

tx0_irq:    ar = dm(stat_flag);
            ar = pass ar;
            if eq rti;
            jump next_cmd;

rx0_irq:    jump input_samples;
            rti; nop; nop;

            rti; nop; nop; nop;
            rti; nop; nop; nop;
            rti; nop; nop; nop;

/***************
 Code Start:
****************/
sys_init:
            i0 = ^rx_buf;
            l0 = %rx_buf;
            i1 = ^tx_buf;
            l1 = %tx_buf;
            i3 = ^init_cmds;
            l3 = %init_cmds;
            m1 = 1;
```

(listing continues on next page)

12 Hardware Interfacing

```
/*********************
    sport0 setup:
        multichannel enable, ext. sclk, MFD=1,
        32 words, ext. rfs, DTYPE 00, 16 bits
*********************/
        ax0 = b#1000011000001111;
        dm(0x3ff6) = ax0;


/*********************
    Multichannel enable setup:
        Rx: 3, 4, 5, 19, 20, 21
        Tx: 0, 1, 2, 16, 17, 18
*********************/
        ax0 = b#0000000000000111;
        dm(0x3ff9) = ax0;
        dm(0x3ffa) = ax0;

        ax0 = b#0000000000000111;
        dm(0x3ff7) = ax0;
        dm(0x3ff8) = ax0;


/*********************
    SPORT Autobuffer Setup
        Rx: I0, M1
        Tx: I1, M1
*********************/
        ax0 = b#0000001010000111;
        dm(0x3ff3) = ax0;

        ax0 = b#0001000000000000;
        dm(0x3fff) = ax0;               /* SPORT0 Enabled,
                                        SPORT1 = FI, FO, IRQs */
start_setup:
        ifc = b#000000111111;
        nop;

        ax0 = 1;
        dm(stat_flag) = ax0;

        imask = b#010000;               /* enable tx0 interrupt */

        ax0 = dm(i1,m1);
        tx0 = ax0;
check_init:
        ax0 = dm(stat_flag);            /* wait for entire init */
        af = pass ax0;                  /* buffer to be sent to */
        if ne jump check_init;          /* the codec           */
check_aci:
        ax0 = dm(rx_buf);               /* once initialized, wait */
        ay0 = b#0000000000000010;       /* for codec to come out  */
        ar = ax0 and ay0;               /* of autocalibration     */
        if ne jump check_aci;
```

588

```
        idle;

        ay0 = 0xbf3f;
        ax0 = dm(init_cmds+6);
        ar = ax0 AND ay0;
        dm(tx_buf) = ar;                    /* unmute left DAC */
        idle;

        ax0 = dm(init_cmds+7);
        ar = ax0 AND ay0;
        dm(tx_buf) = ar;                    /* unmute right DAC */
        idle;

        ifc = b#000000111111;
        nop;
        imask = b#001000;                   /* enable rx0 interrupt */

/***********************************************************************/
/* Main Loop: talkthru                                                */
/***********************************************************************/
talkthru:idle;
        jump talkthru;

/***********************************************************************/
input_samples:
        ena sec_reg;
        ax1 = dm(rx_buf+1);                 /* L channel input */
        mx1 = dm(rx_buf+2);                 /* R channel input */
        dm(tx_buf+2) = mx1;
        dm(tx_buf+1) = ax1;
        rti;

next_cmd:
        ena sec_reg;
        ax0 = dm(i3,m1);
        dm(tx_buf) = ax0;
        ax0 = i3;
        ay0 = ^init_cmds;
        ar = ax0 - ay0;
        if gt rti;
        ax0 = 0x8000;
        dm(tx_buf) = ax0;
        ax0 = 0;                            /* remove MCE if done initialization */
        dm(stat_flag) = ax0;                /* reset status flag */
        rti;

.endmod;
```

Listing 12.3 ADSP-2101/AD1847 Talk-Through Routine (TALK_47.DSP)

12 Hardware Interfacing

Listing 12.4 is an ADSP-2101/AD1847 demonstration routine. This program was written to demonstrate several features of the AD1847 when it is connected to the ADSP-2101 EZ-LAB® Demonstration Board.

```
.module/ram/abs=0          ad1847;

/****************************************************************************
  ADSP-2101 -> AD1847: Talkthru Interface

This code provides the necessary initializations and setup to allow for
communication between the AD1847 SoundPort codec and the ADSP-2100 family
Serial Port 0.

This interface code was written around an AD1847 connected to an ADSP-2101
EZ-Lab board through the J2 SPORT Connector.

There are two modes of operation of this interface:

1) If the board is RESET, the codec is initialized for an
   8kHz sampling rate of stereo PCM data.

2) If the FLAG_IN button is held at reset, then the code enters its AD1847
   configuration mode.

Codec attributes that can be changed are: line or microphone input source,
input gain, and the sample rate. Do not press the IRQ2 button too fast when
changing the sample rate as the codec needs time to perform its autoclibration
procedure for each sampling rate.

Initialization:
        SPORT1 must be configured as flags and interrupts.
        irq2 must be enabled.

Operation:
        Press and hold      <FLAG> button
        Press and release   <IRQ2> button
        Release             <FLAG> button to enter the setup routine

        Push <IRQ2> button to toggle between Line_1 and Line_2 input
        Push <FLAG> button to go to the next state

        Push <IRQ2> button to change the input gain, (8 levels)
        Push <FLAG> button to go to the next state

        Push <IRQ2> button to change the sample rate, (16 steps)
        Push <FLAG0> button to exit the setup routine
```

```
Line Input Gain Default:      level 0 = 0dB
Sample Rates:
          (1)      8 (default)     (9)      5.5125
          (2)      16              (10)     11.025
          (3)      27.42857        (11)     18.9
          (4)      32              (12)     22.05
          (5)      N/A             (13)     37.8
          (6)      N/A             (14)     44.1
          (7)      48              (15)     33.075
          (8)      9.6             (16)     6.615
```

This module can be used without modification for the ADSP-2101, ADSP-2103, and ADSP-2115. Use with the ADSP-2111, ADSP-2171, or ADSP-21msp5x processors would require modification of the interrupt vector table and any instructions relating to IFC, IMASK, or any other interrupt-related structure.

This interface is not recommended for the ADSP-2105.

```
*************************************************************************/

.var/dm/ram/circ        rx_buf[3];              /* Status + L data + R data */
.var/dm/ram/circ        tx_buf[3];              /* Cmd + L data + R data    */
.init tx_buf:           0xc000, 0x0000, 0x0000; /* Initially set MCE        */

.var/dm/ram/circ        init_cmds[13]; /
/*****************************************************************/
/*         Initial codec setup:                         */
/* 0:  Left Input Control: 0 gain, Line 1 input         */
/* 1:  Right Input Control: 0 gain, Line 1 input        */
/* 2:  Left Aux #1 Input Control: Muted                 */
/* 3:  Right Aux #1 Input Control: Muted                */
/* 4:  Left Aux #2 Input Control: Muted                 */
/* 5:  Right Aux #2 Input Control: Muted                */
/* 6:  Left DAC Control: 0 attenuation, muted           */
/* 7:  Right DAC Control: 0 attenuation, muted          */
/* 8:  Data Format: XTAL1, 8kHz sampling, stereo,       */
/*             16-bit linear PCM                        */
/* 9:  Interface Config: Playback enabled, ACAL allowed */
/* 10: Pin Control: CLKOUT active, XCTL1/0 LO           */
/* 12: Misc. Info: Transmit on 0,1,2, 32-word frame     */
/* 13: Digital Mix Control: DME Disabled, 0 attenuation */
/*****************************************************************/
```

(listing continues on next page)

12 Hardware Interfacing

```
        .init init_cmds:
                    0xc000,
                    0xc100,
                    0xc280,
                    0xc380,
                    0xc480,
                    0xc580,
                    0xc680,
                    0xc780,
                    0xc850,
                    0xc909,
                    0xca00,
                    0xcc40,
                    0xcd00;

    .var/dm     stat_flag;
    .var/dm     gain_state;
    .var/dm     sampling_state;
    .var/dm     chng_state;

reset_vect: jump setup_mode; nop; nop; nop;

irq2_svc:   toggle FLAG_OUT; si = 1; dm(chng_state) = si; rti;

tx0_irq:    ar = dm(stat_flag);
            ar = pass ar;
            if eq rti;
            jump next_cmd;

rx0_irq:    jump input_samples;
            rti; nop; nop;

            rti; nop; nop; nop;
            rti; nop; nop; nop;
            rti; nop; nop; nop;

/*********************
   Code Start:
*********************/
setup_mode:
            set flag_out;
            icntl = b#00100;               /* edge sensitive IRQ2 */
            ax0 = 0x0018;
            dm(0x3fff) = ax0;
            nop;
            nop;
            if FLAG_IN jump sys_init;

            ifc = b#000000111111;
            ax0 = 0;
            dm(chng_state) = ax0;

            imask = b#100000;
```

```
wait_irq:    ar = dm(chng_state);              /* wait for IRQ2 */
             ar = pass ar;
             if eq jump wait_irq;
             imask = 0x0000;

wait_rel_FO:if NOT FLAG_IN jump wait_rel_FO;   /* wait for FLAG_IN release */

             ax0 = 0;
             dm(chng_state) = ax0;

             jump set_codec;

sys_init:
             i0 = ^rx_buf;
             l0 = %rx_buf;
             i1 = ^tx_buf;
             l1 = %tx_buf;
             i3 = ^init_cmds;
             l3 = %init_cmds;

             m1 = 1;

/***********************
    sport0 setup: multichannel enable, ext. sclk, MFD=1,
          32 words, ext. rfs, DTYPE 00, 16 bits
*********************/
             ax0 = b#1000011000001111;
             dm(0x3ff6) = ax0;

/*********************
    Multichannel enable setup:
          Rx: 3, 4, 5, 19, 20, 21
          Tx: 0, 1, 2, 16, 17, 18
********************/
             ax0 = b#0000000000000111;
             dm(0x3ff9) = ax0;
             dm(0x3ffa) = ax0;

             ax0 = b#0000000000000111;
             dm(0x3ff7) = ax0;
             dm(0x3ff8) = ax0;
```

(listing continues on next page)

12 Hardware Interfacing

```
/**********************
   SPORT Autobuffer Setup
         Rx: I0, M1
         Tx: I1, M1
**********************/
         ax0 = b#0000001010000111;
         dm(0x3ff3) = ax0;

         ax0 = b#0001000000000000;
         dm(0x3fff) = ax0;                    /* SPORT0 Enabled,
                                              SPORT1 = FI, FO, IRQs */

start_setup:
         ifc = b#000000111111;
         nop;

         ax0 = 1;
         dm(stat_flag) = ax0;

         imask = b#010000;                    /* enable tx0 interrupt */

         ax0 = dm(i1,m1);
         tx0 = ax0;

check_init:
         ax0 = dm(stat_flag);                 /* wait for entire init */
         af = pass ax0;                       /* buffer to be sent to */
         if ne jump check_init;               /* the codec            */

check_aci:
         ax0 = dm(rx_buf);                    /* once initialized, wait */
         ay0 = b#0000000000000010;            /* for codec to come out  */
         ar = ax0 and ay0;                    /* of autocalibration     */
         if ne jump check_aci;

         idle;

         ay0 = 0xbf3f;
         ax0 = dm(init_cmds+6);
         ar = ax0 AND ay0;
         dm(tx_buf) = ar;                     /* unmute left DAC */
         idle;

         ax0 = dm(init_cmds+7);
         ar = ax0 AND ay0;
         dm(tx_buf) = ar;                     /* unmute right DAC */
         idle;

         ifc = b#000000111111;
         nop;
         imask = b#001000;                    /* enable rx0 interrupt */
```

```
/***********************************************************************/
/* Main Loop: talkthru                                                 */
/***********************************************************************/

talkthru:
        idle;
        jump talkthru;

/***********************************************************************/

input_samples:
        ena sec_reg;
        ax1 = dm(rx_buf+1);             /* L channel input */
        mx1 = dm(rx_buf+2);             /* R channel input */
        dm(tx_buf+2) = mx1;
        dm(tx_buf+1) = ax1;
        rti;

next_cmd:
        ena sec_reg;
        ax0 = dm(i3,m1);
        dm(tx_buf) = ax0;
        ax0 = i3;
        ay0 = ^init_cmds;
        ar = ax0 - ay0;
        if gt rti;
        ax0 = 0x8000;
        dm(tx_buf) = ax0;
        ax0 = 0;                        /* remove MCE if done initialization */
        dm(stat_flag) = ax0;            /* reset status flag */
        rti;

/****************************************************************************
 Codec configuration section
 ***************************************************************************/
set_codec:
        set FLAG_OUT;
        ax0 = 0;
        dm(gain_state) = ax0;
        dm(sampling_state) = ax0;

        IFC = b#000000111111;
        nop;
        imask = b#100000;               /* enable IRQ2 interrupt     */
```

(listing continues on next page)

12 Hardware Interfacing

```
/****************
 Input Selection
 ****************/

set_input:
        if NOT FLAG_IN jump gain_wait;
        sr0 = dm(chng_state);
        af = pass sr0;
        if eq jump set_input;

        ax0 = dm(init_cmds);
        ay0 = 0xc080;
        ar = ax0 XOR ay0;           /* if already set for line 1 in, */
        if ne jump set_to_2;        /* switch to line 2              */
        ax0 = 0xc000;
        dm(init_cmds) = ax0;
        ax0 = 0xc100;
        dm(init_cmds+1) = ax0;
        ax0 = 0;                    /* reset flag for IRQ2 */
        dm(chng_state) = ax0;
        jump set_input;

set_to_2:
        ax0 = 0xc080;
        dm(init_cmds) = ax0;
        ax0 = 0xc180;
        dm(init_cmds+1) = ax0;
        ax0 = 0;                    /* reset flag for IRQ2 */
        dm(chng_state) = ax0;
        jump set_input;

/****************
 Gain Selection
 ****************/

gain_wait:
        if NOT FLAG_IN jump gain_wait;      /* software switch de-bounce */

set_gain:
        if NOT FLAG_IN jump sampling_wait;
        sr0 = dm(chng_state);
        af = pass sr0;
        if eq jump set_gain;

        ax1 = dm(gain_state);
        ar = pass ax1;
        if gt jump next1;
```

```
          ax0 = 0xfff0;                  /* set gain = 0 for both inputs */
          ay0 = dm(init_cmds);
          ar = ax0 AND ay0;
          dm(init_cmds) = ar;
          ay0 = dm(init_cmds+1);
          ar = ax0 AND ay0;
          dm(init_cmds+1) = ar;
          ax0 = 1;
          dm(gain_state) = ax0;
          jump set_gain;

next1:    ax1 = dm(gain_state);
          ay0 = 1;
          ar = ax1 - ay0;
          if gt jump next2;

          ax0 = 0x0002;                  /* set gain = 2 = 3dB */
          call gain_adjust;
          jump set_gain;

next2:    ax1 = dm(gain_state);
          ay0 = 2;
          ar = ax1 - ay0;
          if gt jump next3;

          ax0 = 0x0006;                  /* set gain = 4 = 6dB */
          call gain_adjust;
          jump set_gain;

next3:    ax1 = dm(gain_state);
          ay0 = 3;
          ar = ax1 - ay0;
          if gt jump next4;

          ax0 = 0x0002;                  /* set gain = 6 = 9dB */
          call gain_adjust;
          jump set_gain;

next4:    ax1 = dm(gain_state);
          ay0 = 4;
          ar = ax1 - ay0;
          if gt jump next5;

          ax0 = 0x000e;                  /* set gain = 8 = 12dB */
          call gain_adjust;
          jump set_gain;
```

(listing continues on next page)

12 Hardware Interfacing

```
next5:      ax1 = dm(gain_state);
            ay0 = 5;
            ar = ax1 - ay0;
            if gt jump next6;

            ax0 = 0x0002;                    /* set gain = 10 = 15dB */
            call gain_adjust;
            jump set_gain;

next6:      ax1 = dm(gain_state);
            ay0 = 6;
            ar = ax1 - ay0;
            if gt jump next7;

            ax0 = 0x0006;                    /* set gain = 12 = 18dB */
            call gain_adjust;
            jump set_gain;

next7:      ax0 = 0x0002;                    /* set gain = 13 = 21dB */
            call gain_adjust;
            ax0 = 0;
            dm(gain_state) = ax0;
            jump set_gain;

/*————
 gain adjust
————*/
gain_adjust:
            ay0 = dm(init_cmds);
            ar = ax0 XOR ay0;
            dm(init_cmds) = ar;
            ay0 = dm(init_cmds+1);
            ar = ax0 XOR ay0;
            dm(init_cmds+1) = ar;
            ay0 = dm(gain_state);
            ar = ay0+1;
            dm(gain_state) = ar;
            ax0 = 0;
            dm(chng_state) = ax0;
            rts;
```

```
/************************
 Sampling Rate Selection
************************/
sampling_wait:
        if NOT FLAG_IN jump sampling_wait;

set_sampling:
        if NOT FLAG_IN jump done_config;
        sr0 = dm(chng_state);
        af = pass sr0;
        if eq jump set_sampling;

        ax0 = 0;
        dm(chng_state) = ax0;

set_rate:
        ar = dm(sampling_state);
        ar = pass ar;
        if gt jump to_16;

to_8:   ar = 1;                         /* buffer initialized for 8 kHz */
        dm(sampling_state) = ar;
        jump set_sampling;

to_16:  ax0 = dm(sampling_state);
        ay0 = 1;
        ar = ax0 - ay0;
        if gt jump to_27;

        ax1 = 0xc852;                   /* set for 16.0 kHz */
        dm(init_cmds+8) = ax1;
        ax0 = 2;
        dm(sampling_state) = ax0;
        jump set_sampling;

to_27:  ax0 = dm(sampling_state);
        ay0 = 2;
        ar = ax0 - ay0;
        if gt jump to_32;

        ax1 = 0xc854;                   /* set for 27.42857 kHz */
        dm(init_cmds+8) = ax1;
        ax0 = 3;
        dm(sampling_state) = ax0;
        jump set_sampling;
```

(listing continues on next page)

12 Hardware Interfacing

```
to_32:     ax0 = dm(sampling_state);
           ay0 = 3;
           ar = ax0 - ay0;
           if gt jump to_na;

           ax1 = 0xc856;                    /* set for 32.0 kHz */
           dm(init_cmds+8) = ax1;
           ax0 = 4;
           dm(sampling_state) = ax0;
           jump set_sampling;

to_na:     ax0 = dm(sampling_state);
           ay0 = 4;
           ar = ax0 - ay0;
           if gt jump to_na2;

           ax0 = 5;
           dm(sampling_state) = ax0;
           jump set_sampling;

to_na2:    ax0 = dm(sampling_state);
           ay0 = 5;
           ar = ax0 - ay0;
           if gt jump to_48;

           ax0 = 6;
           dm(sampling_state) = ax0;
           jump set_sampling;

to_48:     ax0 = dm(sampling_state);
           ay0 = 6;
           ar = ax0 - ay0;
           if gt jump to_96;

           ax1 = 0xc85c;                    /* set for 48.0 kHz */
           dm(init_cmds+8) = ax1;
           ax0 = 7;
           dm(sampling_state) = ax0;
           jump set_sampling;

to_96:     ax0 = dm(sampling_state);
           ay0 = 7;
           ar = ax0 - ay0;
           if gt jump to_55;

           ax1 = 0xc85e;                    /* set for 9.6 kHz */
           dm(init_cmds+8) = ax1;
           ax0 = 8;
           dm(sampling_state) = ax0;
           jump set_sampling;
```

```
to_55:     ax0 = dm(sampling_state);
           ay0 = 8;
           ar = ax0 - ay0;
           if gt jump to_11;

           ax1 = 0xc851;                    /* set for 5.5125 kHz */
           dm(init_cmds+8) = ax1;
           ax0 = 9;
           dm(sampling_state) = ax0;
           jump set_sampling;

to_11:     ax0 = dm(sampling_state);
           ay0 = 9;
           ar = ax0 - ay0;
           if gt jump to_18;

           ax1 = 0xc853;                    /* set for 11.025 kHz */
           dm(init_cmds+8) = ax1;
           ax0 = 10;
           dm(sampling_state) = ax0;
           jump set_sampling;

to_18:     ax0 = dm(sampling_state);
           ay0 = 10;
           ar = ax0 - ay0;
           if gt jump to_22;

           ax1 = 0xc855;                    /* set for 18.9 kHz */
           dm(init_cmds+8) = ax1;
           ax0 = 11;
           dm(sampling_state) = ax0;
           jump set_sampling;

to_22:     ax0 = dm(sampling_state);
           ay0 = 11;
           ar = ax0 - ay0;
           if gt jump to_37;

           ax1 = 0xc857;                    /* set for 22.05 kHz */
           dm(init_cmds+8) = ax1;
           ax0 = 12;
           dm(sampling_state) = ax0;
           jump set_sampling;
```

(listing continues on next page)

12 Hardware Interfacing

```
to_37:    ax0 = dm(sampling_state);
          ay0 = 12;
          ar = ax0 - ay0;
          if gt jump to_44;

          ax1 = 0xc859;                  /* set for 37.8 kHz */
          dm(init_cmds+8) = ax1;
          ax0 = 13;
          dm(sampling_state) = ax0;
          jump set_sampling;

to_44:    ax0 = dm(sampling_state);
          ay0 = 13;
          ar = ax0 - ay0;
          if gt jump to_33;

          ax1 = 0xc85b;                  /* set for 44.1 kHz */
          dm(init_cmds+8) = ax1;
          ax0 = 14;
          dm(sampling_state) = ax0;
          jump set_sampling;

to_33:    ax0 = dm(sampling_state);
          ay0 = 14;
          ar = ax0 - ay0;
          if gt jump to_66;

          ax1 = 0xc85d;                  /* set for 33.075 kHz */
          dm(init_cmds+8) = ax1;
          ax0 = 15;
          dm(sampling_state) = ax0;
          jump set_sampling;

to_66:    ax0 = 0;
          dm(sampling_state) = ax0;

          ax1 = 0xc85f;                  /* set for 6.615 kHz */
          dm(init_cmds+8) = ax1;
          jump set_sampling;


done_config:
          toggle FLAG_OUT;
          jump sys_init;
.endmod;
```

Listing 12.4 ADSP-2101/AD1847 Demonstration Routine (DEMO_47.DSP)

Hardware Interfacing 12

12.3 INTERFACING DRAMS WITH THE ADSP-2100 FAMILY

As new algorithms for digital signal processors are developed, the memory requirements for these applications will continue to grow. Not only will these applications require more memory, but they will require increased efficiency for data storage and retrieval. Examples of these new applications include digital processing for two- and three-dimensional graphic images and speech storage for voice mail systems and digital telephone answering machines.

Two common storage options for these applications are Static Random Access Memories (SRAMs) and Dynamic Random Access Memories (DRAMs).

The functional difference between an SRAM and a DRAM is how the devices store data. In an SRAM, data is stored in transistors that hold their value until you overwrite them with new data. The DRAM stores data in capacitors that gradually loose their charge and, without refreshing, will loose the data.

Size is another important difference. Capacitors are significantly smaller than transistors, so DRAMs have a higher bit density.

Although a DRAM package is smaller than the equivalent memory size in an SRAM package, you should consider the following factors when you decide which memory device to use:

- Capacitor leakage.
  Over time, DRAM capacitors "leak" or lose current and they must be "refreshed" periodically.

- Multiplexed-addressing.
  To take advantage of the higher bit density of the DRAM, the capacitor cells are accessed using multiplexed-addressing. This reduces the required number of address pins because you can use each pin twice: once to address a memory row and a second time to address a memory column. Although this makes the package smaller, it complicates memory addressing.

- Availability, price, and performance.
  DRAMs are readily available in 1 Mbit, 4 Mbit, and larger capacities, while SRAMs are typically available in 64 Kbyte (512 Kbit) and 128 Kbyte (1 Mbit) ranges. DRAMs are usually less expensive than SRAMs, but DRAM access times are significantly slower than SRAM access times.

603

12 Hardware Interfacing

- Program and hardware simplicity.
  The SRAM has a simple microprocessor interface. With N address lines, you can address 2^N sequential locations in SRAM. You do not need additional software overhead (to refresh) when your program reads or writes to memory. Each read or write requires only a single DSP instruction. When using the ADSP-2100 family, each instruction takes one clock cycle. If necessary, wait states can be programmed for addressing slower memories.

  When you use a DSP to address DRAM, you need additional hardware or software to handle the multiplexed row and column addressing and the refresh or precharge requirements. Many systems with DRAM use a hardware-intensive solution, like a DRAM controller.

  As an alternative, you can move the control functions into the DSP software. This is a good implementation for systems requiring large memory spaces and cost efficient solutions. The chip count is reduced at the expense of decreased software throughput.

Although DSPs were designed to interface with SRAMs, in the right application, DRAMs can provide a larger, more cost-effective storage medium.

This chapter presents a software solution for addressing DRAM with the ADSP-2100 family. To implement this solution, the ADSP-2101 EZ-LAB® Evaluation Board was used as the development platform. A simple DRAM interface board was designed for this application. The DRAM interface board connects to the EZ-LAB through the expansion connector to let the EZ-LAB access the interface board's bank of memory. Figure 12.1 is a block diagram of the test system. The subroutines included at the end of this chapter were verified on this interface board and the ADSP-2101 EZ-LAB®.

Hardware Interfacing 12

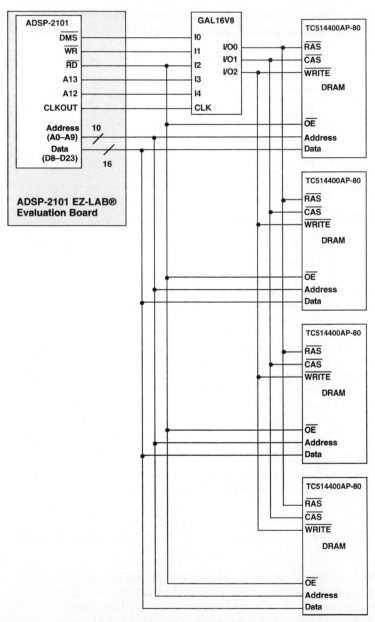

Figure 12.1 Functional Block Diagram Of DRAM Interface Test System

12 Hardware Interfacing

Depending on your application and DRAM selection, you may have to modify the subroutines and the PAL equations included in this chapter.

Table 12.1 lists the components used in the test system described above.

ADSP-2101 EZ-LAB Evaluation Board	1
1 M x 4 DRAM, PN TC514400AP-80	4
DRAM, 80 ns access time	
Programmable Logic Device,	
PN GAL16V8-15	1
Ribbon Cable Connector,	1
60-Pin, Straight wire-wrap	
Wire-wrap board	1

Table 12.1 Test System Components

When you design a DRAM memory system for an ADSP-2100 family DSP, you must consider the following points:

- DRAM configuration

- Multiplexed memory addressing

- DSP and DRAM control signals

- DSP to DRAM Interface timing

- DRAM Memory access modes

- DRAM Refresh

12.3.1 DRAM Configuration

Originally, DRAMs were organized in one-bit widths, such as 256 K x 1 or 1 M x 1. Since ADSP-2100 family DSPs are 16-bit fixed point processors, it would have taken 16 DRAMs to store data memory. As DRAM chip organization evolves, larger memories and wider widths are becoming available. DRAMs are available up to 4 M and larger with 4-bit and 16-bit widths.

You must make a trade-off between system cost, power consumption, and available board space when you choose DRAMs. Larger and wider DRAMs are more expensive and consume more power than smaller DRAMs, but you need fewer chips for the same amount of memory.

Hardware Interfacing   12

For this application, 1 M x 4 DRAMs (available from several manufactures) were chosen. These four DRAMs provide 1 M x 16 data memory.

12.3.2   Multiplexed Memory Addressing

To address a 1 M x 4 DRAM, you need ten address lines. With ten multiplexed address lines, you can address each row (2^{10}, or 1024 rows) and each column (2^{10}, or 1024 columns) in the DRAM. This is equal to 1024 x 1024, or 1 M, of addressable, 4-bit locations.

The ADSP-2100 family processors have 14 address lines for data memory. In an SRAM this can provide up to 2^{14}, or 16,384 (16 K) addressable data memory locations. In this application, you only need ten address lines to address 4, 1 M x 4 DRAMs; this system uses A0–A9 of the ADSP-2101 (see Figure 12.1). Since the remaining address lines are not used to address memory, they could be left unconnected, but this system uses two available address lines (A12 and A13) as control lines for the DRAMs.

12.3.3   DSP & DRAM Control Signals

To interface a DRAM to the DSP, compare the control signals and timing diagrams for both parts. Although this section has a brief overview of DSP and DRAM timing, you must determine the most effective use of the ADSP-2100 family control signals to control the reads and writes to DRAM in your application.

The DSP uses several control lines to access data memory: $\overline{\text{RD}}$, $\overline{\text{WR}}$, $\overline{\text{DMS}}$. This application uses $\overline{\text{DMS}}$ (Data Memory Select) to differentiate between a program and data memory access because the program memory (PM) and data memory (DM) address and data lines are multiplexed off-chip.

12.3.3.1 DSP Read/Write Timing

Figure 12.2 shows the read and write timing for the DSP. The read cycle (or write cycle) begins when the processor puts the address on the data memory address (DMA) bus and asserts $\overline{\text{DMS}}$. The $\overline{\text{RD}}$ (or $\overline{\text{WR}}$) signal is then asserted. Data is placed on the data bus within a specified time, then $\overline{\text{RD}}$ (or $\overline{\text{WR}}$) is deasserted. Finally, $\overline{\text{DMS}}$ is deasserted, ending the memory access.

12 Hardware Interfacing

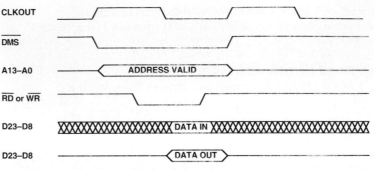

Figure 12.2 DSP Read/Write Timing

12.3.3.2 DRAM Read/Write Timing

The DRAM uses the control signals $\overline{\text{RAS}}$ (Row Address Strobe), $\overline{\text{CAS}}$ (Column Address Strobe), $\overline{\text{WRITE}}$, and $\overline{\text{OE}}$ (Output Enable).

A DRAM read cycle (see Figure 12.3 for the read cycle timing) starts when the falling edge of $\overline{\text{RAS}}$ strobes the row address into the DRAM. The falling edge of $\overline{\text{CAS}}$ strobes the column address into the DRAM and, after an access delay, enables the output buffer. The $\overline{\text{WRITE}}$ signal must stay high before and after the falling edge of $\overline{\text{CAS}}$. The read cycle ends when the $\overline{\text{RAS}}$ and $\overline{\text{CAS}}$ lines are brought high.

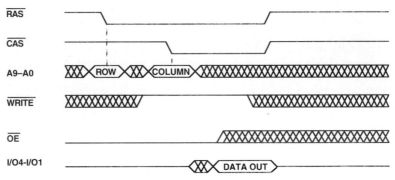

Figure 12.3 DRAM Read Cycle Timing

Another read cannot occur until the precharge time is met after $\overline{\text{RAS}}$ and $\overline{\text{CAS}}$ are brought high. There are minimum pulse-widths and setup and hold times associated with all control lines.

Hardware Interfacing 12

The DRAM write cycle is similar to the read cycle, except the $\overline{\text{WRITE}}$ line is held low. There are two basic write cycles for DRAMs: the early-write cycle and the delayed-write cycle. In the early-write cycle, the $\overline{\text{WRITE}}$ line is asserted before the assertion of $\overline{\text{CAS}}$; in the delayed-write cycle, it is asserted after. For this DSP interface, use the delayed-write (or output enabled write) cycle.

A DRAM delayed-write cycle (write cycle timing shown in Figure 12.4) also starts when the falling edge of $\overline{\text{RAS}}$ strobes the row address into the DRAM. The falling edge of $\overline{\text{CAS}}$ strobes the column address into the DRAM. Then, the falling edge of $\overline{\text{WRITE}}$ latches the data into the DRAM. There are minimum pulse-widths and setup and hold times associated with these control lines as well.

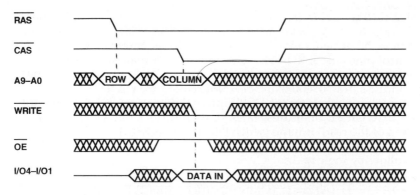

Figure 12.4 DRAM Delayed-Write (Output Enabled) Cycle Timing

12.3.3.3 $\overline{\text{RAS}}$ Generation

Because the falling edge of $\overline{\text{RAS}}$ latches the row address into the DRAM, the address must be valid before $\overline{\text{RAS}}$ is deasserted.

To drive $\overline{\text{RAS}}$, use a combination of the $\overline{\text{DMS}}$, $\overline{\text{RD}}$, and address lines A13 and A12. A read from data memory locations 0x1000 to 0x13FF drives $\overline{\text{RAS}}$ low and latches the row address from 0x000 to 0x3FF. To drive $\overline{\text{RAS}}$ high again, read from any data memory address in the range 0x2000 to 0x27FF.

12 Hardware Interfacing

Use the following logic to drive $\overline{RAS}$:

When (A13 = 0), (A12 = 1), ($\overline{DMS}$ = 0), and ($\overline{RD}$ = 0),
$\overline{RAS}$ = 0;

or, when (A13 = 1), (A12 = 0), ($\overline{DMS}$ = 0), and ($\overline{RD}$ = 0),
$\overline{RAS}$ =1;

otherwise, $\overline{RAS}$ = $\overline{RAS}$;

You can implement the above logic discretely, or in a PAL, such as the GAL16V8 used in this application.

12.3.3.4 $\overline{CAS}$ Generation

The generation of $\overline{CAS}$ must occur after the generation of $\overline{RAS}$. Since the negative transition of $\overline{CAS}$ latches the column address into the DRAM, the address must be present before $\overline{CAS}$ goes low. To generate $\overline{CAS}$, use the same control pins ($\overline{DMS}$, $\overline{RD}$, $\overline{WR}$, A13, and A12). A read or write to the data memory addresses 0x3000 to 0x33FF drives $\overline{CAS}$ low and latches the column address from 0x000 to 0x3FF. (Note a read from data memory 0x3800 or higher reads from the DSP's internal data memory and does not assert the external $\overline{RD}$ or address lines.) $\overline{CAS}$ returns to logic high at the completion of the read or write (when $\overline{DMS}$ returns high).

Use the following logic to drive $\overline{CAS}$:

When (A13 = 1), (A12 = 1), ($\overline{DMS}$ = 0), (($\overline{RD}$ = 0)#($\overline{WR}$ =0))),
then $\overline{CAS}$ = 0;

otherwise $\overline{CAS}$ = 1;

The actual read or write from the DSP occurs when $\overline{CAS}$ is low. To meet the write timing requirements of the DRAM and the DSP, $\overline{CAS}$ must be held low for two clock cycles. To accomplish this, use the external data memory wait states of the processor.

Data memory has several configurable wait states. The address range for $\overline{CAS}$ (0x3000–0x33FF) is configured by setting DWAIT3 in the data memory wait state control register. For this application, set DWAIT3 to 1.

Hardware Interfacing   12

12.3.3.5 $\overline{WRITE}$ & $\overline{OE}$ Generation

The DRAM latches data on the falling edge of $\overline{WRITE}$, while the DSP latches data on the rising edge. To ensure that the data from the DSP is still valid during the high-to-low transition of $\overline{WRITE}$, the $\overline{WR}$ signal must be delayed one clock cycle. By delaying $\overline{WR}$ one cycle to create $\overline{WRITE}$, you create a falling edge to latch valid data into the DRAM (see Figure 12.6). You can use a D flip-flop or PAL with registers to delay $\overline{WR}$.

The $\overline{RD}$ output of the DSP can be run directly into the $\overline{OE}$ input of the DRAM.

12.3.4 DSP To DRAM Interface Timing

This section describes the specific interface timing required between the DSP and DRAM.

12.3.4.1 DRAM Read Timing

To read from the DRAM, the $\overline{OE}$ signal of the DRAM is tied to the DSP's $\overline{RD}$ line. Data is latched into the DSP on the rising edge of the $\overline{RD}$ signal and this matches the availability of data from the DRAM. Figure 12.5 shows the timing for generating the $\overline{RAS}$ and $\overline{CAS}$ signals to read from DRAM.

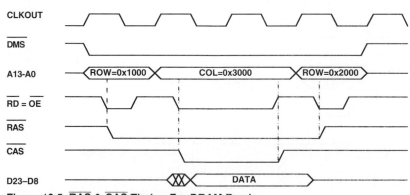

Figure 12.5 $\overline{RAS}$ & $\overline{CAS}$ Timing For DRAM Read

To accomplish the read cycle, execute a dummy read to a DM location (0x1000 to 0x13FF) to select a row and generate $\overline{RAS}$. Follow this cycle with a read from external DM (0x3000 to 0x33FF), which selects a column, makes data available on bus lines, and latches it on the DSP. To finish the read cycle, perform a dummy read to a DM location (0x2000) to drive the $\overline{RAS}$ line high again.

12 Hardware Interfacing

Implemented in ADSP-2100 family assembly code, a DRAM read requires four DSP cycles. A fifth cycle (nop) is needed when consecutive reads are performed to assure $\overline{\text{RAS}}$ precharge timing is met. The following example illustrates a read from row address 0xABC and column address 0xDEF:

```
DRAM_read:
    ax0=DM(0x1ABC);           /* dummy read sets !RAS, ROW addr */
    ay0=DM(0x3DEF);           /* !CAS set, ay0=DRAM data */
    ax0=DM(0x2000);           /* dummy read to deselect !RAS */
    nop;                      /* necessary for precharge time */
```

This may appear to be excessive overhead for a single memory read. You can achieve faster reads by using a DRAM with Fast-Page addressing (described later in this chapter).

12.3.4.2 DRAM Write Timing

A write cycle is similar to a read cycle. For a write to the DRAM, the DSP's $\overline{\text{WR}}$ line must be delayed to create the DRAM $\overline{\text{WRITE}}$ signal. The DRAM $\overline{\text{WRITE}}$ signal is generated from a clocked D-flip flop within the GAL16V8 with the ADSP-2101 $\overline{\text{WR}}$ signal as input. The clock is obtained from DSP's CLKOUT signal. Figure 12.6 shows the timing for generating the $\overline{\text{RAS}}$ and $\overline{\text{CAS}}$ signals to write to DRAM.

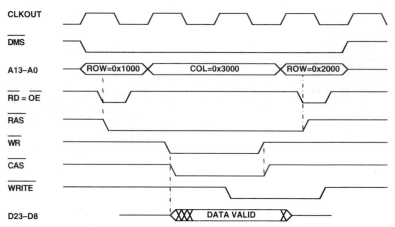

Figure 12.6 $\overline{\text{RAS}}$ & $\overline{\text{CAS}}$ Timing For DRAM Write

Hardware Interfacing 12

For a delayed-write (or output-enable write), the falling edge of $\overline{\text{WRITE}}$ latches the data into the DRAM.

In ADSP-2100 family assembly code, a DRAM write requires four DSP cycles. A fifth cycle (nop) is needed when consecutive writes are performed to assure $\overline{\text{RAS}}$ precharge timing is met. The following example shows you how to write to row address 0xABC and column address 0xDEF:

```
DRAM_write:
    ax0=DM(0x1ABC);          /* dummy read sets !RAS, ROW addr */
    DM(0x3DEF)=ay0;          /* !CAS set, ay0 written to DRAM */
    ax0=DM(0x2000);          /* dummy read to deselect !RAS */
    nop;                     /* necessary for precharge time */
```

12.3.5   Memory Access Modes
DRAMs support several memory access modes to help reduce memory access times. The access mode that you select depends on the modes supported by the particular DRAM used in your application. Memory access modes include: Page Mode, Enhanced or Fast Page Mode, Static Column Mode, and Nibble or Ripple Mode. This section describes two modes: page mode and fast page mode.

12.3.5.1 Page Mode
Page mode is the simplest memory access for a DRAM, but it also takes the longest time. To read one memory location requires a row and then a column access. A "page" is equivalent to a row.

Page-access provides quick access to memory locations in a page and can be accomplished by any DRAM. Page mode starts as a normal access when $\overline{\text{RAS}}$ is driven low. A memory access is accomplished by asserting $\overline{\text{CAS}}$ and performing a read or write. While keeping $\overline{\text{RAS}}$ low, you can access any other column location in the page by again asserting $\overline{\text{CAS}}$. Since $\overline{\text{RAS}}$ remains low, the $\overline{\text{RAS}}$ precharge time is saved, which results in a faster access speed.

12.3.5.2 Enhanced Or Fast Page Mode
Fast page mode requires a DRAM that supports this feature. It is similar to page-mode access because you can access multiple columns within one page. However, it is faster than normal page mode because the column access is started as soon as the new address is placed on the DRAM address input. This saves the $\overline{\text{CAS}}$ precharge time.

12 Hardware Interfacing

Memory access in fast page mode is possible using the available signals. In this mode, memory reads or writes to dynamic memory are executed by leaving the $\overline{RAS}$ signal low after the row latch occurs. You can continuously fill columns in the same row. The only limitation is that $\overline{RAS}$ has a maximum pulse width timing requirement (for example 200 µs).

Multiple reads or writes are performed much faster if done within the same page. The following example shows you how to read three locations in one row;

```
DRAM_page_reads:
      ax0=DM(0x1ABC);             /* dummy read sets !RAS, ROW addr */
      ay0=DM(0x3123);             /* !CAS set, ay0=DRAM data */
      ay1=DM(0x3456);             /* !CAS set, ay0=DRAM data */
      ax1=DM(0x3789);             /* !CAS set, ay0=DRAM data */
      ax0=DM(0x2000);             /* dummy read to deselect !RAS */
      nop;                        /* necessary for precharge time */
```

12.3.6   DRAM Refresh

The DRAM stores data in capacitor cells. Since capacitors lose their charge, or "leak", over time, each cell of the DRAM must be refreshed periodically to maintain adequate voltage levels. Each memory read actually causes the capacitors to discharge slightly, so DRAMs have built-in write-back features that follow a read. This write-back feature is called a "precharge". The precharge occurs after the read when both $\overline{RAS}$ and $\overline{CAS}$ have returned high. It is part of the read cycle timing and you must account for it in timing analysis.

The read/write-back of the DRAM recharges the capacitor cells. The DRAM architecture refreshes every column cell when a row is accessed. Therefore, to accomplish a refresh of all memory cells, it is only necessary to read every DRAM row within the specified refresh period of your DRAM.

The $\overline{RAS}$-only refresh is the simplest type of refresh operation and is supported by all DRAMs. To accomplish an $\overline{RAS}$-only refresh, the $\overline{RAS}$ line is brought low one cycle for each row of the DRAM while the $\overline{CAS}$ remains high.

On the DRAM interface, this is done by reading from external data memory twice. One read sends $\overline{RAS}$ low and latches the row address, the other read brings $\overline{RAS}$ back high. CAS ($\overline{DMS}$) remains high through the entire $\overline{RAS}$ cycle. The following example shows you one method to refresh DRAM.

```
DRAM_refresh:
    ax0=DM(0x1ABC);              /* dummy read sets !RAS, ROW addr */
    ax0=DM(0x2000);              /* dummy read to deselect !RAS */
```

Note: You may have to use a NOP instruction at the end of the refresh loop if the processor's instruction rate is less than 100 ns. The NOP instruction is necessary to meet the $\overline{RAS}$ precharge time and varies with DRAM access time (the DRAMs used in this application need a minimum $\overline{RAS}$ precharge time of 60 ns).

Some DRAMs have other refresh methods built-in. These include the hidden refresh, the CAS-before-RAS refresh, and refresh with scrubbing. For simplicity, this application concentrates on the RAS-only refresh.

12.3.7   DRAM Refresh Timing

Typical DRAM refresh periods are 4–16 ms. Depending on your application, you can refresh all the rows at one time, called a burst refresh, or interleave refreshes with normal memory accesses, called an interleaved refresh.

If you use a 1 M deep DRAM with a 16 ms refresh period, you must access each of the 1024 rows during the 16 ms interval. The overall time needed to refresh the DRAM is the same regardless of whether you choose burst or interleaved refreshes. To refresh 1024 rows, you need 1024 x 2 cycles since it takes one cycle to assert $\overline{RAS}$ and one cycle to deassert $\overline{RAS}$. For a DSP running at 10 MHz, there is a 100 ns cycle time. So, your 2048 cycles will take 2048 x 100 ns, or about 0.2 ms.

If your application can afford to pause for 0.2 ms every 16 ms, then the burst method is the simplest. If not, consider breaking the 0.2 ms into more manageable pieces, for example ten refreshes of 0.02 ms each.

Typical real-time applications are based on some periodic input or cycle. You can use this timing to determine when to perform the refresh. Otherwise, you must generate the period interrupt yourself to ensure that refresh occurs to prevent data loss. With ADSP-2100 family DSPs, you can use the internal timer to generate this periodic input. If the timer is not available, use an external interrupt through $\overline{IRQ2}$. If the external interrupt is not available, use a serial port to create a periodic interrupt even if the port is not being used for communication.

12 Hardware Interfacing

The following code example refreshes all 1M DRAM locations within 16 ms. Since a 1M DRAM consists of 1024 rows x 1024 columns, you need to access 1024 rows within the 16 ms. The code uses the DSP's internal timer to create a periodic interrupt that refreshes 256 rows every 4 ms. For this example, assume the DSP is running at 16.67 MHz (60 ns clock cycle).

To refresh every 4 ms using 60 ns cycles, you need to create a timer interrupt every 4 ms/60 ns, or 66,667 clock cycles. If you set TSCALE to 2, TCOUNT is decremented every third clock cycle. Setting TCOUNT and TPERIOD to 22,222 or 0x56CE generates a timer interrupt every 66,666 cycles.

```
Timer_initialization:
        ax0=0x2                      /* TSCALE=2,  */
        dm(0x3ffb)=ax0;
        ax0=0x56CE                   /* TCOUNT=22,222 */
        dm(0x3ffc)=ax0;
        ax0=0x56CE                   /* TPERIOD=22,222 */
        dm(0x3ffd)=ax0;
```

During a timer interrupt, the timer interrupt service routine calls the following refresh subroutine.

```
/* Refresh for 256 rows

bank1: 000-0FF
bank2: 100-1FF
bank3: 200-2FF
bank4: 300-3FF    */

DRAM_timer_refresh:
        ax0=DM(bank);            /* determine which bank to refresh */
        ay0=0x3;                 /* set ay0 for masking */
        ay1=1;                   /* set ay1 for incrementing */
        ar=ax0+ay0;              /* update for next bank */
        ar=ar and ay1;           /* mask upper bits (bank = 0-3) */
        DM(bank)=ar;             /* store new bank for next refresh */
        sr=lshift ar by 8 (lo);  /* left shift bank by 8 */
        M0=SR0;
        I0=0x1000;               /* start of RAS addresses */
        modify(I0,M0);           /* offset to start of bank */
        M1=1;
        CNTR=256;
```

```
  do refresh until ce;
      ax0=DM(I0,M1);      /* dummy read sets !RAS, ROW addr */
      ax0=DM(0x2000);     /* dummy read to deselect !RAS */
      refresh: nop;       /* end of loop (nop for precharge) */
  rts;
```

12.3.8   EZ-LAB Implementation

To illustrate a DRAM interface to a DSP, this application uses the ADSP-2101 EZ-LAB evaluation board and a DRAM expansion card. The 2101 EZ-LAB has an ADSP-2101 processor and 64k x 8 EPROM. The EPROM is used only for booting the internal program RAM of the DSP. No additional data memory is on the board, however all data, address, and control lines are available through the connector.

To add DRAM to the EZ-LAB, the expansion card is connected to the EZ-LAB with a ribbon cable (see Figure 12.7). The DRAM interface card has one programmable logic device (GAL16V8), and four 1M x 4 DRAMs, (TC514400). Since the ADSP-2101 is a 16-bit fixed point DSP, the application needs four 1M x 4 DRAMs for 16-bit wide memory. The 16V8 PAL provides the glue logic to create the DRAM control signals, $\overline{RAS}$, $\overline{CAS}$, and $\overline{WRITE}$.

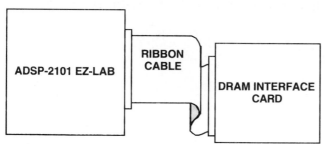

Figure 12.7 EZ-LAB/DRAM Interface Board Connection

12 Hardware Interfacing

12.3.9 DRAM Program Listings

This section contains several program listings that were tested on the
application described in this chapter.

```
.module/ram/boot=0DRAM_read;
/*
File Name:   DRAMRD.DSP
Version:     Version 0.00
Purpose:     Conducts a random read from DRAM
Calling parameters:
             setras: points to read location for RAS deassertion
             i3: points to the row being accessed
             i2: points to the column being accessed
Return Values
             ax1: contains data stored in location (row = i3, column = i2)
Registers affected:
             ax0, ax1, i2
Computation Time:
             5 cycles
             Random read from DRAM = 4 cycles ( dwait3 = 1)
*/

.external setras;
.entry read;

read: ax0 = dm(i3, m3);                  /* select row for DRAM access */
      ax1 = dm(i2, m2);                  /* read from column pointed by i2 */
      ax0 = dm(setras);                  /* deassert RAS */
      rts;

.endmod;
```

Listing 12.5 DRAM Read Program

```
.module/ram/boot=0DRAM_write;
/*
File Name:    DRAMWR.DSP
Version:      Version 0.00
Purpose:      Conducts a random write to DRAM
Calling parameters:
              setras: points to read location for RAS deassertion
              i3: points to the row being accessed
              i2: points to the column being accessed
              mx0: contains the data that requires to be written
Return Values:
        data in mx0 stored in location (row = i3, column = i2)
Registers affected:
              ax0, i2
Computation Time:
              5 cycles
              Random (OE controlled write) to DRAM = 4 cycles ( dwait3 = 1)
*/

.external setras;
.entry write;

write: ax0 = dm(i3, m3);      /* read selects row to be accessed */
       dm(i2, m2) = mx0;             /* write to column pointed by i2 */
       ax0 = dm(setras);            /* deassert RAS */
       rts;

.endmod;
```

Listing 12.6 DRAM Write Program

12 Hardware Interfacing

```
.module/ram/boot=0DRAM_refresh;
/*
File Name:   DRAMREF.DSP
Version:     Version 0.00
Purpose:     Conduct memory accesses to DRAM for refresh
Calling Parameters:
             refcntr: has the number of rows to refresh
             setras: points to the read location for RAS deassertion
             i6: contains the address of the first row to refresh
Return Values:
             Refresh of the number of rows specified by refcntr
Registers Affected:
             ax0, i6
Computation Time:
             Total execution time = 3 + cntr(3) cycles;
             RAS-only refresh per row = 3 cycles
*/

.external refcntr;
.external setras;
.external row;
.entry refresh;

refresh: cntr = dm(refcntr);               /* number of rows to refresh */
      do rasonly until ce;
      ax0 = dm(i6, m6);                     /* refresh row i6 */
      ax0 = dm(setras);                     /* deassert RAS */
rasonly: nop;                               /* inserted to meet RAS precharge */
      rts;

.endmod;
```

Listing 12.7 DRAM Refresh Program

Hardware Interfacing　12

```
.module/boot=0/abs=0     DRAM_test;
/*
File Name:    DRAMTST.DSP
Version:      Version 0.00
Purpose:      Assembly source code for Dynamic memory test. This code was
              written for the purpose of testing the DRAM interface board using
              the EZ-ICE emulator. It accomodates testing of refresh as well as
              independence of address lines.

              It is designed with two sections, the first fills all of the
              external DRAM and the second section reads and tests stored values
              for errors.

              This routine also sustains the DRAM storage using a timer
              implemented burst refresh every 16ms.
Return Values:
              buffer of 256 locations in internal data memory of error values
              followed by the actual data values that should have been read back
              from the DRAMs.
*/

.const rows        = 1024;                 /* number of rows */
.const cols        = 1024;                 /* number of columns */
.const ref_rows    = 1024;
.const maxfill     = 0xffff;               /* maximum fill value */
.var/dm/ram/seg=int_dm nbr_err, err_row, err_col, err_ov, buffer, refcntr;
.var/dm/ram/seg=row_deassert setras;
.var/dm/ram/seg=row_range row;
.var/dm/ram/seg=col_range col;
.global row, setras, refcntr;

/* _____ Interrupt vectors _____ */
     jump main; nop; nop; nop;             /* reset interrupt */
     rti; nop; nop; nop;                   /* irq2 */
     rti; nop; nop; nop;                   /* sport0 transmit */
     rti; nop; nop; nop;                   /* sport0 receive */
     rti; nop; nop; nop;                   /* sport1 transmit or irq1 */
     rti; nop; nop; nop;                   /* sport1 receive or irq0 */
     i6 = ^row; m6 = 0;
     call refresh; rti;                    /* timer expired */

/* _____ External functions _____ */
.external refresh, write, read;
```

(listing continues on next page)

12 Hardware Interfacing

```
/* _____ M a i n _____ */
main: call setup;
      ax0 = 0;                           /* initialize DM variables */
      dm(nbr_err) = ax0;      /* number of errors encountered */
      dm(err_ov) = ax0;              /* number of errors plus nbr_err */
      dm(err_row) = ax0;      /* saves row location of one error */
      dm(err_col) = ax0;      /* saves col loc. of one error */
      ifc = 0x3f; nop;               /* clear all pending interrupts */
      icntl = b#00111;               /* interrupts are edge sensitive*/
      ax0 = ref_rows;
      dm(refcntr) = ax0;      /* initialize refresh counter */
      i3 = ^row;
      m3 = 1;
      l3 = 0;                        /* init. row pointer */
      i2 = ^col;
      m2 = 1;
      l2 = 0;                        /* init. column pointer */
      i4 = ^buffer;
      m4 = 1;
      l4 = 0x200;                    /* init. error buffer */
      imask = b#000001;              /* enable timer irq*/
      mstat = b#0100000;      /* begin decrementing */

/* Fill routine - This routine fills all of memory with values in the range of
   0 to 0xffff using random writes (non-fast-page mode) so that address
   line errors can be debugged. */

      ax0 = dm(setras);             /* deassert RAS initially */
      cntr = rows;                  /* initialize variables */
      mx0 = 0;                      /* for nested fill loop */
      ay1 = 1;                      /* increment value for fill counter*/
      m3 = 0;
      mr1 = 0;
      do rowfill until ce;
            i2 = ^col;                   /* reset column pointer */
            cntr = cols;                 /* reset cntr to number of colums */
            do colfill until ce;
            imask = 0;                   /* disable timer during write */
                  call write;
                  imask = 1;
                  ax1 = maxfill;
                  ay0 = mx0;             /* mx0 is the fill counter */
                  ar = ay0 + 1;         /* compare with max. fill value */
                  af = ax1 - ay0;
                  if eq ar = pass af;    /* if mx0 = 0xffff, reset to 0 */
            colfill: mx0 = ar;
rowfill: modify(i3,m2);                   /* increment row pointer */
```

```
/* Check routine - reads stored values and compares them to actual filled
      values, if a difference is encountered in the read value it is
      accumulated in the error routine, which stores actual and erroneous
      values. */

reread: i3 = ^row;                      /* reset row pointer */
        i2 = ^col;                      /* reset column pointer */
        mr0 = 0;                        /* mr0 = check counter */
        cntr = rows;                    /* cntr = number of rows */
        do rowread until ce;
              i2 = ^col;
              cntr = cols;
              do colread until ce;
                    imask = 0;                /* disable timer during read */
                    call read;
                    imask = 1;
                    ay0 = mr0;
                    ar = dm(nbr_err);
                    af = ax1 - ay0;
                    if ne call error;
                    ax1 = maxfill;
                    ar = ay0 + 1;
                    af = ax1 - ay0;
                    if eq ar = pass af;     /* if mr0 = 0xffff, reset it to 0 */
        colread: mr0 = ar;
        rowread: modify(i3,m2);                /* increment row pointer */

        sr1 = dm(nbr_err);              /* sr1 + mr0 = number of errors */
        mr0 = dm(err_ov);
        ar = sr1;
        ar = pass ar;                          /* this tests refresh ability */
        if eq jump reread;              /* continually read until error */

wait: idle;                            /* set breakpoint here in emulation */
      jump wait;
/* _____ End Main _____ */

/* _____ Subroutines _____ */
error: ay1 = maxfill;                   /* if an error occurs */
       ar = dm(nbr_err);
       af = ar - ay1;                          /* add it to nbr_err or err_ov */
       if ne jump not_ov;
       ay1 = 1;
       ax1 = dm(err_ov);
       ar = ax1 + ay1;
       dm(err_ov) = ar;
       jump cont;
```

(listing continues on next page)

```
not_ov: ay1 = 1;
        ar = ar + ay1;
        dm(nbr_err) = ar;                /* save 256 error values */

cont:   dm(i4, m4) = ax1;                /* saves error value followed by */
        dm(i4, m4) = ay0;                /* actual value in internal DM */
        dm(err_row) = i3;                /* saves last error row and */
        dm(err_col) = i2;                /* column location */
        rts;

setup:  ax0 = 0x0003;
        dm(0x3ffb) =ax0;                 /* TSCALE */
        ax0 = 0xc350;                    /* 50,000 for 16ms per interrupt */
        dm(0x3ffc) =ax0;                 /* TCOUNT */
        dm(0x3ffd) =ax0;                 /* TPERIOD */
        ax0=0x0200;
        dm(0x3ffe)=ax0;                  /* dWait3 = 1 WS */
        ax0=0x0000;
        dm(0x3ffa)=ax0;                  /* No Sport functions enabled */
        dm(0x3ff9)=ax0;
        dm(0x3ff8)=ax0;
        dm(0x3ff7)=ax0;
        dm(0x3ff6)=ax0;
        dm(0x3ff5)=ax0;
        dm(0x3ff4)=ax0;
        dm(0x3ff3)=ax0;
        dm(0x3ff2)=ax0;
        dm(0x3ff1)=ax0;
        dm(0x3ff0)=ax0;
        dm(0x3fef)=ax0;
        dm(0x3fff)=ax0;
        rts;

.endmod;
```

Listing 12.8 DRAM Test Program

Hardware Interfacing 12

```
.module/boot=0/abs=0    DRAM_record;
/*
File Name:   DRAMRCRD.DSP
Version:     Version 0.00
Purpose:     Assembly source code for u-law companded speech sample
             storage and playback. This routine enables approximately
             2 minutes and 11 seconds of storage on the DRAM interface
             board. The maximum storage time possible is twice that
             mentioned above using u-law companding (8-bit word length).
Return Values:
             IRQ2 - switch between record and forward playback
             FLAG_IN - switch between foward and backward playback.
             Note: pressing FLAG_IN in record mode causes backward
             playback.
             record mode - flag_out is low
             playback mode - flag_out is high
*/
.const  rows     = 0x400;     /* # of rows and columns on 1M DRAM */
.const  cols     = 0x400;
.const  lastrow  = 0x13ff;
.const  lastcol  = 0x33ff;
.const  ref_rows = 0x10;
.var/dm/ram/seg=int_dm       lastr, lastc, sflag, flagin, refcntr; .var/dm/ram/
seg=row_deassert  setras;    /* location for RAS deselection */
.var/dm/ram/seg=row_range    row;       /* row selection range */ .var/dm/ram/
seg=col_range     col;       /* column selection range */
.global row, setras, refcntr;

/* _____ Interrupt vectors _____ */
      jump main; nop; nop; nop;             /* reset interrupt */
      jump changestate; nop; nop; nop;    /* irq2 */
      rti; nop; nop; nop;                   /* sport0 transmit */
      jump record; nop; nop; nop;           /* sport0 receive */
      rti; nop; nop; nop;                   /* sport1 transmit or irq1 */
      rti; nop; nop; nop;                   /* sport1 receive or irq0 */
      rti; nop; nop; nop;                   /* timer expired */

/* _____ External functions _____ */
.external refresh, read, write;

/* _____ Interrupt handlers _____ */
changestate: ax0 = 0;                     /* pressing IRQ2 causes state change */
      dm(flagin) = ax0;                   /* clear backward playback mode */
      ax0 = dm(sflag);                    /* if 0 make 1 = record mode */
      ar = pass ax0;
      if ne jump set_s1;            /* sflag = 0 = playback mode */
      ax0 = 0x0001;                       /* sflag = 1 = record mode */
      dm(sflag) = ax0;
      rti;
```

(listing continues on next page)

625

12 Hardware Interfacing

```
set_s1: ax0 = 0;                    /* if 1 make 0 = playback mode */
        dm(sflag) = ax0;
        dm(lastr) = i3;             /* save row location for wraparound */
        dm(lastc) = i2;             /* save column location for wraparound */
        rti;

record: ax0 = dm(flagin);           /* if flagin = 1 playback backwards */
        ar = pass ax0;
        if ne jump yalp;
        ax0 = dm(sflag);            /* if sflag = 0 playback forwards */
        ar = pass ax0;
        if eq jump play;
        reset flag_out;             /* flag_out = low */
        mx0 = rx0;
        tx0 = mx0;
        call write;                 /* read a sample and write it to DRAM */
        call inc_loc;               /* increment column/row pointer */
        ax0 = i3;                   /* compare inc'd location with */
        ay0 = lastrow;              /* ...the last row and column */
        ar = ax0 - ay0;
        if le jump refr;
        ax0 = i2;
        ay0 = ^col;
        ar = ax0 - ay0;
        if ne jump refr;
        ax0 = 0;                    /* if the last row and column available */
        dm(sflag) = ax0;            /* has been reached save the current */
        ax0 = lastrow;              /* row and column location for */
        dm(lastr) = ax0;            /* wraparound */
        ax0 = lastcol;
        dm(lastc) = ax0;
        jump refr;

play:   call read;                  /* playback forward */
        tx0 = ax1;                  /* read from DRAM and output sample */
        call inc_loc;
        ax0 = dm(lastr);            /* compare to last column and row, */
        ay0 = i3;                   /* if so set to row and column 0 */
        ar = ax0 - ay0;
        if ne jump refr;
        ax0 = dm(lastc);
        ay0 = i2;
        ar = ax0 - ay0;
        if ne jump refr;
        i3 = ^row;
        i2 = ^col;
        jump refr;

yalp:   call read;                  /* playback backward */
```

Hardware Interfacing   12

```
        tx0 = ax1;                    /* read from DRAM and output sample */
        ax0 = i2;                     /* decrement the colunm/row pointer */
        ay0 = ^col;
        ar = ax0 - ay0;
        if ge jump done;
        i2 = lastcol;
        modify(i3, m2);

done:   ax0 = i3;                     /* compare to row and column 0, if so */
        ay0 = ^row;                   /* reset to last row and last column */
        ar = ax0 - ay0;
        if ge jump refr;
        ax0 = i2;
        ay0 = lastcol;
        ar = ax0 - ay0;
        if ne jump refr;
        i3 = dm(lastr);
        i2 = dm(lastc);

refr:   call refresh;
        ar = i6;
        ay1 = lastrow;                /* if last row to refresh has been */
        af = ar - ay1;                /* reached reset it to row 0 */
        if lt jump cont;
        i6 = ^row;

cont:   rti;

/* _____ M a i n _____ */
main:   call setup;                   /* initialize sport registers */
        ax0 = 0;                      /* initialize state variables */
        dm(sflag) = ax0;
        dm(flagin) = ax0;
        ax0 = ref_rows;               /* initialize refresh counter */
        dm(refcntr) = ax0;
        i6 = ^row;
        m6 = 1; l6 = 0;               /* init. refresh row pointer */
        i3 = ^row;
        m3 = 0; l3 = 0;               /* init. row pointer */
        i2 = ^col;
        m2 = 1; l2 = 0;               /* init. column pointer */
        ax0 = dm(setras);             /* initially deassert RAS */
        ifc = 0x3F; NOP;
        icntl = b#00111;              /* interrupts are edge sensitive */
        imask = b#100000;             /* enable irq2 */
        idle;                         /* wait for irq2 */
```

(listing continues on next page)

12 Hardware Interfacing

```
state_1: reset flag_out;        /* flag_out = low (indicate record) */
        ax0 = 0;
        dm(flagin) = ax0;
        i3 = ^row;
        i2 = ^col;
        m2 = 1;
        imask = b#101000;               /* enable irq2 and sp0 receive */

wait1: idle;                            /* receive until flag_in or irq2 */
        ax0 = dm(sflag);                /* or last column and row */
        ar = pass ax0;                  /* check mode (record/playback) */
        if eq jump playback;
        if not flag_in jump f1;         /* if not_flagin play backwards */
        jump wait1;

f1:     if not flag_in jump f1;         /* first debounce flag_in */
        dm(lastr) = i3;                 /* save row location for wraparound */
        dm(lastc) = i2;                 /* save column location for wraparound */
        jump kabyalp;

playback: set flag_out;                 /* flag_out = high (indicate playback) */
        ax0 = 0;
        dm(flagin) = ax0;
        m2 = 1;                         /* reset modify value to + 1 */
        i3 = ^row;                      /* reset to row and col 0 */
        i2 = ^col;

wait2: idle;                            /* wait for irq2 or flag_in */
        ax0 = dm(sflag);
        ar = pass ax0;                  /* if sflag = 0 = record */
        if ne jump state_1;
        if not flag_in jump f3;         /* if FLAG_IN play backwards */
        jump wait2;

f3:     if not flag_in jump f3;         /* first debounce flag_in */

kabyalp: set flag_out;                  /* set up for backward play */
        m2 = -1;                        /* modify for column = -1 */
        ax0 = 0;                        /* signify playback mode */
        dm(sflag) = ax0;
        ax0 = 1;                        /* signify play backwards mode */
        dm(flagin) = ax0;
        i3 = dm(lastr);                 /* init. to last row and col. location */
        i2 = dm(lastc);

wait3: idle;                            /* wait for irq2 or flag_in */
        if not flag_in jump f5;         /* switch back to playback */
        ax0 = dm(sflag);
        ar = pass ax0;                  /* if sflag = 0, record again */
```

```
        if ne jump state_1;
        jump wait3;

f5:     if not flag_in jump f5;        /* first debounce flag_in */
        jump playback;
/* _____ End Main _____ */

/* _____ Subroutines _____ */
inc_loc:    ax0 = i2;                   /* increment column and/or row pointer */
        ay0 = lastcol;
        ar = ax0 - ay0;
        if le jump done1;
        i2 = ^col;
        modify(i3, m2);

done1: rts;

setup: ax0 = 0;
        dm(0x3ffb) =ax0;                /* TSCALE */
        dm(0x3ffc) =ax0;                /* TCOUNT */
        dm(0x3ffd) =ax0;                /* TPERIOD */
        ax0=0x0200;
        dm(0x3ffe)=ax0;                 /* Dwait3 = 1 WS */
        ax0=0x0000;
        dm(0x3ff9)=ax0;                 /* Disable Receive Multichannels */
        dm(0x3ffa)=ax0;
        dm(0x3ff7)=ax0;                 /* Disable Transmit Multichannels */
        dm(0x3ff8)=ax0;
        ax0=0x6b27;                     /* Multichannel disabled */
        dm(0x3ff6)=ax0;                 /* Int. gen serial clock */
                                        /* Receive frame sync required, width 0 */
                                        /* Transmit frame sync required, width 0 */
                                        /* Int trans, receive frame sync enabled */
                                        /* u-law companding, 8 bit word length */
        ax0=0x0002;
        dm(0x3ff5)=ax0;                 /* Generate 2.048 MHz serial clock */
        ax0=255;
        dm(0x3ff4)=ax0;                 /* Divide by 256 for 8KHz sampling rate */
        ax0=0x0000;
        dm(0x3ff3)=ax0;                 /* SPORT0 AUTOBUFF disabled */
        dm(0x3ff2)=ax0;                 /* SPORT1 CNTL disabled */
        dm(0x3ff1)=ax0;                 /* SPORT1 timer not used */
        dm(0x3fef)=ax0;                 /* SPORT1 AUTOBUFF disabled */
        ax0 = 0x1000;
        dm(0x3fff)=ax0;                 /* SPORT0 enabled, No PM Wait States */
                                        /* BOOT Wait State 0, BOOT page 0   */

        rts;

.endmod;
```

Listing 12.9 DRAM Speech Sample Record/Playback Program

12 Hardware Interfacing

12.3.10 DRAM Interfacing References

Analog Devices:
ADSP-2100 Family User's Manual
ADSP-2100 Family EZ-Tool Manual
ADSP-2100 Family Assembler Tools & Simulator Manual
Digital Signal Processing Laboratory
      Using the ADSP-2101 Microcomputer
ADSP-2101 and ADSP-21msp50 data sheets

Data I/O Corp.,
Abel Design Software User's Manual, 1990.

Driscoll, Frederick F.,
      "Dynamic Refresh", *Interfacing the 68000 Microprocessor*
            pgs. 358-65, 173-177.

Clements, Alan,
      "Designing Dynamic Read/Write RAM Systems", Microprocessor
            Systems Design, pg. 275-93.

Steve Gumm, Carl T. Dreher,
      "Unraveling the Intricacies of Dynamic RAMs", Dynamic RAMs Part 1,
            pg. 162.

Toshiba, Signetics, SGS Thomson data sheets.

Hardware Interfacing 12

12.4   LOADING AN ADSP-2101 PROGRAM VIA THE SERIAL PORT

For many DSP applications, it is desirable to have a DSP processor under the control of a host computer. In these situations, the host computer would download a program for the DSP to execute. The ADSP-2101 provides two serial ports suitable for program download from a host computer. This section note details the ADSP-2101 monitor program for downloading from a serial port. The monitor program itself would be booted from EPROM or other boot memory. While this example uses serial port zero, the principal could be extended to download via a memory-mapped parallel port.

12.4.1   A Monitor

The task of the host computer is to download a series of instructions to the ADSP-2101 for execution. The ADSP-2101 receives the incoming instructions, loads them into program memory and when all instructions have been received, executes them. Prior to and during the download from the host, the ADSP-2101 executes a monitor program. This monitor activates the serial port, receives the instructions and places them in program memory for execution.

The ADSP-2101 instruction is twenty-four bits wide but many hosts, including eight-bit processors, more readily handle byte-wide data. Since the serial port can accommodate serial words from three to sixteen bits in length, byte-length data words are easily received.

Whenever a program memory write occurs, the sixteen most significant bits are supplied by the source register, explicitly named in the instruction, and the eight LSBs are supplied by the PX register. The basic tactic of the monitor program is to assemble the two most significant bytes in a data register (using the Shifter) and load PX explicitly with the least significant byte. A program memory write then writes the correct twenty-four bit instruction.

In addition to the transfer of instructions through the serial port into program memory, the monitor program must also know when the download is complete and execution can begin. A straight forward method is to count the number of instructions sent to the serial port. A count value is sent to the ADSP-2101 before the first instruction. This is the count of the instructions to follow. After each instruction is downloaded, the count can be decremented.

12 Hardware Interfacing

The downloaded program must avoid overwriting the monitor program while the monitor executes. The last instruction of the monitor program is identified by a global label which also identifies the beginning of the available space for downloaded code. The monitor program must be linked with the downloaded program so that the downloaded program makes the correct address references including the reference to this global label.

The indirect addressing capabilities of the Data Address Generators on the ADSP-2101 make it easy to cycle through the correct sequential locations starting with the label.

The final concern is the interrupt table. If the downloaded program is interrupt-driven, the interrupt table (program memory H#0000 to H#001C) must contain valid instructions for servicing expected interrupts.

There are several ways to do this. First, the monitor program itself could contain the valid interrupt table for the program to be downloaded. This assumes that the interrupt structure of the downloaded program is known when the monitor program is created. Second, the interrupt table may be downloaded through the serial port just as the rest of the program is. The DAG can loaded with the start address of the interrupt table and the instructions can be loaded, but you may not overwrite the interrupt being used to receive the data on the serial port until all instructions have been received.

The monitor program example does not load an interrupt table. The best approach is dependent on your application.

12.4.2   Implementation

The first task of the monitor program is to setup and enable the serial port. Serial ports on the ADSP-2101 are extremely flexible in terms of framing options, word lengths and timing. The ADSP-2101 serial ports may receive the frame synch and serial clock from the host processor or generate them internally.

As the program is downloaded from a host computer, the ADSP-2101 looks to the host for serial port information. That is, the serial port frame synchronization and serial port clock are supplied by the host computer. For purposes of illustration, the code that appears at the end of this section uses normal framing and external receive frame synchronization. For externally generated serial clocks the ADSP-2101 can support frequencies up to the processor instruction rate.

Hardware Interfacing 12

The flow for the monitor program is shown in Figure 12.8.

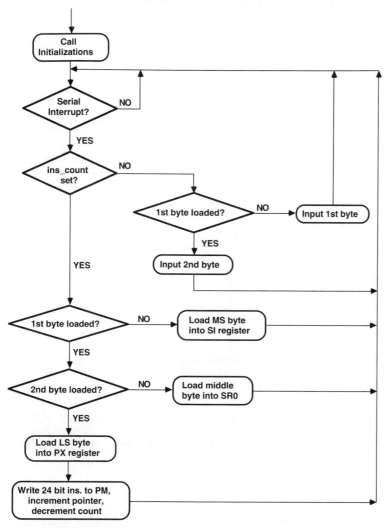

Figure 12.8 Boot Program Flow Diagram

12 Hardware Interfacing

Once the serial ports are enabled, the monitor program waits for a serial port interrupt signifying that a serial word has been received. The first two serial words received are the instruction count. As the serial word is eight bits, two serial port words make up the instruction count. The separate bytes of the instruction count are combined in the shifter and loaded into data memory. This count represents the number of instructions to be downloaded from the host and does not include the interrupt table. The interrupts are handled automatically, as the interrupt table has a fixed length.

With the count downloaded, the ADSP-2101 is ready to accept instructions through the serial port. Instructions are downloaded a byte at a time just as the instruction count was. The most significant byte is first. It is loaded into the SI register and the byte count ("count") is decremented. The middle byte of the instruction is loaded into the SR0 register. These two bytes are combined in the shifter with the results residing in the SR0 register. Once again the byte count is decremented. Finally, the least significant byte is loaded into the PX register. Now that all three bytes are loaded into registers on the ADSP-2101, the downloaded instruction can be written to program memory.

When all is downloaded, a jump to the new downloaded program is all that is necessary to begin execution.

The monitor program is shown in Listing 12.10.

A monitor program initializes the serial port and receives instructions, writing them into program memory, then beginning execution. This method of booting is useful when the ADSP-2101 is under the control of a host computer or controller. Any size program may be downloaded (up to the full addressing capability of the ADSP-2101) with this particular method of implementation. Only the program memory used by the monitor program (eighty-six instructions) cannot be loaded. That space could, however, be used for program memory data storage.

Hardware Interfacing 12

```
.Module/ram/BOOT=0          serial_boot_monitor;
.Var/dm                     count;              {counts bytes}
.VAR/DM                     ins_count;          {counts instructions}
.GLOBAL                     code_start;         {end of monitor space}


              JUMP restarter; NOP; NOP; NOP;  {restart vector}
              RTI; NOP; NOP; NOP;             {IRQ2 not used}
              RTI; NOP; NOP; NOP;             {sport0 TX not used}
              JUMP serial; NOP; NOP; NOP;     {sport0 RX }
              RTI; NOP; NOP; NOP;             {sport1 TX not used}
              RTI; NOP; NOP; NOP;             {sport1 RX not used}
              RTI; NOP; NOP; NOP;             {no timer used}


restarter:    CALL initializations;
wait_loop:    IDLE;
              JUMP wait_loop;


initializations:  I4 = H#3ff3;               {pointer to mem map reg}
                  I5 = PM(^code_start);      {pointer to start of prog}
                  I6 = 0000;                 {pointer to interrupt tab}
                  M4 = 0;
                  M5 = 1;

                  DM(count) = 1;             {count val for # of bytes}

                  DM(I4,M5) = 0;             {disable autobuffer}
                  DM(I4,M5) = 0;             {no frame divide modulus}
                  DM(I4,M5) = 0;             {no clk divide modulus}
                  DM(I4,M5) = H#2007;        {extrnl RFS & SCLK, no compand}
                                             {SLEN 8, no multichannel}
                  DM(H#3fff) = H#1000;       {enable sport0}

                  DM(ins_count) = H#FFFF;

                  imask = 8;                 {sport0 rec interrupt only}
                  RTS;

serial:       AY1 = DM(ins_count);
              AR = PASS AY1;
              IF GT JUMP next_instruction;   {get next instruction}
              IF LT JUMP load_word_count;    {get number of instructions}
              IF EQ IMASK=0;                 {done; turn off interrupts}
              JUMP code_start;               {start downloaded program}

{load the count, that is, the number of instructions to be downloaded}
{this happens in two bytes  The most significant byte first}
```

(listing continues on next page)

635

12 Hardware Interfacing

```
load_word_count:    AY0 = DM(count);                      {is this 1st or 2nd byte}
                    AR = PASS AY0;
                    IF NE JUMP first_byte;
                    IF EQ JUMP second_byte;

first_byte:         SI = RX0;                              {first byte decrem. count}
                    AR = AY0 - 1;
                    DM(count) = AR;
                    RTI;

second_byte:        SR0 = RX0;                             {second byte...}
                    SR = SR OR LSHIFT SI BY 8 (LO);        {put two bytes together}
                    DM(ins_count) = SR0;                   {store in ins_count}
                    DM(count) = 3;                         {load count for ins.}
                    RTI;
```

{load the next instruction. Instructions are 24 bits long and appear}
{at the serial port in 8 bit fragments. The most significant byte 1st}

```
next_instruction:   AX0 = 2;                               {decide which byte is due}
                    AY0 = DM(count);
                    AR = AX0 - AY0;

                    IF LT JUMP most_sig_byte;
                    IF EQ JUMP middle_byte;
                    IF GT JUMP least_sig_byte;

most_sig_byte:      SI = RX0;                              {load MS byte into SI}
                    AR = AY0 - 1;                          {decrement count}
                    DM(count) = AR;
                    RTI;

middle_byte:        SR0 = RX0;                             {load Middle into SR}
                    SR = SR OR LSHIFT SI BY 8 (LO);        {put MS and middle together}
                    AR = AY0 - 1;                          {decrement count}
                    DM(count) = AR;
                    RTI;

least_byte:         PX = RX0;                              {put LS byte into PX}
                    PM(I5,M5) = SR0;                       {write SR0 into PM}
                                                           {PX provides 8 LS bits}
                    DM(count) = 3;                         {reset byte count}
                    AR = AY1 - 1;                          {decrement ins count}
                    DM(ins_count) = AR;
                    RTI;

code_start:         NOP;

.ENDMOD;
```

Listing 12.10 Monitor Program Listing

Hardware Interfacing   12

12.5   MEMORY INTERFACING FOR THE ADSP-2105

In order to utilize the full potential of the ADSP-2105, several techniques for memory allocation and development tools usage will be discussed. The techniques and examples described here can also be applied to any of the processors in the ADSP-2100 family.

This section will describe several example systems, along with the correct development tool set-ups to implement them. Listings of example architecture descriptions, module declarations, and linker-output map files will be provided for each example. A description of C-compiler use for each case will follow the examples.

12.5.1   Example System1:
Using Boot Pages For Program Memory

An ADSP-2105 system may contain up to eight pages of boot memory. ADSP-2105 boot pages are 1K long, as opposed to ADSP-2101/2111 boot pages which are 2K long.

To use boot pages, the following steps should be taken.

Builder	In your .SYS file, use the ADSP2105 directive. Define BOOT ROM pages as needed for your system.
Assembler	Define each module as bootable. (.MODULE / boot=x, where x=0-7)
Linker	Use linker as you normally would.
Splitter	Use the **-bs**[1024] option. This sets the generated boot page size to 1024.

The example system provided uses all eight boot pages. Each module performs a dummy task and then boots in the next page. Note that in the .map file (*ex1.map*), the data memory declared for each overlaps. This is because each boot page is treated as a separate run-time context by the processor. Each boot page is an entirely different program with no relation to any other program; thus memory space is reused for each boot page. If there is a need to share variables between boot pages, the variables must be defined as **static** variables when declared. This forces the linker to place these variables in high memory and not to overwrite them when booting in a new page.

12 Hardware Interfacing

12.5.2 Example System 2:
Booting With The -loader Option (RAM Initialization)

Like Example 1, this system uses the concept of boot pages to store the executable code. What is different in this system is the automatic initialization of external PM RAM as well as internal and external DM RAM. This initialization is done through the use of the **-loader** option when using the Prom Splitter (see *ADSP-2100 Family Development Software, Release 3.1 Release Note* for a description of the -loader option operation).

In order to use the **-loader** option, the following steps must be taken.

Builder   In your .SYS file, define internal PM and DM RAM segments, as well as any desired external PM and DM RAM segments. Define BOOT ROM pages (but don't assign any modules to it). Boot page 0 will be filled with the splitter-generated loader code. User code will be placed in subsequent boot pages (1 and higher if needed).

Assembler   DO NOT use the "BOOT = xx" qualifier in any module description. (Modules may, however, be defined to reside in internal PM RAM segments.)

Linker   Use linker as you normally would. Linker will place modules according to their definition and the architecture description.

Splitter   Use **-loader** and **-bs**[1024]
              (and -bb[ ] if needed) options. The splitter will create a bootable image file which will load program memory with the appropriate modules, and data memory with appropriate values, according to their definitions.
Do not use the -pm, -dm, or -bm switches when using the -loader option.

The boot-page qualifiers are removed from the module declarations because, on chip reset, the **loader** option causes special code to be booted into internal PM RAM which copies the ROM's contents into the appropriate RAM spaces. The executable code will be written into internal PM RAM when the initialization procedure is complete. This operation is transparent to the user and system operation on startup will appear to be the same as for Example 1, except that RAM will be initialized automatically. The linker output file (.BNM image file) should be used to

Hardware Interfacing   12

burn boot memory PROMs. The bootable image file will contain multiple pages. The -loader routine will be placed on boot page 0. User modules will be placed in ROM starting on boot page 1.

12.5.3   Example System 3:
Using Internal & External PM RAM For Code

If your particular application requires code that is longer than the 1K internal PM RAM limit of the ADSP-2105, do not despair. The 2100 family PROM splitter **-loader** option will allow these types of programs to be used with the ADSP-2105. As described in the previous section, the **loader** option will initialize external memory. This feature can be used to initialize external PM RAM with executable code while still using the internal PM RAM to its fullest extent.

System development should follow the following steps.

Builder	In your .SYS file, define sufficient internal and external PM RAM space. Define BOOT ROM pages (but don't assign any modules to it) sufficient for the **-loader** code and your code.
Assembler	DO NOT use the "BOOT = xx" qualifier in any module description. (Modules may, however, be defined to reside in internal PM RAM segments.)
Linker	Use linker as you normally would. Linker will place modules according to their definition and the architecture description.
Splitter	Use **-loader** and **-bs**[1024] options. The splitter will create a bootable image file which will load program memory with the appropriate module according to their definitions.

Do not use the -pm, -dm, or -bm switches when using the -loader option.

This strategy is especially useful if your application software can be subdivided into *time-critical* and *non-time-critical modules*. The time-critical modules should be set to reside in the internal PM segment, while non-time-critical modules can be placed in external PM RAM.

639

12 Hardware Interfacing

12.5.4 Example System 4:
Using External PM ROM

Another useful method for partitioning code too big to fit in one page of memory is to use a combination of boot memory and external ROM. To accomplish this both internal and external ROM modules must be defined to reside in the same boot page. Since boot page size is limited to 1K, defining the modules this way forces the linker to place one module externally while keeping it in the same run-time context.

To implement this system, the following steps should be taken.

Builder	In your .SYS file, use the ADSP2105 directive. Define the necessary boot page. Define an appropriate PM ROM segment.
Assembler	Define each module to be bootable from boot page 0 (/boot=0).
Linker	Use linker as you normally would.
Splitter	Run the splitter twice--once to generate the BOOT ROM and once to generate the program memory ROM.

The example system provided uses one page of boot memory along with 1K of internal program ROM. The program ROM module (*ex4_2.dsp*) is declared to reside in program ROM **and** is defined with the /boot=0 directive. This qualifier insures *ex4_2.dsp* is included in the same run-time context as the module defined to reside in the boot ROM (*ex4_1.dsp*).

12.5.5 Hardware Implications

The ADSP-2105 allows several boot EPROM configurations to be used. Refer to the Memory Interface chapter of the *ADSP-2100 Family User's Manual* for an in-depth discussion of the boot memory interface.

For direct plug-in compatibility with an ADSP-2101, use the following connections (for 27512).

```
ADSP-2105 A0 - A13     =>  EPROM A0 - A13
ADSP-2105 D22 - D23    =>  EPROM A14 - A15
```

This allows for eight 1K boot pages on 2K boundaries.

Hardware Interfacing 12

Since the ADSP-2105 only uses 1K boot pages, it is possible to access all eight boot pages while using a smaller EPROM. To accomplish this, boot pages can be placed on 1K boundaries by using the Prom Splitter's **-bb**[1024] option and the following hardware connections (for 27256).

ADSP-2105 A0 - A11	=>	EPROM A0 - A11
ADSP-2105 A12	=>	**no connection**
ADSP-2105 A13	=>	EPROM A12
ADSP-2105 D22 - D23	=>	EPROM A13 - A14

This connection scheme is also useful with a 27512, where code for another processor can utilize the upper half of the EPROM.

Note: Using the -bb[1024] option and the above hardware connections is not compatible with the ADSP-2101. On boot-up, the ADSP-2101 will always load 2K worth of information from the EPROM. Therefore, this configuration cannot be used with any ADSP-2101 In-Circuit Emulator™ (ICE) if booting the ICE from the target system. To use an ICE in these systems, remove the target EPROM and download the executable code into the ICE's overlay memory. The same caveat holds true when simulating the bootable code in the ADSP-2101 simulator using the 'LR' command.

12.5.6   Use Of The C-Compiler With ADSP-2105 Systems

If your application software is derived from C code, the following options and strategies should be employed when using the C-compiler for the above example.

Example 1   Use the **-b#** option when compiling to assign each module to the appropriate boot page. For example, typing **cc21** *filename* **-b0** would assign the module *filename* to boot page 0. Also, modify the *run_hdr* program so that it will be assigned to the correct boot page (ex. MODULE/boot=0 run_hdr;). When re-assembling the modified *run_hdr* program, ALWAYS use the **-c** (case sensitivity) option.

Example 2   DO NOT use the compiler **-b#** option. The **-loader** option of the PROM splitter automatically generates the correct boot image file. Compile the module as you would normally.

12 Hardware Interfacing

Example 3
For memory usage considerations, the **-lpm** option can be used. This places literals into PM rather than the default, DM. This is not a requirement but may come in handy if memory space is tight. Again, when using the **-loader** option, DO NOT use the **-b#** option when compiling.

Example 4
Modify the *run_hdr* program to reside in boot page 0 (see Example 1). For module 1 (INTERNAL), compile with the **-b0** option. For module 2 (EXTERNAL), compile with both the **-b0** and **-crom** options. This will assign both modules to the same run-time context and places PM modules into ROM segments.

12.5.7   Linking Modules Generated By The C-Compiler

For Examples 1, 3, and 4, use the **-p** option when linking. This places the run-time libraries on the correct boot pages. For Example 2, link without this option.

12.5.8   Additional Suggestions

If you find that you have memory limitations, such as the stack having no room to grow (examine the .MAP file generated by the linker's **-x** option), use the **-s###** option when you link to define the minimum stack size (default = 1). This is important because the linker allocates internal DM memory first, starting with the stack, until internal DM is filled. If you suspect that the stack will grow past its default boundaries, such that it may start to overwrite DM variables, give yourself some room using the **-s###** option. The linker will reserve the specified space and then proceed to fill internal memory until space is exhausted. The linker will then start to fill external DM RAM if it exists. The C-compiler does not allow you to explicitly define variables in external memory. External memory is allocated only after internal memory is filled.

Hardware Interfacing 12

12.5.9 About The Example Programs

The disk provided with this book contains sample architecture, assembler, and C files for each example system. Batch files are also provided to create final bootable image files for each example system so that their operation can be examined in the simulator or emulator. Please examine these files and feel free to use them as a framework for your own systems.

Example 1	System file	**ex1.sys**
	Source modules	**ex1_1.dsp, ex1_2.dsp, ex1_3.dsp, ex1_4.dsp,**
		ex1_5.dsp, ex1_6.dsp, ex1_7.dsp, ex1_8.dsp
	Batch file	**makeex1.bat**
	C source modules	**ex1_1.c, ex1_2.c, ex1_3.c, ex1_4.c, ex1_5.c, ex1_6.c, ex1_7.c, ex1_8.c**
	C batch file	**makeex1c.bat**

Each source module for Example 1 resides on a separate boot page. Each module performs a dummy task and then boots in the next page. Note in the **.map** file (**ex1.map**) that the data memory variables declared in each module overlap. This occurs because each boot page is considered to be a separate run-time context by the processor system.

In the C modules, the construct

*(short *)Sys_Ctrl_reg = BOOTx

is used to boot in page x.

Example 2	System file	**ex2.sys**
	Source module	**ex2_1.dsp**
	Batch file	**makeex2.bat**
	C source module	**ex2_1.c**
	C batch file	**makeex2c.bat**

The source module for Example 2 declares and initializes two buffers stored in RAM and then performs a dummy task. The PROM Splitter creates *two* boot pages (0 and 1). The **loader** code is placed on boot page 0 and the source code on boot page 1. In the simulator you will see the **loader** code booted in first. This code initializes the proper memory locations and then boots in the source module.

12 Hardware Interfacing

Example 3	System file	**ex3.sys**
	Source module	**ex3_1.dsp, ex3_2.dsp**
	Batch file	**makeex3.bat**
	C source module	**ex3_1.c, ex3_2.c**
	C batch file	**makeex3c.bat**

One source module in Example 3 is defined to reside in *internal* PM RAM, the other in *external* PM RAM. When viewing the C example, the code is loaded starting with internal memory, until it is filled, and then will begin placing code in external memory, if needed. It is not possible to explicitly place modules in internal or external memory in C. If this feature is desired, the C-compiler-generated assembly modules must be hand modified for explicit placement.

Example 4	System file	**ex4.sys**
	Source module	**ex4_1.dsp, ex4_2.dsp**
	Batch file	**makeex4.bat**
	C source module	**ex4_1.c, ex4_2.c**
	C batch file	**makeex4c.bat**

The program ROM module (**ex4_2.dsp**) is declared to reside in both program ROM *and* on the boot page. Note that in contrast to Example 1, there is no data memory overlap, since both modules are part of the same run-time context.

Hardware Interfacing 12

12.5.10   Appendix: Example System 1

ex1.sys

```
.SYSTEM   example1;
.ADSP2105;
.MMAP0;
.seg/rom/boot=0               boot_page_0[1024];        {2105 1K boot size}
.seg/rom/boot=1               boot_page_1[1024];        {2105 1K boot size}
.seg/rom/boot=2               boot_page_2[1024];        {2105 1K boot size}
.seg/rom/boot=3               boot_page_3[1024];        {2105 1K boot size}
.seg/rom/boot=4               boot_page_4[1024];        {2105 1K boot size}
.seg/rom/boot=5               boot_page_5[1024];        {2105 1K boot size}
.seg/rom/boot=6               boot_page_6[1024];        {2105 1K boot size}
.seg/rom/boot=7               boot_page_7[1024];        {2105 1K boot size}
.seg/PM/ram/abs=0/code/data   int_pm[1024];             {2105 1K int pm}
.seg/DM/ram/abs=h#3800/data   int_dm[512];              {2105 int dm}
.endsys;
```

ex1_1.dsp

```
.module/ram/boot=0          ex1_module_1;
#include <def2105.h>

{this module is loaded into boot memory page 0}

.var/dm/ram     var_mod1_1[100];        {module 1 dm variable 1}
.var/dm/ram     var_mod1_2[100];        {module 1 dm variable 2}

{code section of module 1: this code boots module 2}

ex1_pg0:                    ax0=1;
                            ax0=2;
                            ax0=3;
                            ax0=4;
                            ax0=5;
                            ax0=6;
                            ax0=7;
                            ax0=8;
                            ax0=9;
                            ax0=10;
                            ax0=0x240;
                            dm(Sys_Crtl_Reg)=ax0;         {boot page 1}
```

(example system continues on next page)

12 Hardware Interfacing

makeex1.bat

```
bld21 ex1
asm21 -cp ex1_1
asm21 -cp ex1_2
asm21 -cp ex1_3
asm21 -cp ex1_4
asm21 -cp ex1_5
asm21 -cp ex1_6
asm21 -cp ex1_7
asm21 -cp ex1_8
ld21 ex1_1 ex1_2 ex1_3 ex1_4 ex1_5 ex1_6 ex1_7 ex1_8 -a ex1 -e ex1 -g -x
spl21 ex1 ex1 -bs 1024
```

Index ■

A

AD1847 585. *See also* SoundPort interfaces
AD1849 572, 578. *See also* SoundPort interfaces
Adaptive equalizer 84
  architectures 105
  complex 106
  complex filter listing 109
  decision-directed adaptation 115
  flowchart 112
  history 98
  least mean squared (LMS) algorithm 84, 109
  LMS routine 111
  performance index 105
  practical considerations 119
  real 106
  sampling rates 107
  theory of 102
  training sequence 114
Adaptive predictor 296, 297
Adaptive quantizer 296, 297
Adaptive sample rate decimation 218
ADPCM (G.721) 293
  implementation 300
  receiver 298
  subroutines 301
  transmitter 294
ADSP-2100 family
  ADSP-2101 architecture 7
  base architecture 4
  data memory address (DMA) bus 6
  data memory data (DMD) bus 6
  overview 1
  program memory address (PMA) bus 6
  program memory data (PMD) bus 6
  result (R) bus 7
ADSP-2101 1. *See also* ADSP-2100 family
ADSP-2101 architecture 7. *See also* ADSP-2100 family
  on-chip memory 7
  programmable timer 9
  serial ports ("SPORTs") 8

Index

ADSP-2111 2. *See also* ADSP-2100 family
ADSP-2111 architecture 9. *See also* ADSP-2100 family
   host interface port (HIP) 9
ADSP-21msp50/5x 2. *See also* ADSP-2100 family
ADSP-21msp5x architecture 10. *See also* ADSP-2100 family
   analog interface 10
   codec 10
ALU 2, 4. *See also* ADSP-2100 family, base architecture
Answer mode descrambler 23. *See also* V.32
Answer mode scrambler 23. *See also* V.32
APCM inverse quantization 219
APCM quantization 218
Arithmetic/logic unit 2, 4
Assembly language 11
Auto-correlation 209
Autobaud 561, 564

B

Barrel shifter 2, 4. *See also* ADSP-2100 family, base architecture
Baud rate 561, 564
Bilinear transformation 514
Biquad filter 525. *See also* IIR biquad filter
Boot pages 637. *See also* Memory interfacing
Booting through the serial port 631. *See also* Loading through SPORT
Buses 6. *See also* ADSP-2100 family

C

Call mode descrambler 23. *See also* V.32
Call mode scrambler 22. *See also* V.32
Cepstral coefficients 338
Channel equalization 101
Circular buffers 544. *See also* Modulo addressing
Comfort noise insertion (CNI) 205. *See also* GSM
Companding
   A-law 293
   µ-law 293
Continuous phase frequency-shift keyed modulation 120
   flow diagram 122
   implementation 121
   listing 125
   methodology 120
Control systems 503. *See also* Digital control systems
Convolutional code 37. *See also* Trellis coding

Index

D

Data address generator 5
Data scrambler 18, 127, 152
DCT 443. *See also* Discrete cosine transform
Decoder 220, 294. *See also* GSM
  higher sub-band decoder 298
  lower sub-band decoder 299
Descrambler 22. *See also* V.32
  programs 25
Development system
  Assembler 13
  C Compiler 14
  EZ-ICE® 13
  EZ-LAB® 13
  Linker 13
  PROM Splitter 14
  Simulator 14
  System Builder 13
Digital control system *See also* N'th order digital controllers
  ADSP-2101-based actuator controller 506
  benchmarks 522
  hardware implementation 505
  model 504
  overview 503
  PID controller 511
  software implementation 507
Digital mobile radio (DMR) 205. *See also* GSM
Digital tone detection 481. *See also* Goertzel algorithm
Discrete cosine transform (DCT) 443
  benchmark times 447
  comparison with Fourier transform 445
  computational methods 449
  indirect computation 449
  listings 455
  transform coefficients 444
  two-dimensional DCT 448
  zig-zag scanning of DCT coefficients 453
Double-precision IIR biquad 531, 534
Double-precision multiply routine 529

Index

DRAM
    CAS 608, 610
    fast page mode 613
    listings 618
    multiplexed memory addressing 607
    OE 611
    page mode 613
    RAS 608, 609
    read timing 611
    read/write timing 607, 608
    refresh 614
    refresh timing 615
    WRITE 611
    write timing 612
Dynamic time warping 333, 342
    time warping boundaries 344

E

Echo cancellation
    algorithm 82
    benchmarks 97
    flowchart for LMS stochastic gradient algorithm 85
    frequency offset compensation 88
    implementation of LMS algorithm 84
    LMS adaptive filter block diagram 84
    LMS stochastic gradient code listing 88
    telephone channel block diagram 81
Encoder 207, 294. *See also* GSM
    circular buffering 300
    higher sub-band encoder 296
    lower sub-band encoder 296
End point detection 337
Equalizer 105. *See also* Adaptive equalizer
EZ-LAB® Evaluation Board 604, 617

F

Filter coefficients 158

G

Goertzel algorithm
    benchmarks 489
    detection time 484
    frequency resolution 484
    Goertzel coefficients 488
    leakage loss 483
    single tone detection 484, 485
    symbol detection 484

Index

Group special mobile (GSM) 205
  benchmarks 222
  decoder 220
  encoder 207
  listings 223
  long term prediction (LTP) 207
  speech codec 205
  voice activity detection 205

H

Hardware interfacing 571
  AD1847 585
  AD1849 572, 578
  DRAM 603
  memory 603
  SoundPorts 571
Hilbert transform 90
  implementation 91
  listing 96
Hou's fast discrete cosine algorithm 449. *See also* Discrete cosine transform
Human speech production
  excitation 330
  filtering 330

I

IIR biquad filters 525. *See also* Multiprecision filters
  biquad filter subroutine 525
  filter characteristics 549
  optimized 16-bit biquads 544
  optimized basic biquad filter routine 546
  second-level optimized biquad filter routine 548
  second-order biquad 525
Image compression
  applications of 443
Index registers 454
Interfacing 571. *See also* Hardware interfacing
Inverse adaptive quantizer 298, 299
Isolated word recognition 340

L

Least mean squared (LMS) algorithm 84
Levinson-Durbin recursion 175
Linear prediction
  filter 157
  filter coefficients 158
  speech synthesis filter 158

Index

Linear predictive coding (LPC) 205, 207, 331
    2.4 kbits/s 163
    7.8 kbits/s 159
    analysis 336, 338
    parameters 158
    predictor coefficients 338
    synthesis 158
-loader
    -loader option 638
Loading through SPORT
    flow diagram 633
    implementation 632
    listing 636
Logarithmic-area-ratios (LAR)
    coding of 212
    decoding of 212
    quantization 212
    transformation of 214
Long term analysis filtering 216
    cross-correlation 216
    long term correlation lag 216
Long term prediction (LTP) 207. *See also* GSM
Long term synthesis filtering 217, 221
LPC
    2.4 kbits/s 163
    7.8 kbits/s 159
    parameters 158
    synthesis 158
LPC analysis 336, 338

M

Memory buses 6. *See also* ADSP-2100 family
Memory interfacing
    -loader option 638
    boot pages for program memory 637
    DRAM 603
    hardware implications 640
    SRAM 603, 637
    use of the C-Compiler 641
    using external PM ROM 640
    using internal & external PM RAM for code 639
Modem 17
Modulation 75, 140. *See also* Modem
Modulo addressing 544
Monitor 631. *See also* Loading through SPORT

Index

Multiplier/accumulator (MAC) 2, 4. *See also* ADSP-2100 family, base architecture
Multiprecision filters 527. *See also* IIR biquad filters
   double-precision IIR biquad 531, 534
   double-precision multiply routine 529
   half, double-precision IIR biquad subroutine 539
   half, triple-precision IIR biquad subroutine 543
   multiprecision multiplication 528
   optimized double-precision IIR biquad subroutine 537

N

Notch filter
   example 520
   routine 522
N'th order digital controllers
   analog-controller-based digital design 513
   bilinear transformation 514
   fourth-order direct form controller 517
   implementation 517
   second-order biquad section 515
   state-space design 515
   structures 515
Nyquist theory 482

O

Optimized filters 544. *See also* IIR biquad filters

P

Parameters for LPC
   excitation 158
   filter coefficients 158
   gain 158
PID controller
   backward difference 509
   design 508
   implementation 511
   Kd 509
   Ki 509
   Kp 509
   routine 512
Pre-emphasis filtering 208
Predictor coefficients 338
Program sequencer 5. *See also* ADSP-2100 family, base architecture
Pulse shape filter routine 19, 138

Index

Q

Quadrature amplitude modulation (QAM)
  block diagram 76
  demodulator block diagram 77
  demodulator code listing 80
  implementation 78
  methodology 75
Quadrature mirror filter 294
  receive 300
  transmit 294

R

Raised cosine filter 32
  formula 33
  implementation 33
  rolloff factor 33
Random number generator routine 139
Recognition 329. *See also* Speech recognition
Recognition library 331
Reflection coefficients
  coefficients 210
Regular pulse excitation (RPE) 207. *See also* GSM
  adaptive sample rate decimation 218
  APCM inverse quantization 219
  APCM quantization 218
  encoding 217
  reconstructed short term residual signal 219
  weighting filter 217
Rounding (for GSM)
  multiply with rounding 206
RS.232 552. *See also* UART

S

Sampling frequency for digital tone detection
  good sampling frequency 485
  maximum error-squared value 482
  Nyquist theory 482
  selecting sampling frequency 482
Schur recursion 209, 210, 336
Scrambler 21. *See also* V.32
  programs 25
Serial port
  loading the ADSP-2101 631
  using as UART 551

Index

Shifter 4. *See also* ADSP-2100 family, base architecture
Short term analysis filtering 213, 215
Short term synthesis filtering 220
Signal modulation routine 75, 140
signal_map 46
Software tools 13. *See also* Development system
SoundPort interfaces
  ADSP-2101/AD1847 585
  ADSP-2101/AD1847 demonstration routine 602
  ADSP-2101/AD1847 talk-through routine 589
  ADSP-2105/AD1849 578
  ADSP-2105/AD1849 talk-through routine 584
  ADSP-2111/AD1849 572
  ADSP-2111/AD1849 talk-through routine 577
Speech codec 205. *See also* GSM
Speech compression
  ADPCM 293
  GSM 205
  LPC 157
Speech recognition
  demonstration shell 346
  executive shell 346
  filter 331
  hardware implementation 349
  K-Nearest Neighbor 341
  recognition library 331
  recognition phase 330, 333
  software implementation 334
  speaker dependent systems 330, 332
  speaker independent systems 330, 332
  theory of 330
  training phase 330, 332
Speech synthesis filter 158
SRAM 637
Sub-band ADPCM (G.722) 293. *See also* ADPCM
  subroutine descriptions 301

Index

T

Timer interrupt routine 561
Tone detection applications *See also* Goertzel algorithm
   DTMF 481, 484
   remotely controlled equipment 481
Trellis encoding 37
  block diagram 38
  convolutional encoder 41
  convolutional encoder block diagram 41
  convolutional encoder routine 45
  differential encoder 40
  differential encoder lookup table 40
  implementation 39
  Trellis encoder program 43

U

UART
  autobaud 561, 564
  baud rate 561, 564
  character echo example 567
  example system configuration 552
  full-duplex operation 551
  initialization routine 553
  interrupts 552
  receive character routine 563
  RS-232 line driver 551
  timer 552
  timer interrupt routine 553
  transmit character routine 563
Unvoiced sound 157, 158
Unvoiced speech 330

V

V.27 ter
  4-point V.27 ter constellation 129
  4-point V.27 ter phase changes 128
  8-point V.27 ter constellation 128
  8-point V.27 ter phase changes 127
  data acquisition routine 134
  data scrambler 18, 127, 152
  data scrambler routine 135
  IQ generator routine 137
  main V.27 ter routine 133
  pulse shape filter routine 19, 138
  random number generator routine 139
  signal modulation routine 75, 140
  transmitter 126

Index

V.29
  8-point V.29 phase changes 141
  constellation 142, 143
  data acquisition routine 151
  data scrambler routine 152
  IQ generator routine 154
  main V.29 routine 150
V.32
  adaptive equalization 98. *See also* Adaptive equalizer
  adaptive equalizer 20
  descrambler 21, 22
  differential decoder 21
  echo cancellation 81
  programs 25
  pulse shape filters 19. *See also* Raised cosine filter
  quadrature amplitude modulation (QAM) 75
  recommendation 17
  scrambler 18, 21
  signal mapping 19
  transmitter 18
  Viterbi decoder 21
  Viterbi decoding 47
Viterbi algorithm 50
Viterbi decoding. *See* V.32
  algorithm 50
  implementation 52
  last surviving path 55
  listing 74
  shortest path 53, 55
  state diagram 48
  state table 48
  trellis diagram 49
Vocal fold vibration 157
Vocal tract 157
Voice activity detection (VAD) 205. *See also* GSM
Voice recognition 330. *See also* Speech recognition
Voiced sound 157, 158
Voiced speech 330

W

Weighing filter 217
Wide-band ADPCM 293. *See also* ADPCM (G.722)

Z

Zig-zag scanning of DCT coefficients 453